Sustainable Integrated Clean Environment for Human & Nature

Sustainable Integrated Clean Environment for Human & Nature

Editor

Shervin Hashemi

MDPI • Basel • Beijing • Wuhan • Barcelona • Belgrade • Manchester • Tokyo • Cluj • Tianjin

Editor
Shervin Hashemi
Yonsei University
Korea

Editorial Office
MDPI
St. Alban-Anlage 66
4052 Basel, Switzerland

This is a reprint of articles from the Special Issue published online in the open access journal *Sustainability* (ISSN 2071-1050) (available at: https://www.mdpi.com/journal/sustainability/special_issues/clean_environment_human_nature).

For citation purposes, cite each article independently as indicated on the article page online and as indicated below:

LastName, A.A.; LastName, B.B.; LastName, C.C. Article Title. *Journal Name* **Year**, *Volume Number*, Page Range.

ISBN 978-3-0365-1882-4 (Hbk)
ISBN 978-3-0365-1881-7 (PDF)

Cover image courtesy of Shervin Hashemi.

Contents

About the Editor

Shervin Hashemi is a researcher at the Institute for Environmental Research at Yonsei University, Seoul, Republic of Korea. His research expertise is on providing the provision of water and sanitation in remote areas. He invented the Resource Circulated Sanitation system, for which he won the Leaving No One Behind Innovation Award 2019 and Energy Globe Republic of Korea National Award 2019. He also developed the Sanitation Sustainability Index to evaluate the technical, economic, and social sustainability of the local sanitation systems. In 2021, he was nominated for the IWA Global Water Award, which showcases the exceptional individuals who have made a significant contribution to the world wherein water is wisely managed through their innovative leadership and practices. Currently, he is associated with the journal Sustainability (ISSN 2071-1050) as a guest editor.

Preface to "Sustainable Integrated Clean Environment for Human & Nature"

The interaction between humans and nature must be taken into consideration when it comes to technical, economic, and social sustainable development. Considerate interactions between humans and nature can help the current generation to overcome the barriers and difficulties of sustainable development, such as pandemics, and provide a brighter future for our next generation.

In this regard, answering the following key questions can clarify the way forward:

1. What is the current state of the environment? Is it clean?
2. How can we make our environment clean and suitable for humans as well as nature?
3. How can we keep our environment clean through sustainable practices?

As a step toward finding the answers to the abovementioned questions, this book is a collection of articles that are published in the Special Issue of the journal *Sustainability* (ISSN 2071-1050), entitled "Sustainable Integrated Clean Environment for Human and Nature".

The book contains a collection of 20 valuable contributions, including 17 research articles, 2 review papers, and an editorial covering the following subjects:

1. COVID-19 and the sustainability of a clean environment for humans and nature: visions, challenges, and solutions
2. Clean technologies and nature-based approaches, including environmental remediation and resource circulation
3. Global sanitation, hygiene, and public health issues
4. Economic approaches, including the development of economic models, life cycle assessment, and the circular economy
5. Social awareness and effective education on human rights for procuring clean air and water

Through including the latest studies in the abovementioned fields, this book addresses the technicians, economists, social activists, and decision makers who are concerned about clean environment concepts for sustainable development of the current and next-generation through respectful interactions between humans and nature.

I wish to acknowledge the generous support of the journal *Sustainability* and publisher MDPI. I also thank the authors who contributed through submitting their valuable works to the Special Issue "Sustainable Integrated Clean Environment for Human and Nature" as well as the reviewers and editorial staff at the journal *Sustainability* for their time and effort in reviewing, proofreading, and publishing the articles included in this collection. I am grateful to Eng. Farid Hashemi for his kind support and encouragement. This book and the Special Issue were edited during the peak of the COVID-19 pandemic. I express my deepest sympathies to the people worldwide who faced losses due to this pandemic.

Shervin Hashemi
Editor

sustainability MDPI

Editorial

Perspectives on Sustainable Integrated Clean Environment for Human and Nature

Shervin Hashemi

Institute for Environmental Research, Yonsei University College of Medicine, 50-1 Yonsei-ro, Seodae-mun-gu, Seoul 03722, Korea; shervin@yuhs.ac

check for updates

Citation: Hashemi, S. Perspectives on Sustainable Integrated Clean Environment for Human and Nature. *Sustainability* **2021**, *13*, 4150. https://doi.org/10.3390/su13084150

Received: 6 April 2021
Accepted: 7 April 2021
Published: 8 April 2021

Publisher's Note: MDPI stays neutral with regard to jurisdictional claims in published maps and institutional affiliations.

1. Introduction

The term "sustainability" is generally used to describe an ideal condition in which the Earth's biosphere and human civilization can safely interact and coexist [1]. Accordingly, it is essential to have a comprehensive vision of the requirements for a clean environment sustainable by humans and nature. Such a vision can be useful as it encompasses all aspects related to human life that interact with nature, so that human civilization can develop while respecting and protecting nature.

However, there are limited scientific approaches to achieving this vision. To address this knowledge gap, in December 2019, the author proposed a Special Issue of the journal *Sustainability* (ISSN 2071-1050) entitled "Sustainable Integrated Clean Environment for Human & Nature", to gather the missing pieces of this vision among multidisciplinary scientific–technical, economic, and social approaches over a one-year period. The Special Issue attempts to answer the following key questions:

1. What is the current condition of our environment? Is it clean and pollution-free?
2. How can we make our environment clean and suitable for humans and nature?
3. How can we sustainably keep our environment clean?

The articles in the Special Issue contribute to one or more of the following fields:

- Actions for developing countries;
- Clean technologies;
- Economic approaches;
- Environmental remediation;
- Global water security issues;
- Hazardous substances;
- Human–nature interactions;
- Nature-based solutions—from theories and laboratories to fields and actions;
- New policies;
- Public health;
- Risk assessment and modeling;
- Sanitation/hygiene;
- Social awareness/human rights—concepts of leaving no one behind, pilot social studies, reactions from different societies;
- Sustainable water, wastewater, and waste treatment, including pilot scaled case studies;
- Resource circulation—conservation and recycling.

2. List of Contributions

Over 30 manuscripts were submitted for consideration for the Special Issue, and all of them underwent the journal's rigorous review process. In total, 19 papers, including 17 research articles and 2 review papers, were finally accepted for publication and inclusion in this Special Issue. These are listed below:

1. Al-Gheethi, A.; Al-Sahari, M.; Abdul Malek, M.; Noman, E.; Al-Maqtari, Q.; Mohamed, R.; Talip, B.A.; Alkhadher, S.; Hossain, M.S. Disinfection Methods and Survival of SARS-CoV-2 in the Environment and Contaminated Materials: A Bibliometric Analysis. *Sustainability* **2020**, *12*, 7378. https://doi.org/10.3390/su12187378.
2. Amnuaylojaroen, T.; Inkom, J.; Janta, R.; Surapipith, V. Long Range Transport of Southeast Asian PM2.5 Pollution to Northern Thailand during High Biomass Burning Episodes. *Sustainability* **2020**, *12*, 10049. https://doi.org/10.3390/su122310049.
3. Blazy, R.; Hrehorowicz-Gaber, H.; Hrehorowicz-Nowak, A.; Płachta, A. The Synergy of Ecosystems of Blue and Green Infrastructure and Its Services in the Metropolitan Area—Chances and Dangers. *Sustainability* **2021**, *13*, 2103. https://doi.org/10.3390/su13042103.
4. Coccia, M. How (Un)sustainable Environments Are Related to the Diffusion of COVID-19: The Relation between Coronavirus Disease 2019, Air Pollution, Wind Resource and Energy. *Sustainability* **2020**, *12*, 9709. https://doi.org/10.3390/su12229709.
5. Gopalakrishnan, Y.; Al-Gheethi, A.; Abdul Malek, M.; Marisa Azlan, M.; Al-Sahari, M.; Radin Mohamed, R.M.S.; Alkhadher, S.; Noman, E. Removal of Basic Brown 16 from Aqueous Solution Using Durian Shell Adsorbent, Optimisation and Techno-Economic Analysis. *Sustainability* **2020**, *12*, 8928. https://doi.org/10.3390/su12218928.
6. Hashemi, S. Sanitation Sustainability Index: A Pilot Approach to Develop a Community-Based Indicator for Evaluating Sustainability of Sanitation Systems. *Sustainability* **2020**, *12*, 6937. https://doi.org/10.3390/su12176937.
7. Hassan, A.; G Almatar, M.; Torab, M.; Allen, C.D. Environmental Urban Plan for Failaka Island, Kuwait: A Study in Urban Geomorphology. *Sustainability* **2020**, *12*, 7125. https://doi.org/10.3390/su12177125.
8. Kärenlampi, P.P. Diversity of Carbon Storage Economics in Fertile Boreal Spruce (Picea Abies) Estates. *Sustainability* **2021**, *13*, 560. https://doi.org/10.3390/su13020560.
9. Li, Y.; Wang, X.; Dong, X. Delineating an Integrated Ecological and Cultural Corridor Network: A Case Study in Beijing, China. *Sustainability* **2021**, *13*, 412. https://doi.org/10.3390/su13010412.
10. Mofijur, M.; Rizwanul Fattah, I.M.; Saiful Islam, A.B.M.; Uddin, M.N.; Ashrafur Rahman, S.M.; Chowdhury, M.A.; Alam, M.A.; Uddin, M..A. Relationship between Weather Variables and New Daily COVID-19 Cases in Dhaka, Bangladesh. *Sustainability* **2020**, *12*, 8319. https://doi.org/10.3390/su12208319.
11. Muliadi, F.N.A.; Halmi, M.I.E.; Wahid, S.B.A.; Gani, S.S.A.; Mahmud, K.; Shukor, M.Y.A. Immobilization of Metanil Yellow Decolorizing Mixed Culture FN3 Using Gelling Gum as Matrix for Bioremediation Application. *Sustainability* **2021**, *13*, 36. https://doi.org/10.3390/su13010036.
12. Muliadi, F.N.A.; Halmi, M.I.E.; Wahid, S.B.A.; Gani, S.S.A.; Zaidan, U.H.; Mahmud, K.; Abd Shukor, M.Y. Biostimulation of Microbial Communities from Malaysian Agricultural Soil for Detoxification of Metanil Yellow Dye; a Response Surface Methodological Approach. *Sustainability* **2021**, *13*, 138. https://doi.org/10.3390/su13010138.
13. Nur-E-Alam, M.; Hoque, M.N.; Ahmed, S.M.; Basher, M.K.; Das, N. Energy Engineering Approach for Rural Areas Cattle Farmers in Bangladesh to Reduce COVID-19 Impact on Food Safety. *Sustainability* **2020**, *12*, 8609. https://doi.org/10.3390/su12208609.
14. Rahman, M.; Thill, J.-C.; Paul, K.C. COVID-19 Pandemic Severity, Lockdown Regimes, and People's Mobility: Early Evidence from 88 Countries. *Sustainability* **2020**, *12*, 9101. https://doi.org/10.3390/su12219101.
15. Roblek, V.; Thorpe, O.; Bach, M.P.; Jerman, A.; Meško, M. The Fourth Industrial Revolution and the Sustainability Practices: A Comparative Automated Content Analysis Approach of Theory and Practice. *Sustainability* **2020**, *12*, 8497. https://doi.org/10.3390/su12208497.

16. Scharf, M.; Heide, L.; Grahle, A.; Syré, A.M.; Göhlich, D. Environmental Impact of Subsidy Concepts for Stimulating Car Sales in Germany. *Sustainability* **2020**, *12*, 10037. https://doi.org/10.3390/su122310037.
17. Tanveer, M.; Hassan, S.; Bhaumik, A. Academic Policy Regarding Sustainability and Artificial Intelligence (AI). *Sustainability* **2020**, *12*, 9435. https://doi.org/10.3390/su12229435.
18. White, I.; Falkland, T.; Kula, T. National Versus Local Sustainable Development Plans and Island Priorities in Sanitation: Examples from the Kingdom of Tonga. *Sustainability* **2020**, *12*, 9379. https://doi.org/10.3390/su12229379.
19. Woo, B.L.; Lim, M.K.; Park, E.Y.; Park, J.; Ryu, H.; Jung, D.; Ramirez, M.J.; Yang, W. Characteristics of Non-Smokers' Exposure Using Indirect Smoking Indicators and Time Activity Patterns. *Sustainability* **2020**, *12*, 9099. https://doi.org/10.3390/su12219099.

3. Content and Significance of Contributions

Shortly after the Special Issue was announced, the world faced one of the most serious outbreaks of severe acute respiratory syndrome coronavirus 2 (SARS-CoV-2) [2]. In view of humanity's current lifestyle, the coronavirus disease (COVID-19) pandemic has shown the fragility of our concept of sustainability [3]. Consequently, the importance of the concept of "sustainable, clean environment" became clear to individuals around the world.

The COVID-19 pandemic has demonstrated the need for a crucial reconsideration of the need for a sustainable, clean environment for humans and nature. The provision of safe water, sanitation, hygiene, and waste management is essential for preventing virus transmission, and protecting human health has become increasingly critical. Understanding the technical, social, and economic concepts of the COVID-19 pandemic and recognizing challenges can lead to solutions and sustainable management approaches during and after the pandemic.

In response to the COVID-19 pandemic, contributions 1, 4, 10, 13, and 14 cover a wide range of topics related to the sustainability of an integrated clean environment. These contributions are useful in highlighting the novel meaning of sustainability after the COVID-19 pandemic.

Accordingly, the 4th Sustainability Webinar, "COVID-19 and the Sustainability of Clean Environment for Human & Nature: Visions, Challenges, and Solutions", was proposed to provide a global forum for ordinary individuals and decision-makers to understand the vision, challenges, and solutions for sustaining a clean environment for humans and nature during and after the COVID-19 pandemic. The webinar included presentations from four experts, including scholars who made relevant contributions to the Special Issue. A webinar recording is available online at https://sustainability-4.sciforum.net/ (accessed on 29 March 2021) [4].

Along with the medical approaches for overcoming the COVID-19 pandemic, implementing appropriate green technologies may be considered among the highest priorities for sustainable development during and after the COVID-19 pandemic.

The Special Issue contains two contributions on the topic of sustainable sanitation. Contribution 6 introduces the Sanitation Sustainability Index, which is an integrated, community-based indicator for evaluating the sustainability of sanitation systems, which will be implemented in specific communities. The highlighted characteristics of this newly developed index are the consideration of a wider range of sanitation sustainability parameters, including acceptability and public health indicators, and the ability of the index to evaluate sanitation systems before implementation [5].

Another notable contribution is presented by White et al. (contribution 18), who discussed national versus local sustainable development plans and island priorities in sanitation. This article provides examples from the Kingdom of Tonga and presents an analysis of the priorities given to water and sanitation in top-down national sustainable development strategies and nation-wide, bottom-up, village-level community development plans. This contribution has been commended by the editorial office of the journal *Sustainability* and has been recognized as a feature paper.

Contributions 3, 7, and 17 of the Special Issue present discussions on sustainable urban planning, green infrastructure, and academic policies. Contributions 2 and 19 discuss indoor and outdoor air quality from both technical and public health perspectives. In addition, contributions 15 and 16 present intriguing studies that discuss industrial concerns, including impacts on the environment and sustainable development policies. A significant number of studies, including contributions 5, 8, 9, 11, and 12, discuss different aspects of ecological concerns regarding environmental sustainability, including the technical approaches for the treatment or bioremediation of environmental contaminants.

4. Conclusions

The Special Issue "Sustainable Integrated Clean Environment for Human & Nature" is a collection of articles that investigate the technical, economic, and social shortcomings of the current approaches toward a sustainable integrated clean environment for humans and nature in developing countries; these articles also provide innovative solutions aimed at overcoming the challenges of balancing the current approaches to enhance their sustainability.

Consideration of these studies could be helpful for decision-makers seeking to develop efficient strategies leading to the provision of new and effective sustainability policies by considering both the social development of humans and the need to respect and protect nature sustainably.

Funding: This work received no external funding.

Institutional Review Board Statement: Not applicable.

Informed Consent Statement: Not applicable.

Data Availability Statement: Not applicable.

Acknowledgments: The author wishes to acknowledge the generous support of the journal *Sustainability* and the publisher MDPI. The author also thanks those who contributed papers to this Special Issue, as well as reviewers and the authorial staff at the journal *Sustainability*, particularly Liv Zhou, for their time and effort in reviewing, improving, and publishing the articles included in this Special Issue. The author is grateful to Eng. Farid Hashemi for his kind support and encouragement. This Special Issue was edited during the peak of the COVID-19 pandemic. The author would like to express his deepest sympathies to people worldwide who have faced losses due to this pandemic.

Conflicts of Interest: The author declares no conflict of interest.

References

1. Kuhlman, T.; Farrington, J. What is Sustainability? *Sustainability* **2010**, *2*, 3436–3448. [CrossRef]
2. Gorbalenya, A.E.; Baker, S.C.; Baric, R.S.; de Groot, R.J.; Drosten, C.; Gulyaeva, A.A.; Haagmans, B.L.; Lauber, C.; Leontovich, A.M.; Neuman, B.W. The species Severe acute respiratory syndrome-related coronavirus: Classifying 2019-nCoV and naming it SARS-CoV-2. *Nat. Microbiol.* **2020**, *5*, 536–544.
3. Barbier, E.; Burgess, J. Sustainability and Development after COVID-19. *World Dev.* **2020**, *135*, 105082. [CrossRef] [PubMed]
4. Sustainability Webinar | COVID-19 and the Sustainability of Clean Environment for Human & Nature: Visions, Challenges, and Solutions. Available online: https://sustainability-4.sciforum.net/ (accessed on 29 March 2021).
5. Narzetti, D.A.; Marques, R.C. Access to Water and Sanitation Services in Brazilian Vulnerable Areas: The Role of Regulation and Recent Institutional Reform. *Water* **2021**, *13*, 787. [CrossRef]

sustainability

MDPI

Article

COVID-19 Pandemic Severity, Lockdown Regimes, and People's Mobility: Early Evidence from 88 Countries

Md. Mokhlesur Rahman [1,2], Jean-Claude Thill [3,*] and Kamal Chandra Paul [4]

1 Department of Urban and Regional Planning, Khulna University of Engineering & Technology, Khulna 9203, Bangladesh
2 INES Program, The William States Lee College of Engineering, The University of North Carolina at Charlotte, 9201 University City Blvd, Charlotte, NC 28223, USA; mrahma12@uncc.edu
3 Department of Geography and Earth Sciences and School of Data Science, The University of North Carolina at Charlotte, 9201 University City Blvd, Charlotte, NC 28223, USA
4 Department of Electrical and Computer Engineering, The William States Lee College of Engineering, The University of North Carolina at Charlotte, 9201 University City Blvd, Charlotte, NC 28223, USA; kpaul9@uncc.edu
* Correspondence: jfthill@uncc.edu; Tel.: +1-704-687-5973

Received: 2 October 2020; Accepted: 30 October 2020; Published: 1 November 2020

Abstract: This study empirically investigates the complex interplay between the severity of the coronavirus pandemic, mobility changes in retail and recreation, transit stations, workplaces, and residential areas, and lockdown measures in 88 countries around the world during the early phase of the pandemic. To conduct the study, data on mobility patterns, socioeconomic and demographic characteristics of people, lockdown measures, and coronavirus pandemic were collected from multiple sources (e.g., Google, UNDP, UN, BBC, Oxford University, Worldometer). A Structural Equation Modeling (SEM) framework is used to investigate the direct and indirect effects of independent variables on dependent variables considering the intervening effects of mediators. Results show that lockdown measures have significant effects to encourage people to maintain social distancing so as to reduce the risk of infection. However, pandemic severity and socioeconomic and institutional factors have limited effects to sustain social distancing practice. The results also explain that socioeconomic and institutional factors of urbanity and modernity have significant effects on pandemic severity. Countries with a higher number of elderly people, employment in the service sector, and higher globalization trend are the worst victims of the coronavirus pandemic (e.g., USA, UK, Italy, and Spain). Social distancing measures are reasonably effective at tempering the severity of the pandemic.

Keywords: COVID-19; lockdown; social distancing; mobility; SEM

---

1. Introduction

The novel coronavirus, also known as Coronavirus Disease 2019 (COVID-19), first emerged in Wuhan (P.R. China) in late fall 2019 and has now spread to 213 countries around the globe [1]. The World Health Organization (WHO) declared COVID-19 a pandemic on 11 March 2020, considering its outbreak in many countries [2]. As of now, more than 44 million people have been infected by this highly infectious disease and over 1.1 million people have died [1,3]. The current fatality rate among closed cases is about 10%, which speaks volume about the sheer severity of the pandemic. The increasing number of coronavirus cases and deaths poses challenges to the healthcare system, economic development, supply chain, education, and travel pattern of the people [4]. Figure 1 represents the increasing number of coronavirus infection cases and deaths (in million) in the world from 31 December 2019 to 29 October 2020. To control the spread of COVID-19, governments have implemented travel bans through national lockdown, stay-at-home order, restriction on mass gathering and non-essential travel, which further affected people's mobility and social distancing practices.

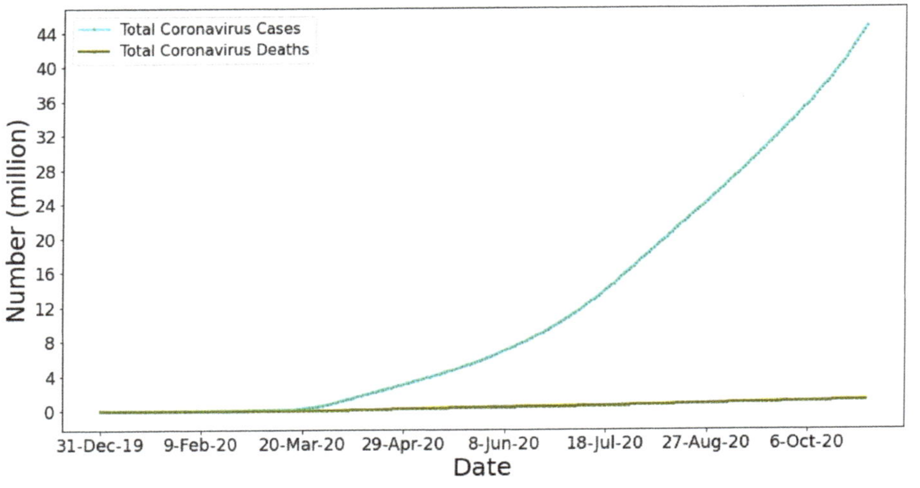

Figure 1. Total number of coronavirus infection cases and deaths in the world [5].

This study mainly aims at unraveling the complex relationships between the incidence of the pandemic, lockdown measures on populations and their social distancing and mobility behaviors in the early stage of the pandemic. The following three research questions are form the core of the understanding these interrelated relationships:

(1) What are the impacts of different lockdown measures on reducing people's mobility patterns and the severity of the pandemic?
(2) What are the effects of socioeconomic and institutional arrangements and dispositions on population mobility and on the pandemic severity?
(3) What are the consequences of pandemic severity on the condition of social distancing of the people?

The impacts of COVID-19 on public health have been discussed in many previous papers [6–9]. This disease is imposing tremendous pressure on the health care system [8]. Besides, COVID-19 is affecting the mental health of people in the form of mass fear, panic, and uncertainties [10–12]. Because of the escalation of the pandemic, there has been a huge increase in the personal stockpiling of necessary goods (e.g., food staples, toilet paper, cleaning supply) which is unsettling the balance in the demand and supply of consumer goods [6].

Many researchers have investigated the impacts of COVID-19 on the global economy [4,6,9,13,14]. Globally, stock markets collapsed by 50%. As COVID-19 threw millions out of work in the US, it caused an unemployment rate soaring to 14.7% in April 2020, which is the highest rate since the Great Depression [15,16]. The US Congress passed a $2 trillion coronavirus aid package to help businesses and workers. Global annual GDP is expected to contract by 3–4%. With the COVID-19 outbreak, a massive freeze in the industrial and logistical infrastructure caused a devastation throughout the global economy. Many investors moved towards safer investments because of the fear of a worldwide recession [9]. Meanwhile, the global supply chain has been deeply disrupted. About 940 of the Fortune 1000 companies have reported a supply chain disruption due to COVID-19 [17]. A simulation study observed that changes in opening and closing time of the facilities due to the coronavirus pandemic are affecting supply chain performance [18]. However, considering the sharp economic downturn, influence on reducing mobility and thus control the severity of the pandemic. Social distancing also has direct impact on reducing pandemic severity, although the effect is rather marginal. The socioeconomic and institutional factors of a country significantly affect pandemic severity. However, pandemic severity, socioeconomic and institutional factors have no significant impacts on social distancing.

2. Materials and Methods

2.1. Data

To test and validate the conceptual model in Figure 2, data were collected from multiple sources (Table 1). The data sources include Google, the United Nations (UN), United Nations Development Program (UNDP), Worldometer, Oxford University, Hofstede, The Fraser Institute, KOF Swiss Economic Institute, and BBC.

Table 1. Description of the variables and data sources.

Variable	Description	Source
RR	Percentage change of mobility in retail and recreation trips	[38]
TS	Percentage change of mobility in transit stations trips	[38]
WP	Percentage change of mobility in workplaces trips	[38]
RD	Percentage change of mobility in residential trips	[38]
l_case	Total coronavirus infection cases per 1 million population	[1]
l_death	Total coronavirus deaths per 1 million population	[1]
NL	National lockdown	[39]
WPC	Workplace closing	[40]
SH	Stay-at-home order	[40]
SI	Stringency index [1]	[40]
FS	Percentage of female smokers	[40]
AGE65	Percentage of the population age 65 and older	[40]
MA	Median age	[41]
EI	Average of years of schooling vs. expected years of schooling	[41]
AE	Percentage of the population employed in agriculture	[42]
SE	Percentage of the population employed in services	[42]
HE	Percentage of health expenditure to total GDP	[42]
IDV	Individualism versus Collectivism emphasis [2]	[43]
KOFGI	KOF Globalization Index [3]	[44]

[1] A composite index considering all implemented lockdown measures. The score of this index ranges from 0 to 100. A high score indicates the strictest measures and low score indicates loose measure. [2] This indicator measures the degree of interdependence among the members of a society. The score ranges from 0 to 100. A low score indicates collective culture and higher interdependence among the members and conversely a high score indicates Individualist culture and a low level of interdependence. [3] A composite index that indicates openness to trade and capital flows considering economic, social and political aspects. The score of the index ranges from 0 to 100. A high score denotes a highly globalized country and a low score indicates poorly globalized country.

Google prepared a COVID-19 Community Mobility Report to help policymakers and public health professionals to understand changes in mobility in responses to lockdown measures (e.g., travel ban, work-from-home, shelter-in-place, restriction on public gathering) [38,45]. This report shows how visits and length of stay at different places, such as retail and recreation (e.g., restaurant, café, shopping center, theme park), workplaces (i.e., place of work), transit stations (e.g., subway stations, seaport, taxi stand, rest area), residential areas (i.e., place of residence), parks (e.g., public park, national forest), grocery stores and pharmacies (e.g., supermarket, convenience store, drug store) changed as of April 17 compared to a baseline value, with a potential to reduce the impact of COVID-19 pandemic. The baseline value is the median value of the corresponding week during the 5-week period from 3 January to 6 February 2020. The data were collected from the Google account holders who have turned on their travel location history. This study uses mobility changes in retail and recreation, workplaces, transit stations, and residential areas for 88 countries (Figure 3). Due to the ambiguity of the nature of visits and trips to grocery stores and pharmacies and the inconsistent definition of parks across countries (i.e., only include public parks), mobility changes in these two points of interest (POIs) were excluded from the study.

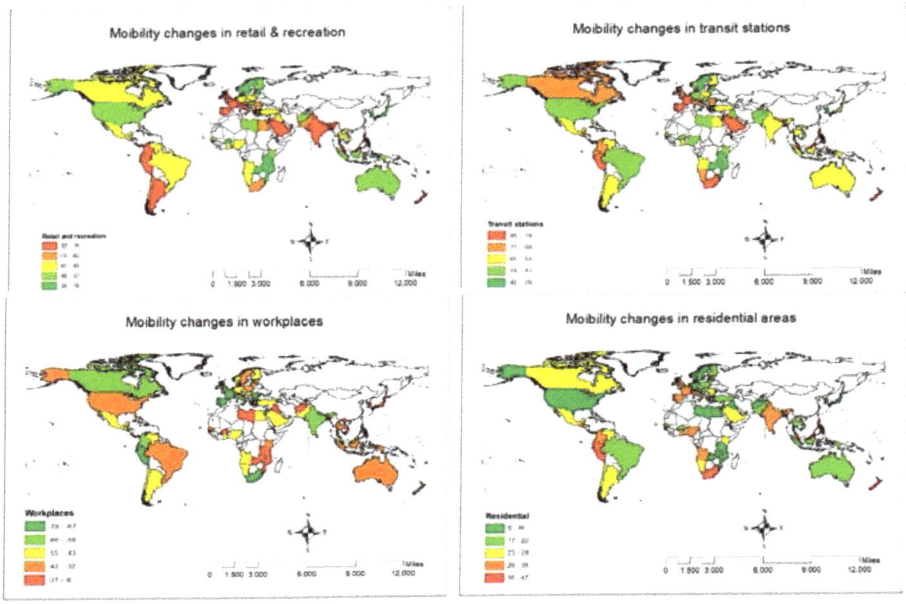

Figure 3. Mobility changes in POIs.

The total number of coronavirus infection cases and death cases as of April 17 were collected from Worldometer [1]. They collect data from thousands of sources around the world, analyze and validate them in real-time, and provide COVID-19 live statistics. To flatten the curve of COVID-19, governments issued different lockdown measures for part or whole country to restrict all non-essential movements. Data related to lockdown measures were collected from Dunford, Dale [39] and Oxford [40]. This study also collected socioeconomic (e.g., age, education, employment sector) and institutional context (e.g., individualism versus collectivism, globalization index) data to investigate their impacts on coronavirus infection cases and deaths, lockdown measures, and travel patterns (Table 1). After collecting data for 88 countries, they were integrated to build a complete dataset and conduct this study. Table 1 lists the variables that were included in the final model. A complete list of variables that were tested in the SEM framework to achieve the final model is provided in Appendix A.

Descriptive statistics of 19 different social distancing measures, lockdown variables, coronavirus infection cases and deaths, socioeconomic, and institutional context variables of all 88 countries are reported in Table 2. They are included in the statistical model as dependent variables, independent variables, mediators, and control variables.

Table 2. Descriptive statistics of the variables (N = 88).

Variable	Unit	Min	Max	Mean	SD
RR	%	−92	−18	−59.41	18.20
TS	%	−95	−20	−60.91	15.06
WP	%	−78	−6	−48.41	16.86
RD	%	7	47	24.19	8.78
l_case	Cases/1M pop	0.69	8.64	4.98	2.10
l_death	Deaths/1M pop	0.69	6.11	2.25	1.52
NL	Dummy (1, 0)	0	1	0.59	0.49
WPC	Dummy (1, 0)	0	1	0.83	0.38
SH	Dummy (1, 0)	0	1	0.67	0.47
SI	Index	38.22	100	82.07	13.73
FS	%	0.2	35.3	13.02	10.05
AGE65	%	1.14	27.05	11.09	6.88
MA	Year	16.7	48.4	33.68	8.95
EI	Index	0.3	1	0.72	0.16
AE	%	0.1	73.2	16.72	18.51
SE	%	21.1	87.6	61.79	15.83
HE	%	2.4	17.1	6.95	2.74
IDV	Index	6	91	40.02	22.95
KOFGI	Index	38.2	91.3	71.82	12.96

2.2. Statistical Model

SEM is used to investigate the causal relationships between socioeconomic and institutional factors, lockdown variables, coronavirus infection and death rates, and social distancing measures. This multivariate statistical technique is a common method to investigate complex relationships between dependent variables, independent variables, mediators, and latent dimensions. Many researchers have used SEM to investigate the factors that affect travel behaviors (e.g., mode choice, trip purpose, travel distance), for instance [46–49]. SEM consists of regression analysis, factor analysis, and path analysis to explore interrelationships between variables. It is a confirmatory technique where an analyst tests a model to check consistency between the existing theories and the nature of constructs.

Based on Exploratory Factor Analysis (EFA) and extant theories, latent dimensions are created to reduce dimensions and easily understand the data and represent underlying concepts. The following four latent dimensions are created based on the observed data:

(1) Social distancing measures: TS, RR, WP, and RD
(2) Pandemic severity: l_case and l_death
(3) Lockdown measures: NL, WPC, SH, and SI
(4) Socioeconomics and institutional factors: MA, AGE65, KOFGI, AE, SE, HE, FS, EI, and IDV

Moreover, a path diagram is constructed to graphically represent interdependencies of the independent variables, mediators, and dependent variables in the model specification. Finally, a set of fit indices (e.g., Chi-square, CFI, TLI, RMSEA, SRMR) are estimated to establish goodness-of-fit of the model.

3. Results

3.1. Calibrated Model

The model is calibrated using the SEM Builder within STATA 15 [50]. The maximum likelihood estimation method is used to calculate the coefficients. The overall structure of the model with direct standardized coefficients is depicted in Figure 4. The final structure of the model includes interactions between dependent and independent variables through mediators.

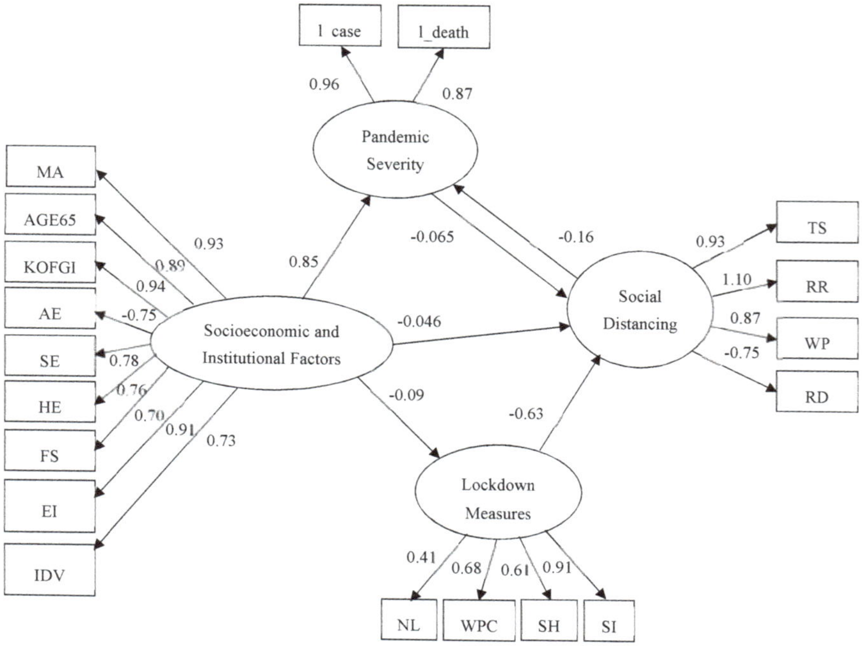

Figure 4. The calibrated model with direct standardized effects.

The fit of the calibrated model is evaluated based on several goodness-of-fit statistics (Table 3). The Chi-square statistic of the estimated model is 261.331. A lower value of the Chi-square indicates a better-fit model. Other fit statistics confirm that the estimated model is satisfactory. Thus, by all accounts, the goodness-of-fit of the estimated SEM is within the acceptable range and is quite satisfactory, which validates the use of this model [46,47].

Table 3. Goodness-of-fit Statistics.

Fit Statistic	Value
Chi-square	261.331
Chisq/df	2.026
RMSEA (Root mean squared error of approximation)	0.108
CFI (Comparative fit index)	0.920
TLI (Tucker-Lewis index)	0.894
SRMR (Standardized root mean squared residual)	0.099

3.2. *Standardized Direct Effects*

Table 4 reports on the standardized coefficients by pair of variables in the model, including the direction of the modeled effect. These coefficients indicate the direct impacts of the socioeconomic and institutional factors, on the dependent variables of lockdown measures, pandemic severity and social distancing measures, and the direct interactions between and among dependent, independent, and latent variables. However, this table does not represent any indirect effects of independent variables through mediators. Table 4 also reports the standard error, z-value, and probability level (*P*-value) of the estimates. Most of the coefficients are statistically significant at the 0.001 level. However, a few coefficients with a P-value greater than 0.001 are retained in the model to preserve the overall representation of the relationships.

Table 4. Estimated standardized direct effects (N = 88).

Variables	Std. Coef.	Std. Err.	z	P > z
Social Distancing <——— Pandemic severity	−0.065	0.220	−0.300	0.767
Social Distancing <——— Lockdown measures	−0.626	0.093	−6.760	0.000
Social Distancing <——— Socioeconomic & institutional	−0.046	0.204	−0.230	0.821
Pandemic severity <——— Social distancing	−0.160	0.086	−1.860	0.063
Pandemic severity <——— Socioeconomic & institutional	0.847	0.037	22.680	0.000
Lockdown measures <—— Socioeconomic & institutional	−0.090	0.115	−0.790	0.432
MA <——— Socioeconomic & institutional	0.925	0.020	46.660	0.000
IDV <——— Socioeconomic & institutional	0.734	0.050	14.720	0.000
HE <——— Socioeconomic & institutional	0.762	0.053	14.400	0.000
FS <——— Socioeconomic & institutional	0.701	0.054	12.910	0.000
EI <——— Socioeconomic & institutional	0.911	0.021	43.400	0.000
SE <——— Socioeconomic & institutional	0.777	0.046	16.920	0.000
AE <——— Socioeconomic & institutional	−0.746	0.048	−15.400	0.000
AGE65 <——— Socioeconomic & institutional	0.887	0.023	38.130	0.000
KOFGI <——— Socioeconomic & institutional	0.936	0.017	54.230	0.000
RR <——— Social distancing	1.133	0.109	10.380	0.000
TS <——— Social distancing	0.933	0.038	24.760	0.000
WP <——— Social distancing	0.868	0.046	18.800	0.000
RD <——— Social distancing	−0.751	0.054	−13.980	0.000
l_case <——— Pandemic severity	0.963	0.021	45.320	0.000
l_death <——— Pandemic severity	0.872	0.031	28.310	0.000
NL <——— Lockdown measures	0.413	0.099	4.180	0.000
WPC <——— Lockdown measures	0.681	0.066	10.390	0.000
SH <——— Lockdown measures	0.614	0.076	8.120	0.000
SI <——— Lockdown measures	0.911	0.047	19.390	0.000

Four latent dimensions are created to understand social distancing, pandemic severity, lockdown measures, and socioeconomic and institutional characteristics. Now we discuss the model results by observing the relationships between latent dimensions and observed independent variables:

Social Distancing: This latent dimension is created from four observed variables: TS, RR, WP, and RD. It is the only dependent latent factor that represents the level of mobility changes of the people at transit stations, retail and recreation facilities, workplaces, and residences. Social distancing is positively associated with changes in the use of transit stations (0.933), retail and recreation facilities (1.133), and workplaces (0.868). In contrast, social distancing is negatively associated with residences (−0.751). Moreover, social distancing is negatively associated with pandemic severity (−0.160). All other things being held equal, a one-unit change in social distancing reduces pandemic severity by 0.16 units by reducing people's mobility, and consequently the risk of exposure to other individuals infected by the COVID-19 virus. Thus, increasing social distancing reduces the severity of the coronavirus pandemic (i.e., number of infection cases, and deaths). However, the relationship is marginally significant at a *P*-value of 0.063.

Pandemic Severity: This endogenous latent dimension is measured by two observed variables: l_case and l_death. Pandemic severity is positively associated with both of the observed variables (l_case: 0.963 and l_death: 0.876). In contrast, pandemic severity is negatively associated with social distancing, which implies that increasing severity of the pandemic reduces mobility in transit stations, retail and recreation, and workplaces and increases the time spent at one's residence. However, the association is not statistically significant (p-value: 0.767).

Lockdown measures: This endogenous latent factor is estimated by using four observed variables: NL, WPC, SH, and SI. The Lockdown latent dimension is positively associated with all of the measures (NL: 0.413, WPC: 0.681, SH: 0.614, and SI: 0.911) taken by governments to bring about social distancing and control the pandemic. Furthermore, lockdown measures are negatively associated with social distancing (−0.626). Thus, adopting strict lockdown measures (e.g., restriction on public gathering, workplace closing, and stay-at-home order) significantly reduces mobility at transit stations, retail and recreation facilities, and workplaces and increases time spent near one's home, all of which entailing people to stay home and avoid unnecessary travel.

Socioeconomic and institutional factors: This is the only exogenous latent dimension in the model. It comprises nine observed variables: MA, AGE65, KOFGI, AE, SE, HE, FS, EI, and IDV. Socioeconomics and institutional factors are positively associated with median age (0.925), elderly people (0.887), level of globalization (0.936), employment in the service sector (0.777), expenditure on health (0.762), female smokers (0.701), level of education (0.911), and the degree of interdependence in the society (0.734). Conversely, it is negatively associated with employment in the agricultural sector (−0.746). This latent dimension can therefore be interpreted as an indicator of urbanity and modernity. Moreover, socioeconomic and institutional factors are positively associated with pandemic severity (0.847) and negatively associated with lockdown measures (−0.090) and social distancing (−0.046). Thus, one unit change in socioeconomic and institutional factors leads to an increase in pandemic severity by 0.847 unit through increases in the number of elderly people, level of globalization, employment in the service sector, and reduction in employment in the agricultural sector. In contrast, one unit change in socioeconomic and institutional factors lead to a decrease in lockdown measures and in social distancing by 0.090 units and 0.046 units, respectively, by encouraging people to be more considerate of their impact on the rest of society. However, the impacts of socioeconomic and institutional factors on lockdown measures and social distancing are very minor and statistically non-significant at P-value 0.05.

3.3. Estimated Standardized Total Effects

It is important to analyze the total effect of latent factors on social distancing, pandemic severity, and lockdown measures considering their indirect effects, which remain unrevealed in the path diagram (Figure 4). Table 5 details the standardized total effects of latent factors on each of the observed variables of social distancing, pandemic severity, and lockdown regime.

Table 5. Total effects on social distancing and pandemic severity.

Latent Factor	Social Distancing				Pandemic Severity		Lockdown Measures			
	TS	RR	WP	RD	l_case	l_death	NL	WPC	SH	SI
Pandemic severity	−0.061	−0.075	−0.057	0.049	-	-	-	-	-	-
Lockdown measures	−0.591	−0.717	−0.550	0.475	0.098	0.088	-	-	-	-
Socioeconomic and institutional factors	−0.042	−0.051	−0.039	0.034	0.823	0.746	−0.037	−0.061	−0.055	−0.082
Social distancing	-	-	-	-	−0.156	−0.141	-	-	-	-

Taking into account both direct and indirect effects, the analysis reveals that pandemic severity, lockdown measures, and socioeconomic and institutional factors reduce mobility at transit stations, retail and recreation centers, and workplaces and increase residential mobility. However, lockdown measures have much stronger and significant effects on all four social distancing aspects than pandemic severity and socioeconomic and institutional factors. In addition, the SEM analysis shows that lockdown and socioeconomic and institutional factors magnify pandemic severity while social distancing reduces pandemic severity. However, the impacts of socioeconomic and institutional factors are higher and statistically significant than lockdown measures and social distancing. Thus, lockdown measures are important to persuade people to stay home and maintain social isolation, and socioeconomic and institutional variables of urbanity and modernity substantially increase the severity of coronavirus pandemic. The table also indicates that only socioeconomic and institutional factors have direct impacts on the lockdown regime. However, the impacts are very insignificant.

Considering the complex relationships on hand, SEM extracts direct and indirect effects of variables and latent dimensions on social distancing, pandemic severity, and lockdown regime (Table 6). Direct and indirect impacts allow us to comprehend the core causes of social distancing and pandemic severity in different countries. Observing the direct and indirect effects, we understand that the direct effects of different latent factors on social distancing and pandemic severity is higher and significant compared to indirect effects. In some cases, indirect effects are statistically insignificant and often trivially small (Table 6). Thus, overall, the direct effects of latent factors represent the total effects without any mitigating or amplifying indirect effects. Results articulated in Table 6 illustrates that lockdown measures directly reduce people's mobility while socioeconomic and institutional factors increase the severity of the pandemic to a greater extent. Socioeconomic and institutional factors only have direct effects on lockdown, without any indirect influence on it.

Table 6. Direct, indirect, and total effects on social distancing, pandemic severity, and lockdown regime.

Latent Factor	Social Distancing			Pandemic Severity			Lockdown		
	Direct	Indirect	Total	Direct	Indirect	Total	Direct	Indirect	Total
Pandemic severity	−0.065	−0.001	−0.066	-	0.011	0.011	-	-	-
Lockdown measures	−0.626	−0.007	−0.633	-	−0.101	−0.101	-	-	-
Socioeconomic and institutional factors	−0.046	0.001	−0.045	0.847	0.007	0.855	−0.090	-	−0.090
Social distancing	-	0.011	0.011	−0.160	−0.002	−0.162	-	-	-

4. Discussion and Conclusions

COVID-19 has become a piercing issue and its numerous negative impacts on public health, economy, lifestyle, and wellbeing of populations are striking policymakers to come up with some solutions. To this end, this study provides significant contributions by empirically investigating the root causes of the interplay between mobility changes, pandemic severity, and lockdown regimes in 88 countries in the early stage of the pandemic. To perform this study, data were collected from multiple sources. An SEM was developed to find out the complex relationships among the observed variables and latent dimensions. Results from the SEM exhibit that different lockdown measures have significant repercussions to maintain social distancing. However, pandemic severity and socioeconomic and institutional context factors have no significant impact to sustain social distancing practices. The results also explain that socioeconomic and institutional context factors have significant effects on increasing pandemic severity. Elderly people, globalization, and employment in the service sector are primarily responsible for a higher number of coronavirus infection cases and deaths in many countries (e.g., USA, UK, Italy, and Spain). Moreover, social distancing is reasonably able to reduce the severity of coronavirus pandemic, although the impacts are marginal at the granularity of national populations (−0.162). It is also understood that lockdown measures affect the socio-economic context of the countries along with reducing the severity of the coronavirus pandemic. People are adjusting their lifestyle and travel pattern to cope with the new circumstances. Individuals and industries are adopting new alternative strategies to keep pace with the global trend. New possibilities are emerging in the world (i.e., a greater use of information and communication technologies in business and personal life) to flourish in a new environment. Thus, it is surmised that this new normal situation is shaping the personal and business world in such a way to keep moving and meet every demand of the people.

Several policy implications can be drawn from this analysis. An effective way to maintain social distancing is to implement strict lockdown measures. Rather than putting into effect casual stay-at-home recommendations and piecemeal efforts, comprehensive and strict lockdown measures are indispensable to maintain social distancing that can reduce coronavirus infection cases and deaths in a meaningful way. However, since globalization is a reality in the modern era, imposing strict restrictions on people and freight movement within and outside country boundaries is detrimental to the economy and to business partnership. Thus, alternative strategies (e.g., e-shopping, application of information technology) should be undertaken by the authorities to ensure the safe transfer of the people and freight from origin to destination and continue international trade during crisis times.

Despite making significant and timely contributions, the strengths of this study are bound by a few cautionary remarks. First, the Google mobility report was prepared based on data collected from Google account users who turned on their travel location history setting [38]. Thus, it may not represent the true travel behaviors of the general population. More generally, the sway that data quality may have on the conclusions of the analysis should not be brushed off. Data quality may be a concern because health outcome variables are difficult to measure with good accuracy as the pandemic unfolds. Furthermore, international studies are notoriously difficult to conduct due to the heterogeneous adherence to data quality standards in different national contexts. Second, data were collected from multiple sources and integrated to perform the analysis. Thus, it is very challenging to make consensus and consistent policy decisions that can be applied generally. Thirdly, to deal with the ambiguous definition of trips, a comparative analysis of essential versus non-essential travels can be performed based on a recent dataset on changes in the visits to non-essential venues (e.g., restaurants, department stores, and cinemas) published by Unacast [35]. Finally, this study has been conducted at the coarse geographic resolution of countries. Thus, a future study at a finer scale would provide more insights on the interplay between social distancing, pandemic severity, and lockdown regimes. In addition, we propose to pursue further research at the interface of mobility changes, pandemic severity, and lockdown regimes as the COVID-19 pandemic continues to afflict populations around the globe. As more complete time series become available and as the pandemic will have eased into other phases, the stability of our model, or alternatively its dynamic properties, will be critically important to assess to better prepare the world for future pandemics under changing socio-politico-medical contexts. However, this horizon maybe 6, 9, or even 12 months away. We believe the present analysis and results achieved here have value as they stand, as they capture the reality of the pandemic a few months after its global onset.

Author Contributions: Data curation, M.M.R., J.-C.T. and K.C.P.; Formal analysis, M.M.R., J.-C.T. and K.C.P.; Investigation, M.M.R., J.-C.T. and K.C.P.; Methodology, M.M.R. and J.-C.T.; Software, M.M.R.; Supervision, J.-C.T.; Validation, J.-C.T.; Visualization, M.M.R. and K.C.P.; Writing—original draft, M.M.R., J.-C.T. and K.C.P.; Writing—review & editing, M.M.R., J.-C.T. and K.C.P. All authors have read and agreed to the published version of the manuscript.

Funding: This research did not receive any specific funding from agencies in public, private, and non-profit organizations.

Conflicts of Interest: The authors declare no conflict of interest. The authors are responsible for the contents of this paper.

Appendix A. List of Variables Tested in the Model

Table A1. Description of the variables considered in the study.

Variable	Description	Measure	Source
l_case	Total infection cases per 1 million population	Cases/1M	Worldometer
l_death	Total deaths per 1 million population	Deaths/1M	Worldometer
Case_Mar20	Total number of infection cases on March 20	#	Worldometer
Case_May15	Total number of infection cases on May 15	#	Worldometer
Wcase_Pre	Weekly change of infection cases before April 17	#	Worldometer
Wcase_post	Weekly change of infection cases after April 17	#	Worldometer
Death_Mar20	Total number of deaths on March 20	#	Worldometer
Death_May15	Total number of deaths on May 15	#	Worldometer
Wdeath_Pre	Weekly change of death counts before April 17	#	Worldometer
Wdeath_post	Weekly change of death counts after April 17	#	Worldometer
RR	Percentage change of mobility to retail and recreation POIs compared to baseline	%	Google
TS	Percentage change of mobility to transit stations POIs compared to baseline	%	Google
WP	Percentage change of mobility to workplaces compared to baseline	%	Google
RD	Percentage change of time spent at home compared to baseline	%	Google
PODEN	Population density	Pop/km2	UN
MFR	Male-female ratio	Ratio	UN
GDPGR	Annual GDP growth rate	%	UN
GDPP	GDP per capita	$	UN
AE	Percentage of population employed in agriculture	%	UN
IE	Percentage of population employed in industry	%	UN
SE	Percentage of population employed in services	%	UN
UR	Percentage of unemployed people in the workforce	%	UN
CPI	Consumer Price Index	Index	UN
HE	Percentage of health expenditure to total GDP	%	UN
HP	Number of physicians per 1000 population	#	UN
FA	Percentage of forested areas to country land area	%	UN
Tourist	Number of tourist/visitor arrivals at national borders (000)	#	UN
EI	Average years of schooling (adults) and expected years of schooling (children)	Index	UNDP
LR	Percentage of population aged 15 and above who can read and/or write	%	UNDP
ExIm	Percentage of exports and imports of goods and services to total GDP	%	UNDP
FDI	Percentage of additional long-term and short-term capital of total GDP	%	UNDP
Remit	Percentage of earning by migrants to total GDP	%	UNDP
IU	Percentage of people with access to the internet	%	UNDP
MU	Mobile phone subscriptions per 100 people	#	UNDP
MR	Ratio of difference between in-migrants and out-migrants to the average population per 1000 people	Ratio	UNDP
MA	Median age of the people	Year	UNDP
DR	Number of elderly (65 and older) per 100 working age (15–64)	Ratio	UNDP
UR	Percentage of people living in urban areas	%	UNDP
Temp	Average temperature in April (DC: Degree Centigrade)	DC	Weatherbase
Rainfall	Average precipitation in April	Mm	Weatherbase

Table A1. *Cont.*

Variable	Description	Measure	Source
DE1	Days elapsed since the lockdown to April 17	Days	Different sources
DE2	Days elapsed between first case and imposing social distancing measures	Days	Different sources
PDI	Power Distance Index	Index	Hofstede
IDV	Individualism Versus Collectivism	Index	Hofstede
MAS	Masculinity Versus Feminity	Index	Hofstede
UAI	Uncertainty Avoidance Index	Index	Hofstede
LTO	Long Term Orientation versus Short Normative Orientation	Index	Hofstede
IVR	Indulgence Versus Restraint	Index	Hofstede
EFS	Economic Freedom Score	Score	The Fraser Institute
KOFGI	KOF Globalisation Index (2017)	Index	KOF Swiss Economic Institute
Regime	Regime classification (2008)	Type	Xavier Marquez
Agereg	Age in years of the current regime as classified by regime (2008)	Year	Xavier Marquez
FC	Date of the first reported infection case	Date	Worldometer
LRL	Localized recommended lockdown (1 = Yes, 0 = No)	Dummy (1, 0)	BBC
LL	Localized lockdown (1 = Yes, 0 = No)	Dummy (1, 0)	BBC
NRL	National recommended lockdown (1 = Yes, 0 = No)	Dummy (1, 0)	BBC
NL	National lockdown (1 = Yes, 0 = No)	Dummy (1, 0)	BBC
Asia	Country located in Asia continent (1 = Yes, 0 = No)	Dummy (1, 0)	Wikipedia
Africa	Country located in Africa continent (1 = Yes, 0 = No)	Dummy (1, 0)	Wikipedia
Europe	Country located in Europe continent (1 = Yes, 0 = No)	Dummy (1, 0)	Wikipedia
North America	Country located in North America continent (1 = Yes, 0 = No)	Dummy (1, 0)	Wikipedia
South America	Country located in South America continent (1 = Yes, 0 = No)	Dummy (1, 0)	Wikipedia
Australia	Country located in Australia continent (1 = Yes, 0 = No)	Dummy (1, 0)	Wikipedia
ME	Country located in Middle East (1 = Yes, 0 = No)	Dummy (1, 0)	Wikipedia
SC	Closing of schools and universities (1 = required closing, 0 = no measure/recommended closing)	Dummy (1, 0)	Oxford University
WPC	Closing of workplaces (1 = required closing, 0 = no measure/recommended closing)	Dummy (1, 0)	Oxford University
CPE	Cancellation of public events (1 = required canceling, 0 = no measure/recommended canceling)	Dummy (1, 0)	Oxford University
RG	Restriction on gatherings (1 = restriction on gathering-less than 100 people, 0 = no restrictions/restriction on very large gathering-above 100 people)	Dummy (1, 0)	Oxford University
CPT	Closing of public transport (1 = required closing, 0 = no measure/recommended closing)	Dummy (1, 0)	Oxford University
SH	Stay-at-home order (1 = required not to leave, 0 = no measure/recommended not to leave)	Dummy (1, 0)	Oxford University
RIM	Restriction on internal movement between cities/regions (1 = required not to travel, 0 = no measure/recommended not to travel)	Dummy (1, 0)	Oxford University
RIT	Restriction on international travel (1 = ban/quarantine arrivals, 0 = no restriction/screening arrivals)	Dummy (1, 0)	Oxford University
CP	Direct cash payments to people who lose their jobs or cannot work (1 = income support from govt., 0 = no income support	Dummy (1, 0)	Oxford University

Table A1. *Cont.*

Variable	Description	Measure	Source
FO	Freezing financial obligations for households (1 = debt/contract relief, 0 = no measures)	Dummy (1, 0)	Oxford University
ES	Announced economic stimulus spending	$	Oxford University
PIC	Public campaigns (1 = public information campaign, 0 = public officials urging caution)	Dummy (1, 0)	Oxford University
TP	Testing policy (1 = comprehensive testing policy, 0 = no/limited testing policy)	Dummy (1, 0)	Oxford University
CT	Contact testing after positive diagnosis (1 = comprehensive testing policy, 0 = no/limited testing policy)	Dummy (1, 0)	Oxford University
EIH	Short term spending on healthcare system (e.g., hospitals, masks)	$	Oxford University
IV	Public spending on Covid-19 vaccine development	$	Oxford University
SI	Government Response Stringency Index (0 to 100, 100 = strictest)	Index	Oxford University
AGE65	Percentage of population aged 65 and older	%	Oxford University
AGE70	Percentage of population aged 70 and older	%	Oxford University
DB	Percentage of people with diabetes	%	Oxford University
FS	Percentage of female smokers	%	Oxford University
MS	Percentage of male smokers	%	Oxford University
HB	Hospital beds per 1000 people	Beds/1k people	Oxford University
PD	Presidential democracy	Dummy (1, 0)	Oxford University
SPD	Mixed (semi-presidential) democracy	Dummy (1, 0)	Oxford University
ParD	Parliamentary democracy	Dummy (1, 0)	Oxford University
CD	Civilian dictatorship	Dummy (1, 0)	Oxford University
MD	Military dictatorship	Dummy (1, 0)	Oxford University
RD	Royal dictatorship	Dummy (1, 0)	Oxford University

References

1. Worldometer. Reported Cases and Deaths by Country, Territory, or Conveyance. Available online: https://www.worldometers.info/coronavirus/#countries (accessed on 30 April 2020).
2. CDC. *New ICD-10-CM code for the 2019 Novel Coronavirus (COVID-19)*; CDC: Atlanta, Georgia, USA, 2020.
3. Hopkins, J. COVID-19 Dashboard Johns Hopkins Coronavirus Resource Center. Available online: https://coronavirus.jhu.edu/map.html (accessed on 30 April 2020).
4. Evans, O. Socio-economic impacts of novel coronavirus: The policy solutions. *Bizecons Q.* **2020**, *7*, 3–12.
5. Ritchie, E.O.-O.H.; Beltekian, D.; Mathieu, E.; Hasell, J.; Macdonald, B.; Giattino, C.; Roser, M. Coronavirus (COVID-19) Cases. Available online: https://ourworldindata.org/coronavirus (accessed on 30 April 2020).
6. Abodunrin, O.; Oloye, G.; Adesola, B. Coronavirus pandemic and its implication on global economy. *Int. J. Arts Lang. Bus. Stud.* **2020**, *4*, 13–23.
7. Craig, A.T.; Heywood, A.E.; Hall, J. Risk of COVID-19 importation to the Pacific islands through global air travel. *Epidemiol. Infect.* **2020**, *148*, e71. [CrossRef] [PubMed]
8. Ma, T.; Heywood, A.; MacIntyre, C.R. Travel health risk perceptions of Chinese international students in Australia—Implications for COVID-19. *Infect. Dis. Health* **2020**. [CrossRef]
9. Igwe, P.A. *Coronavirus with Looming Global Health and Economic Doom*; African Development Institute of Research Methodology: Enugu North, Nigeria, 2020.
10. Samuel, J.; Ali, G.G.M.N.; Rahman, M.M.; Esawi, E.; Samuel, Y. COVID-19 Public Sentiment Insights and Machine Learning for Tweets Classification. *Information* **2020**, *11*, 314. [CrossRef]
11. Samuel, J.; Rahman, M.; Ali, G.; Samuel, Y.; Pelaez, A. Feeling Like It is Time to Reopen Now? COVID-19 New Normal Scenarios based on Reopening Sentiment Analytics. *IEEE Access* **2020**, *8*, 142173–142190. [CrossRef]
12. Rahman, M.; Ali, G.; Li, X.J.; Paul, K.C.; Chong, P.H. Twitter and Census Data Analytics to Explore Socioeconomic Factors for Post-COVID-19 Reopening Sentiment. *arXiv* **2020**, arXiv:2007.00054. [CrossRef]
13. Ayittey, F.K.; Ayittey, M.K.; Chiwero, N.B.; Kamasah, J.S.; Dzuvor, C. Economic impacts of Wuhan 2019-nCoV on China and the world. *J. Med. Virol.* **2020**, *92*, 473–475. [CrossRef] [PubMed]

14. Iacus, S.M.; Natale, F.; Satamaria, C.; Spyratos, S.; Vespe, M. Estimating and Projecting Air Passenger Traffic during the COVID-19 Coronavirus Outbreak and its Socio-Economic Impact. *arXiv* **2020**, arXiv:2004.08460. [CrossRef]
15. Long, H.; Dam, A.V. U.S. Unemployment Rate Soars to 14.7 Percent, the Worst Since the Depression Era. The Washington Post. Available online: https://www.washingtonpost.com/business/2020/05/08/april-2020-jobs-report/?utm_campaign=wp_post_most&utm_medium=email&utm_source=newsletter&wpisrc=nl_most. (accessed on 8 May 2020).
16. Economics, T. United States Unemployment Rate: 2021–2022 Forecast. Available online: https://tradingeconomics.com/united-states/unemployment-rate#:~{}:text=In%20the%20long%2Dterm%2C%20the,according%20to%20our%20econometric%20models (accessed on 30 April 2020).
17. Sherman, E. Fortune 2020. In *Fortune*; Fortune Media Group Holdings: New York, NY, USA, 2020; p. 2020.
18. Ivanov, D. Predicting the impacts of epidemic outbreaks on global supply chains: A simulation-based analysis on the coronavirus outbreak (COVID-19/SARS-CoV-2) case. *Transp. Res. Part E: Logist. Transp. Rev.* **2020**, *136*, 101922. [CrossRef] [PubMed]
19. Liu, H.; Bai, X.; Shen, H.; Pang, X.; Liang, Z.; Liu, Y. Synchronized Travel Restrictions across Cities can be Effective in COVID-19 Control. *medRxiv* **2020**. [CrossRef]
20. Zheng, R.; Xu, Y.; Wang, W.; Ning, G.; Bi, Y. Spatial transmission of COVID-19 via public and private transportation in China. *Travel Med. Infect. Dis.* **2020**, 101626. [CrossRef] [PubMed]
21. Adekunle, A.I.; Meehan, M.; Alvarez, D.R.; Trauer, J.; McBryde, E. Delaying the COVID-19 epidemic in Australia: Evaluating the effectiveness of international travel bans. *medRxiv* **2020**. [CrossRef] [PubMed]
22. Bryant, P.; Elofsson, A. Estimating the Impact of Mobility Patterns on COVID-19 Infection Rates in 11 European Countries. *Medrxiv* **2020**. [CrossRef]
23. Flaxman, S.; Mishra, S.; Gandy, A.; Unwin, H.J.T.; Mellan, T.A.; Coupland, H.; Whittaker, C.; Zhu, H.; Berah, T.; Eaton, J.W. Estimating the effects of non-pharmaceutical interventions on COVID-19 in Europe. *Nature* **2020**, *584*, 257–261. [CrossRef]
24. Fang, H.; Wang, L.; Yang, Y. *Human Mobility Restrictions and the Spread of the Novel Coronavirus (2019-nCoV) in China*; 26906; National Bureau of Economic Research: Cambridge, MA, USA, 2020.
25. Aleta, A.; Hu, Q.; Ye, J.; Ji, P.; Moreno, Y. A data-driven assessment of early travel restrictions related to the spreading of the novel COVID-19 within mainland China. *medRxiv* **2020**. [CrossRef]
26. Anzai, A.; Kobayashi, T.; Linton, N.M.; Kinoshita, R.; Hayashi, K.; Suzuki, A.; Yang, Y.; Jung, S.-m.; Miyama, T.; Akhmetzhanov, A.R.; et al. Assessing the Impact of Reduced Travel on Exportation Dynamics of Novel Coronavirus Infection (COVID-19). *J. Clin. Med.* **2020**, *9*, 601. [CrossRef]
27. Chang, M.-C.; Kahn, R.; Li, Y.-A.; Lee, C.-S.; Buckee, C.O.; Chang, H.-H. Modeling the impact of human mobility and travel restrictions on the potential spread of SARS-CoV-2 in Taiwan. *medRxiv* **2020**. [CrossRef]
28. Chinazzi, M.; Davis, J.T.; Ajelli, M.; Gioannini, C.; Litvinova, M.; Merler, S.; Piontti, A.P.y.; Mu, K.; Rossi, L.; Sun, K.; et al. The effect of travel restrictions on the spread of the 2019 novel coronavirus (COVID-19) outbreak. *Science* **2020**, *368*, 395–400. [CrossRef]
29. Devi, S. Travel restrictions hampering COVID-19 response. *Lancet* **2020**, *395*, 1331–1332. [CrossRef]
30. Jin, H.; Liu, J.; Cui, M.; Lu, L. Novel coronavirus pneumonia emergency in Zhuhai: Impact and challenges. *J. Hosp. Infect.* **2020**, *104*, 452–453. [CrossRef]
31. Linka, K.; Peirlinck, M.; Costabal, F.S.; Kuhl, E. Outbreak dynamics of COVID-19 in Europe and the effect of travel restrictions. *Comput. Methods Biomech. Biomed. Eng.* **2020**, 1–8. [CrossRef] [PubMed]
32. Qiu, Y.; Chen, X.; Shi, W. Impacts of social and economic factors on the transmission of coronavirus disease (COVID-19) in China. *Medrxiv* **2020**. [CrossRef]
33. Tian, H.; Li, Y.; Liu, Y.; Kraemer, M.; Chen, B.; Cai, J.; Li, B.; Xu, B. Early evaluation of the Wuhan City travel restrictions in response to the 2019 novel coronavirus outbreak. *Medrxiv* **2020**, 7. [CrossRef]
34. Jittrapirom, P.; Tanaksaranond, G. *An Exploratory Survey on the Perceived Risk of COVID-19 and Travelling*; SocArXiv, University of Maryland: College Park, MD, USA, 2020.
35. Engle, S.; Stromme, J.; Zhou, A. Staying at Home: Mobility Effects of COVID-19. *SSRN J.* **2020**. [CrossRef]
36. Gao, S.; Rao, J.; Kang, Y.; Liang, Y.; Kruse, J. Mapping County-Level Mobility Pattern Changes in the United States in Response to COVID-19. *SSRN J.* **2020**. [CrossRef]
37. Rodríguez-Morales, A.J.; MacGregor, K.; Kanagarajah, S.; Patel, D.; Schlagenhauf, P. Going global–Travel and the 2019 novel coronavirus. *Travel Med. Infect. Dis.* **2020**, *33*, 101578. [CrossRef]

38. Google. *COVID-19 Community Mobility Report*; Google: Mountain View, CA, USA, 2020.
39. Dunford, D.; Dale, B.; Stylianou, N.; Lowther, E.; Ahmed, M.; De la Torres Arenas, I. *Coronavirus: The World in Lockdown in Maps and Charts*; BBC: London, UK, 2020.
40. University of Oxford. *Coronavirus Government Response Tracker*; University of Oxford: Oxford, UK, 2020.
41. UNDP. *Global Human Development Indicators*; UNDP: New York, NY, USA, 2019.
42. UN. *Popular Statistical Tables, Country (Area) and Regional Profiles*; UN: New York, NY, USA, 2019.
43. Hofstede. *National Culture: Country Comparison*; Hofstede Insights: Helsinki, Finland, 2020.
44. Gygli, S.; Haelg, F.; Potrafke, N.; Sturm, J.-E. The KOF globalisation index–revisited. *Rev. Int. Organ.* **2019**, *14*, 543–574. [CrossRef]
45. Aktay, A.; Bavadekar, S.; Cossoul, G.; Davis, J.; Desfontaines, D.; Fabrikant, A.; Gabrilovich, E.; Gadepalli, K.; Gipson, B.; Guevara, M. Google COVID-19 community mobility reports: Anonymization process description (version 1.0). *arXiv* **2020**, arXiv:2004.04145.
46. Najaf, P.; Thill, J.-C.; Zhang, W.; Fields, M.G. City-level urban form and traffic safety: A structural equation modeling analysis of direct and indirect effects. *J. Transp. Geogr.* **2018**, *69*, 257–270. [CrossRef]
47. Wang, D.; Lin, T. Built environment, travel behavior, and residential self-selection: A study based on panel data from Beijing, China. *Transportation* **2019**, *46*, 51–74. [CrossRef]
48. Golob, T.F. Structural equation modeling for travel behavior research. *Transp. Res. Part B Methodol.* **2003**, *37*, 1–25. [CrossRef]
49. Zong, F.; Yu, P.; Tang, J.; Sun, X. Understanding parking decisions with structural equation modeling. *Phys. A Stat. Mech. Appl.* **2019**, *523*, 408–417. [CrossRef]
50. Stata Cooperation. *Stata 15*; Stata Cooperation: College Station, TX, USA, 2017.

sustainability

MDPI

Review

Disinfection Methods and Survival of SARS-CoV-2 in the Environment and Contaminated Materials: A Bibliometric Analysis

Adel Al-Gheethi [1,*], Mohammed Al-Sahari [1], Marlinda Abdul Malek [2,*], Efaq Noman [3,4], Qais Al-Maqtari [5], Radin Mohamed [1], Balkis A. Talip [4], Sadeq Alkhadher [1] and Md. Sohrab Hossain [6]

1 Micropollutant Research Centre (MPRC), Faculty of Civil Engineering & Built Environment, Universiti Tun Hussein Onn Malaysia, Parit Raja 86400, Johor, Malaysia; mohammedalsahari@gmail.com (M.A.-S.); maya@uthm.edu.my (R.M.); sadeqalkhather@yahoo.com (S.A.)
2 Institute of Sustainable Energy, Universiti Tenaga Nasional, Selangor 43000, Malaysia
3 Department of Applied Microbiology, Faculty of Applied Science, Taiz University, Taiz 00967, Yemen; eanm1984@gmail.com
4 Faculty of Applied Sciences and Technology, University Tun Hussein Onn Malaysia (UTHM), Pagoh Higher Education Hub, KM 1, Jalan Panchor, Panchor 84000, Johor, Malaysia; balkis@uthm.edu.my
5 Department of Biology, Faculty of Sciences, Sanaa University, Sanaa 00967, Yemen; qais.almaqtari@gmail.com
6 School of Industrial Technology, University Sains Malaysia (USM), 11800 USM, Penang, Malaysia; sohrab@usm.my
* Correspondence: adel@uthm.edu.my or adelalghithi@gmail.com (A.A.-G.); Marlinda@uniten.edu.my (M.A.M.); Tel.: +60-19-332-2775 or +60-3-8921-2020 (M.A.M.); Fax: +60-3-8921-2116 (M.A.M.)

Received: 22 July 2020; Accepted: 20 August 2020; Published: 9 September 2020

Abstract: The presence of SARS-CoV-2 in sewage and water resources has been used as an indication for the possible occurrence of the virus among communities and for its potential of transmission among humans through the surrounding environment or water resources. In order to reduce the transmission of SARS-CoV-2, contaminated surfaces should be disinfected frequently by using an effective disinfectant. The present review discusses a bibliometric analysis of the global SARS-CoV-2 research and focuses mainly on reviewing the efficiency of the most traditional disinfection technologies. The disinfection methods reviewed include those for hospitals' or medical facilities' wastewater, contaminated surfaces, and contaminated masks. The elimination of the virus based on the concept of sterility assurance level (SAL) is also discussed. In addition, the chemical disinfectants that are currently used, as well as their temporary efficiency, are also reviewed. The different technologies that are globally used for disinfection processes during the COVID-19 pandemic are shown. However, more advanced technologies, such as nanotechnology, might have more potential for higher inactivation effectiveness against SARS-CoV-2.

Keywords: disinfection; COVID-19; sterility assurance level; medical masks; inactivation; UV irradiation

1. Introduction

The genus of coronavirus is composed of at least three genetically and antigenically distinct groups of CoV that cause mild to severe enteric, respiratory, or systemic disease in domestic and wild animals, poultry, rodents, and carnivores, as well as mild colds in humans. SARS-CoV-2, from the genus of coronavirus, is an enveloped and positive-sense virus with single-stranded RNA. However, SARS-CoV-2 has a high potential to cause respiratory or systemic illnesses in humans.

The zoonotic transmission of this virus has been reported in several regions, including a severe acute respiratory syndrome CoV (SARS-CoV) infection, which was transmitted from bats and the Himalayan palm civet (*Paguma larvata*) in 2002–2003 in China, and the Middle East respiratory syndrome CoV (MERS-CoV), which has been transmitted from dromedary camels (*Camelus dromedaries*) in the Arabian Peninsula since 2012. SARS-CoV-2 causes self-limiting upper- and lower-respiratory infections in people with immunocompetence. It has been detected that SARS-CoV-2 has 89% similarity in the nucleotide sequences to bat SARS-like-CoVZXC21, while it has 82% similarity to the human SARS-CoV genome [1–3].

One of the main serious problems with SARS-CoV-2 is the transmission route. It has been revealed that the virus has a clear transmission route through respiratory droplets and survives for more than one day in the environment. It is hypothesized that there is a risk of exposure to the virus on environmental surfaces, which has not yet been proven. Thus, indoor and outdoor surfaces should be disinfected frequently [4,5]. This study presents a bibliometric analysis of global SARS-CoV-2 research and focuses mainly on reviewing the efficiency of the most traditional disinfection technologies on contaminated surfaces, medical masks, and hospital wastewater, which have a higher probability of having greater loads of the virus. The elimination of the virus based on the concept of sterility assurance level is also discussed.

2. Bibliometric Analysis of SARS-CoV-2 in the Literature

Literature data were collected from the SCOPUS database by restricting the search to all published documents from 1 January, 2019 to 24 June, 2020 under the keywords "SARS-CoV-2" AND "Transmission Route" OR "Genomic Map" OR "Pathogenicity" OR "Symptoms" OR "Therapeutics" OR "Survival" OR "Environment", from which 2000 documents were obtained. The bibliometric analysis used various bibliometric ranking indices, including the top journals, countries, and the most frequent keywords in the obtained documents. The worldwide distribution of documents was used to clarify the connections and relationships between terms, keywords, and countries as well as to establish a rainbow density map of bibliographic coupling and countries of publication in this area, as presented in Figures 1–3. In the current review, the top countries are shown in the bibliometric map of the published documents and their links among countries (Figure 1). A circle's size indicates a country's number of publications, while the color of the circle refers to the utilized keywords. In the same sense, the connected lines between countries express the citations and the link strengths between the publications of each country. It was noted that most research conducted on SARS-CoV-2 was performed in the USA, followed by China and Italy. The *Journal of Medical Virology* and *Journal of Medical Hypotheses* were the top journals that published high numbers of SARS-CoV-2 articles (Figure 2). The articles addressed the infection among females, the isolation and purification of the virus for further studies, genetic mapping, and the symptoms of the infections, such as fever (Figure 3).

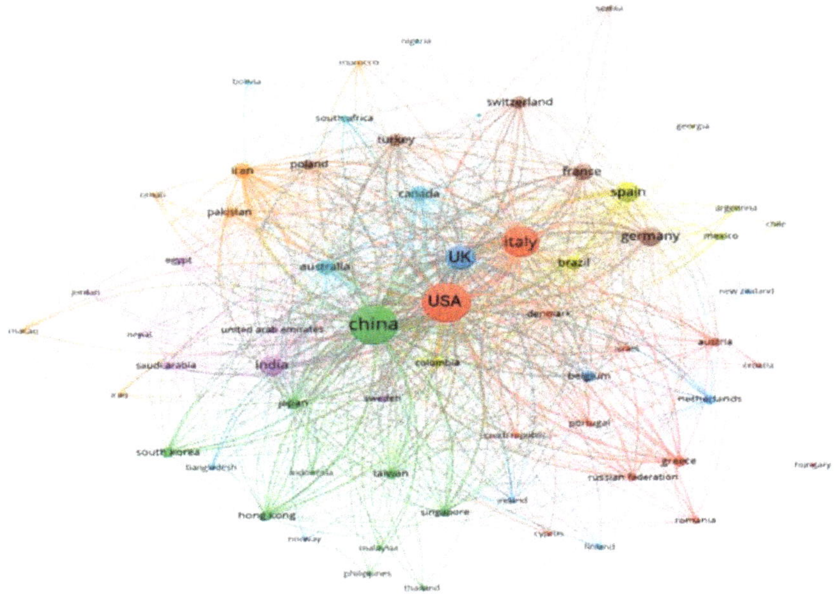

Figure 1. Bibliographic map of the publisher countries of SARS-CoV-2 documents listed in the SCOPUS database (2019 to 22/07/2020).

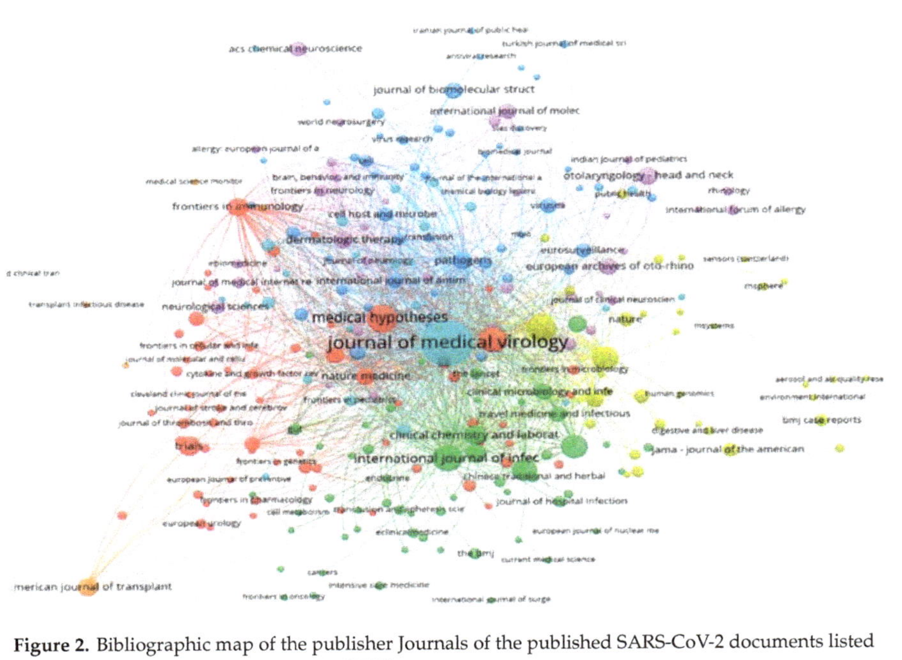

Figure 2. Bibliographic map of the publisher Journals of the published SARS-CoV-2 documents listed in the SCOPUS database (2019 to 22/07/2020).

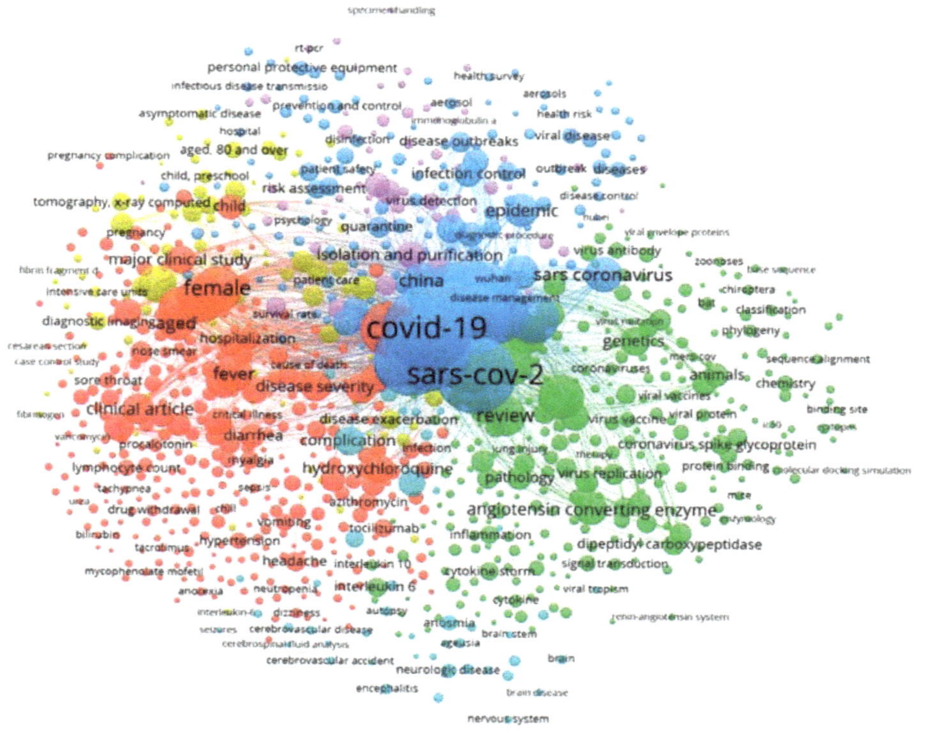

Figure 3. Bibliographic map of the keywords used in the published SARS-CoV-2 documents listed in the SCOPUS database (2019 to 22/07/2020).

The main findings of the reviewed articles are summarized in Figure 4. In order to focus on the literature within the study scope, all reviews were categorized into five sections, including the sources of the SARS-CoV-2 pandemic, pathogenicity and symptoms, transmission, diagnostic and genome organization, and therapeutics that are associated with the research objectives.

The life cycle of the virus is presented in Figure 5. The virus enters into the cell cytoplasm through the ACE2 receptors available on the surface of the human cells. The virus genome is multiplied inside the cells and each of the new genomes is enveloped by the cell membrane and then released to infect a new cell.

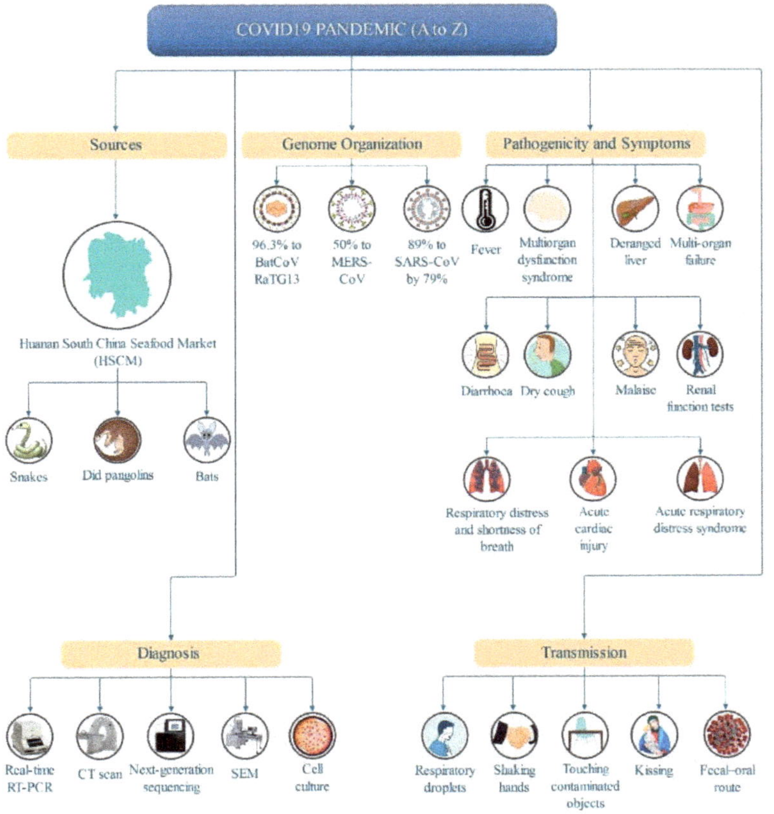

Figure 4. Main findings of SARS-CoV-2 documents listed in the SCOPUS database (2019 to 22/07/2020).

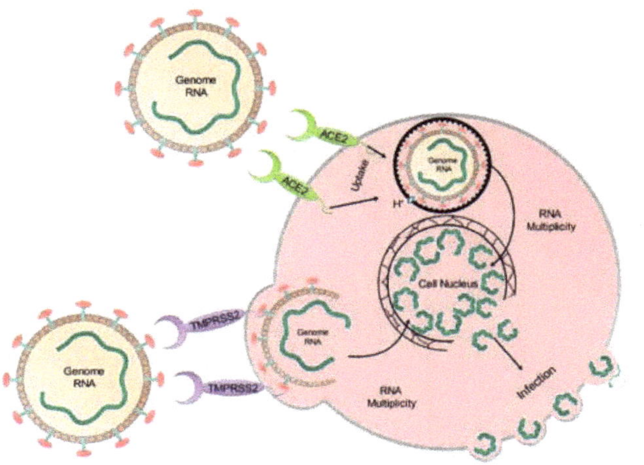

Figure 5. SARS-CoV-2 cycle in the human cells.

3. SARS-CoV-2 in Wastewater

Hospitals' and medical facilities' wastewater may contain significant numbers of microbial communities including pathogenic bacteria and viruses. With the emergence of the novel coronavirus, scientists are now facing challenges in understanding its potential faecal-oral transmission and its association with the environmental pollutants originating from the disposal of hospital wastewater. It has been indicated by researchers in the literature that the pathogenicity and virulence factors of the infectious agents lie in their ability to survive in the environment as well as their potential to transmit into humans and animals without a host cell [6]. Hospital and medical facilities' effluents should be subjected to an efficient disinfection process before final disposal into the environment. There are two factors that should be considered for any disinfection process, namely the biological indicators and the assessment procedure. The selection of an indicator for evaluating the efficiency of the disinfection process depends on the nature of the inactivation process, including chemical or physical processes. The response of indicator organisms reflects the behavior of infectious agents. For instance, bacteriophages are used as an indicator for human viral pathogens [7].

Although there are no scientific reports on the presence of SARS-CoV-2 in sewage, a number of studies have reported high levels of viral shedding on fecal samples derived from SARS-CoV-2 patients [8,9]. In addition, several news reports in France and the Netherlands have reported the presence of SARS-CoV-2 gene fragments in sewage and water surfaces [9].

4. Treatment of Hospital Wastewater Contaminated with SARS-CoV-2

Many of the treatment processes used in treating wastewater focused mainly on inactivating pathogenic bacteria that are harmful to aquatic life. However, recent studies have revealed a detectable number of pathogens that remained in effluents even after they underwent sewage treatment [10]. Chlorine is one of the most common disinfectants used in hospitals and medical facilities. In addition to being effective in killing most microbial communities, this detergent has a negative effect on the environment, which may be harmful to aquatic life if these contaminants enter the streams. Chlorination may also lead to the production of nitrosodimethylamine (NDMA), which is known to be carcinogenic to humans. The chlorination process can react with organic matter, forming another carcinogenic compound called trihalometanes (THMs) [11]. Conventional filtration techniques employed in wastewater treatment are not highly effective in removing micropollutant such as viruses. In addition, some microbes were found to be resistant to certain types of chemical materials. These findings are consistent with the study reported by Al-Gheethi et al. [12], which showed the presence of viable microbial cells even after going through the treatment process.

The use of solar-based disinfection (SODIS) technology is a promising approach, especially in the process of disinfection of water and wastewater. This technology is suitable based on the availability of high solar radiation, and is characterized by low resources cost and sustainability [13]. Meanwhile, the application of nanotechnology in wastewater treatment has also been reported in the literature [14]. For instance, Noman et al. [15] investigated the inactivation of antibiotic-resistant *Escherichia coli* (Gram-negative) and *Staphylococcus aureus* (Gram-positive) seeded in greywater by bimetallic bio-nanoparticles. The bimetallic nanoparticles (Zn/Cu NPs) were biosynthesized in secondary metabolite of a novel fungal strain identified as *Aspergillus iizukae* EAN605. This study revealed the high efficiency of Zn/Cu NPs in inhibiting the growth of *E. coli and S. aureus*. The inactivation mechanism showed that the bacterial cells were inactivated due to the damage in the cell wall structure. This is due to the degradation of carbohydrates and protein structures of the bacterial cell wall. Analysis using Fourier transform infrared spectroscopy (FTIR) confirmed that destruction has occurred in the C–C bond of the functional groups found in the bacterial cell wall. The combination between SODIS and nanotechnology might provide a novel process of disinfection in inactivating human viruses. ZnO are the most common nanoparticles used for wastewater disinfection and these particles were reported to be more effective under sunlight, which might increase the antiviral activity [16,17]. Wastewater treatment plants are usually operated all days of the week for 24 hours. In many places located far from the Equator, solar radiation is very limited in winter and the solar radiation is depleted during night; hence, SODIS is not a suitable method to employ in these areas. In addition, adequate UV radiation cannot penetrate the surface of wastewater effectively. Therefore, if UV radiation is used, treatment should be performed using a reactor, where wastewater is mixed and drained under UV lamps. Nonetheless, Kibbee and Örmeci [18] stated that peracetic acid (PAA) with low-pressure ultraviolet (LP-UV) radiation can inactivate Coxsackievirus B3 (CVB3) in municipal wastewater.

Sterility assurance level (SAL) is a term used to describe the killing efficiency of a treatment process, where the treatment process is very effective if the SAL is very low, as described by ISO 13408-1 [19], and log reduction (accepted by United States Environmental Protection Agency USEPA) is a term that is commonly used for assessing the efficiency of the disinfection processes. SAL is usually expressed as 10^{-n}. A 10^{-3} or 10^{-6} value is most often used for sterilization depending on the initial concentration of the pathogen. Log reduction is calculated as (10^{-1}), representing a 90% reduction in microbial population. The process of disinfection with a 6 log reduction (10^{-6}) indicates that the microbial population decreases from one million (10^{6}) to almost zero, or a reduction of 99.9999%. A kill rate of 99.99% is expressed in the form of a 4 log reduction.

In order to confirm that the removal of treated effluent is safe, the growth of inactive bacteria should be determined as an indicator for inactive viruses. Therefore, samples should be isolated and cultured on enrichment media. These cells are considered killed if no growth can be seen in the culture medium after the incubation period. However, the storage conditions of disinfected samples may affect the ability of microbial cells to resuscitate. Al-Gheethi et al. [20] revealed that *Salmonella* spp. *S. aureus* and *E. faecalis* were resuscitated in sewage samples treated with solar disinfection (SODIS) for 6 h and stored at 37 °C for 4 days. The survival of cells may occur due to the insufficiency of the disinfection process to damage the cell walls of the bacteria and in this case the pathogen growth potential (PGP) bioassay needs to be carried out.

5. Disinfection of Sars-Cov-2 in Contaminated Surfaces

The efficiency of several disinfectants against microbes depends on the ability to create an irreversible destruction of microbial cells. Thus, several methods can be used including the process of physical and chemical disinfection.

However, the concept of chemical disinfection depends on the inhibition of a critical metabolic and anabolic pathways that cause microbial activity to be irreversibly inhibited. For some types of microbes, the cell has an alternative metabolic pathway, which can cause the cell to remain active even after disinfection. On the other hand, the physical disinfection is dependent on the destruction of the cytoplasm. This damage is irreversible when microbial cells do not have the potential to recover to normal activity due to damage to the cells' enzymes.

Many studies have been conducted to better understand the mechanism of action of antibacterial disinfectants. By contrast, similar studies on viruses are relatively sparse. Viruses have no metabolic activity. Therefore, studies related to whether the disinfectant agent used can inactivate prions are still limited. However, a study has reported that lipophilic disinfectants and chaotropic agents have little difficulty in reacting to a normal lipid bilayer [19]. The veridical efficiency of chemical disinfection may act by denaturing virus envelops, which are derived from portions of host cell membranes (phospholipids and proteins). Nonetheless, the denaturation of the virus envelop may not cause damage to the virus genome. Thus, many chemical disinfectants, such as alcohol (70%), chlorine, soap, and detergent may be inadequately effective in inactivating the virus genome. Physical disinfection, such as temperature and pressure, would be more effective due to its ability to destruct both the viral envelope and the genome.

Another limitation in the use of chemical disinfectants is their stability in the environment. The inactivation of viruses on contaminated surfaces by using alcohol or detergents may work immediately. However, the ability of disinfectants to kill viruses on the surface in outdoor areas, which are exposed to pollutants, is limited. This is because the coverage rate of detergents is limited, and it is possible that the detergent is not evenly distributed. Alcohol and other disinfectants loss their activity against the viruses or any microbes after a few minutes due to high evaporation rates, especially in tropical countries. It has been demonstrated that the natures of chemical disinfectants play an important role in determining their efficacy. This process is influenced by environmental parameters, such as temperature and relative humidity.

In the chlorination of water and wastewater, there is an amount of residual chlorine that constitutes an important safeguard against the risk of subsequent microbial contamination after treatment [20]. The diluted chlorine is also effective to inactivate the virus. The diluted chlorine should be freshly prepared due to its sensitivity to light, which may cause it to lose its effectiveness [21]. Residual chlorine is not available for surface disinfection. Thus, outdoor surfaces, such as in public transport, are recommended to be disinfected frequently. For this reason, there is an urgency to find a disinfectant that contains stable active ingredients, which can be used to frequently combat surface contamination, particularly in open public areas. SARS-CoV-2 has the ability to survive for longer periods (possibly up to 24 hours) compared to chemical disinfectants that disperse on contaminated surfaces. This is because chemical disinfectants have short stability. By contrast, nanoparticles have a longer stability in the environment. This has been proven by a study of the use of nanoparticles against the H1N1 virus in different environments [17].

NPs biosynthesis represents an alternative to chemical and physical synthesis processes. For example, the surface of fungal cells consists of proteins, enzymes, and reducing components that produce reducing agents, such as napthoquinones and anthraquinones [22]. Therefore, the use of fungi as stabilizing/reducing agents is very effective in producing NPs. Moreover, fungi have a significant morphology, as they can produce various types of intracellular enzymes for the development of metal and metal oxide NPs. Various species of fungi have been used in synthesizing NP metals due to their high binding capacity and ability to bioaccumulate metals as well as their intracellular uptake. The biosynthesis of NPs using fungi is better than other microorganisms because fungi are easy to isolate, grow fast, and can be cultured in the laboratory [15].

ZnO-NPs with a size of less than 100 nm may have the potential to inactivate SARS-CoV-2 (with a size of 400 nm) and physically destroy the viral genome. In addition, ZnO-NPs plays a dual role as an anti-adhesive and antimicrobial agent simultaneously. It also has the ability to stabilize on the surface under different humidity and ambient conditions [23]. Because SARS-CoV-2 poses a high risk of infection, studies on the effectiveness of nano-disinfectants have been initiated using bacteriophage as a modal organism.

6. Disinfection of Disposable Medical Face Masks Contaminated with SARS-CoV-2

The use of disposable masks such as respirators facial filters (N95), medical face masks, and other face masks (such as non-medical, cloth, or barrier masks) during the SARS-CoV-2 pandemic has increased the amount of waste. A WHO report [24] states that approximately 89 million medical masks are needed each month in dealing with this pandemic. Accordingly, the disposal of this mask to the environment, especially among public users, without undergoing the process of disinfection may contribute to the spread of the virus. Several countries have adopted regulation for reuse of disposable masks, as studied by Rubio-Romero et al. [25]. These regulations are used to ensure that the disinfection process is carried out in accordance with the mandated procedure. The choice of disinfection process depends on the type of mask and the species of the virus.

It has been reported that the survival period of the COVID-19 virus ranges from few hours (72 h) to several days, which might reach up to 9 days [26,27]. The most common disinfectants used are ethanol (62–71%), hydrogen peroxide (0.5%) and sodium hypochlorite (0.1%). These disinfectants have the ability to inactivate the virus within 1 minute, since the inactivation process depends on the oxidation reactions. Heat sterilization may be more efficient in disinfection. However, it is not suitable for direct use because most masks are sensitive to high temperatures [28]. Some references have recommended to keep the mask for at least five days before reuse. However, all masks that have been used will eventually be disposed of into the environment, which may cause virus transmission. Therefore, it is important to find an effective and safe disinfection process before disposing or reuse.

Disinfection methods are categorized into two types, namely chemical methods (by using H_2O_2, bleach, chlorine dioxide, soap solutions, alcohol, ozone or ethylene oxide) and physical methods (heat, steam autoclave, dry air, microwaves, gamma irradiation or UV irradiation) [29]. To date, many studies have focused on non-thermal disinfection, which depends on high pressure (which may reach to 40 MPa), such as supercritical carbon dioxide (SC-SO_2). This technology is more suitable than others because there is no toxic by-product generated, as in the case of chemical methods [30]. However, the SC-CO_2 instrument is very expensive and not applicable for personal use. Medical masks are not classified as clinical waste unless they have been contaminated with blood. In addition, more than 90% of these masks are used by the public for non-medical use. Thus, these masks are not subject to clinical waste management. Chemical disinfectants are more suitable for individual use. Due to these limitations, treatment methods using UV radiation have been widely used compared to other methods.

Studies in the literature have revealed that UV radiation (pulsed xenon ultraviolet (PX-UV)) is effective in inactivating the Ebola virus on glass carriers and Polyethylene terephthalate (PET) materials [31]. O'Hearn et al. [32] concluded that ultraviolet germicidal irradiation (UVGI) with a dose of 40,000 J/m^2 was effective for disinfecting N95 masks. The main advantage of disinfection using UVGI is that it does not degrade the polymers if a double 36 W lamp is used and the inactivation process occurs within 148 s from the time of exposure. According to Rubio-Romero et al. [25], radiation through 254 mm UV light has high efficacy for disinfection during the COVID-19 pandemic. ultraviolet light (UV-C) is also used as one of the disinfection methods. However, one of the disadvantages of its use is the inability to reach the inner layer of the masks. However, the use of a pair of UV lamps (above and below the mask) can enhance the inactivation process. Furthermore, the efficiency of UV in inactivating and killing viruses must be confirmed based on the concept of sterility assurance level (SAL).

7. Survival of SARS-CoV-2 in the Air

Many studies have been conducted to investigate the survivability of SARS-CoV-2 in aerosols. These studies have revealed that the survivability of the virus depends on the moisture and size of droplets. Larger droplets, which can only travel through aerosols at a short distance, settle down very quickly. Nonetheless, as saliva droplets have a longer drying time, viable viral cells present in the saliva droplets can easily be transmitted through direct contact. It has been reported that the survival period of the virus in aerosols and on a surface ranges from 3 h to several days, depending on the type of the surface, as presented in Table 1 [33–35].

Table 1. Survival period of Sars-Cov-2 in aerosol and on different surfaces.

Surface	Survival Time
Aerosols	3 h
Plastic	3 days
Stainless steel	2–5 days
Cardboard	8 h
Paper	5 min to 5 days
Glass	4–5 days
Polyvinyl chloride (PVC)	5 days
Silicon rubber	5 days
Surgical gloves (latex)	5 days
Polyfluorotetraethylene (PTFE)	5 days

The ability of SARS-CoV-2 to survive for different periods of time on different surfaces depends on the properties of a surface. For example, the ability to absorb moisture plays an important role in the adhesion of viral particles. The virus is transmitted from an infected person to the air with a droplet adhesion mechanism that depends on the droplet size (aerosol), where smaller droplets are more easily suspended in air than larger droplets.

8. Conclusions

The presence of SARS-CoV-2 in water and wastewater indicates the potential for the spread of the virus in the environment. However, there is no evidence to show that DNA fragments detected in water and wastewater are pathogenic. In addition, the survival of the virus on surfaces requires effective disinfection to ensure that the virus has become inactive.

Author Contributions: Conceptualization, A.A.-G. and E.N.; methodology, Q.A.-M. and M.A.-S.; software, A.A.-G. and M.A.-S.; validation, R.M., M.A.M., Q.A.-M. and B.A.T.; investigation, S.A.; resources; data curation, M.S.H.; writing—original draft preparation, A.A.-G.; writing—review and editing, R.M. and M.A.M.; visualization, E.N.; supervision, R.M. and M.A.M.; project administration, A.A.-G.; funding acquisition, M.A.M. All authors have read and agreed to the published version of the manuscript.

Funding: The research was funded by UNITEN RMC Internal Research Grant: RJO 10517919/iRMC/Publication.

Acknowledgments: The Authors would like to acknowledge Ministry of Higher Education Malaysia (MoHE) under the Fundamental Research Grant Scheme, (FRGS) FRGS/WAB05/UTHM/02/5.

Conflicts of Interest: The authors declare no conflict of interest.

References

1. Woo, P.C.; Lau, S.K.; Chu, C.M.; Chan, K.H.; Tsoi, H.W.; Huang, Y.; Wong, B.H.; Poon, R.W.; Cai, J.J.; Luk, W.K.; et al. Characterization and complete genome sequence of a novel coronavirus, coronavirus HKU1, from patients with pneumonia. *J. Virol.* **2005**, *79*, 884–895. [CrossRef] [PubMed]
2. Chan, J.F.; Kok, K.H.; Zhu, Z.; Chu, H.; To, K.K.; Yuan, S.; Yuen, K.Y. Genomic characterization of the 2019 novel human-pathogenic coronavirus isolated from a patient with atypical pneumonia after visiting Wuhan. *Emerg. Microbes Infect.* **2020**, *9*, 221–236. [CrossRef] [PubMed]
3. Liu, Y.; Rosenfield, E.; Hu, M.; Mi, B. Direct observation of bacterial deposition on and detachment from nanocomposite membranes embedded with silver nanoparticles. *Water Res.* **2013**, *47*, 2949–2958. [CrossRef] [PubMed]
4. Gralinski, L.E.; Menachery, V.D. Return of the Coronavirus: 2019-nCoV. *Viruses* **2020**, *12*, 135. [CrossRef]
5. Wang, M.; Cao, R.; Zhang, L.; Yang, X.; Liu, J.; Xu, M.; Shi, Z.; Hu, Z.; Zhong, W.; Xiao, G. Remdesivir and chloroquine effectively inhibit the recently emerged novel coronavirus (2019-nCoV) in vitro. *Cell Res.* **2020**, *4*, 1–3. [CrossRef]
6. Ceustermans, A.; De Clercq, D.; Aertsen, A.; Michiels, C.; Geeraerd, A.; Van Impe, J.; Coosemans, J.; Ryckeboer, J. Inactivation of *Salmonella* Senftenberg strain W 775 during composting of biowastes and garden wastes. *J. Appl. Microbiol.* **2007**, *103*, 53–64. [CrossRef]
7. Dias, E.; Ebdon, J.; Taylor, H. The application of bacteriophages as novel indicators of viral pathogens in wastewater treatment systems. *Water Res.* **2018**, *129*, 172–179. [CrossRef]
8. Hindson, J. COVID-19: Faecal–oral transmission? *Nat. Rev. Gastroenterol. Hepatol.* **2020**, *17*, 259. [CrossRef]
9. Chen, Y.; Chen, L.; Deng, Q.; Zhang, G.; Wu, K.; Ni, L.; Yang, Y.; Liu, B.; Wang, W.; Wei, C.; et al. The Presence of SARS-CoV-2 RNA in Feces of COVID-19 Patients. *J. Med. Virol.* **2020**, *92*, 833–840. [CrossRef]
10. Spongberg, A.L.; Witter, J.D. Pharmaceutical compounds in the wastewater process stream in Northwest Ohio. *J. Sci. Environ.* **2008**, *397*, 148–157. [CrossRef]
11. Kumari, M.; Gupta, S.K.; Mishra, B.K. Multi-exposure cancer and non-cancer risk assessment of trihalomethanes in drinking water supplies—A case study of Eastern region of India. *Ecotoxicol. Environ. Saf.* **2015**, *113*, 433–438. [CrossRef] [PubMed]
12. AL-Gheethi, A.A.; Ismail, N.; Lalung, J.; Talib, A.; Kadir, M.O.A. Reduction of Faecal Indicators and Elimination of Pathogens from Sewage Treated Effluents by Heat Treatment. *Casp. J. Appl. Sci. Res.* **2013**, *2*, 39–55.
13. Gomez-Couso, H.; Fontan-Sainz, M.; Sichel, C.; Fernandez-Ibanez, P.; Ares-Mazas, E. Efficacy of the solar water disinfection method in turbid waters experimentally contaminated with Cryptosporidium parvum oocysts under real field conditions. *Trop. Med. Int. Health* **2009**, *14*, 620–627. [CrossRef] [PubMed]

14. Athirah, A.; Al-Gheethi, A.A.S.; Noman, E.A.; Mohamed, R.M.S.R.; Kassim, A.H.M. Centralised and decentralised transport systems for greywater and the application of nanotechnology for treatment processes. In *Management of Greywater in Developing Countries*; Springer: Cham, Switzerland, 2019; pp. 227–244.
15. Noman, E.; Al-Gheethi, A.; Talip, B.A.; Mohamed, R.; Kassim, A.H. Inactivating pathogenic bacteria in greywater by biosynthesized Cu/Zn nanoparticles from secondary metabolite of Aspergillus iizukae; optimization, mechanism and techno economic analysis. *PLoS ONE* **2019**, *14*. [CrossRef]
16. Siddiqi, K.S.; Husen, A. Fabrication of metal nanoparticles from fungi and metal salts: Scope and application. *Nano Res. Lett.* **2016**, *11*, 98. [CrossRef]
17. Ghaffari, H.; Tavakoli, A.; Moradi, A.; Tabarraei, A.; Bokharaei-Salim, F.; Zahmatkeshan, M.; Farahmand, M.; Javanmard, D.; Kiani, S.J.; Esghaei, M.; et al. Inhibition of H1N1 influenza virus infection by zinc oxide nanoparticles: Another emerging application of nanomedicine. *J. Biomed. Sci.* **2019**, *26*, 70. [CrossRef]
18. Kibbee, R.; Örmeci, B. Peracetic acid (PAA) and low-pressure ultraviolet (LP-UV) inactivation of Coxsackievirus B3 (CVB3) in municipal wastewater individually and concurrently. *Water Res.* **2020**, *183*, 116048. [CrossRef]
19. ISO. *International Standard Organisation. Aseptic Processing of Health Care Products, Part 1, General Requirements*; International Standard ISO 13408-1 1998; International Standard Organisation: Geneva, Switzerland, 1998.
20. Al-Gheethi, A.A.; Efaq, A.N.; Bala, J.D.; Norli, I.; Abdel-Monem, M.O.; Kadir, M.A. Removal of pathogenic bacteria from sewage-treated effluent and biosolids for agricultural purposes. *Appl. Water Sci.* **2018**, *8*, 74. [CrossRef]
21. Trajano, D.; Dias, E.; Ebdon, J.; Taylor, H. Limitations of chlorine disinfection of human excreta: Implications for Ebola disease control. In Proceedings of the 39th WEDC International Conference, Kumasi, Ghana, 11–15 July 2016.
22. Siddiqi, K.S.; ur Rahman, A.; Husen, A. Properties of zinc oxide nanoparticles and their activity against microbes. *Nanoscale Res. Lett.* **2018**, *13*, 1–13. [CrossRef]
23. Liu, Z.; Magal, P.; Seydi, O.; Webb, G. Understanding Unreported Cases in the 2019-Ncov Epidemic Outbreak in Wuhan, China, and the Importance of Major Public Health Interventions. *Biology* **2020**, *9*, 50. [CrossRef]
24. WHO. Shortage of Personal Protective Equipment Endangering Health Workers Worldwide. 2020. Available online: https://www.who.int/news-room/detail/03-03-2020-shortage-of-personal-protective-equipment-endangering-health-workers-worldwide (accessed on 3 April 2020).
25. Rubio-Romero, J.C.; del Carmen Pardo-Ferreira, M.; García, J.A.T.; Calero-Castro, S. Disposable masks: Disinfection and sterilization for reuse, and non-certified manufacturing, in the face of shortages during the COVID-19 pandemic. *Saf. Sci.* **2020**, *129*, 104830. [CrossRef] [PubMed]
26. Günter, K.; Todt, D.; Pfaender, S.; Steinmann, E. Persistence of coronaviruses on inanimate surfaces and their inactivation with biocidal agents. *J. Hosp. Infect.* **2020**, *104*, 246–251.
27. Van Doremalen, N.; Bushmaker, T.; Morris, D.H.; Holbrook, M.G.; Gamble, A.; Williamson, B.N.; Tamin, A.; Harcourt, J.L.; Thornburg, N.J.; Gerber, S.I.; et al. Aerosol and surface stability of SARS-CoV-2 as compared with SARS-CoV-1. *N. Engl. J. Med.* **2020**, *382*, 1564–1567. [CrossRef] [PubMed]
28. Rowan, N.J.; Laffey, J.G. Challenges and solutions for addressing critical shortage of supply chain for personal and protective equipment (PPE) arising from Coronavirus disease (COVID19) pandemic–Case study from the Republic of Ireland. *Sci. Total Environ.* **2020**, *725*, 138532. [CrossRef] [PubMed]
29. Mohapatra, S. Sterilization and disinfection. In *Essentials of Neuroanesthesia*; Academic Press: Cambridge, MA, USA, 2017; pp. 929–944.
30. Efaq, A.N.; Rahman, N.N.N.A.; Nagao, H.; Al-Gheethi, A.; Shahadat; Kadir, M.O.A. Supercritical carbon dioxide as non-thermal alternative technology for safe handling of clinical wastes. *Environ. Process.* **2015**, *2*, 797–822. [CrossRef]
31. Jinadatha, C.; Simmons, S.; Dale, C.; Ganachari-Mallappa, N.; Villamaria, F.C.; Goulding, N.; Tanner, B.; Stachowiak, J.; Stibich, M. Disinfecting personal protective equipment with pulsed xenon ultraviolet as a risk mitigation strategy for health care workers. *Am. J. Infect. Control* **2015**, *43*, 412–414. [CrossRef]
32. Katie, O.; Gertsman, S.; Sampson, M.; Webster, R.; Tsampalieros, A.; Ng, R.; Gibson, J.; Lobos, A.-T.; Acharya, N.; Agarwal, A.; et al. Decontaminating N95 masks with Ultraviolet Germicidal Irradiation (UVGI) does not impair mask efficacy and safety: A Systematic Review. *J. Hosp. Infect.* **2020**, *106*, 163–175.

33. Fiorillo, L.; Cervino, G.; Matarese, M.; D'Amico, C.; Surace, G.; Paduano, V.; Fiorillo, M.T.; Moschella, A.; La Bruna, A.; Romano, G.L.; et al. COVID-19 Surface Persistence: A Recent Data Summary and Its Importance for Medical and Dental Settings. *Int. J. Environ. Res. Public Health* **2020**, *17*, 3132. [CrossRef]
34. Wang, J.; Zhou, M.; Liu, F. Reasons for healthcare workers becoming infected with novel coronavirus disease 2019 (COVID-19) in China. *J. Hosp. Infect.* **2020**, *105*, 100–101. [CrossRef]
35. Warnes, S.L.; Little, Z.R.; Keevil, C.W. Human coronavirus 229E remains infectious on common touch surface materials. *mBio* **2015**, *6*. [CrossRef]

sustainability

MDPI

Article

How (*Un*)sustainable Environments Are Related to the Diffusion of COVID-19: The Relation between Coronavirus Disease 2019, Air Pollution, Wind Resource and Energy

Mario Coccia

CNR—National Research Council of Italy, Research Institute on Sustainable Economic Growth, Collegio Carlo Alberto, Via Real Collegio, 30-10024 Moncalieri (Torino), Italy; mario.coccia@cnr.it; Tel.: +39-011-6824-915

Received: 6 October 2020; Accepted: 17 November 2020; Published: 20 November 2020

Abstract: The pandemic caused by novel coronavirus disease 2019 (COVID-19) is generating a high number of cases and deaths, with negative effects on public health and economic systems. One of the current questions in the contemporary environmental and sustainability debate is how high air pollution and reduced use of renewable energy can affect the diffusion of COVID-19. This study endeavors to explain the relation between days of air pollution, wind resources and energy, and the diffusion of COVID-19 to provide insights into sustainable policy to prevent future epidemics. The statistical analysis here focuses on a case study of Italy, one of the first countries to experience a rapid increase in confirmed cases and deaths. The results reveal two main findings: (1) cities with high wind speed and high wind energy production have a lower number of cases of COVID-19 in the context of a more sustainable environment; (2) cities located in hinterland zones with high air pollution, low wind speed and less wind energy production have a greater number of cases and total deaths. The results presented here suggest that the pandemic caused by novel coronavirus (SARS-CoV-2) and future epidemics similar to COVID-19 cannot be solved only with research in medicine but the solution also needs advanced capabilities and technologies for supporting sustainable development based on the reduction of air pollution and increase of production in renewable energy to improve air quality and as a consequence public health.

Keywords: air pollution; wind energy; renewable energy; COVID-19; coronavirus disease; SARS-CoV-2; sustainable development; cleaner production; sustainable technologies

1. Introduction and Related Works

Within environmental and sustainability science, new and relatively unexplored topics are continually emerging, such as factors determining the diffusion of novel coronavirus disease 2019 (COVID-19), which generates a severe respiratory disorder and is causing the deaths of many individuals worldwide [1–3]. Manifold studies suggest a possible relation between *un*sustainable environments with high air pollution and the spread of COVID-19 [3–5]. In particular, populations living in regions with high levels of air pollution have a high probability of developing respiratory disorders because of the likely negative effects of particulate matter commingled with infective agents, such as SARS-CoV-2 [3,6–9]. In fact, scholars state that a high level of air pollution can increase the viral infectivity and lethality of the COVID-19 [10]. In addition, atmosphere with low wind prevents the dispersion of air pollutants, which seem to be one of the determinants of higher incidence of COVID-19 in some European regions, such as North Italy [11]. The study by van Doremalen et al. [12] revealed that in China, viral agents of SARS-CoV-2 may be suspended in the air for several minutes and this finding can explain higher number of cases and deaths related to COVID-19 in manifold countries, such as the USA, India, Brazil, Mexico, etc. [13]. In general, several studies present the hypothesis that atmospheric pollution associated with certain climatological factors (such as, high humidity and low wind speed) may support a longer permanence of viral particles in the air, fostering the rapid spread of COVID-19 within polluted regions [5,11,12,14,15]. In this context, Martorell-Marugán et al. [16] suggest new tools for supporting increased data analysis and statistical capabilities that allow users to explore trends and associations between critical environmental data and the spread of COVID-19 in society. In particular, Megahed and Ghoneim [17] argue that the COVID-19 pandemic has transformed the built environment because of the fear of infection; as a result, appropriate architecture and urbanism can reduce potential risks or stop the spread of infectious diseases by designing a healthy and sustainable built environment. In addition, Rainisch et al. [18] show that a lockdown strategy or combination of non-pharmaceutical interventions may reduce air pollution and negative effects on public health, lowering the demand for hospitalization and intensive care units. In other words, these scholars argue that strong containment and mitigation policies can improve air quality and mitigate the substantial morbidity and mortality caused by the COVID-19 pandemic [18].

In order to extend previous studies on critical aspects of the transmission dynamics of SARS-CoV-2 [3–5], the goal of this paper is to analyze the relation between cases, air pollution and *un*sustainable environments with low renewable wind energy production (and high air pollution) that can explain some vital relationships determining the spread of COVID-19 and negative effects on public health. This study has the potential to support long-term policies directed towards fostering sustainability by maintaining and sustaining environmental quality both at national and at global level through biological, eco-friendly methods and the dissemination of knowledge on various topics and technologies relating to renewable energy systems (such as wind energy technology) in order to reduce environmental pollution and biodegradation and foster bioremediation, which could be one of the strategies for preventing future epidemics similar to COVID-19. In fact, Ferrannini et al. [19] point out that industrial policy should facilitate sustainable structural change in modern economies towards human and social development by investing in renewable resources and by designing innovation policies directed to environmental benefits in the period post COVID-19 pandemic crisis.

2. Research Questions, Research Setting and Study Design

The research questions of this study are as follows:

What is the relation between cases of COVID-19, air pollution and *un*sustainable environments with low renewable wind energy production?

What is the impact of this relationship on the spread of COVID-19 and on public health?

In the context of these problems in environmental and sustainable sciences, this study focuses on a case study of Italy, one of the first countries to accumulate a high number of deaths from COVID-19. The sample is constituted by fifty-five cities that are provincial capitals in Italy ($N = 55$). Epidemiological data of COVID-19 are from Ministero della Salute [20]; data of air pollution are from Regional Agencies for Environmental Protection [21]; climatological information are from meteorological stations of the Italian provinces under study [22]. Data of the density of population are from the Italian National Institute of Statistics [23], and finally, data concerning the production of wind energy per Italian region are from an Italian transmission operator named Terna [24,25].

The measures for statistical analyses are as follows:

- *Air pollution.* Total days exceeding the limits set for PM_{10} or for ozone in 2018 per Italian provincial capital. Days of air pollution are a major factor that affects the environment and public health [3,21]. Moreover, by using 2018 as the baseline year for air pollution data, we separate out the effects of COVID-19. Experimental results reveal that $PM_{2.5}$ and PM_{10} have a strong correlation in atmospheric pollution [26,27].
- *Spread of COVID-19.* Number of confirmed cases in March–April 2020 (during the first wave of this pandemic).

- *Climatological information*. Average wind speed in km/h in February–March 2020.
- *Interpersonal contact*. Population density of cities (individual/km^2) in 2019.
- *Sustainable environment*. Production of renewable wind energy with power in MegaWatt (MW) of overall wind farms in Italian regions in January 2020.

Descriptive statistics of measures just mentioned are performed by categorizing Italian provincial capitals into groups, as follows.

Renewable wind energy production is used to categorize:

- cities with *high wind energy production* (seven regions in Italy have 94% of national production of wind energy);
- cities with *low wind energy production* (regions that have 6% of national production of wind energy, a proxy for a less sustainable environment).
- Days of air pollution is used to categorize:
- Cities with *high number of days of air pollution* (>100 days per year exceeding the limits set for PM_{10} or for ozone);
- Cities with *low number of days of air pollution* (≤100 days per year exceeding the limits set for PM_{10} or for ozone).

Correlation and regression analyses verify the relationships between the variables under study. In particular, regression analysis considers the number of cases across Italian provincial capitals (dependent variable y) as a linear function of the explanatory variable x of total days exceeding the limits set for PM_{10} (i.e., air pollution).

The specification of linear relationship is a *log-log* model:

$$log\ y_t = \alpha + \beta\ log\ x_{t-1} + u \tag{1}$$

α = constant; β = coefficient of regression; u = error term.

An alternative model (1) applies as explanatory variable the density of population per km^2, considering the categorization of cities according to the days of air pollution and their location in regions with high or low intensity of wind energy production. The ordinary least squares (OLS) method is applied for estimating the unknown parameters of linear models (1). Statistical analyses are performed with the Statistics Software SPSS® version 26.

3. Results

The main results of the analysis are:

- Regions with a high intensity of wind-based renewable energy and low air pollution experience lower spread of COVID-19 within society;
- Cities with high air pollution and low production of wind energy have a very high number of cases in environments with high average density of population and low average speed of wind;
- Cities in regions with lower levels of wind energy production demonstrate a high positive correlation between days of air pollution and cases of COVID-19;
- In cities with little days of air pollution, an increase of 1% in the density of population, it increases the expected number of cases by around 0.25%, whereas in cities with many days of air pollution, an increase of 1% in the density of population, it increases the expected number of cases by around 0.85%;
- The percentage of cases and total deaths weighted with population of Italian regions reveals that around 74.50% of cases and around 81% of deaths caused by COVID-19 in Italy occur in regions with many days of air pollution and with low production of renewable energy based on wind resources.

In particular, the findings of the statistical analyses are described in the following tables and figures. First of all, the wind energy production in Italy per region is shown in Table 1.

Table 1. Wind energy production in Italy per region in January 2020 and energy production compared to demand in 2018.

Italian Regions	Number Wind Farms	Power (MW)	Total Deficit (−) or Surplus (+) of Energy Production (from Renewable and Traditional Resources) Compared to Demand in 2018 [24]
Abruzzo	47	264.23	−17.6%
Basilicata	1413	1300.12	+10.6%
Calabria	418	1125.77	+179.0%
Campania	619	1734.61	−44.2%
Emilia Romagna	72	44.85	−28.9%
Friuli Venezia Giulia	5	0.01	−5.50%
Lazio	69	70.94	−22.0%
Liguria	33	56.83	−15.7%
Lombardia	10	0.05	−34.6%
Marche	51	19.24	−67.80
Molise	79	375.87	+94.5%
Piemonte	18	23.82	+12.8%
Puglia	1176	2570.12	+55.8%
Sardegna	595	1105.34	+33.6%
Sicilia	884	1904.10	−18.1%
Toscana	126	143.01	−25.1%
Trentino Alto Adige	10	0.39	+66.9%
Umbria	25	2.09	−42.70
Valle d'Aosta	5	2.59	+208.3%
Veneto	18	13.43	−48.1%

Source: [24,25], last report available.

Table 1 shows that seven regions in Italy, mainly Southern regions (Molise, Puglia, Calabria, Basilicata, Campania, Sicilia and Sardegna), constitute 94% of the total wind energy production. These regions have at least 1 GigaWatt (GW) of power, with the Puglia region (South-East Italy) having the highest value of 2.5 GW. In particular, Italy had in January 2020 around 5645 wind farms, with almost 7000 wind turbines of various power sizes: above 10 MW (MegaWatt) of power, there were 313 plants for a total power of 9.07 GW. The most relevant power class ranges from 20 to 200 kW (kiloWatt), with 3956 systems having a total power of approximately 234 MW. As mentioned, the Puglia region generates the largest share of wind power in Italy—24.8% of the total, with 92 factories generating over 10 MW of power.

Table 2 shows that regions in Northern Italy mainly produce energy from thermal power stations using fossil fuels, liquefied natural gas coal and crude oil (average value is 48.33%).

Table 2. Main energy sources in Northern Italy, with low wind and solar energy, 2018.

Regions of North Italy	Thermal Power Stations Using Fossil Fuels, Liquefied Natural Gas Coal and Crude Oil %	Wind and Solar Farms %
Emilia Romagna	60.2	7.3
Friuli Venezia Giulia	72.1	5.2
Liguria	76.5	3.7
Lombardia	46.7	3.2
Piemonte	74.0	6.6
Trentino Alto Adige	20.3	6.1
Valle d'Aosta	4.00	2.4
Veneto	32.8	6.2
Arithmetic mean	48.33	5.09
Stand. Deviation	27.13	1.78

Source: [24], last report available.

The results in Table 3 suggest that cities in regions with high production of wind energy (94% of total) have a very low number of cases of COVID-19 (in March and April 2020, first wave of COVID-19), whereas cities located in regions with a low intensity of wind energy production (6% of total), within an *un*sustainable environment (in terms of low production of wind energy and many days of air pollution), have a very high number of cases. In fact, Table 3 also shows that cities in regions with low production of wind energy (6% of total) have higher number of days of air pollution than cities with a high production of wind energy (around 79 polluted days vs. 48 polluted days exceeding PM_{10} or ozone per year). This preliminary result suggests that regions with a high intensity of wind-based renewable energy and low air pollution also experience a lower spread of COVID-19 within society.

Table 3. Descriptive statistics of Italian province capitals according to intensity of wind energy production.

	Cities in Regions with 94% of Wind Energy Production $N = 5$		**Cities in Regions with 6% of Wind Energy Production** $N = 50$	
Variables	Mean	Std. Deviation	Mean	Std. Deviation
Days exceeding limits set for PM_{10} or ozone, 2018	48.0	30.27	79.44	41.7
Cases, 17th March 2020	59.8	90.84	475.58	731.11
Cases, 7th April 2020	505.6	646.12	2119.68	2450.71
Cases, 27th April 2020	708.2	949.19	3067.67	3406.67
Density inhabitants/km^2, 2019	2129	3384.1	1385.76	1489.31
Wind km/h, February–March 2020	14.6	5.45	8.1	3.08

In order to confirm this result, Table 4 focuses on air pollution of cities: in particular, cities with high number of days of air pollution (greater than 100 days exceeding limits set for PM_{10} or ozone per year) and low production of wind energy accumulated a very high number of cases in March and April 2020, in environments with high average density of population and low average speed of wind (atmospheric stability) [11,14,28].

Table 4. Descriptive statistics of Italian provincial capitals according to days of air pollution.

	Cities with High Air Pollution: >100 Days Exceeding Limits Set for PM_{10} $N = 20$		**Cities with Low Air Pollution: <100 Days Exceeding Limits Set for** PM_{10} $N = 35$	
Variables	Mean	Std. Deviation	Mean	Std. Deviation
Days exceeding limits set for PM_{10} or ozone, 2018	125.25	13.4	48.77	21.37
Cases, 17th March 2020	881.7	1010.97	184.11	202.76
Cases, 7th April 2020	3650	3238.82	1014.63	768.91
Cases, 27th April 2020	4838.05	4549.41	1637.21	1292.26
Density inhabitants/km^2, 2019	1981.4	1988.67	1151.57	1466.28
Wind km/h, February–March 2020	7.67	2.86	9.28	4.15

Table 5 shows results of variables under study considering cities with the same density of population: Italian provincial capitals with high average density of people per km^2 (i.e., >1000 inhabitant/km^2, mostly those bordering large urban conurbations, such as the cities of Brescia, Bergamo and Cremona, close to Milan, the second most populous city in Italy after Rome Capital) had higher numbers of COVID-19 cases. These cities, located in hinterland zones, have also high number of days of air pollution and low average wind speed (and, as a consequence, low production of renewable energy based on wind resources).

Table 5. Descriptive statistics of Italian provincial capitals according to density of population.

	Density of Population HIGH > 1000 Inhabitant/km^2 N = 25 Cities		Density of Population LOW < 1000 Inhabitant/km^2 N = 30 Cities	
Variables	Mean	Std. Deviation	Mean	Std. Deviation
Days exceeding limits set for PM_{10} or ozone, 2018	91.24	40.24	64.37	39.25
Cases, 17th March 2020	665.08	919.7	248.37	386.95
Cases, 7th April 2020	2967.44	3092.46	1144.2	1065.99
Cases, 27th April 2020	4195.42	4333.91	1727.55	1491.47
Density inhabitants/km^2, 2019	2584.4	2000.63	510.77	282.11
Wind km/h, February–March 2020	7.99	2.79	9.28	4.41

Table 6 shows that cities in regions with less than 6% wind energy production demonstrated a high positive correlation between days of air pollution and cases of COVID-19 on 17th March ($r = 0.69$, p-value < 0.01), 7th April ($r = 0.55$, p-value < 0.01) and 27th April 2020 ($r = 0.36$, p-value < 0.01). Instead, in regions with a high intensity of wind energy production, the results are not significant.

Table 6. Correlation.

	Cities in Regions with 94% of Wind Energy Production	Cities in Regions with 6% of Wind Energy Production
	Log days exceeding limits set for PM_{10} or ozone, 2018	Log days exceeding limits set for PM_{10} or ozone, 2018
Log cases 17th March 2020 Pearson Correlation	0.81	0.69 **
Log cases 7th April 2020 Pearson Correlation	0.74	0.55 **
Log cases 27th April 2020 Pearson Correlation	0.69	0.36 **

Note: ** Correlation is significant at the 0.01 level (2-tailed).

4. Discussion of a Likely Relation between *Un*sustainable Environments and the Spread of COVID-19

This study analyzed data on COVID-19 cases alongside factors associated with an *un*sustainable environment given by the level of atmospheric pollution and the production of renewable wind energy. Results suggest that environments with high air pollution and low production of renewable wind energy may have accelerated the spread of COVID-19 in Northern Italian cities, leading to a higher number of cases and deaths. In particular, cities with little wind and frequently high numbers of days of air pollution—exceeding safe levels of ozone or particulate matter—had higher numbers of COVID-19 cases and deaths [3,11,14].

In fact, Table 7 suggests that days of air pollution, in areas with low production of wind energy, can explain the number of cases of COVID-19 in society. In particular,

- Cities in regions with 94% of wind energy production yield non-significant results because of the low number of statistical units in the sample under study;
- Instead, in cities with a mere 6% of wind energy production, an increase of 1% in air pollution, measured by days exceeding the limits set for PM_{10}, the expected number of cases increases by around 0.92% (p-value < 0.001).

Table 7. Parametric estimates of the relationship of log cases on log air pollution considering groups of cities in regions with high or low production of wind energy.

	Cities in Regions with 94% of Wind Energy Production	Cities in Regions with 6% of Wind Energy Production
Constant α	0.70	3.39 ***
(St. Err.)	(2.64)	(0.85)
Coefficient β	1.34	0.92 ***
(St. Err.)	(0.70)	(0.20)
R^2 (St. Err. of Estimate)	0.55 (0.86)	0.31 (0.82)
F	3.65	21.28 ***

Note: Explanatory variable: log days exceeding limits set for PM_{10} or ozone 2018; Dependent variable: log cases on 7th April 2020; *** *p*-value < 0.001.

Figure 1 shows a visual representation of regression lines: cities demonstrating higher production of renewable wind energy in Italy tend to have a low number of cases driven by days of air pollution. In order to confirm these findings, Table 8 considers cities with high and low number of days of air pollution.

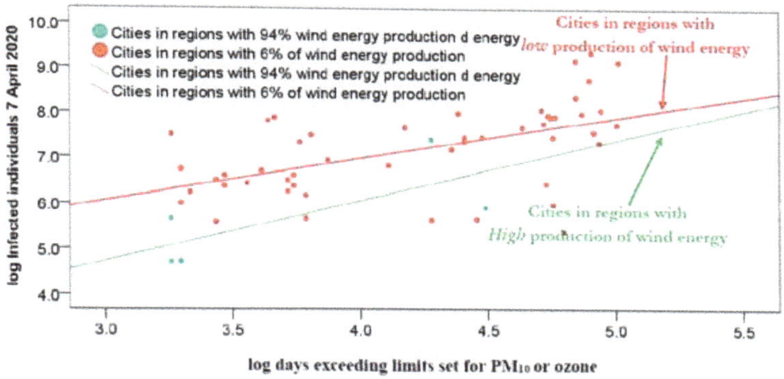

Figure 1. Regression lines of log cases on log air pollution according to production of wind energy. *Note*: This result suggests that the spread of COVID-19 increases with air pollution in regions with low production of wind energy (in general, *un*sustainable environments).

Table 8. Parametric estimates of the relationship of log cases (7th April 2020) on log density of population (inhabitants/km^2 in 2019), considering groups of cities with high and low number of days of air pollution.

	Cities with Low Air Pollution	Cities with High Air Pollution
Constant α	4.976	1.670
(St. Err.)	(0.786)	(1.491)
Coefficient β	0.252 *	0.849 ***
(St. Err.)	(0.120)	(0.205)
R^2 (St. Err. of Estimate)	0.119	0.488
F	17.168 ***	4.457 *

Note: Explanatory variable: log density of population (inhabitants/km^2) in 2019; Dependent variable: log cases on 7th April 2020; *** *p*-value < 0.001; * *p*-value < 0.05.

Table 8 reveals that in cities with little days of air pollution, an increase of 1% in the density of population, it increases the expected number of cases by around 0.25% (*p*-value = 0.042), whereas, in cities with many days of air pollution, an increase of 1% in the density of population, it increases the expected number of cases by around 0.85% (*p*-value < 0.001). Figure 2 shows regression lines: regions with many days of air pollution generate an atmosphere rich in air pollutants that, associated with low wind speed, can commingle with viral agents and lead to poor air quality creating an habitat fostering a higher diffusion of COVID-19 [3,11,14,28]. Overall, then, empirical evidence here suggests a relation of association between the variables under study rather than a relation of dependence because of manifold confounding factors that influence both the dependent variables and independent variables.

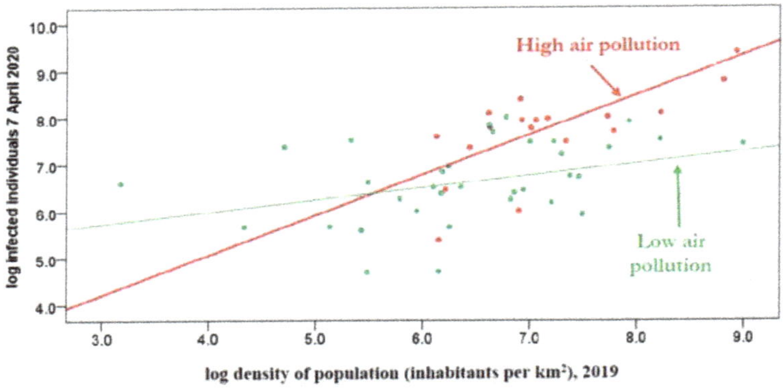

Figure 2. Regression line of log cases on log population density, considering groups of cities with high or low number of days of air pollution. Note: This result reveals that the spread of COVID-19 is higher in cities with high air pollution.

In addition, if we consider regions with high/low number of days of air pollution, using arithmetic mean of days exceeding the limits set for PM_{10} or ozone in cities, the percentage of cases and total deaths weighted with the population of these regions reveals that around 74.50% of cases and around 81% of deaths from COVID-19 in Italy occur in regions with many days of air pollution and with low production of renewable energy based on wind resources.

These findings here provide valuable insight into the geo-environmental factors and types of energy resources and energy production that may affect the environment and may accelerate the spread of COVID-19 and similar infectious diseases within society. In short, the main results of the study, based on a case study of the COVID-19 outbreak in Italy, are that cities with high production of wind energy and a more sustainable environment, associated with low air pollution, experience reduced spread of COVID-19 and a lower number of deaths (during the first wave of COVID-19 pandemic). In fact, the results reveal that Italian cities in Southern regions with high production of wind energy and low air pollution (also for high speed of wind), which result in a cleaner environment and better air quality, likely explain a lower impact of COVID-19 in society. By contrast, cities in Northern Italy, with high air pollution and low wind speed (and, as a consequence, low wind energy production), experience the stagnation of air pollution in the atmosphere which, commingled with viral agents, can facilitate the spread of COVID-19 and similar infectious diseases [11]. Hence, COVID-19 experienced low spread in regions with low air pollution and high production of renewable (wind) energy (more *sustainable environments*). From the perspective of sustainable science, Xu et al. [29] found that winds significantly aid the dissipation of fine particulate matter (PM), and high concentrations of fine PM only persisted for a very short time and dissipated after several hours. The role of wind speed and direction, temperature and humidity is critical for urban ventilation and quality of air [30]. Considering the benefits of wind as resource that can reduce air pollution and, as a consequence, improve environmental sustainability, Gu et al. [31] argue that a strategy to enhance air quality in cities is improving the exchange of air between areas within and above the urban canopy with appropriate urban planning and management [17]. In fact, the health and economic benefits associated with national and local reduction of air pollution are now rarely contested [3]. In this context, Cui et al. [32] show that reductions in ambient air pollution can decrease mortality and related morbidity, with manifold economic benefits in society.

Overall, then, environments with little wind and high air pollution can sustain, in fall and winter seasons, the stagnation of air pollution, which, associated with viral agents, seems to have facilitated the spread of COVID-19 in regions, such as Northern Italy [3,11,14,28]. Regions have to apply long-term sustainable polices directed towards reducing air pollution and supporting the production of renewable energy [33]. Ibn-Mohammed et al. [34] suggest that the lessons learned from the COVID-19 pandemic should be used to support sustainable development goals directed towards more resilient and low-carbon economic systems. In particular, policymakers should support a sustainable model of development based on the circular economy in order to balance the complex equation of generating profit with minimal environmental harm [35–37]. DeWit et al. [38], to reduce the risk of exposure of Japanese cities to new infectious diseases [4], point out that the goal of protecting public health has to be associated with the promotion of accelerated decarbonization and a resilient and sustainable development. In fact, Sharifi and Khavarian-Garmsir [39] argue that the lessons learned from this health crisis for post-COVID urban planning should be directed towards improvements in air and water quality to create cities with more resilient and sustainable environments for the benefit of human society [17]. Finally, Wells ([40], p. 29) states that "the COVID 19 pandemic can act as a catalytic event in which the legitimacy and efficacy of existing economic and political structures will be challenged and reshaped, and hence is an opportunity to redefine the ecological burdens our activities create". In this context, countries should introduce organizational, product and process innovations to support sustainable development for coping with future threats of novel infectious diseases, also through the expansion of sustainable technologies for resource-efficiency optimization and for renewable energy production, as well as for water purification, for material recycling and material recovery from complex products to enable low-impact production and/or use in the environment and society [41–49].

5. Conclusions and Outlook

The results here suggest that, among Italian provincial capitals, the number of COVID-19 cases was higher in *un*sustainable environments, i.e., cities with >100 days per year exceeding the limits set for PM_{10} or ozone and cities with a low average intensity of wind energy production. In fact, in hinterland cities (mostly those bordering large urban conurbations) with high air pollution coupled with low wind speed, the average number of cases in April 2020 more than doubled, and the same social and health issues are ongoing in the autumn–winter season of 2020–2021. The findings here also provide valuable insight into the understanding of geo-environmental and industrial factors that may accelerate the spread of COVID-19 in *un*sustainable environments. In this context, a proactive strategy to cope with future epidemics should concentrate on reducing the levels of air pollution in polluted cities with sustainable policies concerning traffic reduction, higher renewable energy production, increase of manufacturing processes with reduced energy, water and material footprints, as well as growing application of efficient materials and technologies for producing sustainable energy, devices and systems [4,42,50].

However, these conclusions are of course tentative because there are several challenges to such studies that can only capture certain aspects of the ongoing complex relations between atmospheric pollution, diffusion of viral infectivity, production of renewable energy and other factors of socioeconomic and environmental systems. Therefore, this study encourages further investigations into vital aspects of the spread of COVID-19 and other viral agents in highly industrialized areas to design appropriate sustainable policies that can reduce air pollution, improve air quality and control the spread of new waves of COVID-19 pandemic and novel infections similar to COVID-19, reducing the total impact on public health and also economic system [51–53]. To conclude, in the presence of regions with high air pollution and with low renewable (wind) energy production that deteriorate environment, damage public health and facilitate the spread of infectious diseases, a comprehensive strategy to prevent future epidemics similar to COVID-19 must be also designed by considering recommendations of sustainable and environmental sciences for cleaner production and higher utilization of energy from renewable resources and sustainable technologies for improving the environment, economic system and public health in the long run.

Funding: This research received no external funding.

Conflicts of Interest: The author declares that he has no known competing financial interests or personal relationships that could have appeared to influence the work reported in this paper. No funding was received for this study.

References

1. Alwan, N.A.; Burgess, R.A.; Ashworth, S.; Beale, R.; Bhadelia, N.; Bogaert, D.; Dowd, J.; Eckerle, I.; Goldman, L.R.; Greenhalgh, T.; et al. Scientific consensus on the COVID-19 pandemic: We need to act now. *Lancet* **2020**, *396*, e71–e72. [CrossRef]
2. Andersen, L.M.; Harden, S.R.; Sugg, M.M.; Runkle, J.D.; Lundquist, T.E. Analyzing the spatial determinants of local Covid-19 transmission in the United States. *Sci. Total Environ.* **2021**, *754*, 142396. [CrossRef]
3. Coccia, M. Factors determining the diffusion of COVID-19 and suggested strategy to prevent future accelerated viral infectivity similar to COVID. *Sci. Total Environ.* **2020**, *729*, 138474. [CrossRef] [PubMed]
4. Coccia, M. An index to quantify environmental risk of exposure to future epidemics of the COVID-19 and similar viral agents: Theory and practice. *Environ. Res.* **2020**, *191*, 110155. [CrossRef] [PubMed]
5. Frontera, A.; Martin, C.; Vlachos, K.; Sgubin, G. Regional air pollution persistence links to COVID-19 infection zoning. *J. Infect.* **2020**, *81*, 318–356. [CrossRef]
6. Anenberg, S.C.; Achakulwisut, P.; Brauer, M.; Moran, D.; Apte, J.S.; Henze, D.K. Particulate matter-attributable mortality and relationships with carbon dioxide in 250 urban areas worldwide. *Sci. Rep.* **2019**, *9*, 1–6. [CrossRef]
7. Ash'Aari, Z.H.; Aris, A.Z.; Ezani, E.; Kamal, N.I.A.; Jaafar, N.; Jahaya, J.N.; Manan, S.A.; Saifuddin, M.F.U. Spatiotemporal Variations and Contributing Factors of Air Pollutant Concentrations in Malaysia during Movement Control Order due to Pandemic COVID-19. *Aerosol Air Qual. Res.* **2020**, *20*, 2047–2061. [CrossRef]
8. He, H.; Shen, Y.; Jiang, C.; Li, T.; Guo, M.; Yao, L. Spatiotemporal Big Data for $PM_{2.5}$ Exposure and Health Risk Assessment during COVID-19. *Int. J. Environ. Res. Public Health* **2020**, *17*, 7664. [CrossRef]
9. Srivastava, A. COVID-19 and air pollution and meteorology-an intricate relationship: A review. *Chemosphere* **2021**, *263*, 128297. [CrossRef]
10. Gupta, A.; Bherwani, H.; Gautam, S.; Anjum, S.; Musugu, K.; Kumar, N.; Anshul, A.; Kumar, R. Air pollution aggravating COVID-19 lethality? Exploration in Asian cities using statistical models. *Environ. Dev. Sustain.* **2020**, 1–10. [CrossRef]
11. Coccia, M. The effects of atmospheric stability with low wind speed and of air pollution on the accelerated transmission dynamics of COVID-19. *Int. J. Environ. Stud.* **2020**, 1–27. [CrossRef]
12. Van Doremalen, N.; Bushmaker, T.; Morris, D.; Holbrook, M.G.; Gamble, A.; Williamson, B.N.; Tamin, A.; Harcourt, J.L.; Thornburg, N.J.; Gerber, S.I.; et al. Aerosol and Surface Stability of SARS-CoV-2 as Compared with SARS-CoV-1. *N. Engl. J. Med.* **2020**, *382*, 1564–1567. [CrossRef] [PubMed]
13. Center for System Science and Engineering at Johns Hopkins 2020. Coronavirus COVID-19 Global Cases. Available online: https://gisanddata.maps.arcgis.com/apps/opsdashboard/index.html#/bda7594740fd40299423467b48e9ecf6 (accessed on 4 October 2020).
14. Coccia, M. How do low wind speeds and high levels of air pollution support the spread of COVID-19? *Atmos. Pollut. Res.* **2020**. [CrossRef] [PubMed]
15. Tzampoglou, P.; Loukidis, D. Investigation of the Importance of Climatic Factors in COVID-19 Worldwide Intensity. *Int. J. Environ. Res. Public Health* **2020**, *17*, 7730. [CrossRef]
16. Martorell-Marugán, J.; Villatoro-García, J.A.; García-Moreno, A.; López-Domínguez, R.; Requena, F.; Merelo, J.J.; Lacasaña, M.; Luna, J.D.D.; Díaz-Mochón, J.J.; Lorente, J.A.; et al. DatAC: A visual analytics platform to explore climate and air quality indicators associated with the COVID-19 pandemic in Spain. *Sci. Total Environ.* **2021**, *750*, 141424. [CrossRef]
17. Megahed, N.A.; Ghoneim, E.M. Antivirus-built environment: Lessons learned from Covid-19 pandemic. *Sustain. Cities Soc.* **2020**, *61*, 102350. [CrossRef]

18. Rainisch, G.; Undurraga, E.A.; Chowell, G. A dynamic modeling tool for estimating healthcare demand from the COVID19 epidemic and evaluating population-wide interventions. *Int. J. Infect. Dis.* **2020**, *96*, 376–383. [CrossRef]
19. Ferrannini, A.; Barbieri, E.; Biggeri, M.; Di Tommaso, M.R. Industrial policy for sustainable human development in the post-Covid19 era. *World Dev.* **2021**, *137*, 105215. [CrossRef]
20. Ministero della Salute 2020. Covid-19—Situazione in Italia. Available online: http://www.salute.gov.it/portale/nuovocoronavirus/dettaglioContenutiNuovoCoronavirus.jsp?lingua=italiano&id=5351&area=nuovoCoronavirus&menu=vuoto (accessed on 1 April 2020).
21. Legambiente 2019. Mal'aria 2019, il Rapporto Annuale Sull'Inquinamento Atmosferico Nelle Città Italiane. Available online: https://www.legambiente.it/malaria-2019-il-rapporto-annuale-annuale-sullinquinamento-atmosferico-nelle-citta-italiane/ (accessed on 28 March 2020).
22. Il Meteo 2020. Medie e Totali Mensili. Available online: https://www.ilmeteo.it/portale/medie-climatiche (accessed on 28 March 2020).
23. ISTAT 2020. The Italian National Institute of Statistics-Popolazione residente al 1 gennaio. Available online: http://dati.istat.it/Index.aspx?DataSetCode=DCIS_POPRES1 (accessed on 18 October 2020).
24. Terna. *Statistiche Regionali 2018*; Terna SpA: Roma, Italy, 2018.
25. Terna 2020. Fonte Rinnovabili. Wind Energy. Available online: https://www.terna.it/it/sistema-elettrico/dispacciamento/fonti-rinnovabili (accessed on 20 May 2020).
26. Kong, L.; Xin, J.; Zhang, W.; Wang, Y. The empirical correlations between PM2.5, PM10 and AOD in the Beijing metropolitan region and the PM2.5, PM10 distributions retrieved by MODIS. *Environ. Pollut.* **2016**, *216*, 350–360. [CrossRef]
27. Zhou, X.; Cao, Z.; Ma, Y.; Wang, L.; Wu, R.; Wang, W. Concentrations, correlations and chemical species of PM2.5/PM10 based on published data in China: Potential implications for the revised particulate standard. *Chemosphere* **2016**, *144*, 518–526. [CrossRef]
28. Rashed, E.A.; Kodera, S.; Gomez-Tames, J.; Hirata, A. Influence of Absolute Humidity, Temperature and Population Density on COVID-19 Spread and Decay Durations: Multi-Prefecture Study in Japan. *Int. J. Environ. Res. Public Health* **2020**, *17*, 5354. [CrossRef] [PubMed]
29. Xu, J.; Zhu, F.; Wang, S.; Zhao, X.; Zhang, M.; Ge, X.; Wang, J.; Tian, W.; Wang, L.; Yang, L.; et al. A preliminary study on wind tunnel simulations of the explosive growth and dissipation of fine particulate matter in ambient air. *Atmos. Res.* **2020**, *235*, 104635. [CrossRef]
30. Yuan, M.; Song, Y.; Huang, Y.; Shen, H.; Li, T. Exploring the association between the built environment and remotely sensed PM2.5 concentrations in urban areas. *J. Clean. Prod.* **2019**, *220*, 1014–1023. [CrossRef]
31. Gu, K.; Fang, Y.; Qian, Z.; Sun, Z.; Wang, A. Spatial planning for urban ventilation corridors by urban climatology. *Ecosyst. Health Sustain.* **2020**, *6*, 1747946. [CrossRef]
32. Cui, L.; Zhou, J.; Peng, X.; Ruan, S.; Zhang, Y. Analyses of air pollution control measures and co-benefits in the heavily air-polluted Jinan city of China, 2013–2017. *Sci. Rep.* **2020**, *10*, 5423. [CrossRef]
33. Wang, Z.; Zhu, Y. Do energy technology innovations contribute to CO_2 emissions abatement? A spatial perspective. *Sci. Total Environ.* **2020**, *726*, 138574. [CrossRef]
34. Ibn-Mohammed, T.; Mustapha, K.; Godsell, J.; Adamu, Z.; Babatunde, K.; Akintade, D.; Acquaye, A.; Fujii, H.; Ndiaye, M.; Yamoah, F.; et al. A critical analysis of the impacts of COVID-19 on the global economy and ecosystems and opportunities for circular economy strategies. *Resour. Conserv. Recycl.* **2021**, *164*, 105169. [CrossRef]
35. Coccia, M. The relation between price setting in markets and asymmetries of systems of measurement of goods. *J. Econ. Asymmetries* **2016**, *14*, 168–178. [CrossRef]
36. Coccia, M. Why do nations produce science advances and new technology? *Technol. Soc.* **2019**, *59*, 101124. [CrossRef]
37. Coccia, M. Theories of Development. In *Global Encyclopedia of Public Administration, Public Policy, and Governance*; Springer Science and Business Media LLC.: Berlin/Heidelberg, Germany, 2019; pp. 1–7.
38. Dewit, A.; Shaw, R.; Djalante, R. An integrated approach to sustainable development, National Resilience, and COVID-19 responses: The case of Japan. *Int. J. Disaster Risk Reduct.* **2020**, *51*, 101808. [CrossRef]
39. Sharifi, A.; Khavarian-Garmsir, A.R. The COVID-19 pandemic: Impacts on cities and major lessons for urban planning, design, and management. *Sci. Total Environ.* **2020**, *749*, 142391. [CrossRef]

40. Wells, P.; Abouarghoub, W.; Pettit, S.; Beresford, A. A socio-technical transitions perspective for assessing future sustainability following the COVID-19 pandemic. *Sustain. Sci. Pract. Policy* **2020**, *16*, 29–36. [CrossRef]
41. Coccia, M. A taxonomy of public research bodies: A systemic approach1. *Prometheus* **2005**, *23*, 63–82. [CrossRef]
42. Coccia, M. Measuring the impact of sustainable technological innovation. *Int. J. Technol. Intell. Plan.* **2009**, *5*, 276. [CrossRef]
43. Coccia, M. Sources of technological innovation: Radical and incremental innovation problem-driven to support competitive advantage of firms. *Technol. Anal. Strat. Manag.* **2017**, *29*, 1048–1061. [CrossRef]
44. Coccia, M. The Fishbone diagram to identify, systematize and analyze the sources of general purpose technologies. *J. Adm. Soc. Sci.* **2017**, *4*, 291–303. [CrossRef]
45. Coccia, M. The origins of the economics of Innovation. *J. Econ. Soc. Thought* **2018**, *5*, 9–28. [CrossRef]
46. Coccia, M. Sources of disruptive technologies for industrial change. *Ind. Riv. Econ. Politica Ind.* **2017**, *38*, 97–120. [CrossRef]
47. Coccia, M. Varieties of capitalism's theory of innovation and a conceptual integration with leadership-oriented executives: The relation between typologies of executive, technological and socioeconomic performances. *Int. J. Public Sect. Perform. Manag.* **2017**, *3*, 148–168. [CrossRef]
48. Coccia, M. Theorem of not independence of any technological innovation. *J. Econ. Bibliogr.* **2018**, *5*, 29–35. [CrossRef]
49. Coccia, M. Deep learning technology for improving cancer care in society: New directions in cancer imaging driven by artificial intelligence. *Technol. Soc.* **2020**, *60*, 101198. [CrossRef]
50. Coccia, M. *Comparative Critical Decisions in Management. Global Encyclopedia of Public Administration, Public Policy, and Governance*; Farazmand, A., Ed.; Springer: Berlin/Heidelberg, Germany, 2020. [CrossRef]
51. McLennan, A.K.; Hansen, A.K.K.; Ulijaszek, S.J. Health and medicine cannot solve COVID-19. *Lancet* **2020**, *396*, 599–600. [CrossRef]
52. Ou, Y.; West, J.J.; Smith, S.J.; Nolte, C.G.; Loughlin, D.H. Air pollution control strategies directly limiting national health damages in the US. *Nat. Commun.* **2020**, *11*, 1–11. [CrossRef] [PubMed]
53. Coccia, M. *Scientific Data of the Principal Factors Determining the Diffusion of COVID-19 in Italy*; Version 1; Mendeley Data; Mendeley-Elsevier: London, UK, 2020. [CrossRef]

sustainability

MDPI

Article

Relationship between Weather Variables and New Daily COVID-19 Cases in Dhaka, Bangladesh

M. Mofijur [1,2,*], I.M. Rizwanul Fattah [1,*], A.B.M. Saiful Islam [3], M.N. Uddin [4], S.M. Ashrafur Rahman [5], M.A. Chowdhury [6], Md Asraful Alam [7,*] and Md. Alhaz Uddin [8]

1 Faculty of Engineering and Information Technology, University of Technology Sydney NSW 2007, Australia
2 Mechanical Engineering Department, Prince Mohammad Bin Fahd University, Al Khobar 31952, Saudi Arabia
3 Department of Civil and Construction Engineering, College of Engineering, Imam Abdulrahman Bin Faisal University, Dammam 31451, Saudi Arabia; asislam@iau.edu.sa
4 Department of Electrical and Electronic Engineering, Northern University Bangladesh, Dhaka 1213, Bangladesh; mnuddin@nub.ac.bd
5 Biofuel Engine Research Facility, Queensland University of Technology (QUT), Brisbane, QLD 4000, Australia; s2.rahman@qut.edu.au
6 Department of Mechanical Engineering, Dhaka University of Engineering and Technology (DUET), Gazipur 1707, Bangladesh; asad@duet.ac.bd
7 School of Chemical Engineering, Zhengzhou University, Zhengzhou 450001, China
8 Department of Civil Engineering, College of Engineering, Jouf University, Sakaka 42421, Saudi Arabia; malhaz@ju.edu.sa
* Correspondence: MdMofijur.Rahman@uts.edu.au (M.M.); IslamMdRizwanul.Fattah@uts.edu.au (I.M.R.F.); alam@zzu.edu.cn (M.A.A.)

Received: 2 August 2020; Accepted: 1 October 2020; Published: 9 October 2020

Abstract: The present study investigated the relationship between the transmission of COVID-19 infections and climate indicators in Dhaka, Bangladesh, using coronavirus infections data available from the Institute of Epidemiology, Disease Control and Research (IEDCR), Bangladesh. The Spearman rank correlation test was carried out to study the association of seven climate indicators, including humidity, air quality, minimum temperature, precipitation, maximum temperature, mean temperature, and wind speed with the COVID-19 outbreak in Dhaka, Bangladesh. The study found that, among the seven indicators, only two indicators (minimum temperature and average temperature) had a significant relationship with new COVID-19 cases. The study also found that air quality index (AQI) had a strong negative correlation with cumulative cases of COVID-19 in Dhaka city. The results of this paper will give health regulators and policymakers valuable information to lessen the COVID-19 spread in Dhaka and other countries around the world.

Keywords: COVID-19; climate indicators; air quality; environment; pandemic; coronavirus

1. Introduction

Coronavirus is a major pathogen affecting the respiratory system of humans [1,2]. On 31 December 2019, an unspecified etiological outbreak from Wuhan, Hubei, China, was reported to the World Health Organization (WHO) [3]. The novel disease, later named COVID-19 caused by SARS-COV-2, spread quickly to other countries around the globe [4]. In light of the rising danger, WHO declared COVID-19 as an international public health emergency (PHEIC) [5]. SARS-COV-2 has already infected people in most countries around the world. As of 25 September 2020, the incidence of SARS-COV-2 infections reached the figure of more than 32 million, with more than 213 countries and regions affected by the pandemic [6]. It has been reported that in the worst-case scenario, COVID-19 patients can have trouble breathing and can have pneumonia, which results in death [7].

Clinical diagnosis has identified that COVID-19 patients have similar indications to other coronavirus affected patients, e.g., Middle East respiratory syndrome (MERS) and severe acute respiratory syndrome (SARS) [8]. The initial indication of SARS-COV-2 infection is cough, fever, short breath, and in the late stage, it can damage the kidneys, cause pneumonia, and unexpected death. Worldwide experience has shown that COVID-19 can affect all age groups with varying recovery rates among those groups [9]. The vulnerability of the elderly (>80 years of age) is high, with a fatality rate of ~22% of cases infected by COVID-19 [1].

In Bangladesh, the first coronavirus cases were identified on 8 March 2020, in three young patients (two male and one female). The two male patients had returned from Italy, and the female was a family member of one of the two males [10]. Bangladesh's infections remained low until the end of March but increased steeply in April. Bangladesh's cases reached 100 on 9 April and exceeded 200 cases in two days, so the time for the caseload to double was two days, which later slowed to three and then five days. On 1 June 2020, the time required for the caseload in Bangladesh to double remained at five days. The spread of infection was similar to that of Italy, France, and South Korea, which resulted in WHO announcing COVID-19 a pandemic [11]. Between 8 March and 1 June 2020, there were 49,534 COVID-19 cases in Bangladesh confirmed by rt-PCR, including 672 related fatalities (Case Fatality Rate 1.36%) according to the Institute of Epidemiology Disease Control And Research (IEDCR). Figure 1 shows the COVID-19 case distribution up until 1 June 2020, in Bangladesh [12]. As of 1 June 2020, the geographical distribution of confirmed reported COVID-19 cases was available on 65% of all cases (32,120 of 49,534). Of these cases, 70.3% (22,576) were from Dhaka division, 16.7% (5361) Chattogram division, 3.1% (996) Mymensingh division, 2.9% (937) Rangpur division, 2.3% (744) Sylhet division, 2.2% (700) Rajshahi division, 1.8% (574) Khulna division, and 0.7% (232) Barisal division [13]. The highest attack rate (AR) was observed in the Dhaka division (524.1/1,000,000), whereas Chattogram division had the second highest AR (159.5/1,000,000), and Mymensingh division had the third highest AR (76.6/1,000,000). As of 1 June 2020, according to IEDCR, a total of 320,199 COVID-19 tests, with the overall positivity rate of 15.5%, were conducted in Bangladesh by 52 laboratories (28 laboratories in Dhaka and 24 laboratories in other divisions of the country).

Huang et al. [9] reported that COVID-19 was transmitted from bat to human, though the transitional host is still unidentified. The transmission of the novel virus is through air droplets from human to human. It has been reported that the temperature, humidity, wind, and precipitation may favour either the spread or the inhibition of epidemic episodes [14]. Bashir et al. [15] reported that the transmission of viruses is influenced by weather conditions and the density of people. Dalziel et al. [16] considered climate indicators as the best predictors because weather conditions play a significant role in the transmission of the coronavirus. Epstein [17] reported that the adverse weather associated with long-term changes in climate made a significant contribution to the West Nile virus spreading in the US and Europe. Even though some research has looked into identifying the link between weather variables and the spread of viruses, there is limited research on the relation between them. Most of these studies focused solely on observing the correlation of temperatures, while overlooking the impact of air quality on COVID-19 dynamics. A previous study by Haque and Rahman [18] studied the relation of temperature and humidity with COVID-19 and found that under a linear regression framework, high temperature and high humidity significantly reduced the transmission. However, they concluded that due to high local transmission, it was unlikely that the pandemic would have diminished by April. The Air Quality Index (AQI) of Bangladesh indicates that Dhaka is the most polluted city, and it is predicted that the weather variables, along with AQI at Dhaka city, are significantly correlated with the new daily COVID-19 cases.

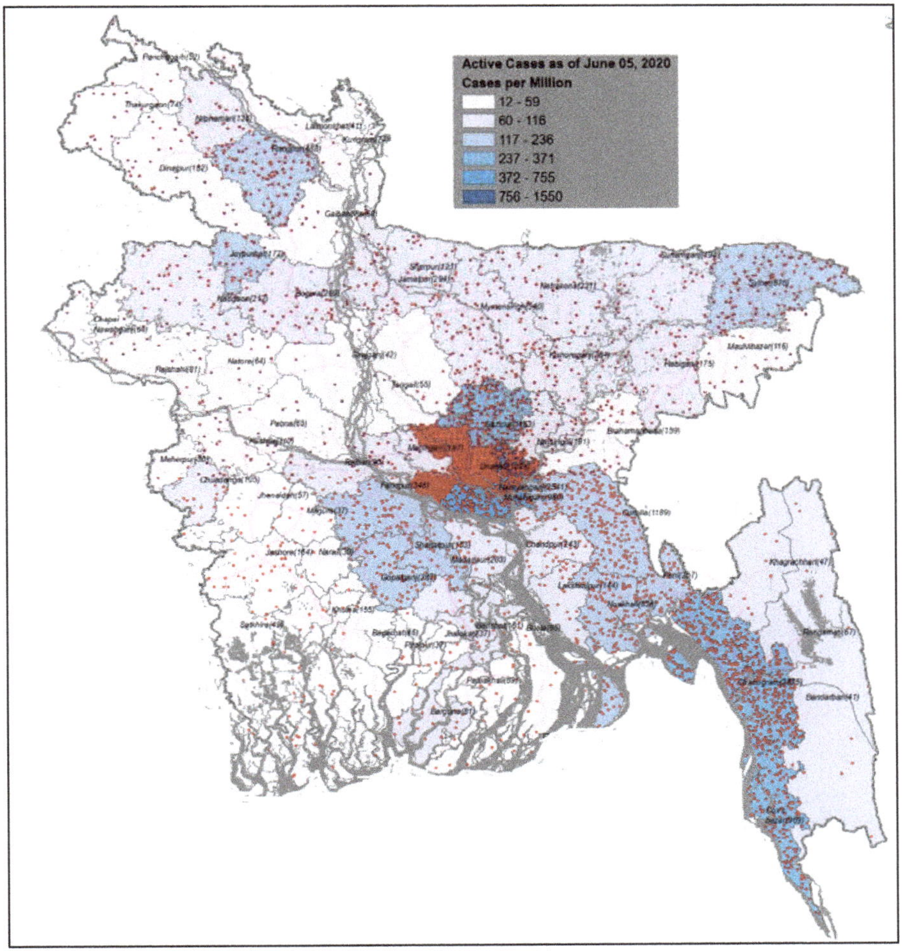

Figure 1. COVID-19 case distribution up until 1 June 2020, in Bangladesh [12].

The trajectory of an epidemic is defined by the reproduction number (R_0), which denotes the average number of secondary cases generated by one primary subject with the whole population susceptible to infection [19]. To consider a disease to be an epidemic, the value of R_0 must be greater than unity, which eventually determines the number of infected people if no intervention is placed. However, as the time elapses, control measures such as non-pharmaceutical interventions (NPIs) (and vaccines or drugs, if available) are introduced as well as the number of people susceptible to infection is exhausted, the effective reproduction number (R) declines. When R falls below unity in value, the epidemic starts to phase out. NPIs, including physical distancing, closure, quarantine, isolation, mask use, and hygiene behaviour, are typically aimed at reducing contact rates in the population [20,21]. It has been reported that the implementation of NPIs generally coincided with a rapid decline in the number of new cases [22]. Several studies have investigated the potential effect of mitigation measures on the dynamics of COVID-19 [19,21,23]. They have suggested that moderate measures could shrink the extent of the epidemic. However, extreme measures would ensure the health system is not overwhelmed, albeit at high economic and social costs [24]. This work will provide expedient information to the policymakers, which can help to lower the present COVID-19 infection rate through the use of NPIs and provide guidance for controlling future pandemics from the onset.

2. Methods

2.1. Study Area

Dhaka is the capital of the People's Republic of Bangladesh. Geographically, Dhaka is situated in South Asia, between 20°30′ to 26°38′ north latitude and 88°01′ to 92°41′ east longitude. It has a population of over 21 million residents. Dhaka has a tropical wet and dry climate. The city has a distinct monsoonal season, with an annual average temperature of 26 °C (79 °F), and monthly means varying between 19 °C (66 °F) in January, and 29 °C (84 °F) in May. Approximately 87% of the annual average rainfall of 2123 mm (83.6 inches) occurs between May and October. Four meteorological seasons are recognised in Dhaka as pre-monsoon (March, April, and May), monsoon (June to September), post-monsoon (October and November), and winter (December, January, and February) [25]. Dhaka is the financial, commercial, and the entertainment capital of Bangladesh. It accounts for up to 35% of Bangladesh's economy. Since its establishment as a modern capital city, the population, area, and the social and economic diversity of Dhaka have grown tremendously. Dhaka is now one of the most densely industrialised regions in the country. Dhaka is a major international city, as it hosts the headquarters of several international cooperations. By the 21st century, it emerged as a megacity, which is now listed as a beta-global city by the Globalization and World Cities Research Network (GaWC).

2.2. Data Collection

The computer-based data on the temporal variation of COVID-19 cases in Dhaka from the period of 1–31 May 2020, were procured from the IEDCR, Bangladesh. Until 30 April, there were a total of 6000 cases recorded (since coronavirus first identified in Bangladesh) in Dhaka and new cases increased steadily from 1 May 2020. The temporal and spatial data on climate indicators (of minimum, maximum, and average temperatures in °C, precipitation in millimetres, and humidity as a percentage) during May 2020 were collected from the Bangladesh Meteorological Department (BMD) [15]. BMD is the only government organisation responsible for monitoring and issuing all types of forecasts and warnings related to weather events. BMD collects the daily weather data for the Dhaka city from Dhaka PBO station (international station number—41923), which is located at 23°46′ N, 90°23′ E, and elevated at 8.45 m. The AQI data was collected from Dhaka US Consulate Air Pollution: Real-time Air Quality Index [26]. The US Embassy air quality monitor measured airborne fine particulate matter on the compound of the Embassy in Dhaka.

2.3. Data Analysis

Data can be analysed in different ways, including using the time series modelling, Spearman, or Pearson correlation coefficient. As the objective of this study was to perform the correlational study of two temporal series of data, the Spearman rank correlation (r_s) test was performed to inspect the correlation between new and total infections with environmental indicators. Spearman's correlation determines the strength and direction of the monotonic relationship between two variables. The coefficient was calculated using the following equation [16]:

$$r_s = 1 - 6\frac{\sum d_i^2}{n(n^2-1)} \quad (1)$$

where d_i is the difference between the x, y pair or the difference between the ranks of the x and y variables; n is the number of coordinate pairs.

3. Results and Discussion

Figure 2a shows the number of daily new infections and total infections statistics for Dhaka from 1 May 2020 to 31 May 2020. It is seen that the daily new cases were not varied steadily. For example, on 6 May, there were 540 new cases, and on 16 May, 407 cases were observed. The maximum new cases (1283) were observed on 20 May. The total confirmed cases increased steadily until the end of the month. On 1 May, the confirmed cases in Dhaka were 6031. By the end of the first week of May, confirmed cases had risen to 8289, by the end of the second week to 11,280, by the end of the third week to 16,528, and by the end of the month to 22,576.

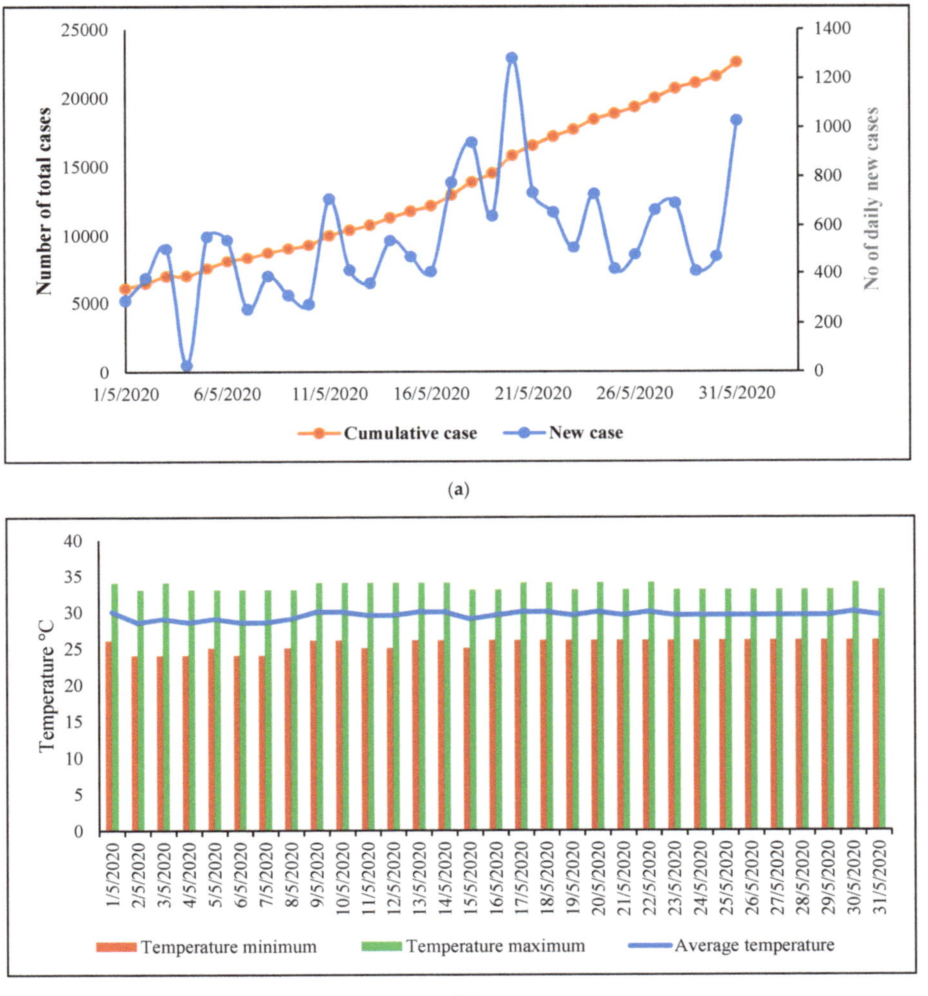

Figure 2. (**a**) Daily new cases and total confirmed cases of COVID-19 in Dhaka, May 2020 [4]. (**b**) Daily maximum, minimum, and average temperatures in Dhaka, May 2020 [14].

The daily temperature changes of Dhaka in May 2020 is depicted in Figure 2b. The lowest daily minimum, maximum, and the average temperature was 24 °C, 33 °C, and 29 °C, respectively, whereas the highest daily minimum, maximum, and the average temperature was 26 °C, 34 °C, and 30 °C, respectively. The average daily lowest and highest rainfall was 2 mm and 20 mm, respectively. The average daily minimum and maximum humidity was 47% and 98%, respectively. The average daily lowest and highest wind speed was 1 km/h and 17 km/h, respectively. The average daily minimum and maximum AQI was 54 and 150, respectively. AQI is inversely related to air quality.

Table 1 shows the correlation of these seven climate indicators with daily new infections and total infections of SARS-COV-2 in Dhaka. Among the seven weather variables, only mean temperature ($r_s = 0.470$; $p = 0.02$) and minimum temperature ($r_s = 0.427$; $p = 0.01$) were positively and significantly correlated with daily new infections. The mean temperature and minimum temperature indicated a significant positive relationship. The reason may be that the viral aerosols last longer in the air due to their smaller sizes at minimal or average temperature, thereby raising the spread of the virus. The AQI indicated a significant negative relationship with total confirmed infections. The reason may be that Dhaka was in full lockdown during the entire month of May so that people's mobility in and out of the region was minimal and pollutant emission was reduced due to absence of public transportation. Maximum temperature, wind speed, rainfall, and humidity were not significantly correlated with the dynamics of coronavirus.

Table 1. Relationship coefficients result between climate indicators and COVID-19 outbreak.

	Climate Variable	New Case	Cumulative Case
Spearman correlation Coefficient (r_s)	Min. temp.	0.427 **	0.732 ***
	Max. temp.	0.156	−0.041
	Avg. temp.	0.472 ***	0.701 ***
	Rainfall	0.233	0.591
	Humidity	0.111	0.013
	Wind	0.287	0.388
	AQI	−0.205	−0.607 ***

*** 1% significance level (two-tailed); ** 5% significance level (two-tailed).

For the present study, the link between climate indicators and the spread of COVID-19 in Dhaka is analysed. Figure 3 shows the most correlation factors on daily new infections and the total number of infections. The analysis determined that the minimum and average temperatures in Dhaka were linked to the spread of COVID-19. Increasing the average temperature by 4 °C gradually lowered the COVID-19 spread. This relationship is supported by the work [27] on the respiratory syncytial virus (RSV) and [28] on SARS viruses. Previous research on the effect of climate indicators on COVID-19 dynamics indicated that temperature served as the key driver for the spread of the coronavirus in China [29], Indonesia [30], and the US [15]. Chen et al. [31] also highlighted that temperature influences the spread of the coronavirus. There are reports of other meteorological indicators affecting the dynamics of the novel coronavirus. It has been reported that humidity and temperature are crucial indicators, which significantly motivate the seasonal transmission of SARS-COV-2 [32]. However, weather parameters affect not only the transmission of the novel virus but also mortality [33]. Poole [34] cited that environment latitude also correlates with the COVID-19 pandemic. Liu et al. [35] also reported that that local weather conditions of lowered temperature, mild diurnal temperature range, and low humidity might favour the transmission.

The findings of this study have been compared with a recent study [30] undertaken in Indonesia which indicated that only mean temperature significantly affected daily cases, with other variables bearing no relationship with the COVID-19 pandemic. In this study, we also found that the average temperature was linked with daily new cases in Bangladesh ($r_s = 0.472$) and that the observed relationship was higher than that found for Indonesia ($r_s = 0.393$). Unlike the Indonesian study [30], this study also considered the effect of AQI on the impact of the outbreak. We found that AQI was strongly related ($r_s = 0.607$; $p = 0.001$) to total cases in Bangladesh.

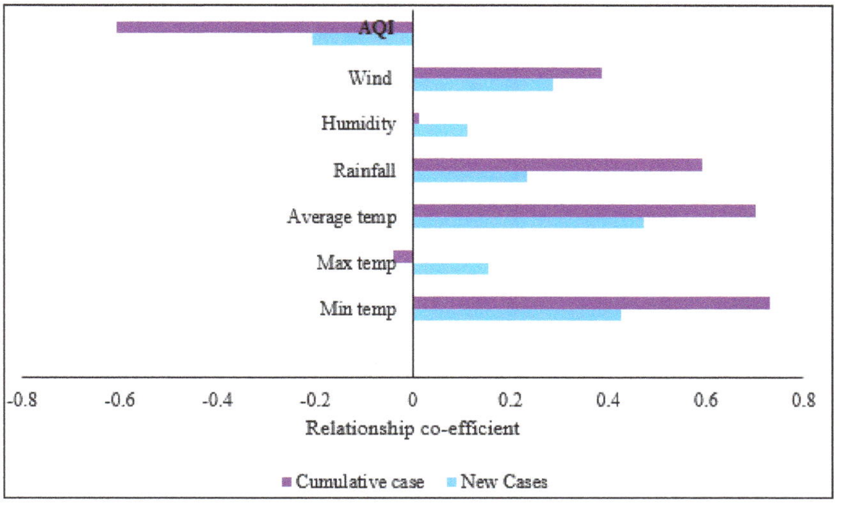

Figure 3. The comparison of r_s among the most indicators.

In Dhaka, the fast spread of COVID-19 was due to the extremely high level of human mobility, along with the weather. Being the capital city, Dhaka is the centre of almost all commercial, economic, and cultural activities of Bangladesh. Many people from regional areas in Bangladesh travel to Dhaka to find a job and develop their financial stability. Another explanation perhaps is that Dhaka has a high population density, allowing very quick transmission of SARS-COV-2. Dhaka has a population of 21 million, growing at 3.6% per year and a density of 121,720 residents per square mile, according to census data. Most of the people are involved either in households or unauthorised work and considerable proportion work in the garment industry (approximately 800,000). Such statistics make Dhaka a perfect epicentre for infection compared to other areas of the country. Zu et al. [36] also reported that population density is one of the causes for the rapid transmission of SARS-COV-2.

In general, the length of the incubation period (duration between exposure to the virus and the onset of symptoms) and the infectiousness period determine the initial spread, infection rate, and the probable duration of an epidemic [19]. Fast detection and isolation of infected cases, as well as social distancing, had shown to produce a more significant restrictive effect on the outbreak by other studies [24]. Davis et al. [37] assessed the influence of several control measures for mitigating the impact of COVID-19 in the UK through modelling and reported that extreme measures were required to bring the epidemic under control due to characteristics of SARS-COV-2. In addition, targeted and robust NPI measures can significantly reduce transmission to the local community, as shown by China [24].

4. Conclusions

Climate indicators play an essential role in determining the COVID-19 outbreak in Dhaka. This study showed a significant correlation between mean temperature, low temperature, and AQI and rapid transmission of COVID-19 infections. The outcomes of this research will provide useful information to lower the incidence rate. In particular, the significant relationship between AQI and the spread of COVID-19 infections indicated the importance of implementing clean environment policies within the country. However, further research on daily pollutant emissions, such as carbon, particulate matter, and nitrogen oxides is necessary as forced confinement due to the pandemic significantly lowered pollutant emissions.

Despite important findings on the impact of weather on the COVID-19 pandemic, some limitations remain. For example, the infection of the novel coronavirus is influenced by many variables, including people's mobility, immunity, hygiene and proximity, the size of gatherings, testing facilities, and health care management. Therefore, further and comprehensive research is necessary to explore data about these variables to give a clearer picture of the dynamics of COVID-19.

Author Contributions: Conceptualisation, M.M. and I.M.R.F.; data curation, M.M. and A.B.M.S.I.; formal analysis, S.M.A.R. and M.A.A.; supervision, M.A.C.; writing—original draft, M.M. and I.M.R.F.; writing—review and editing, M.N.U. and M.A.U. All authors have read and agreed to the published version of the manuscript.

Funding: This research has not been externally funded.

Acknowledgments: Authors would like to acknowledge the 'research development fund' of the School of Information, Systems and Modelling, University of Technology Sydney, Australia.

Conflicts of Interest: The authors declare no conflict of interest.

References

1. Abdullah, S.; Mansor, A.A.; Napi, N.N.L.M.; Mansor, W.N.W.; Ahmed, A.N.; Ismail, M.; Ramly, Z.T.A. Air quality status during 2020 Malaysia Movement Control Order (MCO) due to 2019 novel coronavirus (2019-nCoV) pandemic. *Sci. Total Environ.* **2020**, *729*, 139022. [CrossRef] [PubMed]
2. Kabir, M.T.; Uddin, M.S.; Hossain, M.F.; Abdulhakim, J.A.; Alam, M.A.; Ashraf, G.M.; Bungau, S.G.; Bin-Jumah, M.N.; Abdel-Daim, M.M.; Aleya, L. nCOVID-19 Pandemic: From Molecular Pathogenesis to Potential Investigational Therapeutics. *Front Cell Dev. Biol.* **2020**, *8*, 616. [CrossRef] [PubMed]
3. Li, Q.; Guan, X.; Wu, P.; Wang, X.; Zhou, L.; Tong, Y.; Ren, R.; Leung, K.S.M.; Lau, E.H.Y.; Wong, J.Y.; et al. Early transmission dynamics in Wuhan, China, of novel coronavirus-infected pneumonia. *N. Engl. J. Med.* **2020**, *382*, 1199–1207. [CrossRef] [PubMed]
4. Forster, P.M.; Forster, H.I.; Evans, M.J.; Gidden, M.J.; Jones, C.D.; Keller, C.A.; Lamboll, R.D.; Quéré, C.L.; Rogelj, J.; Rosen, D.; et al. Current and future global climate impacts resulting from COVID-19. *Nat. Clim. Chang.* **2020**. [CrossRef]
5. Sohrabi, C.; Alsafi, Z.; O'Neill, N.; Khan, M.; Kerwan, A.; Al-Jabir, A.; Iosifidis, C.; Agha, R. World Health Organization declares global emergency: A review of the 2019 novel coronavirus (COVID-19). *Int. J. Surg.* **2020**, *76*, 71–76. [CrossRef]
6. Rizwanul Fattah, I.M.; Ong, H.C.; Mahlia, T.M.I.; Mofijur, M.; Silitonga, A.S.; Rahman, S.M.A.; Ahmad, A. State of the Art of Catalysts for Biodiesel Production. *Front. Energy Res.* **2020**, *8*. [CrossRef]
7. Holshue, M.L.; DeBolt, C.; Lindquist, S.; Lofy, K.H.; Wiesman, J.; Bruce, H.; Spitters, C.; Ericson, K.; Wilkerson, S.; Tural, A.; et al. First case of 2019 novel coronavirus in the United States. *N. Engl. J. Med.* **2020**, *382*, 929–936. [CrossRef]
8. Wang, Y.; Wang, Y.; Chen, Y.; Qin, Q. Unique epidemiological and clinical features of the emerging 2019 novel coronavirus pneumonia (COVID-19) implicate special control measures. *J. Med Virol.* **2020**, *92*, 568–576. [CrossRef]
9. Masrur, A.; Yu, M.; Luo, W.; Dewan, A. Space-Time Patterns, Change, and Propagation of COVID-19 Risk Relative to the Intervention Scenarios in Bangladesh. *Int. J. Environ. Res. Public Health* **2020**, *17*, 5911. [CrossRef]
10. Paul, R. Bangladesh Confirms its First Three Cases of Coronavirus. Available online: https://www.reuters.com/article/us-health-coronavirus-bangladesh-idUSKBN20V0FS (accessed on 30 September 2020).
11. Cucinotta, D.; Vanelli, M. WHO declares COVID-19 a pandemic. *Acta Biomed.* **2020**, *91*, 157–160. [CrossRef]
12. Bengal Institute. COVID-19 in Bangladesh: Case Rates. Available online: https://bengal.institute/research/covid19bangladesh/ (accessed on 29 September 2020).
13. World Health Organization (WHO). Bangladesh COVID-19 Situation Report No #14, 1 June 2020. Available online: https://www.who.int/docs/default-source/searo/bangladesh/covid-19-who-bangladesh-situation-reports/who-ban-covid-19-sitrep-14-20200601.pdf? (accessed on 30 September 2020).
14. Cheval, S.; Mihai Adamescu, C.; Georgiadis, T.; Herrnegger, M.; Piticar, A.; Legates, D.R. Observed and Potential Impacts of the COVID-19 Pandemic on the Environment. *Int. J. Environ. Res. Public Health* **2020**, *17*, 4140.

15. Bashir, M.F.; Ma, B.; Bilal; Komal, B.; Bashir, M.A.; Tan, D.; Bashir, M. Correlation between climate indicators and COVID-19 pandemic in New York, USA. *Sci. Total Environ.* **2020**, *728*, 138835. [CrossRef] [PubMed]
16. Dalziel, B.D.; Kissler, S.; Gog, J.R.; Viboud, C.; Bjørnstad, O.N.; Metcalf, C.J.E.; Grenfell, B.T. Urbanisation and humidity shape the intensity of influenza epidemics in US cities. *Science* **2018**, *362*, 75–79. [CrossRef] [PubMed]
17. Epstein, P.R. West Nile virus and the climate. *J. Urban Health* **2001**, *78*, 367–371. [CrossRef] [PubMed]
18. Haque, S.E.; Rahman, M. Association between temperature, humidity, and COVID-19 outbreaks in Bangladesh. *Env. Sci. Policy* **2020**, *114*, 253–255. [CrossRef]
19. Anderson, R.M.; Heesterbeek, H.; Klinkenberg, D.; Hollingsworth, T.D. How will country-based mitigation measures influence the course of the COVID-19 epidemic? *Lancet* **2020**, *395*, 931–934. [CrossRef]
20. Seale, H.; Dyer, C.E.F.; Abdi, I.; Rahman, K.M.; Sun, Y.; Qureshi, M.O.; Dowell-Day, A.; Sward, J.; Islam, M.S. Improving the impact of non-pharmaceutical interventions during COVID-19: Examining the factors that influence engagement and the impact on individuals. *Bmc Infect. Dis.* **2020**, *20*, 607. [CrossRef]
21. Prem, K.; Liu, Y.; Russell, T.W.; Kucharski, A.J.; Eggo, R.M.; Davies, N.; Flasche, S.; Clifford, S.; Pearson, C.A.B.; Munday, J.D.; et al. The effect of control strategies to reduce social mixing on outcomes of the COVID-19 epidemic in Wuhan, China: A modelling study. *Lancet Public Health* **2020**, *5*, e261–e270. [CrossRef]
22. Flaxman, S.; Mishra, S.; Gandy, A.; Unwin, H.J.T.; Mellan, T.A.; Coupland, H.; Whittaker, C.; Zhu, H.; Berah, T.; Eaton, J.W.; et al. Estimating the effects of non-pharmaceutical interventions on COVID-19 in Europe. *Nature*. [CrossRef]
23. Kissler, S.M.; Tedijanto, C.; Goldstein, E.; Grad, Y.H.; Lipsitch, M. Projecting the transmission dynamics of SARS-CoV-2 through the postpandemic period. *Science* **2020**, *368*, 860–868. [CrossRef]
24. Lai, S.; Ruktanonchai, N.W.; Zhou, L.; Prosper, O.; Luo, W.; Floyd, J.R.; Wesolowski, A.; Santillana, M.; Zhang, C.; Du, X.; et al. Effect of non-pharmaceutical interventions to contain COVID-19 in China. *Nature* **2020**, *585*, 410–413. [CrossRef]
25. Bangladesh Meteorological Department (BMD). Meteorological seasons in Dhaka. Available online: http://live.bmd.gov.bd/ (accessed on 29 September 2020).
26. Dhaka US Consulate Air Pollution: Real-time Air Quality Index. Available online: http://aqicn.org/city/bangladesh/dhaka/us-consulate/ (accessed on 4 July 2020).
27. Vandini, S.; Corvaglia, L.; Alessandroni, R.; Aquilano, G.; Marsico, C.; Spinelli, M.; Lanari, M.; Faldella, G. Respiratory syncytial virus infection in infants and correlation with meteorological factors and air pollutants. *Ital. J. Pediatr.* **2013**, *39*, 1. [CrossRef] [PubMed]
28. Tan, J.; Mu, L.; Huang, J.; Yu, S.; Chen, B.; Yin, J. An initial investigation of the association between the SARS outbreak and weather: With the view of the environmental temperature and its variation. *J. Epidemiol. Community Health* **2005**, *59*, 186–192. [CrossRef] [PubMed]
29. Shi, P.; Dong, Y.; Yan, H.; Li, X.; Zhao, C.; Liu, W.; He, M.; Tang, S.; Xi, S. The impact of temperature and absolute humidity on the coronavirus disease 2019 (COVID-19) outbreak-evidence from China. *medRxiv* **2020**. [CrossRef]
30. Tosepu, R.; Gunawan, J.; Effendy, D.S.; Ahmad, L.O.A.I.; Lestari, H.; Bahar, H.; Asfian, P. Correlation between weather and Covid-19 pandemic in Jakarta, Indonesia. *Sci. Total Environ.* **2020**, *725*, 138436. [CrossRef]
31. Chen, B.; Liang, H.; Yuan, X.; Hu, Y.; Xu, M.; Zhao, Y.; Zhang, B.; Tian, F.; Zhu, X. Roles of meteorological conditions in COVID-19 transmission on a worldwide scale. *medRxiv* **2020**. [CrossRef]
32. Sajadi, M.M.; Habibzadeh, P.; Vintzileos, A.; Shokouhi, S.; Miralles-Wilhelm, F.; Amoroso, A. Temperature, Humidity and Latitude Analysis to Predict Potential Spread and Seasonality for COVID-19 (5 March 2020). *SSRN* **2020**. Available online: https://ssrn.com/abstract=3550308 (accessed on 4 July 2020). [CrossRef]
33. Ma, Y.; Zhao, Y.; Liu, J.; He, X.; Wang, B.; Fu, S.; Yan, J.; Niu, J.; Zhou, J.; Luo, B. Effects of temperature variation and humidity on the death of COVID-19 in Wuhan, China. *Sci. Total Environ.* **2020**, *724*, 138226. [CrossRef]
34. Poole, L. Seasonal influences on the spread of SARS-CoV-2 (COVID-19), causality, and Forecastabililty (3-15-2020). *SSRN* **2020**. [CrossRef]
35. Liu, J.; Zhou, J.; Yao, J.; Zhang, X.; Li, L.; Xu, X.; He, X.; Wang, B.; Fu, S.; Niu, T.; et al. Impact of meteorological factors on the COVID-19 transmission: A multi-city study in China. *Sci. Total Environ.* **2020**, *726*, 138513. [CrossRef]

36. Zu, Z.Y.; Jiang, M.D.; Xu, P.P.; Chen, W.; Ni, Q.Q.; Lu, G.M.; Zhang, L.J. Coronavirus Disease 2019 (COVID-19): A Perspective from China. *Radiology* **2020**, 200490. [CrossRef]
37. Davies, N.G.; Kucharski, A.J.; Eggo, R.M.; Gimma, A.; Edmunds, W.J.; Jombart, T.; O'Reilly, K.; Endo, A.; Hellewell, J.; Nightingale, E.S.; et al. Effects of non-pharmaceutical interventions on COVID-19 cases, deaths, and demand for hospital services in the UK: A modelling study. *Lancet Public Health* **2020**, *5*, e375–e385. [CrossRef]

sustainability

MDPI

Article

Energy Engineering Approach for Rural Areas Cattle Farmers in Bangladesh to Reduce COVID-19 Impact on Food Safety

Mohammad Nur-E-Alam [1,*], Mohammad Nasirul Hoque [2], Soyed Mohiuddin Ahmed [3], Mohammad Khairul Basher [4] and Narottam Das [5,6]

1 Electron Science Research Institute, School of Science, Edith Cowan University, Joondalup, WA 6027, Australia
2 Department of Physics, University of Chittagong, Chattogram 4331, Bangladesh; nasirul@cu.ac.bd
3 School of Pharmacy and Life Science, Jiujiang University, Jiujiang 332000, Jiangxi, China; mohiuddinppqc@yahoo.com
4 Institute of Electronics, AERE, Bangladesh Atomic Energy Commission, Dhaka 1349, Bangladesh; khairulcu@gmail.com
5 School of Engineering and Technology, Central Queensland University, Melbourne, VIC 3000, Australia; n.das@cqu.edu.au
6 Centre for Intelligent Systems, School of Engineering and Technology, Central Queensland University, Brisbane, QLD 4000, Australia
* Correspondence: m.nur-e-alam@ecu.edu.au

Received: 21 September 2020; Accepted: 11 October 2020; Published: 17 October 2020

Abstract: This paper reports on the optimization of thin-film coating-assisted, self-sustainable, off-grid hybrid power generation systems for cattle farming in rural areas of Bangladesh. Bangladesh is a lower middle-income country with declining rates of poverty among its 160 million people due to persistent economic growth in conjunction with balanced agricultural improvements. Most of the rural households adopt a mixed farming system by cultivating crops and simultaneously rearing livestock. Among the animals raised, cattle are considered as the most valuable asset for the small-/medium-scale farmers in terms of their meat and milk production. Currently, along with the major health issue, the COVID-19 pandemic is hindering the world's economic growth and has thrust millions into unemployment; Bangladesh is also in this loop. However, natural disasters such as COVID-19 pandemic and floods, largely constrain rural smallholder cattle farmers from climbing out of their poverty. In particular, small- and medium-scale cattle farmers face many issues that obstruct them from taking advantage of market opportunities and imposing a greater burden on their families and incomes. An appropriate measure can give a way to make those cattle farmers' businesses both profitable and sustainable. Optimization of thin-film coating-assisted, self-sustainable, off-grid hybrid power generation system for cattle farming is a new and forward-looking approach for sustainable development of the livestock sector. In this study, we design and optimize a thin-film coating-assisted hybrid (photovoltaic battery generator) power system by using the Hybrid Optimization of Multiple Energy Resources (HOMER, Version 3.14.0) simulation tool. An analysis of the results has suggested that the off-grid hybrid system is more feasible for small- and medium-scale cattle farming systems with long-term sustainability to overcome the significant challenges faced by smallholder cattle farmers in Bangladesh.

Keywords: cattle farming; COVID-19 pandemic; economic point of view; food safety; HOMER; hybrid system; smallholder; thin-film coating

---

1. Introduction

Bangladesh is a small country on the world map; located in south Asia having a land area of 148,460 square kilometers. Apart from the hilly areas, most of the land consists of plains with a strong network of rivers throughout the country. Bangladesh is a small country with a large population, where more than 1085 persons live per square kilometer. Besides industrialization and other commercial infrastructures, agriculture plays the most vital role in the country's economy. Agriculture and livestock not only boost the Gross Domestic Product (GDP), they also ensure food security and help to reduce poverty and unemployment in rural areas. The contribution made by livestock to the GDP for the Fiscal Year 2017–2018 was 1.53% at a constant price, which was 13.6% of the total agricultural sector contribution to GDP [1]. Despite achieving improvement in many aspects of food security, the people of Bangladesh still lack dietary diversification, which leads to nutritional imbalance [2]. There is a great discrepancy between demand and production in the case of protein intake. Milk production for the year 2018–2019 was 99.23 lakh Metric Ton against a demand of 152.02 lakh Metric Ton (one lakh is equivalent to 100,000) [3]. On the other hand, many people in rural areas cannot afford to maintain a balanced diet daily. Sustainable and well-distributed livestock farms in different regions of the country not only improve nutrition but are also preferred assets of investment [4]. Cattle farming has a good economic impact in generating income and food security in rural areas which employ about 20% of the rural labor force [5]. The United Nations Food and Agriculture Organization (FAO) reported that per capita meat consumption in Bangladesh is low (about four kilograms) compared to other neighboring countries such as Pakistan. However, according to the data from the Department of Livestock Services (DLS) Bangladesh, the country's total demand for meat stands at 72.97 lakh tonnes with the requirement of 120 g of meat per head, almost 55% of which comes from cows and goats. Besides the regular beef intake, there is a huge need for cows during one of the biggest festivals (Eid-Ul-Adha) of the Muslim calendar. In the past 20 years, Bangladesh has attempted to become fully dependent on domestic production to meet the extra demand for cattle during the Eid-Ul-Adha celebrations due to a drastic reduction in cattle imports from a neighboring country. Twenty years ago, Bangladesh entered a new era introducing professional micro-scale cattle farms to meet the demand of the local market for milk and meat. According to the report published as in Reference [6], 80 to 90% of the country's cattle production comes from rural farmers with either small- or medium-sized farms. Many people are also currently interested in modern agriculture, including cultivation of fruits, vegetables, fish and cattle as e-commerce has opened a promising platform for the young entrepreneur. Nevertheless, the housing shortage in Bangladesh is a major issue with the result being that many cultivable lands, as well as fields formerly used for open grazing, are nowadays utilized to build new houses and apartments which are reducing the open land areas at an alarming rate in Bangladesh. This is the reason Bangladesh cannot consider an open field cattle farm as found in Australia, New Zealand, or large European countries. Most of the cattle farms (small and medium scale) in Bangladesh either in rural or semi-town areas have very limited land size. Though there are many modern and scientifically advanced tools and equipment widely used in agriculture nowadays in Bangladesh, only a negligible number of farms are to be found using the energy engineering approach to become self-sufficient and sustainable and whilst mitigating the energy costs against the production costs. For the cattle farm, the most challenging part is the meat collection, milk collection, storage and marketing, for which each farm needs high-quality electrical equipment including electric blade, vacuum pump milking parlor systems, and high-performance refrigeration systems. Besides all other costs, such kinds of facilities require a massive budget for energy consumption to regulate as well as maintain adequate temperature, relative humidity, and ventilation. In recent years, energy cost efficiency becomes a key factor for sustainable food and crop storage facilities worldwide, which has stimulated many scientists' enthusiasm to introduce more off-grid, self-sustainable energy-harvesting and saving facilities [7–11]. Hybrid (PV, Wind, Diesel and Battery) power generation systems have recently become very promising for 100% electrification for the rural or remote island areas. A significant number of research works have been conducted worldwide by researchers and scientists in the field of renewable energy (RE) to design and optimize various types of hybrid power generation systems [12–17]. In addition, these types of hybrid power systems can be attractive alternatives for cost-effective power generation in Bangladesh. It is very hard for the government to provide on-going subsidies in power production. Bangladesh has set the goal of becoming a country with zero power supply issues such as daily power cuts, to provide electricity for everyone within the very near future. In support of the country's goal these types of hybrid power systems, together with nano-coated building materials with power-saving features, if used in different infrastructures, can help to reach to the goal of 100% electrification of the country. Renewable power generation and savings based multi-component hybrid systems can also lower carbon emissions to an acceptable level. The exploration of possible and economically viable renewable energy sources is one of the prime interests of the country's policymakers and research scientists to mitigate the energy demand of the country. However, there are several factors involved in the successful development of new renewable-type energy production systems [18–21]. M. E. Karim et al. has conducted a comprehensive study on the government involvements, legal issues, regulatory measures and the policy aspects related to the renewable energy in Bangladesh [22]. They have reported on the significance having of comprehensive policy initiatives, including active and enhancing government participation on developing RE systems, ensuring localization and easily accessible of RE technology, evaluation of relevant legal policy initiatives and others that can shape the country's sustainable development.

At present, the novel coronavirus (COVID-19) outbreak (first spread in Wuhan, China in late 2019) has resulted in a worldwide severe pandemic that has already disrupted human life and civilization. This coronavirus has already caused a huge number of deaths and becomes a major threat for mostly older people who are suffering from diabetes, cardiovascular disease, cancer, and chronic respiratory syndrome [23]. The world's leading scientists, chemists, microbiologists, engineers, doctors, and other health professionals are working very hard to find proper solutions for this new virus by understanding the nature of COVID-19. There are many predictions available online nowadays about the microstructural properties and weather- and region-dependent nature of COVID-19; however, getting an active vaccine against this virus requires a significant amount of research effort together with on-going worldwide clinical trials. The world is seeking new hope while currently relying on the precautions that have been recommended by the World Health Organization (WHO), including social and physical distancing to control the rapid spread of this virus [24]. The outbreak of COVID-19 has overturned not only human life but also the global economy by seriously affecting all kinds of industries including food, agriculture, readymade garments, leather, building and infrastructure materials, energy and power sectors, science and technology and others all over the world [25–30]. Many industries and businesses were forced to take actions like job cuts, even in some cases shut down completely. As a result, millions previously in the workforce have become jobless, and severely affecting—directly and indirectly—developing countries like Bangladesh. As COVID-19 can transfer easily from one affected person to others (i.e., community transmission), it becomes a great concern how, as well as how long is needed, to get back to normal life. According to the proposal jointly made by WHO and FAO for food safety including multi-agency co-operation, proper inspections in different stages of production can mitigate significantly the spread of this corona-virus but to achieve this is a very much challenging task [31,32]. In these circumstances, the risk of contamination can be minimized and kept under control only if the local demand for foods and protein supply can be met by the local growers and suppliers. In Bangladesh, the well-being of the economic cycle as well as of a large number of the population is mainly dependent on good agricultural and industrial prospects. All agricultural and industrial sectors including garment, leather and industries requiring construction materials are major candidates where continuous electricity supply is needed for maintaining a non-disruptive production cycle for the well-being of the economy. However, currently, these sectors are facing severe problems during the COVID-19 crisis and maintaining a continuous electricity supply will continue to be a major issue for many smallholder farmers in the rural farming and cultivation areas in post-COVID-19 Bangladesh [33–42]. Besides, the national grid electricity supply, solar photovoltaics (PVs)-assisted power generation can help all industries and factories to maintain their continuous production cycle. Therefore, designing new and modification of existing infrastructures that can generate and save energy will be the most effective approach for the well-being of post-COVID-19 Bangladesh and will assist in returning the country to a normal lifestyle. In addition, this energy engineering can be a model for future modern farming in any country of the world and, in the near future, potentially can help to diminish the current rise in unemployment due to the COVID-19 pandemic by making all the farms and food industries active and fully productive again.

In this article, we design and optimize a hybrid power generation system that can be applied together with thin-film, low-emissivity (Low-E) coating-assisted, energy-saving features for self-sustainable off-grid or semi-off-grid infrastructure for cattle farmers in rural areas. In Bangladesh, grid connection is not available throughout the country, and there is no alternative of a micro-grid to ensure the power availability to any new or existing infrastructures for farming and food storage. However, the implementation of modern scientific processes can increase the production levels by up to 50%, whereas if the conventional power sources are used, the fuel cost will have a negative impact on profit margins. The organization of this paper is as follows: Section 2 briefly describes the materials and methods, where we describe the types of farms targeted and ways of power optimization. Section 3 describes in detail the means of engineering of energy generation and savings including the components that are considered in this work. Section 4 presents the optimized simulation results, and Section 5 includes a short discussion based on the outcome of the hybrid energy engineering optimization, followed by a conclusion in Section 6.

2. Materials and Methods

2.1. Types of Farm

Figure 1 shows the images of mostly available (typical house type) infrastructure in Bangladesh for limited land size cattle farms together with the schematic diagram of a proposed cattle house containing hybrid power generation systems, in conjunction with energy-saving, thin-film-coated roof materials. A thin-film coating, mainly the low-E coating-assisted roof material, can filter daytime sun irradiation. The low-E-coated glass becomes a modern day building material towards the development of net-zero building infrastructures. Low-E coatings are designed to filter the visible spectrum whilst reflecting the infrared (IR) spectrum, thus reducing the heat entering the building [43–47]. Most of the commercially available low-E coatings are glass-based; however, several companies have developed low-E-type coatings on very thin polycarbonate (clear plastic substrate) materials that can be retrofitted to any transparent material, including glass or Perspex. In our proposed cattle house (Figure 1b), rather than a clear polycarbonate roof material, we are introducing a low-E-type coated semitransparent roof material which will allow only the visible light and thus will not only save lighting costs but also reduce the heat by reflecting the IR spectrum (which is mainly heat) during the day time. The effectiveness of thin-film coatings on energy savings is detailed in the Section 3.4.

In order to study the economic feasibility of energy engineering aspects, various types of cattle farms categorized by size, were considered such as small, medium and large farms holding the land size of 0.5 to 2.0, 2.1 to 5.0, and above 5.0 acres, respectively. However, the details information about the land size and the number maximum animals in one row or column are preciously provided in the following Refs. [48,49]. Table 1 presents the criteria for different types of the farm used in this study.

Figure 1. Images of a cattle farms. Existing typical cattle house in Bangladesh (**a**), and the schematic diagram of a proposed cattle house that contains hybrid power generation systems, together with energy-saving thin-film-coated roof materials (**b**).

Table 1. Criteria of different types of cattle farm (in terms of the number of cows).

Particular	Unit	Herd Size		
		Small	Medium	Large
Area of the farm	Acre	0.5 to 2.0	2.1 to 5.0	Above 5.0
Head count of milking cows	Number	5 to 20	21 to 50	Above 50
Head count of milking cows for this study	Number	13	35	100
Shed area	Row shape	Single row	Double row	Separate shed

2.2. *Optimization of a Power-Generating System*

We aimed to find a feasible and sustainable energy generating system that can meet the energy demand in cattle farming for the rural areas in Bangladesh. The PV systems are considered as the best renewable energy resources due to the great advantages that solar cells/module does not require any raw material to feed and have a very low operation and maintenance cost with a long-term operational life. However, the drawback for using solar cell is that it depends on sunlight and when sunlight is not available (such as a rainy and cloudy day or night) the energy extraction is greatly hampered. The PV module can be used only during the daytime with clear sky (i.e., a sunny day) thus confirms the unsuitability of use for the microgrid system. However, a storage device such as the battery can solve the problem by storing the unused excess power and fed to the load when it requires. Therefore, the PV-battery system can be a sustainable solution to meet the demand of a cattle farm. Techno-economic analyses for different energy (single generator, PV-Battery and PV-Generator-Battery hybrid) systems were carried out to find the most convenient and economically feasible microgrid system. Hybrid Optimization of Multiple Energy Resources (HOMER) software, developed by National Renewable Energy Limited (NREL, Boulder, CO, USA) is used to optimize the hybrid power systems. HOMER simulation tool allows comparing various types of systems in combination with capital, maintenance and other types with a given energy resource to meet the system load demand. Figure 2 shows the schematic diagram of HOMER algorithm that used in this work, where a sustainable power source is designed by considering PV cells, storage device, diesel generator, converter and other appropriate devices.

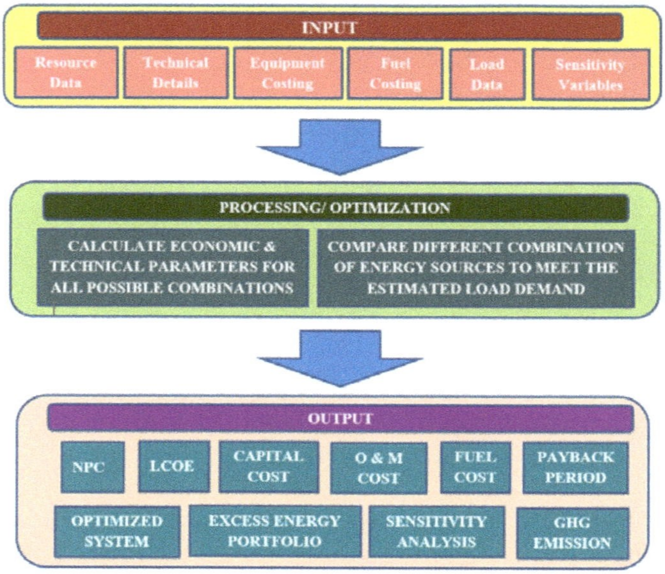

Figure 2. Schematic diagram of Hybrid Optimization of Multiple Energy Resources (HOMER) algorithm.

In this module, the available renewable energy resources and load profile for the system are required to set as input. Besides this, for the cost calculation, the price and operational datasheet need to be supplied for PV, Battery, Converter, generator and other devices. The economic analysis performed using the HOMER software for different types of farms with the payback period and energy engineering are detailed in the next section.

3. Engineering of Energy Generation and Savings

3.1. Design, Components and Optimization of Hybrid Power Generation System

3.1.1. Solar Radiation Data

The solar data available in HOMER pro software was used in this study which was originally extracted from the NASA surface meteorology and Solar energy database. The clearness index defines as the fraction of transmitted solar radiation through the atmosphere to the Earth surface is also an important parameter for HOMER Pro simulation tool. The clearness index is a dimensionless unit which varies from 0 to 1. For a clear sky, the index is close to 1 whereas close to zero for a dense cloud condition. The monthly solar radiation with a clearness index for Bangladesh is shown in Figure 3. It can be seen that the average daily solar radiation is maximum in April while the clearness index is maximum in December. It is because April is the summer when the sun is tilted to the northern hemisphere and at that time the sky becomes cloudy. On the other hand, in winter (December and January), the length of the day is comparatively shorter, and the sun is tilted to the southern hemisphere. However, at this time the sky is clear due to low humidity.

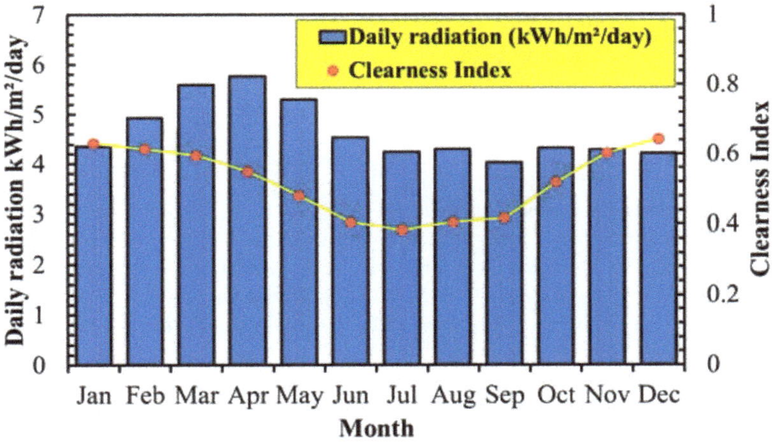

Figure 3. Monthly average solar radiation for a year.

3.1.2. Load Consumption Analysis

Load consumption is the most important part of any power system. The load consumption profile indicates the variation of load demand in different time of the day, month and season. This profile completely depends on the nature of the system. Keown et al. [50] reported that if the temperature rises above 80 °F (26 °C), feed intake by the cattle reduced significantly and thus impacts their growth. Additionally, at 90 °F (32 °C) or above, the milk production reduces by at least 3 to 20%. Therefore, a convenient and comfortable atmosphere is necessary to maintain for the maximum production. For the medium and small farm energy is required only for lighting and ventilation, but for the large farm milk collection and storage device are considered as additional load along with lighting and ventilation system.

For a large cattle farm, the most challenging period is to collect a huge amount of milk and marketing it instantly. For this reason, a large cattle farm should have a milking parlor and refrigeration system. G. Todde et al. shows that the power rating of the vacuum pump for each milking cow is 75 W [51]. A cow has to go to the milking parlor twice a day and on an average, it takes 15–20 min for milking. Therefore, the daily power requirement for milk collection is given by,

$$75 \text{ W} \times 20 \text{ min} \times 2 \text{ times} \times 100 \text{ cows}/1000 \times 60 = 5 \text{ kWh}$$

On the other hand, the liquid milk leaves the cow at 38 °C and requires cooling at 4 °C to avoid milk spoil. To prevent the bacteria growth, the milk should maintain 38 to 10 °C during the first hour after collection and from 10 to 4.4 °C during the second hour [52,53]. Upton et al. [54] reveals that the consumed electricity for milk cooling system is 13.02 Wh/L. If we consider the daily average milk production per cow is 15 L, then the daily power consumption for refrigeration can be calculated as follows:

$$15 \text{ L} \times 100 \text{ cows} \times 13.02 \text{ Wh/L}/1000 \times 24 = 0.81 \text{ kWh}$$

Inside the cowshed, proper lighting and ventilation system required for a healthy environment for the cattle. In Bangladesh, the average temperature is 32 °C in summer. In this study, LED lights and ceiling fans are considered for lighting and ventilation system. Table 2 shows the power rating and appropriate combination for different types of the farm are given below.

Table 2. Power rating, the combination of lighting, and ventilation devices for different types of farm.

Particular	Power Rating (W)	No. of the Unit for Different Types of Farm		
		Small	Medium	Large
LED Light	15	6	16	40
Fan	80	6	16	40

Figure 4 shows the differences in load consumption for three different types of farms, where it can be seen that the cattle farm load stays at the minimum stage in the winter season (November to February). This is because, in the winter season, the cooling system is not necessary for the farm. At night, there is hardly any load for the small- and medium-sized farm except lighting. On the other hand, the large farm has to switch on the refrigerator to store milk and, as a result, it requires a load consumption of more than 30 kW during the winter season. In March and October, the load demand is at a moderate stage. During this time, the temperature at night in such a state that the cooling system is not required for the cowshed. However, during the daytime, fans are needed to be switched on. At the summer season (April to September), there is a huge chance of heatstroke for the cows. In order to stabilize the production of milk, an additional 24 h cooling system needs to be implemented. This 24 h cooling system consisted of a ceiling fan is the major challenge to meet the demand for energy supply. During the summer season, the load demand is maximum for all types of farms due to ensure a congenial environment inside the shed.

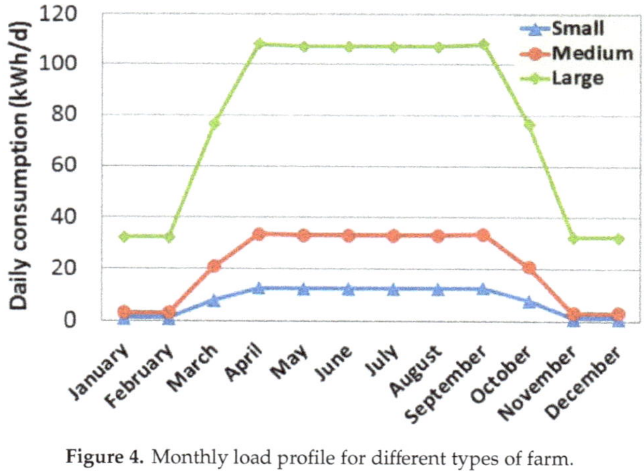

Figure 4. Monthly load profile for different types of farm.

3.2. *System Components*

3.2.1. Solar-PV Module

Polycrystalline silicon PV panels are considered to perform the simulation process in this study. Considering the relation between the orientation of the solar panels and PV output; the PV panels are considered to be placed with south facing and a slope of about 22 degrees with earth surface. In this case, the temperature effect of the solar cell is neglected. The estimated primary cost of PV module is $USD 1720/kWh including transport and installation cost. The replacement cost of the module is $USD 1720/kWh considering the lifetime of about 25 years.

3.2.2. Battery

The battery is a commonly used storage device for electrical energy. We consider a 12 V Lead Acid-type battery for running the simulation optimization. The primary installation cost of this storage equipment is $USD 300/battery, whereas the replacement cost is $USD 3000/battery. The potential technical parameters for the battery are listed in Table 3.

Table 3. List of the technical parameters of the proposed battery.

Criteria	Unit	Values
Nominal voltage	V	12
Nominal capacity	kWh	1
Maximum capacity	Ah	83.4
Capacity ratio	–	0.403
Rate constant	1/hr	0.827
Round trip efficiency	%	80
Maximum charge current	A	16.7
Maximum discharge current	A	24.3
Maximum charge rate	A/Ah	1
Lifetime	Years	10
Throughput	kWh	800
Initial state of charge	%	100
Minimum state of charge	%	20

3.2.3. Converter

A converter is used to convert an electrical signal from alternating current (AC) to direct current (DC) or else vice-versa. A converter is a combination of inverter and rectifier. A PV module produces DC and diesel generator supply AC to serve AC load. Therefore, to synchronize these signals for charging along with supply to the load, a converter is prerequisite for a hybrid power system. The estimated primary installation cost of the converter is $USD 300/kW and the replacement cost is $USD 300/kW. Table 4 presents the summarized technical parameters for the inverter and rectifier module.

Table 4. Summary of technical parameters of the proposed inverter and rectifier module.

Criteria		Unit	Values
Inverter	Lifetime	Year	15
	Efficiency	%	95
Rectifier	Capacity relative to inverter	%	100
	Efficiency	%	95

3.2.4. Diesel Generator

A diesel generator is a conventional power source which converts the chemical energy of diesel into the electric energy. The initial capital cost of the generator is $USD 500/kW with an operational and maintenance cost is $USD 0.030/operational hour. The replacement cost of the diesel generator is $USD 500/kWh. According to the present market value of Bangladesh, the fuel cost is assumed to be $USD 0.77/L (BDT 65). Technical specification of the diesel generator is summarized in Table 5.

Table 5. Summarized technical specification of the diesel generator.

Criteria	Unit	Values
Fuel curve intercept	L/hr	0.838
Fuel curve slope	L/hr/kW	0.236
Lifetime	Hours	

3.3. *Sensitivity Analysis*

A sensitivity variable is a combination of a set of variables for which the HOMER software simulates technical and economic analysis. The incident sunshine is not the same and it varies time to time and place to place from the average value. Again, the levelized cost of energy (LCOE) and net present cost (NPC) depend on load profile and project lifetime. For each sensitivity case, HOMER does search for the cost-effective (lowest cost) system in their respective search limit. In this system, the considered sensitive cases are enlisted in Table 6.

Table 6. List of sensitivity variables.

Scaled Average Load (kWh/day)	Project Lifetime (year)	Solar Scaled Average ($kWh/m^2/day$)
Small (5, 7.97, 10)	10	4.00
Medium (10, 21.24, 30)	15	4.65
Large (50, 77.55, 100)	25	5.00
		5.50
		6.00
		6.50

3.4. Thin-Film Low-E-Type Coatings for Energy Savings

Most of the low-E-type coatings are mainly structured in a sequence of Dielectric/Metal/Dielectric (DMD) type multilayer containing either single or multiple metal layers. The DMD types thin-film multilayers have been studied extensively due to their promising application potentials in various fields including transparent heat regulation (THR). There are some examples of various DMD structure designs and their fabrication technologies are reviewed and listed in Ref [55]. Figure 5 demonstrates the example of clear and semi-transparent polycarbonate roof materials (Figure 5a,b) together with the details explanation of solar radiation spectrum and optical properties of different types of low-E-type coatings (Figure 5c,d). It can be noticed that the polycarbonate roof sheets are, though, suitable for industrial and agricultural purposes; however, they are not capable to filter the solar IR radiation which is mainly heat. The transmission spectral response curves presented in Figure 5d confirms the possible rejection of the IR spectrum mostly over 90%. It is said that 43% of heat is accumulated by the IR spectrum of the sunlight, so by using low-E-type coatings nearly 38% of solar heat can be controlled.

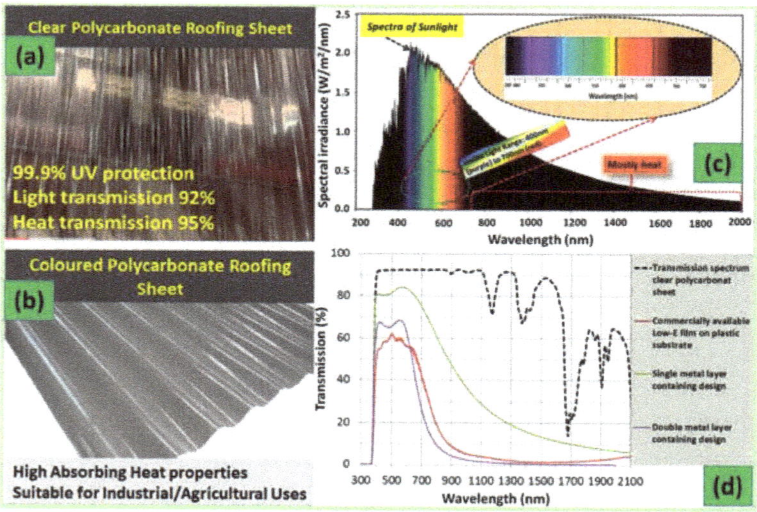

Figure 5. Example of clear and semi-transparent polycarbonate roof materials (**a**,**b**), spectra of sunlight (**c**) and optical properties of different types of low-E-type coatings compare to that of clear polycarbonate roof materials (**d**).

From the spectral response and solar radiation filtering performance (Figure 6), it is very clear that the transparent polycarbonate roof material does not bock any IR spectral radiation of sunlight, while the commercially available low-E-type, thin-film coating and even randomly designed a double metal layer containing DMD type thin-film coating can significantly reduce the radiant heat transmission of sun's spectra into the internal space for any infrastructures. Nur-E-Alam et al., reported that only 29% of solar IR heat (in between 780 to 1700 nm spectral region) can pass through a single metallic nano-composite layer contained DMD type coated glass [55]. Figure 6a shows the standardized spectral power density distribution of solar irradiation (AM 1.5 G) while Figure 6b represents the power density of solar radiation of filtered (through the clear polycarbonate sheet and several thin-film coatings) transmission spectra.

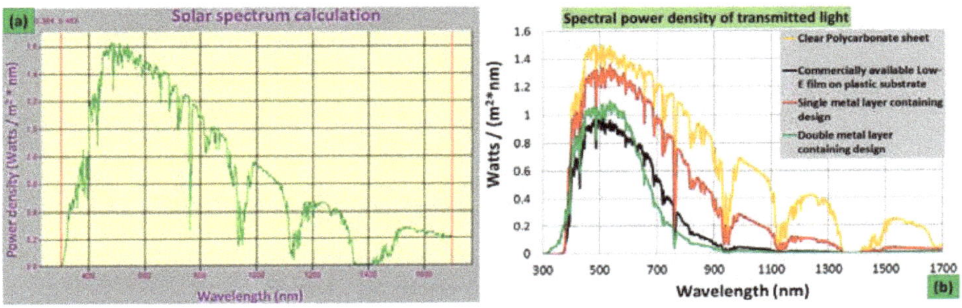

Figure 6. Solar spectrum calculation results in terms of the filtered spectrum of clear polycarbonate roof material and couple low-E-type thin-film coatings.

Overall, features of thin-film coatings clearly indicate that thin-film-coated transparent or semitransparent construction materials (roof or sidewall) can be an additive to save energy for any type of infrastructures for sustainable green earth.

4. Results

In this study based on the volume of the farm, we categorize the cattle farms into three different types, such as small, medium and large. In order to investigate the economic feasibility among the farms, we compare the optimized economic parameters for each type which is optimized by the HOMER pro software. The optimized equipment combination and basic economic parameter for each corresponding farm types are summarized in Table 7.

Table 7. Optimize combination of PV, generator and battery for hybrid system.

Types of the Farm	PV (kW)	Generator (kW)	Battery (Number)	Converter (kW)	NPC ($USD)	Initial Capital ($USD)	O &M ($USD/ Year)	LCOE ($USD)	Payback Period (Years)
Small	1.39	1.40	3	0.774	9703	4227	740	0.451	4.7
Medium	4.06	4.10	7	2.13	26,806	11,767	2032	0.467	4.6
Large	13.10	16.0	40	7.73	104,092	44,922	7995	0.508	3.3

Figure 7 presents the cost analysis for different types of farms with three different energy source systems. It can be seen that the capital cost increases for a three power source component contained (PV, generator and battery) hybrid system compared to the single component (i.e., generator only) power system and a double power source component (i.e., PV and generator) hybrid system. However, the total cost is low for a three-power-source component-contained (PV, generator and battery) hybrid system due to low fuel consumption. It indicates that fuel cost is drastically reduced for PV, generator and battery hybrid system. Besides, this operation and maintenance cost is also low for this system due to reduced operation hour of the generator.

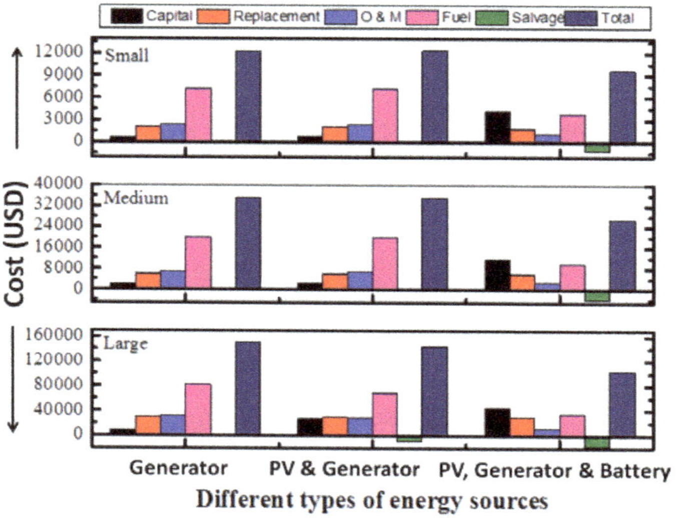

Figure 7. Cost analysis for different types of energy sources.

We have also noticed that the LOCE is smaller for the PV, generator and battery hybrid system compared to that of generator alone or PV and generator system. Comparison of LCOE for different types of farm based on the different combinations of power systems is plotted in Figure 8. The LOCE for generator only and PV and generator hybrid system are almost the same for the small- and medium-sized farms, but for the large farm, the LOCE for PV and generator system is low. On the other hand, the LOCE for PV, generator and battery hybrid system for a small farm is comparatively lower than that of the medium and large farm.

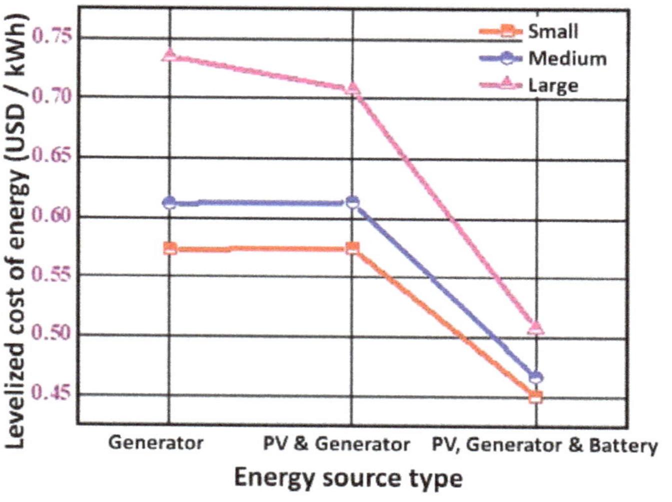

Figure 8. Comparison of the LCOE for different types of farm.

Figure 9 presents a glimpse of global solar radiation and possible output power that could be generated from the solar spectrum. Hourly variation of solar energy, as well as the output power from a solar cell (Figure 9a), states that the power generation from PV plate is directly proportional to the incident global radiation at a particular place. At noon, the global solar radiation is optimum and at that time the production of electricity from the PV cell is optimum as well. The foremost downside of the solar cell is that it can produce energy only at daytime and it reaches its peak for a very short time duration. During the night and a cloudy day, solar module spends its ideal time without producing any energy.

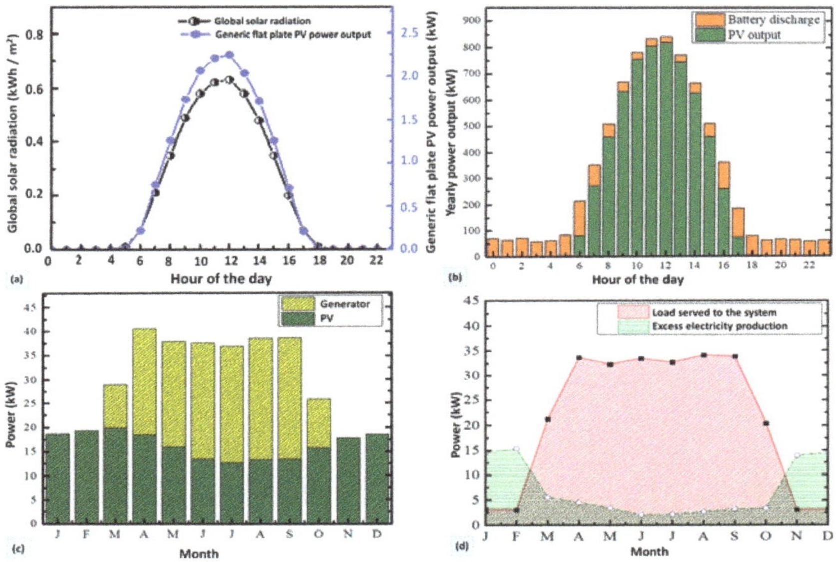

Figure 9. Hourly variation of solar energy as well as the output power from the solar cell (**a**), an hourly variation of the PV output and battery discharge ratio for an optimized system for the medium-sized farm (**b**), monthly variation of PV and generator output pattern for optimized power system for medium-sized farm (**c**), and monthly variation of load demand and excess electricity production for a medium-sized farm (**d**).

Since the solar cell periodically produces energy and it cannot be controlled manually, a storage device can increase the efficiency of the system. From Figure 9b, it is clear that, during the daytime, the power output from the solar cell is optimum and, at night, when the sun is not available, the battery unit fed power to the system as a power source. The storage device can do back up for all time respective to its capacity. As a result, a continuous power source can be ensured by the PV and battery arrangement. Figure 9c shows that there is no need to use the generator at the winter season i.e., during November to February of the year in Bangladesh. During this time, the power generation from the solar cell and battery backup are sufficient to fulfil the load demand. On the other hand, the power generation from the solar cell during July (which is a rainy month in Bangladesh) is minimum (Figure 9c) due to the cloudy sky and solar power is not available as a rainy day as well. However, the load consumption remains at the maximum stage. As a result, to meet the electricity demand, the generator must supply the maximum amount of electricity at the same time. Whilst Figure 9d shows that the PV, generator and battery system produced excess energy during the winter season. At that time, the load demand is minimum and the production of electricity from PV panel excess the load demand. However, from April to October the load demand is maximum and at that time the production of excess electricity is minimum.

Figure 10 represents the average load, solar radiation and project life dependent variation of the LCOE. It can be seen (Figure 10a) that the LCOE decreases for the increases in load consumption for all types of farm. In this study, the baseline load for small-, medium- and large-sized farms are 7.97, 21.24 and 75.8 kWh, respectively. This graph shows that if all other parameters are fixed when the projected load is lower than the baseline load demand then the LCOE increases with decrease in load. On the other hand, if the load demand increase from the baseline load demand, the LCOE decreases with the load. However, the line for below the baseline load is stiffer than the line for higher load. It means that if the load demands decrease, the LCOE increases abruptly. However, if the load demand increases, the LCOE decreases at a slower rate.

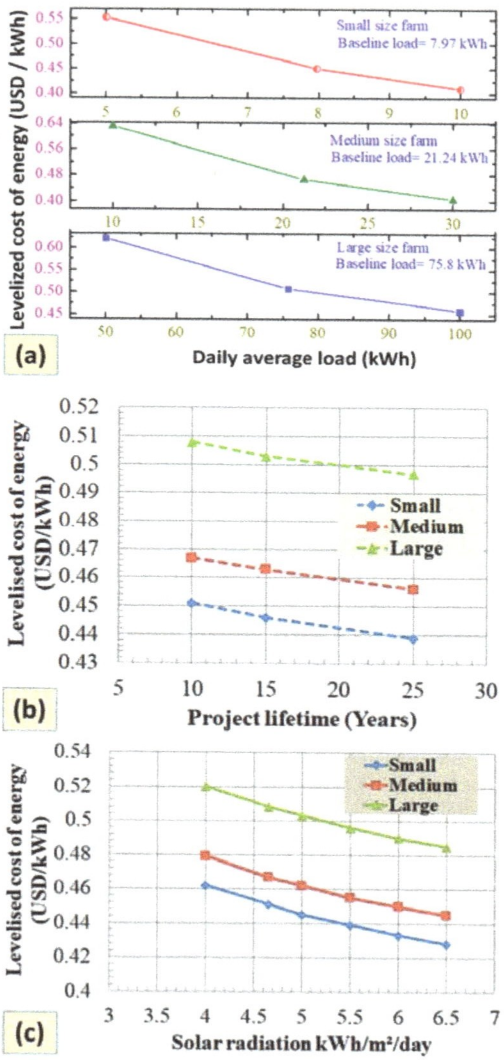

Figure 10. The relation between load and the LCOE for different types of the farm (**a**), a variation of LCOE with projected lifetime for PV, generator and battery hybrid system (**b**), and variation of LCOE with incident solar radiation for PV, generator and battery hybrid system (**c**).

It can be noticed that the LCOE decreases with an increase in project lifetime for all types of farm (Figure 10b). The PV panel is a sustainable energy source, and it can produce energy for a long time without major maintenance thus diminishes the capital cost significantly for a long-term project. The LCOE has a negative correlation with incident solar radiation. On the other hand, it can be seen (Figure 10c) that the LCOE decreases with the increase of incident solar radiation. When the incident solar energy is increasing, the PV output power is also increasing. Since solar energy is a renewable source and it has no fuel cost, the more contribution of PV power over generator will decrease the overall energy cost.

Figure 11 shows that difference of greenhouse gas (CO_2) emission from the generator, and PV and generator type energy sources are negligible for the small and medium farm. However, they have an abrupt decrease in CO_2 emissions for PV, generator and battery hybrid systems. On the other hand, for a large farm, greenhouse gas emissions reduce chronologically for the generator, PV and generator, and PV, generator and battery system. Finally, it can be concluded that based on the greenhouse gas emission, PV, generator and battery hybrid system is more feasible than any other two types of farms.

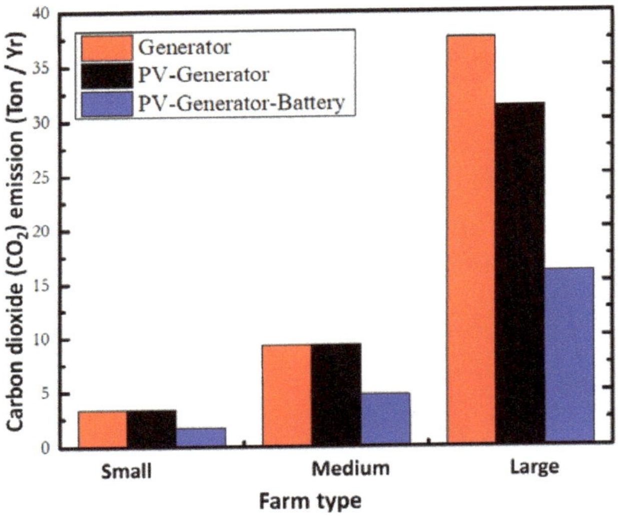

Figure 11. Carbon dioxide emission from different types of the farms for different energy sources.

5. Discussion

From the simulation results, we would like to state the following remarks about this study:

- In this study, the LCOE for small-, medium- and large-sized farms are 0.451, 0.467 and 0.508 $USD/kWh, respectively. It can be concluded here that medium- and small-sized farms are more feasible than the large farms in rural areas for off-grid power supply systems.
- The total cost for PV, generator and battery hybrid system is lower than the generator only or PV and battery system. On the other hand, PV, generator and battery hybrid systems significantly reduce greenhouse gas emissions. Therefore, the PV, generator and battery hybrid system are economically and environmentally feasible for off-grid power generation in rural areas of Bangladesh.
- A cattle farm is more convenient if its size is synchronized with the local market because, if it requires time for trading the milk, it needs to be stored in a refrigerator. As a result of the proposed PV, generator and battery hybrid system may not be ideal due to high load demand.

- ❖ Though the capital cost and initial investment is very high for large farm compare with the small- or medium-type farms, the payback period for small and medium farms are 4.7 and 4.6 years, respectively. If the government provides some subsidies, then the payback period will reduce significantly.
- ❖ According to the current local market, fuel price is $USD 0.77/L which may have a great impact on the LCOE confirmed by several studies, as reported in Refs. [56,57], that is based on the diesel price (0.2–0.5 $USD/L); the LCOE is found in the range of 0.22–0.35 $USD/kWh.

In addition, the energy demand per cow for the small and medium farm is almost the same; however, for large-scale cattle farming, nearly 25% extra energy is required, as shown in Figure 12. This is because of the automation of collecting and storing milk for trading some distance place. As a result, the energy consumption per cow increase and the LOCE increases. Since per capita energy consumption is big for the large scale farm due to storing milk, the small- and medium-scale cattle farming is feasible from the economic point of view where the production is only for the local market.

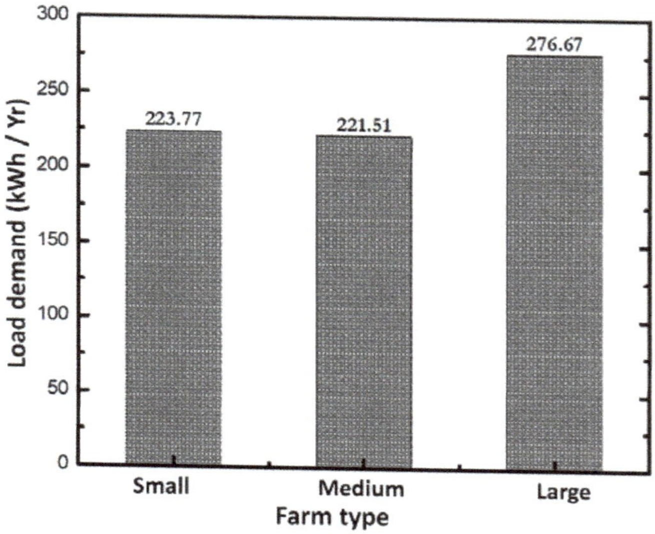

Figure 12. Load demand for a single cow depending on the farm size.

6. Conclusions

We demonstrated the design and optimization of hybrid power system together with the addition of thin-film, Low-E coating assisted power-saving feature for next-generation cultivation and farming system. From the obtained hybrid energy system optimization results, we found the energy payback period for small and medium farms are 4.7 and 4.6 years, respectively; however, if the farm size is larger, this payback period could be 3.5 years. Note that during the hybrid energy system simulation the advantage of thin-film Low-E-type coatings was not accounted for. In summary, we can comment that our study of energy engineering together with the advanced building material coatings will be the model for modern farming in any rural areas, and will help to reduce the carbon emission footprint for any other countries as well.

Author Contributions: Conceptualization, M.N.-E-A.; methodology M.N.-E-A., M.N.H.; software, M.N.-E-A., and M.N.H.; validation, M.N.-E-A., M.N.H., S.M.A., M.K.B., and N.D.; formal analysis, M.N.-E-A., M.N.H., S.M.A., and M.K.B.; investigation, M.N.-E-A., M.N.H.; resources, M.N.-E-A., and M.N.H.; data curation, M.N.-E-A., M.N.H., and M.K.B.; writing—original draft preparation, M.N.-E-A., and M.N.H.; writing—review and editing, M.N.-E-A., M.N.H., S.M.A., M.K.B., and N.D.; visualization, M.N.-E-A., M.N.H., S.M.A., and M.K.B.; supervision,

M.N.-E-A.; project administration, M.N.-E-A. All authors have read and agreed to the published version of the manuscript.

Funding: This research received no external funding.

Conflicts of Interest: The authors declare no conflict of interest.

References

1. Bangladesh Economic Review 2018–2019, Ministry of finance, Government of Bangladesh (GoB). Available online: www.mof.gov.bd (accessed on 20 August 2020).
2. Roy, D.; Dev, D.S.; Sheheli, S. Food Security in Bangladesh: Insight from Available Literature. *J. Nutr. Food Secur.* **2019**, *4*, 66–75. [CrossRef]
3. Livestock Economy at a Glance 2018–19, Department of Livestock Services (DLS), Government of Bangladesh (GOB). Available online: www.dls.gov.bd (accessed on 20 August 2020).
4. Nisbett, N.; Davis, P.; Yosef, S.; Akhtar, N. Bangladesh's story of change in nutrition: Strong improvements in basic and underlying determinants with an unfinished agenda for direct community level support. *Glob. Food Secur.* **2017**, *13*, 21–29. [CrossRef]
5. Bishwajit, G.; Barmon, R.; Ghosh, S. Reviewing the status of agricultural production in bangladesh from a food security perspective. *Russ. J. Agric. Socio-Economic Sci.* **2014**, *25*, 19–27. [CrossRef]
6. The Business Standard report "Decision to import meat will be suicidal". Available online: https://tbsnews.net/bangladesh/decision-import-meat-will-be-suicidal. (accessed on 16 October 2020).
7. Manandhar, A.; Milindi, P.; Shah, A. An overview of the post-harvest grain storage practices of smallholder farmers in developing countries. *Agriculture* **2018**, *8*, 57. [CrossRef]
8. Kumar, D.; Kalita, P. Reducing Postharvest Losses during Storage of Grain Crops to Strengthen Food Security in Developing Countries. *Foods* **2017**, *6*, 8. [CrossRef]
9. Pekmez, H. Cereal storage techniques: A review. *J. Agric. Sci. Technol. B* **2016**, *6*, 67–71. [CrossRef]
10. On-Farm-Cold-Storage. Available online: https://fruit.triforce.cals.wisc.edu/wp-content/uploads/sites/36/2017/02/On-Farm-Cold-Storage.pdf (accessed on 30 August 2020).
11. Thomas, J.A.; Vasiliev, M.; Nur-E-Alam, M.; Alameh, K. Investigation of the optimum illumination spectrum for maximizing the energy savings and yield of Lactuca sativa, L. in glass greenhouses. *Sustainability* **2020**, *12*, 3740. [CrossRef]
12. Patel, K.; Das, N.; Khan, M.M.K. Optimization of hybrid solar, wind and diesel energy systems from economic point of view. In Proceedings of the 29th Australasian Universities Power Engineering Conference (AUPEC), Nadi, Fiji, 26–29 November 2019.
13. Tan, Y.; Meegahapola, L.; Muttaqi, K.M. A review of technical challenges in planning and operation of remote area power supply systems. *Renew. Sustain. Energy Rev.* **2014**, *38*, 876–889. [CrossRef]
14. Linn, S.; Ya, A.Z. Solar/Wind/Diesel hybrid energy system with battery storage for rural electrification. *Int. J. Sci. Eng. Tech. Res.* **2014**, *3*, 2172–2176.
15. Robinson, P.; Gowda, A.C.; Sameer, S.; Patil, S. Development of renewable energy based hybrid system for electricity generation-A case study. *Int. J. Latest Tech. Eng. Manag. Appl. Sci.* **2017**, *6*, 2278–2540.
16. Akikur, R.K.; Saidur, R.; Ping, H.W.; Ullah, K.R. Comparative study of stand-alone and hybrid solar energy systems suitable for off-grid rural electrification: A review. *Renew. Sustain. Energy Rev.* **2013**, *27*, 738–752. [CrossRef]
17. Mohamed, M.A.; Eltamaly, A.M.; Alolah, A.I. Sizing and techno-economic analysis of stand-alone hybrid photovoltaic/wind/diesel/battery power generation systems. *J. Renew. Sustain. Energy* **2015**, *7*, 063128. [CrossRef]
18. Karim, M.E.; Karim, R.; Islam, M.T.; Muhammad-Sukki, F.; Bani, N.A.; Muhtazaruddin, M.N. Renewable Energy for Sustainable Growth and Development: An Evaluation of Law and Policy of Bangladesh. *Sustainability* **2019**, *11*, 5774. [CrossRef]
19. Sustainable and Renewable Energy Development Authority, Department of Power Ministry of Power, Energy and Mineral Resources, Government of the People's Republic of Bangladesh. Available online: http://www.sreda.gov.bd/index.php/site/page/7b9b-49f7-69fb-40fd-45a3-9e6c-b391-7ba5-31f9-13ee (accessed on 9 October 2020).

20. Mollik, S.; Rashid, M.M.; Hasanuzzaman, M.; Karim, M.E.; Hosenuzzaman, M. Prospects, progress, policies, and effects of rural electrification in Bangladesh. *Renew. Sustain. Energy Rev.* **2016**, *65*, 553–567. [CrossRef]
21. Ministry of Power, Energy and Mineral Resources (MPEMR). *Policy Guidelines for Small Power Plants in Private Sector*; MPEMR: Dhaka, Bangladesh, 1996.
22. Bangladesh Energy Regulatory Commission. *Bangladesh Energy Regulatory Commission (Tariff for Roof Top Solar PV Electricity) Regulations, 2016 (Draft)*; Bangladesh Energy Regulatory Commission: Dhaka, Bangladesh, 2016. Available online: http://berc.portal.gov.bd/sites/default/files/files/berc.portal.gov.bd/page/a250b6fc_8bcf_4c96_bb20_3c3de230467a/berc_tariff_regulations_rooftof_solar%28draft%29.pdf (accessed on 9 October 2020).
23. World Health Organization (WHO). 2020. Available online: https://www.who.int/indonesia/news/detail/08-03-2020-knowing-the-risk-for-covid-19#:~{}:text=People%20who%20are%20aged%20over,infected%20with%20the%20virus (accessed on 20 August 2020).
24. World Health Organization (WHO). 2020. Available online: https://www.who.int/emergencies/diseases/novel-coronavirus-2019/advice-for-public/myth-busters?gclid=EAIaIQobChMIsOGxvYTM6gIVyhErCh0swgmPEAAYASAAEgKbM_D_BwE#climate (accessed on 20 August 2020).
25. Trade Set to Plunge as COVID-19 Pandemic Upends Global Economy. Available online: https://www.wto.org/english/news_e/pres20_e/pr855_e.htm (accessed on 31 August 2020).
26. Coronavirus Update: Industry Fast Facts, by IBISWorld. 1 June 2020. Available online: https://www.ibisworld.com/industry-insider/coronavirus-insights/coronavirus-update-industry-fast-facts/ (accessed on 2 September 2020).
27. Live-blog: How the Coronavirus Affects Garment Workers In Supply Chains. Available online: https://cleanclothes.org/news/2020/live-blog-on-how-the-coronavirus-influences-workers-in-supply-chains (accessed on 2 September 2020).
28. COVID-19 Challenges for the Indian Economy: Trade and Foreign Policy Effects. Available online: https://www.eepcindia.org/eepc-download/617-Covid19_Report.pdf (accessed on 2 September 2020).
29. COVID-19: Impacts on Australia's Food and Agribusiness Sector. Available online: https://home.kpmg/au/en/home/insights/2020/03/coronavirus-covid-19-impact-on-food-agribusiness-sector.html (accessed on 2 September 2020).
30. Shimanta, M.L.R.; Gope, H.; Sumaiya, I.J. Readymade Garments Sector and COVID-19 in Bangladesh. *Preprints* **2020**. [CrossRef]
31. World Health Organization (WHO), 2020; COVID-19 and Food Safety: Guidance for Competent Authorities Responsible for National Food Safety Control Systems. 2020. Available online: https://www.who.int/publications/i/item/covid-19-and-food-safety-guidance-for-competent-authorities-responsible-for-national-food-safety-control-systems (accessed on 20 August 2020).
32. COVID-19 and Food Safety: Guidance for competent authorities responsible for national food safety control systems. Available online: http://agriculture.vic.gov.au/agriculture/dairy/energy-in-dairy/solar-panel-systems-for-the-dairy-shed (accessed on 10 August 2020).
33. Agriculture Victoria. Available online: http://agriculture.vic.gov.au/agriculture/dairy/energy-in-dairy/solar-panel-systems-for-the-dairy-shed (accessed on 10 August 2020).
34. Mbow, C.C.; Rosenzweig, L.G.; Barioni, T.G.; Benton, M.; Herrero, M.; Krishnapillai, E.; Liwenga, P.; Pradhan, M.G.; Rivera-Ferre, T.; Sapkota, F.N.; et al. Food Security. In *Climate Change and Land: An IPCC Special Report on Climate Change, Desertification, Land Degradation, Sustainable Land Management, Food Security, and Greenhouse Gas Fluxes in Terrestrial Ecosystems*; Shukla, P.R., Skea, J., Calvo Buendia, E., Masson-Delmotte, V., Pörtner, H.-O., Roberts, D.C., Zhai, P., Slade, R., Connors, S., van Diemen, R., et al., Eds.; IPCC: Geneva, Switzerland, 2019; in press; Available online: https://www.ipcc.ch/srccl/chapter/chapter-5/ (accessed on 10 August 2020).
35. The Business Standard report "The impact of COVID-19 on the dairy industry of Bangladesh". Available online: https://tbsnews.net/thoughts/impact-covid-19-dairy-industry-bangladesh-68569 (accessed on 11 August 2020).

36. World Economic Forum report "How Bangladesh's leaders should respond to the economic threats of COVID-19". Available online: https://www.weforum.org/agenda/2020/04/covid-19-coronavirus-bangladesh/ (accessed on 12 August 2020).
37. Bdnews24.com "Dairy farmers hit hard by coronavirus lockdown in Bangladesh". Available online: https://bdnews24.com/bangladesh/2020/04/13/dairy-farmers-hit-hard-by-coronavirus-lockdown-in-bangladesh (accessed on 10 August 2020).
38. Nur-E-Alam, M. Can Energy Engineering be a Hope for the Post COVID-19 Bangladesh. Innovate July 2020/25. Available online: https://issuu.com/iceaus/docs/innovate-icea_july_2020?fr=sNTE1NzE2OTYyNTA&fbclid=IwAR3ONzUf6ORRz-T-gDRrHOsxJSJvniiREH5BmZ1IZRmyM5Tk1_qrkg-vW90 (accessed on 16 October 2020).
39. Momtaj, I.A.; Nur-E-Alam, M. Sweatshops Workers in post COVID-19. Innovate July 2020/48. Available online: https://issuu.com/iceaus/docs/innovate-icea_july_2020?fr=sNTE1NzE2OTYyNTA&fbclid=IwAR3ONzUf6ORRz-T-gDRrHOsxJSJvniiREH5BmZ1IZRmyM5Tk1_qrkg-vW90 (accessed on 16 October 2020).
40. Begum, M.; Farid, M.S.; Barua, S.; Alam, M.J. COVID-19 and Bangladesh: Socio-economic analysis towards the future correspondence. *Preprints* **2020**, 2020040458. [CrossRef]
41. Poultry Sector Stares at over Tk1,150 Crore Losses. Available online: https://tbsnews.net/economy/industry/poultry-sector-stares-over-tk1150-crore-losses-63472 (accessed on 2 September 2020).
42. Sunny, A.R.; Sazzad, S.A.; Datta, G.C.; Sarker, A.K.; Ashrafuzzaman, M.; Prodhan, S.H. Assessing impacts of COVID-19 on aquatic food system and small-scale fisheries in Bangladesh. *Preprints* **2020**. [CrossRef]
43. Dalapati, G.K.; Kushwaha, A.K.; Sharma, M.; Suresh, V.; Shannigrahi, S.; Zhuk, S.; Masudy-Panah, S. Transparent heat regulating (THR) materials and coatings for energy saving window applications: Impact of materials design, micro-structural, and interface quality on the THR performance. *Prog. Mater. Sci.* **2018**, *95*, 42–131. [CrossRef]
44. Leftheriotis, G.; Yianoulis, P.; Patrikios, D. Deposition and optical properties of optimised ZnS/Ag/ZnS thin films for energy saving applications. *Thin Solid Films* **1997**, *306*, 92–99. [CrossRef]
45. Pracchia, J.A.; Simon, J.M. Transparent heat mirrors: influence of the materials on the optical characteristics. *Appl. Opt.* **1981**, *20*, 251–258. [CrossRef]
46. Kulczyk-Malecka, J.; Kelly, P.; West, G.; Clarke, G.; Ridealgh, J.; Almtoft, K.; Greer, A.; Barber, Z. Investigation of silver diffusion in TiO2/Ag/TiO2 coatings. *Acta Mater.* **2014**, *66*, 396–404. [CrossRef]
47. Vasiliev, M.; Nur-E-Alam, M.; Alameh, K. Highly stable thin-film multilayers for thermal regulation and energy savings in smart cities. In Proceedings of the 2019 IEEE 16th International Conference on Smart Cities: Improving Quality of Life Using ICT, IoT & AI, Charlotte, NC, USA, 6–9 October 2019.
48. Hossen, M.; Hossain, M.; Abedin, M.; Karim; Rume, F. Animal Production Strategies in Southern Region of Bangladesh. *Agriculturists* **2010**, *6*, 77–83. [CrossRef]
49. Expert System for Cattle and Buffalo. Available online: http://www.agritech.tnau.ac.in/expert_system/cattlebuffalo/Housing%20Management%20of%20Cattle%20and%20Buffalo.html#plan (accessed on 23 August 2020).
50. Keown, J.F.; Kononoff, P.J.; Grant, R.J. *How to Reduce Heat Stress in Dairy Cattle. Historical Materials from University of Nebraska-Lincoln Extension*; Nebraska Extension: Lincoln, NE, USA, 2005; p. G05-1582.
51. Todde, G.; Murgia, L.; Caria, M.; Pazzona, A. A multivariate statistical analysis approach to characterize mechanization, structural and energy profile in Italian dairy farms. *Energy Rep.* **2016**, *2*, 129–134. [CrossRef]
52. Alahmer, A.; AlSaqoor, S. Design of a dairy cooling thermal storage supported with secondary refrigeration cooling unit. *IOSR J. Mech. Civ. Eng.* **2014**, *11*, 30–36. [CrossRef]
53. Cengel, Y. *Heat and Mass Transfer: A Practical Approach*, 3rd ed.; McGraw-Hill: New York, NY, USA, 2006.
54. Upton, J.; Humphreys, J.; Koerkamp, P.G.; French, P.; Dillon, P.; De Boer, I.J.M. Energy demand on dairy farms in Ireland. *J. Dairy Sci.* **2013**, *96*, 6489–6498. [CrossRef] [PubMed]
55. Nur-E-Alam, M.; Vasiliev, M.; Alameh, K. *Dielectric/Metal/Dielectric (DMD) Multilayers: Growth and Stability of Ultra-Thin Metal Layers for Transparent Heat Regulation (THR)*; Elsevier BV: Amsterdam, The Netherlands, 2020; pp. 83–112.

56. Aziz, A.S.; Tajuddin, M.F.N.; Adzman, M.R.; Ramli, M.A.M.; Mekhilef, S. Energy Management and Optimization of a PV/Diesel/Battery Hybrid Energy System Using a Combined Dispatch Strategy. *Sustainability* **2019**, *11*, 683. [CrossRef]
57. Abd-El Mageed, H.S. Cost analysis and optimal sizing of PV-Diesel hybrid energy systems. *Am. J. Renew. Sustain. Energy* **2018**, *4*, 47–55.

Publisher's Note: MDPI stays neutral with regard to jurisdictional claims in published maps and institutional affiliations.

sustainability

MDPI

Article

Sanitation Sustainability Index: A Pilot Approach to Develop a Community-Based Indicator for Evaluating Sustainability of Sanitation Systems

Shervin Hashemi

Institute for Environmental Research, Yonsei University College of Medicine, 50-1 Yonsei-ro, Seodaemun-gu, Seoul 03722, Korea; shervin@yuhs.ac

Received: 5 June 2020; Accepted: 25 August 2020; Published: 26 August 2020

Abstract: Evaluating the sustainability of sanitation systems is essential in achieving the sixth sustainable development goal. However, there are only limited number of available evaluation indexes, which are utilized to macroscopically determine a community's sanitation coverage. Consequently, an index is required, which can evaluate different sanitation options for a specific community. In this paper, the sanitation sustainability index (SSI) is suggested as an indicator for evaluating the sustainability of sanitation systems. The SSI has sub-indexes that consider the technical, social, and economic aspects of the sanitation system, and all the variables are dimensionless and heavily dependent on the current state of the community where the sanitation system is going to be implemented. The applicability of the SSI was demonstrated by evaluating the implementation of two onsite sanitation systems, including one septic tank system and one resource-oriented sanitation (ROS) system in South Korea. A sensitivity analysis defined the variables that have significant impact, and the statistical distribution of the SSI for both systems was forecasted. The results showed that for South Korea, which has a profound history of utilizing human waste as fertilizer, utilizing the resource-oriented sanitation system is more sustainable, although it has a lower social sub-index score compared to the septic tank system.

Keywords: resource-oriented sanitation; sanitation sustainability index; SDG6; sustainability assessment; sustainable development goals; sustainable sanitation

1. Introduction

According to UNICEF and WHO [1], the population that utilize safely managed sanitation services increased from 28% to 45% from 2000 up to 2017. Accordingly, 2.1 billion people gained access to at least basic sanitation services. Despite that, 2 billion people still have no access to safe and sustainable sanitation system, 673 million people are practicing open defecation, and 3 billion people still lack basic sanitation; consequently, the sanitation crisis remains a global concern.

The sixth sustainable development goal (SDG6) is designed to solve these global issues before the year 2030 [2]. The objective of SDG6-2 is to provide full coverage of adequate and sustainable sanitation for the entire population. Following this objective, innovative approaches have been applied on new disciplinary understandings of sanitation, to provide sustainable sanitation for low-income regions [3]. Recently, national-scaled programs have been conducted to improve the notion of equity in sanitation to end critical challenges like open defecation [4]. Meanwhile, there are several indicators and indexes that are defined to monitor or evaluate the progress [5–7]. One example is the water, sanitation, and hygiene (WaSH) performance index suggested by Cronk et al. [8], which can be utilized to compare the performances of countries in achieving universal water, sanitation, and hygiene. The elements of this index include water and sanitation access and equity.

The development of these indexes is essential in achieving SDG6, because they present a comparative view of the progress of each country towards SDG6. Additionally, these indexes are useful for undertaking general national-scaled decisions to further progress towards fulfilling SDG6.

In many cases, comparing two different countries based on such macroscopic indexes may be misleading, particularly when there is a notable difference between the human development indexes of those countries [9]. Nevertheless, the economic, social, and technical characteristics of each country are different. Determining an appropriate index, which is flexible, suitable for every community, and reliable for comparison of progress, is crucial. Consequently, the use of an index that is better suited to evaluate sanitation systems for the implementation in a specific community is a suitable approach, and the evaluation should be performed based on the current technical, social, and economic status of the community.

An index proposed by Lundin et al. [10], which evaluates the sustainability of sanitation systems included several technical considerations. However, waste recycling and utilization, along with social and economic indicators, were not considered in that index. Another comprehensive index was suggested by Iribarnegaray et al. [11], which assesses the sustainability of water and sanitation management systems. This index is useful, as it considers the majority of relevant issues of sustainability. However, by developing innovative sanitation systems which can recycle sanitary wastes to be utilized as resources (e.g., fertilizers), there is an emerging need for providing innovative indexes, which can consider the effect of ability for onsite waste recycling on the technical, economic, and social aspects of the community [5]. In particular, economic indicators are required to consider a broader range of issues influencing the cost and probable benefits of implementing sanitation systems in developing countries [12]. Moreover, the insufficient evaluation process for the applicability of an index through limited case studies and demonstration projects is another bottleneck on the way to finding an appropriate sustainability index.

To provide sustainable sanitation for a community, understanding the technical, economic, and social characteristics of the community is critical. Numerous technical approaches toward the provision of sustainable sanitation have been studied [13–17]. These approaches provide a range of options for sanitation, including onsite sanitation systems using water-saving toilets, septic tanks, and resource-oriented sanitation (ROS) systems [18–25]. Moreover, some studies compare different sanitation systems based on economic views [12,26]. Social approaches toward sustainable sanitation were also considered in multiple studies [27–31]. However, there is still a need for a reliable index that can be utilized to determine a suitable sanitation system for a community based on availability, accessibility, and sustainability.

In this paper, the sanitation sustainability index (SSI) is proposed to evaluate the sanitation systems for implementation in any community. The SSI indicates the technical, social, and economic aspects of the sanitation system. The applicability of this index is demonstrated by utilizing it to evaluate the implementation of one septic tank and one ROS system in South Korea.

2. Materials and Methods

2.1. Definition of SSI

Figure 1 presents the SSI structure, which contains technical, economic, and social sub-indexes. Each sub-index has specific dimensionless variables, having values ranging from 0 to 1, with 0.2 intervals, indicating very low, low, medium, high, and very high criteria for evaluating the sustainability of a sanitation system.

Technical and economic sub-indexes are analyzed by simple equations, containing a fraction. The numerator of the fraction is the difference between the specific parameter describing a particular feature of the sanitation system and the current norm of the community regarding the parameter. This value is then divided by the current norm of the community regarding the parameter to indicate how significant this difference is. Hence, the fraction makes the sub-index dimensionless, making it useful to compare two or more systems with each other in any community.

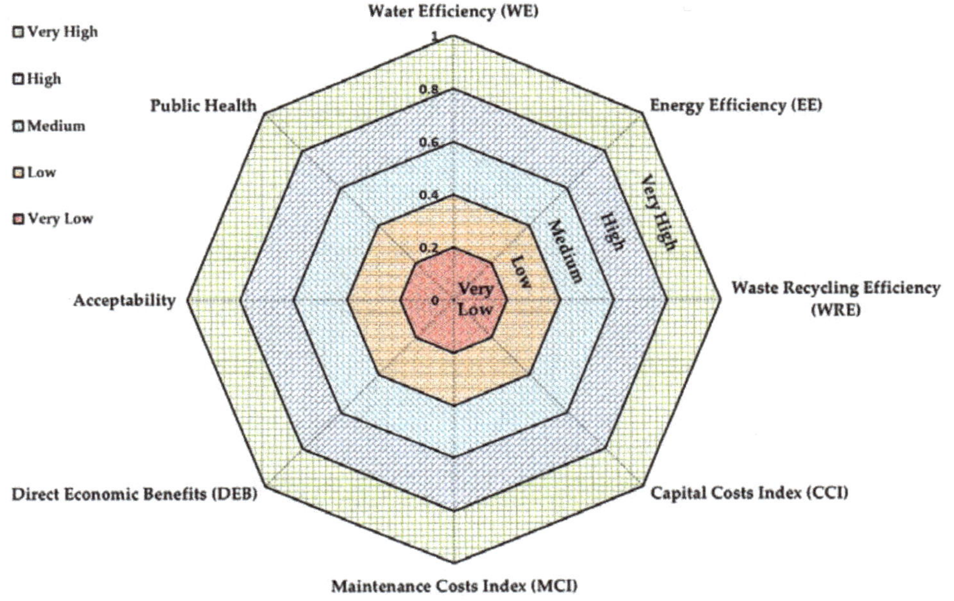

Figure 1. Structure of the sanitation sustainability index and its evaluation criteria.

Water efficiency (WE), energy efficiency (EE), and waste recycling efficiency (WRE) are dimensionless variables assigned to the technical sub-index. The WE variable is calculated by Equation (1), where V_C is the typical amount of water consumption utilized for a single use of the sanitation system, including both water for flushing and other hygienic purposes like handwashing. V_M is the mean of the volume of consumed water in the common sanitation system of the community. When $V_C \geq V_M$, the value of WE should be considered zero.

$$WE = \frac{V_M - V_C}{V_M}, \quad (1)$$

The EE is calculated by Equation (2), where E_C is the energy consumption for treating wastewater made from a single usage of the sanitation system, and E_M is the mean energy consumption for wastewater treatment in common sanitation systems of the community. When $E_C \geq E_M$, the value of EE should be considered zero.

$$EE = \frac{E_M - E_C}{E_M}, \quad (2)$$

WRE is the ratio of safely recycled waste to total waste production. This variable also evaluates the ability of a sanitation system to both treat and utilize produced waste.

For the economic sub-index, capital cost index (CCI), maintenance cost index (MCI), and direct economic benefits (DEBs) are suggested as the dimensionless variables. CCI and MCI are calculated by Equations (3) and (4), respectively. In these equations, the CC_E is the estimated capital cost, and CC_M is the mean capital cost for the common sanitation systems of the community. Similarly, the MC_E is the estimated maintenance cost for a year, and MC_M is the mean maintenance cost for the common sanitation systems of the community. If $CC_E \geq CC_M$, then the value of CCI should be considered as zero. Similarly, if $MC_E \geq MC_M$, then the value of MCI should be considered as zero.

$$\text{CCI} = \frac{\text{CC}_\text{M} - \text{CC}_\text{E}}{\text{CC}_\text{M}}, \quad (3)$$

$$\text{MCI} = \frac{\text{MC}_\text{M} - \text{MC}_\text{E}}{\text{MC}_\text{M}}, \quad (4)$$

DEB is a binary variable. When a sanitation system can produce economic benefits or has the potential that its economic benefit covers the capital and maintenance costs, then DEB should be 1, otherwise, it is 0.

The social sub-index includes acceptability and public health indicators as its relative variables. Both mentioned variables are qualitative, which must be turned into dimensionless quantitative variables through onsite investigations, and acceptability should reflect the cultural sanitation actions in the society. In this case, investigations should be performed by analyzing questionnaires that reflect the opinion of the users after experiencing the system. Therefore, the acceptability variable can be defined as the percentage of people who were satisfied after using the system and would use it again. For newly developed sanitation systems like ROS, user training can significantly increase the acceptability variable.

The public health indicator should reflect the hygiene aspects of the sanitation system. A wide range of considerations including safe treatment, recycling process, reduction of risks of incidence for sanitary diseases (e.g., diarrhea), total hygiene status of the system, especially where there is an interface with users or maintenance technicians, and efficiency in controlling public health concerns, e.g., reducing open defecation cases, are required to define this variable. Although several studies have been conducted on the social aspects of sanitation in different communities, there is no established standard survey to evaluate the acceptability and public health aspects of a sanitation system [4,27–29,32]. Therefore, regardless of the method of investigation, a value of 0 to 1, representing the evaluation criteria, should be assigned for the acceptability and public health variables.

Accordingly, for both acceptability and public health, the index is flexible for any appropriate and possible local application. The scoring can be done by the decision-maker or technician who is applying this index to the results of a local investigation as well as other appropriate data.

After evaluating all the described sub-indexes and variables, the SSI score is calculated as the arithmetic mean of all the values. For a specific community, when two or more sanitation systems are being compared, the system with the higher SSI is concluded to be more sustainable for implementation, and an SSI value closer to 1 reflects higher sustainability for implementation.

2.2. Applicability of SSI for Sanitation System Implementation in South Korea

To demonstrate SSI applicability, the SSI is calculated to evaluate the implementation of two onsite sanitation systems: a septic tank system and an ROS system. The systems are considered for implementation in a suburban residential area of Nowon District, Seoul, Republic of Korea to offer public hygiene services. For both sanitation systems, variables related to technical and economic sub-indexes were assigned based on the characteristics and assumptions of these systems, which were described in a previous study [12]. Table 1 presents the assumptions for calculating the SSI for both sanitation systems, including references that were used for assigning specific values to each parameter. A currency exchange rate of KRW 1100 for USD 1 is used, wherever applicable.

There is insufficient research on the social acceptability and public health issues for sanitation systems in Korea. However, considering the ancient sanitation practices in Korea as described by Han & Hashemi [5], along with other local pilot studies, it might be reasonable to consider a value of 0.7–0.8 for both social acceptability and public health indicators for the ROS system variables [20]. As using septic tank systems is quite common in both urban and rural areas of South Korea, a value of 0.9 was utilized for both the acceptability and public health variables.

A sensitivity analysis for the SSI scores calculated for both sanitation systems was conducted by applying a $\Delta = \pm 10\%$ variation to each parameter. Additionally, to account for the lack of knowledge and availability of data, statistical distributions for each variable were utilized to forecast the statistical distribution of the calculated SSI indexes through Monte Carlo simulation using Oracle Crystal Ball version 11.1.2.4 with 1,000,000 trials. Details of the assumptions and results of this statistical analysis are presented in the Supplementary Material.

Table 1. Assumptions for calculating sanitation sustainability index of septic tank and resource-oriented sanitation systems in South Korea.

Sub-Index	Variable	Unit	Sanitation System	Value (Range: min-max)	References
Technical	V_C	Liter	Septic Tank	8 (4–13)	[5,12,33,34]
			ROS	1 (0–2.5)	
	V_M		-	10	
	E_C	Wh	Septic Tank	5.2 (1.04–10.9)	
			ROS	0.65 (0–2.1)	
	E_M		-	6.5	
	WRE	-	Septic Tank	0.519 (0.135–0.519)	[5,20,35]
			ROS	0.9 (0.7–0.9)	
Economic	CC_E †	US$	Septic Tank	3380.00 (3121.00–9436.00)	[12]
			ROS	4668.18 (2899.22–5002.77)	
	CC_M †		-	6061.00	
	MC_E †		Septic Tank	2240.16 (2118.18–3401.82)	
			ROS	2195.28 (1687.58–2702.92)	
	MC_M †		-	2931.82	
	DEB	-	Septic Tank	0	
			ROS	1	
Social	Acceptability	-	Septic Tank	0.9	[16,27,28]
			ROS	0.75	
	Public Health	-	Septic Tank	0.9	
			ROS	0.75	

† Including laboring costs for initialization and one-year maintenance services.

3. Results

3.1. SSI Applicability for Evaluating the Proposed Implementation of Sanitation Systems in South Korea

Figure 2 presents the results of the SSI sub-indexes evaluation for both mentioned sanitation systems in South Korea. The SSI was quantified as 0.42 and 0.71 for septic tank and ROS systems, respectively. Thus, notable differences between the calculated SSIs were observed, particularly for the ROS system, which shows better sustainability in the technical and economic sub-indexes. As described in previous studies, the ability of ROS systems to recycle waste and their economic benefits is significant [12,20]. According to the Republic of Korea, Ministry of Environment, there is an average recycling rate of 51.9% for wastewater generated from wastewater treatment facilities, such as septic tanks, in South Korea [35]. The wastewater produced in septic tank sanitation systems includes a mixture of urine, feces, hygienic materials (e.g., toilet paper), and flushing water. However, properly designed ROS systems can have a significantly higher WRE, approximately 0.7–0.9, by storing and treating urine and feces separately [5,20].

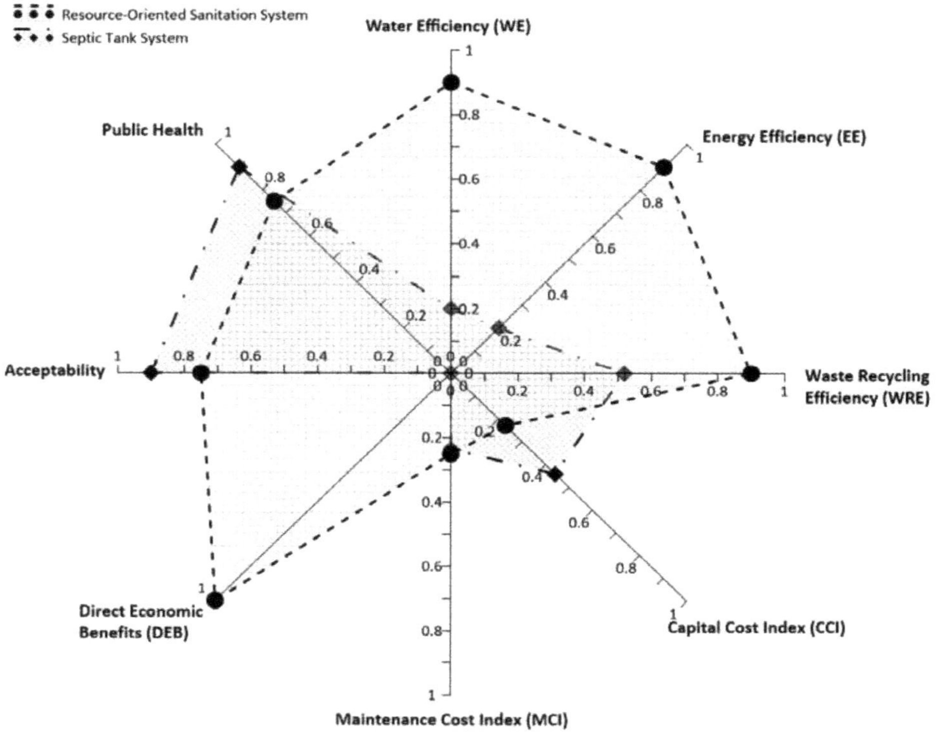

Figure 2. Comparison of sanitation sustainability index for septic tank and resource-oriented sanitation systems in South Korea.

The social sub-index of ROS systems is considered lower than that of a septic tank. Conducting in-depth studies on the social acceptability of sanitation systems in South Korea can provide a better assessment for this sub-index. Conducting safe, natural treatment methods for urine and feces, and providing sustainable operation monitoring, regular maintenance and training for users are beneficial actions for increasing the social sub-index scores of the ROS system.

3.2. Sensitivity Analysis for SSI

Figure 3 presents the sensitivity analysis for the studied sanitation systems. For the septic tank system, after social sub-indexes, the following descending order of variables in terms of their impact on SSI apply: E_C, V_C, MC_E, CC_E, and WRE. However, for the ROS system, DEBs and WRE have the greatest impact, while CC_E, acceptability, public health, and MC_E indicators have notable impacts. The sensitivity of other variables was negligible.

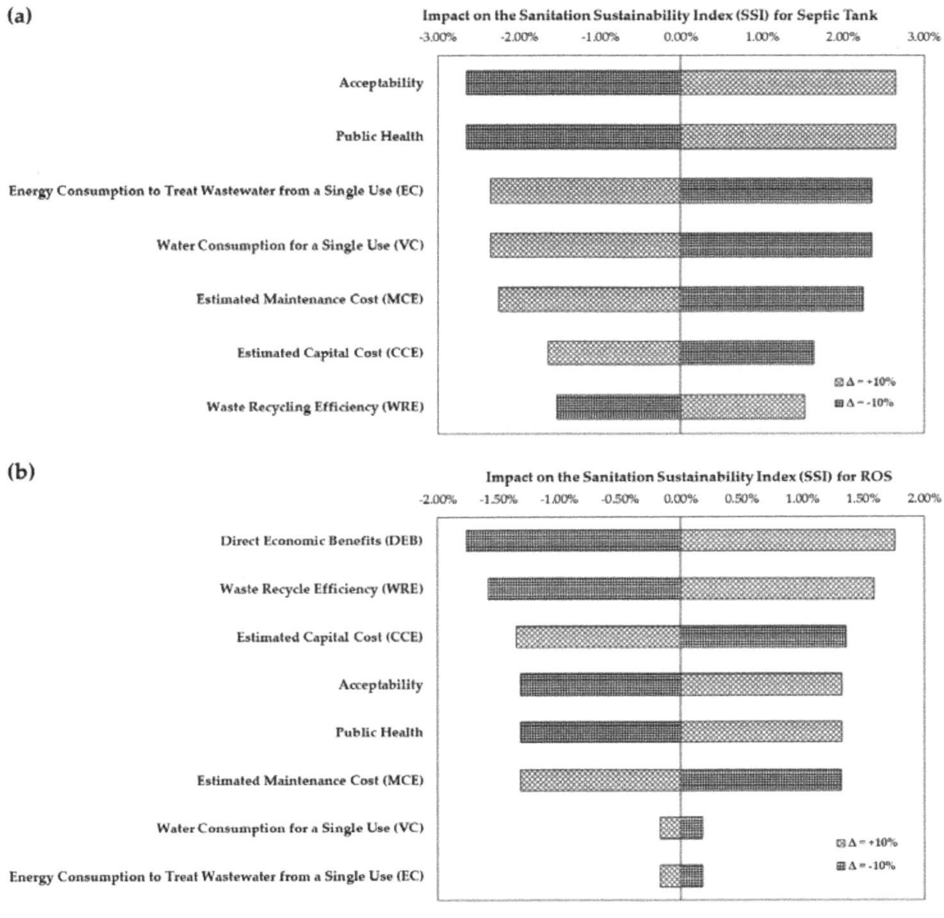

Figure 3. Sensitivity analysis for sanitation sustainability index calculated for (**a**) septic tank and (**b**) resource-oriented sanitation systems in South Korea.

These results indicate that for South Korea, which has a notably long history in recycling human waste as fertilizer, the development of an ROS system for waste recycling and utilization can potentially increase the SSI. Additionally, reducing the CC_E and MC_E by utilizing local labor and materials can also significantly increase the ROS system sustainability.

3.3. Statistical Distribution Forecast for SSI

Figure 4 presents the results of forecasting the SSI statistical distributions for both sanitation systems. Results for 1,000,000 trials yielded a mean of 0.36 (95% CI = 0.24–0.50) and 0.67 (95% CI = 0.55–0.75) for septic tank and ROS systems, respectively. The calculated SSI scores for both systems (0.42 for septic tank system and 0.71 for ROS system) are within the 95% CI, and do not have a significant difference in the forecasted mean. Similar trends were observed for the forecasted statistical distributions of WE, EE, WRE, CCI, and MCI, and these results are presented as Supplementary Material.

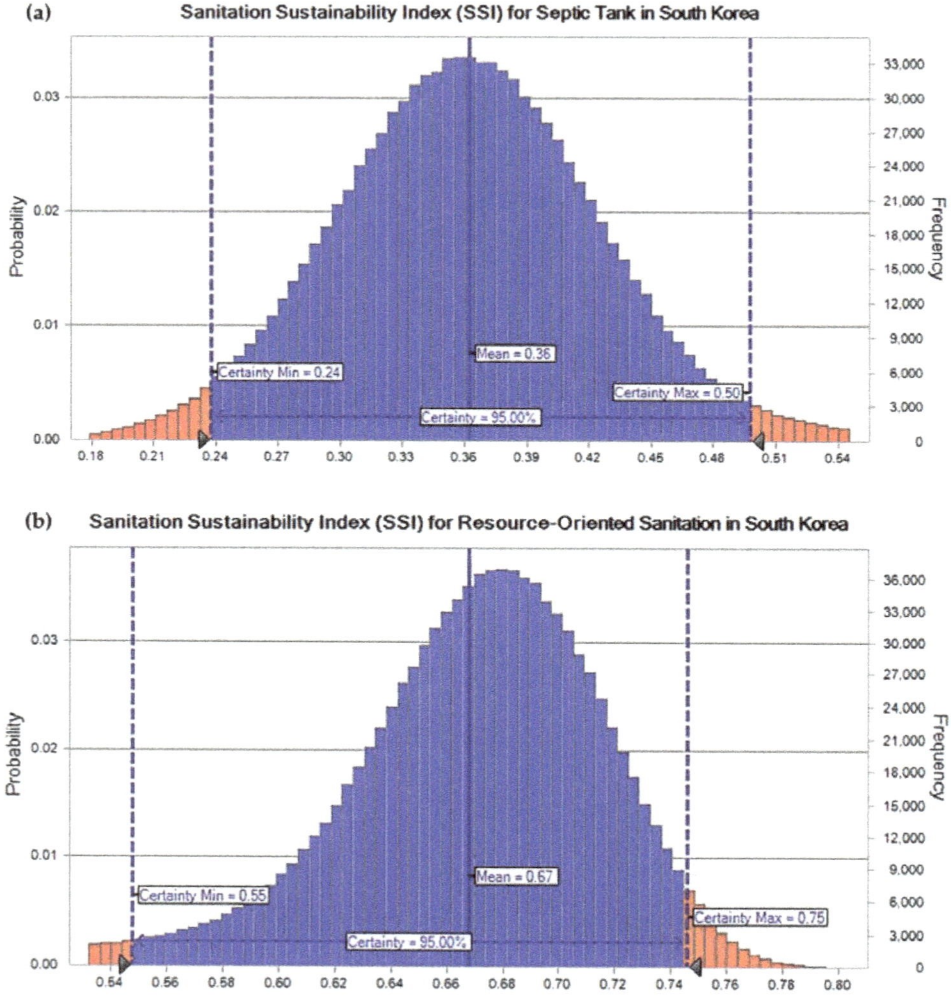

Figure 4. Forecasted statistical distribution of sanitation sustainability index calculated for (**a**) septic tank and (**b**) resource-oriented sanitation systems in South Korea.

The results concluded that the SSI could be a reliable index as a means for evaluating the sustainability of a sanitation system, as well as comparing the sustainability of different sanitation systems.

4. Discussion

The proposed SSI can be considered a tool to evaluate the sustainability of any local sanitation system and its associated infrastructure. It can be used to evaluate the sustainability of a single sanitation system, as well as to compare two or more systems. The index includes sub-indexes, which can provide technical, economic, and social evaluations. Overall, SSI can be a useful tool for sanitation system designers, managers, and developers. Meanwhile, decision-makers can also use SSI to compare different systems to choose the most sustainable one for implementation.

The sub-indexes and their variables are based on the understanding of the sanitation status of the community, where the system is going to be implemented. Accordingly, the scope of the index is narrowed to one specific community. This means that the value of the index is specifically valid for a particular community and may differ between communities. This feature gives sufficient flexibility to the index, rendering it entirely replicable. Nevertheless, in case of availability of sufficient data reflecting the nation-wide sanitation status, SSI can also be used to evaluate sanitation systems at a national level.

The variables of all sub-indexes are also compatible with time. Community-based variables, such as V_M, E_M, CC_M, and MC_M, can be defined based on the predicted changes to the community (e.g., increase in population). Accordingly, the dimensionless index can also provide a vision about the sustainability of the assessed system in the future.

The flexibility and applicability of this index can be compared with that of the one suggested by Iribarnegaray et al., which also evaluates access to clean and safe water [36]. Most of the available indexes proposed to evaluate water and sanitation sustainability are useful in evaluating systems that are already implemented. Although all these indexes are defined with a similar objective, the SSI is used to evaluate the sustainability of sanitation systems before implementation, which can save time, money, and resources. Furthermore, its technical and economic sub-indexes are defined based on logical and simple equations, reflecting differences between the evaluated systems and the current norms of the community. Moreover, these sub-indexes can determine the potential of the system for waste recycling and reuse as well as provisions for economic benefits.

SSI applicability was demonstrated for evaluating the implementation of a septic tank and an ROS system in a suburban area of Seoul. Results indicated higher sustainability for the ROS system compared to the more commonly used septic tank system. Implementations of ROS systems are limited to the demonstration level through limited small projects.

A notable feature of the SSI is its use in determining the fate of wastes as well as the economic and social aspects of a sanitation system. Considering the high potential of an ROS system for waste recycling, it can be concluded that it is entirely sustainable in both technical and economic aspects. The index clearly shows that the implementation of the ROS has a higher capital cost, with the potential for providing direct economic benefits, which can cover both capital and maintenance costs.

With regard to social evaluation, the index assesses the acceptability and provision of public health services for a sanitation system. South Korea, along with some other Asian countries, has a historical background of using source-separated urine and feces as fertilizer [5]. Accordingly, ROS can be quite acceptable if more of them are implemented. Meanwhile, more efficient and hygienic maintenance and waste treatment processes can notably increase the social evaluation score of an ROS system. In addition, conducting a sensitivity analysis can determine the target parameters or sub-indexes of SSI for a specific sanitation system regarding a specific community. Accordingly, giving priority to these parameters might be useful to increase the sustainability of the sanitation system in the specific community.

The reliability of the SSI was examined through Monte Carlo simulation to forecast the statistical distribution of the calculated SSI scores. It became clear that the results were within the 95% CI. Accordingly, the index can be reliably applied to any other community with different settings when the only requirement for the SSI input is relevant local data.

5. Limitations of SSI

Although SSI is proposed to improve sanitation in communities worldwide, the approach described in this study may not be the ideal one.

For example, the economic sub-indexes are based on specific estimations that rely on the current financial status of the community. However, the economic situation of a community can be profoundly affected by unpredicted social events such as a pandemic.

Moreover, the social sub-indexes are heavily dependent on the availability of related studies and databases. Especially for newly developed sanitation systems, there is a demand for previous pilot projects that investigated the acceptability and efficiency of the system in improving public health by reducing the risk of sanitation-related illnesses. The investigation methods depend on the community culture and available resources for the investigation. The role of SSI is to reflect and simplify the results of these investigations. Accordingly, to use SSI, prior local studies on economic and social perspectives of the community are crucial.

6. Conclusions

In this paper, the SSI is proposed as an index to evaluate the sustainability of sanitation systems intended to be implemented in a community. The SSI is simple to calculate, and its variables are flexible, time compatible, and entirely dependent on the characteristics of the community where the system will be implemented. It considers the technical, economic, and social aspects of the system. Therefore, the SSI is a reliable indicator to evaluate different sanitation systems, compare their sustainability, and choose the most suitable type for a specific community.

The applicability of the SSI has been demonstrated for South Korea by evaluating the implementation of a septic tank and an ROS system. A sensitivity analysis was performed to determine the variables that had significant impact. Statistical distribution was forecasted for the SSI of both systems and proved that the calculated SSI is within the 95% confidence interval.

The SSI can evaluate the sustainability of any sanitation system, considering the current water and sanitation condition of the community before implementation. This feature of SSI can make it useful for choosing the best sanitation system for a community, especially in remote and rural areas where access to clean water and sanitation is limited and inadequate.

The index applies to any community, and its prerequisite is relevant technical, economic, and social data. Accordingly, the provision of comprehensive, relevant databases for the community is essential. The lack of such databases causes limitations, especially for calculating the SSI social sub-index. Consequently, the development of a standard survey to determine social acceptability, cultural sanitation aspects, and public health indicators will be advantageous to further develop the SSI. Thus, studies on customizing the SSI and defining new variables or sub-indexes for specific communities will make this index more beneficial.

Supplementary Materials: The following are available online at http://www.mdpi.com/2071-1050/12/17/6937/s1: Sustainability Evaluation Results for Implementation of Septic Tank and Resource-Oriented Sanitation Systems in South Korea.

Funding: This research received no external funding.

Acknowledgments: The author's profound gratitude goes to Eng. Farid Hashemi, for his kind support and efforts. This article was written during the peak of the COVID-19 pandemic. The author would like to express his deepest sympathies to people worldwide who have faced losses due to this pandemic. Sanitation and hygiene have become even more critical now, and the author will continue to contribute toward achieving global sustainable and adequate sanitation, especially for those in remote and rural areas. The author wishes for the day where no one is left behind regarding access to clean water and sanitation and hopes that everyone stays safe and sound.

Conflicts of Interest: The author declares no conflict of interest.

References

1. UNICEF; WHO. *Progress on Household Drinking Water, Sanitation and Hygiene 2000–2017: Special Focus on Inequalities*; United Nations Children's Fund (UNICEF): New York, NY, USA; World Health Organization: New York, NY, USA, 2019; ISBN 978-92-806-5036-5.
2. UN Sustainable Development Goal 6. Available online: https://sustainabledevelopment.un.org/sdg6 (accessed on 24 March 2020).
3. Hyun, C.; Burt, Z.; Crider, Y.; Nelson, K.L.; Prasad, C.S.S.; Rayasam, S.D.G.; Tarpeh, W.; Ray, I. Sanitation for Low-Income Regions: A Cross-Disciplinary Review. *Annu. Rev. Environ. Resour.* **2019**, *44*, 287–318. [CrossRef] [PubMed]

4. Jain, A.; Wagner, A.; Snell-Rood, C.; Ray, I. Understanding open defecation in the age of Swachh Bharat Abhiyan: Agency, accountability, and anger in rural Bihar. *Int. J. Environ. Res. Public Health* **2020**, *17*, 1384. [CrossRef]
5. Han, M.; Hashemi, S. Sanitation revolution: From waste to resource. *Desalin. Water Treat.* **2017**, *91*, 305–310. [CrossRef]
6. Langergraber, G.; Muellegger, E. Ecological Sanitation—A way to solve global sanitation problems? *Environ. Int.* **2005**, *31*, 433–444. [CrossRef] [PubMed]
7. Schwemlein, S.; Cronk, R.; Bartram, J. Indicators for monitoring water, sanitation, and hygiene: A systematic review of indicator selection methods. *Int. J. Environ. Res. Public Health* **2016**, *13*, 333. [CrossRef] [PubMed]
8. Cronk, R.; Luh, J.; Meier, B.M.; Bartram, J. *The Water, Sanitation, and Hygiene Performance Index: A Comparison of Country Performance in Realizing Universal WaSH*; University of North Carolina at Chapel Hill: Chapel Hill, NC, USA, 2015. [CrossRef]
9. UNDP. Human Development Index (HDI)|Human Development Reports. Available online: http://hdr.undp.org/en/content/human-development-index-hdi/ (accessed on 24 March 2020).
10. Lundin, M.; Molander, S.; Morrison, G.M. A set of indicators for the assessment of temporal variations in the sustainability of sanitary systems. *Water Sci. Technol.* **1999**, *39*, 235–242. [CrossRef]
11. Iribarnegaray, M.A.; Copa, F.R.; Gatto D'Andrea, M.L.; Arredondo, M.F.; Cabral, J.D.; Correa, J.J.; Liberal, V.I.; Seghezzo, L. A comprehensive index to assess the sustainability of water and sanitation management systems. *J. Water Sanit. Hyg. Dev.* **2012**, *2*, 205–222. [CrossRef]
12. Hashemi, S.; Boudaghpour, S. Economic Analysis and Probability of Benefit of Implementing Onsite Septic Tank and Resource-Oriented Sanitation Systems in Seoul, South Korea. *Environ. Technol. Innov.* **2020**, *18*, 100762. [CrossRef]
13. Simha, P.; Ganesapillai, M. Ecological Sanitation and nutrient recovery from human urine: How far have we come? A review. *Sustain. Environ. Res.* **2017**, *27*, 107–116. [CrossRef]
14. Hashemi, S.; Han, M.; Kim, T. Optimization of fertilization characteristics of urine by addition of Nitrosomonas europaea bio-seed. *J. Sci. Food Agric.* **2016**, *96*, 4416–4422. [CrossRef]
15. Hashemi, S.; Han, M. Optimizing source-separated feces degradation and fertility using nitrifying microorganisms. *J. Environ. Manag.* **2018**, *206*, 540–546. [CrossRef] [PubMed]
16. Hashemi, S.; Han, M.; Namkung, E. Fate of Fecal Indicators in Resource-Oriented Sanitation Systems Using Nitrifying Bio-Treatment. *Int. J. Environ. Res. Public Health* **2018**, *15*, 164. [CrossRef] [PubMed]
17. Simha, P.; Lalander, C.; Nordin, A.; Vinnerås, B. Alkaline dehydration of source-separated fresh human urine: Preliminary insights into using different dehydration temperature and media. *Sci. Total Environ.* **2020**, *733*, 139313. [CrossRef] [PubMed]
18. Katukiza, A.Y.; Ronteltap, M.; Oleja, A.; Niwagaba, C.B.; Kansiime, F.; Lens, P.N.L. Selection of sustainable sanitation technologies for urban slums—A case of Bwaise III in Kampala, Uganda. *Sci. Total Environ.* **2010**, *409*, 52–62. [CrossRef] [PubMed]
19. Tobias, R.; O'Keefe, M.; Künzle, R.; Gebauer, H.; Gründl, H.; Morgenroth, E.; Pronk, W.; Larsen, T.A. Early testing of new sanitation technology for urban slums: The case of the Blue Diversion Toilet. *Sci. Total Environ.* **2017**, *576*, 264–272. [CrossRef] [PubMed]
20. Hashemi, S.; Han, M. Field evaluation of the fertilizing potential of biologically treated sanitation products. *Sci. Total Environ.* **2019**, *650*, 1591–1598. [CrossRef]
21. Hashemi, S.; Han, M. Control of urine odor in different sanitation practices and its implication on water saving. *J. Water Sanit. Hyg. Dev.* **2017**, *7*, 156–162. [CrossRef]
22. Hashemi, S.; Han, M.; Kim, T. Identification of urine scale problems in urinals and the solution using rainwater. *J. Water Sanit. Hyg. Dev.* **2015**, *5*, 322–329. [CrossRef]
23. Hashemi, S.; Han, M.; Kim, T. The effect of material and flushing water type on urine scale formation. *Water Sci. Technol.* **2015**, *72*, 2027–2033. [CrossRef]
24. Koottatep, T.; Pussayanavin, T.; Polprasert, C. Nouveau design solar septic tank: Reinvented toilet technology for sanitation 4.0. *Environ. Technol. Innov.* **2020**, *19*, 100933. [CrossRef]
25. Flanagan, C.P.; Randall, D.G. Development of a novel nutrient recovery urinal for on-site fertilizer production. *J. Environ. Chem. Eng.* **2018**, *6*, 6344–6350. [CrossRef]

26. Tahulela, A.C.; Ballard, H.H. Developing the Circular Economy in South Africa: Challenges and Opportunities. In *Sustainable Waste Management: Policies and Case Studies*; Ghosh, S., Ed.; Springer: Singapore, 2020; pp. 125–133. [CrossRef]
27. Simha, P.; Lalander, C.; Vinnerås, B.; Ganesapillai, M. Farmer attitudes and perceptions to the re-use of fertiliser products from resource–oriented sanitation systems—The case of Vellore, South India. *Sci. Total Environ.* **2017**, *581–582*, 885–896. [CrossRef] [PubMed]
28. Drangert, J.O.; Nawab, B. A cultural-spatial analysis of excreting, recirculation of human excreta and health—The case of North West Frontier Province, Pakistan. *Health Place* **2011**, *17*, 57–66. [CrossRef] [PubMed]
29. Mariwah, S.; Drangert, J.O. Community perceptions of human excreta as fertilizer in peri-urban agriculture in Ghana. *Waste Manag. Res.* **2011**, *29*, 815–822. [CrossRef] [PubMed]
30. Hashemi, S.; Han, M. Methods for controlling stored urine odor in resource-oriented sanitation. *J. Water Sanit. Hyg. Dev.* **2017**, *7*, 507–514. [CrossRef]
31. Mkhize, N.; Taylor, M.; Udert, K.M.; Gounden, T.G.; Buckley, C.A. Urine diversion dry toilets in eThekwini municipality, South Africa: Acceptance, use and maintenance through users' eyes. *J. Water Sanit. Hyg. Dev.* **2017**, *7*, 111–120. [CrossRef]
32. Ssemugabo, C.; Halage, A.A.; Namata, C.; Musoke, D.; Ssempebwa, J.C. A socio-ecological perspective of the facilitators and barriers to uptake of water, sanitation and hygiene interventions in a slum setting in Kampala, Uganda: A qualitative study. *J. Water Sanit. Hyg. Dev.* **2020**, *10*, 227–237. [CrossRef]
33. Hashemi, S.; Han, M.; Kim, T.; Kim, Y. Innovative Toilet Technologies for Smart and Green Cities. In *True Smart & Green Urban Technologies and Infrastructure Systems, Proceedings of the 8th Conference of the International Forum on Urbanism (IFoU), Incheon, Korea, 22–24 June 2015*; Kim, D., Kim, S., Schuetze, T., Sohn, S., Chelleri, L., Ostermeyer, Y., Tieben, H., Wolfram, M., Eds.; International Forum on Urbanism (IFoU): Delft, The Netherlands, 2015; p. E013. [CrossRef]
34. Guerrini, A.; Romano, G.; Indipendenza, A. Energy efficiency drivers in wastewater treatment plants: A double bootstrap DEA analysis. *Sustainability* **2017**, *9*, 1126. [CrossRef]
35. MOE. *Modernization of the Sewerage System in Korea*; The International Bench Marking Network (IBNET): Washington, DC, USA, 2015. Available online: https://www.ib-net.org/docs/History_of_Korea\T1\textquoterights_sewerage_system_development.pdf (accessed on 24 March 2020).
36. Iribarnegaray, M.A.; Gatto D'Andrea, M.L.; Rodriguez-Alvarez, M.S.; Hernández, M.E.; Brannstrom, C.; Seghezzo, L. From indicators to policies: Open sustainability assessment in the water and sanitation sector. *Sustainability* **2015**, *7*, 14537–14557. [CrossRef]

sustainability

MDPI

Article

National Versus Local Sustainable Development Plans and Island Priorities in Sanitation: Examples from the Kingdom of Tonga

Ian White [1,*], Tony Falkland [2] and Taaniela Kula [3]

1 Fenner School of Environment and Society, Australian National University, Canberra ACT 0200, Australia
2 Island Hydrology Services, 9 Tivey Place, Hughes, Canberra ACT 2605, Australia; tony.falkland@netspeed.com.au
3 Natural Resources Division, Ministry of Lands and Natural Resources, Nuku'alofa, Tonga; tkula@naturalresources.gov.to
* Correspondence: ian.white@anu.edu.au; Tel.: +61-418-262-881

Received: 22 October 2020; Accepted: 9 November 2020; Published: 11 November 2020

Abstract: Sanitation, water supply, and their governance remain major challenges in many Pacific Island countries. National sustainable development strategies (NSDSs) are promoted throughout the Pacific as overarching improved governance instruments to identify priorities, plan solutions, and fulfill commitments to sustainable development. Their relevance to local village-level development priorities is uncertain. In this work we compare national priorities for sanitation in NSDSs with those in village community development plans (CDPs) and with metrics in censuses from the Kingdom of Tonga. Tonga's Strategic Development Frameworks (TSDFI 2011–2014 and TSDFII 2015–2025) were developed to focus government and its agencies on national outcomes. From 2007 to 2016, 136 villages throughout Tonga's five Island Divisions (IDs) formulated CDPs involving separately 80% of women, youth, and men in each village. It is shown that censuses in 2006 and 2016 reveal linked improvements in water supply and sanitation systems but identify IDs with continuing challenges. It is found that sanitation and water are a national priority in TSDFI but are absent from the current TSDFII. In contrast, analysis of CDPs, published just after TSDFII, show in one ID, 53% of villages ranked sanitation as a priority and marked differences were found between IDs and between women, youth, and men. CDPs' sanitation priorities in IDs are shown to mostly correspond to sanitation and water metrics in the censuses, but some reflect impacts of natural disasters. Explanations for differences in sanitation priorities between the national and local development plans, as well as suggestions for improving NSDS processes in island countries, are advanced.

Keywords: sustainable development strategies; community development plans; small island developing states; governance; sanitation; water supply; hygiene; WASH; census results; top-down versus bottom-up; gender and age; SDG6

---

1. Introduction

The combined challenges of climate change and economic development have long been recognized as major threats to the sustainability of small island developing states (SIDS), impacting especially vital water and sanitation services [1–3]. Water supply and sanitation systems remain continuing challenges in the Pacific region which was the only region in the world where there was an increase in open defecation between 2000 and 2015 [4].

National sustainable development strategies (NSDSs) [5] have been seen as an efficient way for SIDS to identify national challenges, plan their solutions, assign resources and responsibilities, and fulfill country commitments to international and regional agreements and programs, particularly the United Nations (UN) Sustainable Development Goals (SDGs) and UN SIDS action programs [3]. Part of their efficiency stems from the fact that they are mainly top-down processes, often encouraged, assisted, and supported by external agencies. The relevance of such governance processes imported into dispersed small island communities in Pacific Island countries (PICs) has been questioned [6]. PICs characteristically have well-developed local institutions, resilient social systems, a sensitivity to environmental change, and a high degree of equity [7]. These strengths make PICs well suited to bottom-up planning processes, but these are generally time consuming [8]. In this work, using a valuable example from a multi-island PIC, the Kingdom of Tonga, comparison is made between the planning priorities given to sanitation in top-down national processes and bottom-up village development planning processes.

Improvements in water, sanitation, and hygiene (WASH) have long been recognized as one of the keys for the continued development of PICs [1–3]. The incidence of diarrheal diseases in PICs is, on average, four to five times higher than in larger countries in the Oceania region [9] and is mainly linked to contaminated drinking water as a result of poor sanitation and hygiene [10]. This linkage was one of the reasons that Oceania as a region did not meet the UN's 2015 Millennium Development targets for water and sanitation [11].

Sanitation in SIDS remains a significant challenge for many interacting reasons. Dispersed island communities, restricted land areas, limited and sometimes brackish water supply, El Niño Southern Oscillation (ENSO)- and Pacific Decadal Oscillation (PDO)-related droughts [12], frequent extreme events, including intense tropical cyclones (TCs) [13], easily polluted surface and limited groundwater, coupled with a wide range of population densities, variable island geologies and geomorphologies, and restricted land and financial resources mean that identifying locally acceptable, safe, appropriate, and affordable sanitation systems with minimum impacts on fragile island environments is problematic. The most difficult are islands which rely on variable rainwater as their only source of water for all purposes [10].

Improved governance and accessible information are seen as central to improved WASH outcomes [14], particularly in SIDS. NSDSs have been promulgated as overarching, efficient, integrated governance instruments to identify and solve national development challenges and address UN SDGs [15]. The South Pacific Kingdom of Tonga presents an opportunity to contrast identified sanitation priorities in top-down national and bottom-up village planning processes. Tonga's national development planning instrument, the Tonga Strategic Development Framework 2015–2025 (TSDFII) [16] was launched in 2015 after three months of senior-level consultations and a three-month review by ministries. TSDFII built on experiences from the Tonga Strategic Development Framework 2011–2014 (TSDFI) [17]. TSDFII and TSDFI, present an integrated vision of the direction the government and its agencies' plan to pursue as well as contribute to Tonga's international and regional commitments.

A much lengthier village level development planning process occurred throughout all Island Divisions (IDs) in Tonga between 2007 and 2016. This created Community Development Plans (CDPs) in rural villages throughout Tonga's five IDs under the Ministry of Internal Affairs (MIA) and was facilitated by a non-government organization, Mainstreaming of Rural Development Innovation Tonga Trust (MORDI TT), with support from donors [18]. CDPs ranked local village priority development issues which were prepared and endorsed by a minimum of 80% of each village community. In the past, only men had been involved in local planning. Breaking from tradition, the CDP process also included women and youth who ranked priorities separately from men. In 2016, 136 CDPs out of Tonga's 151 rural villages, were presented to the then Prime Minister [18]. Currently, 117 of the CDPs are available for analysis [19].

It was shown recently [20] that the high priority given to improving water supply in village CDPs universally throughout all Tonga's five IDs contrasts with the omission of water supply from planned outcomes in TSDFII, despite TSDFII's claim that it addresses UN SDG6 on water and sanitation. Here, the priorities given to sanitation in both TSDFII and in the preceding TSDFI are compared with its ranked priority in the available CDPs. These are also compared with census data from 2006 and 2016 on sanitation type and available water sources at the ID level.

Several indices and indicators have been proposed to monitor performance in WASH [21,22]. The sanitation sustainability index (SSI) has been introduced recently as an indicator for evaluating the sustainability of sanitation systems [23]. The strength of SSI is that it includes integrated measures of the technical, social, environmental, and economic aspects of local community sanitation systems. Unfortunately, in PIC rural villages, even technical information is often absent. What are available usually are national census data on water sources and sanitation types together with often limited hydrological data.

Four questions are addressed in this work:

1. In the absence of detailed information, can census data be used to identify both improvements in sanitation and the diversity of sanitation needs in PICs?
2. Do census data show a link between island sanitation types and island water sources?
3. Do top-down NSDS plans give the same weight to sanitation priorities as those identified in nation-wide bottom-up village development plans and is there an evolution of priorities in NSDSs?
4. Can NSDSs in multi-island countries be improved for sanitation outcomes?

These questions are addressed at the ID level.

Census data on demographics and household water supply sources and sanitation systems from the last two censuses in 2006 and 2016 [24,25] in Tonga are compared at the ID level as well as for the capital area, Greater Nuku'alofa. The relation between sanitation type and water source used are explored. The emphasis given to sanitation in both TSDFI and TSDFII [16,17] is then examined and compared with that given to other infrastructure services. An analysis is presented of the ranked priorities given by women, youth, and men, as well as aggregate village priorities for sanitation in the available CDPs [19] from villages in all IDs. Relationships between CDP priority rankings for sanitation and census sanitation type and water source data are explored. Underlying reasons for differences in top-down versus bottom-up development priorities for sanitation are discussed and options for improving NSDS processes in PICs are given.

2. Materials and Methods

2.1. Study Location

The Kingdom of Tonga's population of close to 101,000 people live in 169 islands clustered in five IDs with approximate land area of 750 km^2. These are dispersed over 700,000 km^2 of the southwestern Pacific Ocean (Figure 1). Its islands border the seismically active Tonga trench. Tonga's western islands are volcanic in origin, but its eastern islands are geologically different, being uplifted coral limestone and sand islands. Most of the eastern islands, including Tongatapu, the largest island and location of the capital Nuku'alofa, have a covering of rich volcanic soil deposited from eruptions in the western islands. Volcanic eruptions, earthquakes, tsunamis, TCs, storm surges, and droughts and floods linked to ENSO and the PDO are frequent natural hazards. Recent TCs that have devastated parts of Tonga include TC Ian (2014), TC Gita (2017), and TC Harold (2020).

There is a gradient in annual rainfall across Tonga varying from 1750 mm in the south to 2300 mm in the north, influenced by proximity to the South Pacific Convergence Zone (SPCZ). Annual rainfall variability is moderate with a mean coefficient of variability of 0.21 for the period 1947 to 2019 and decreases from north to south. There are no long-term trends in available annual rainfall [20], which is consistent with climate change projections for Tonga [12]. The Kingdom has a wetter season from November to April followed by a drier season from May to October, also influenced by the position of the SPCZ [26].

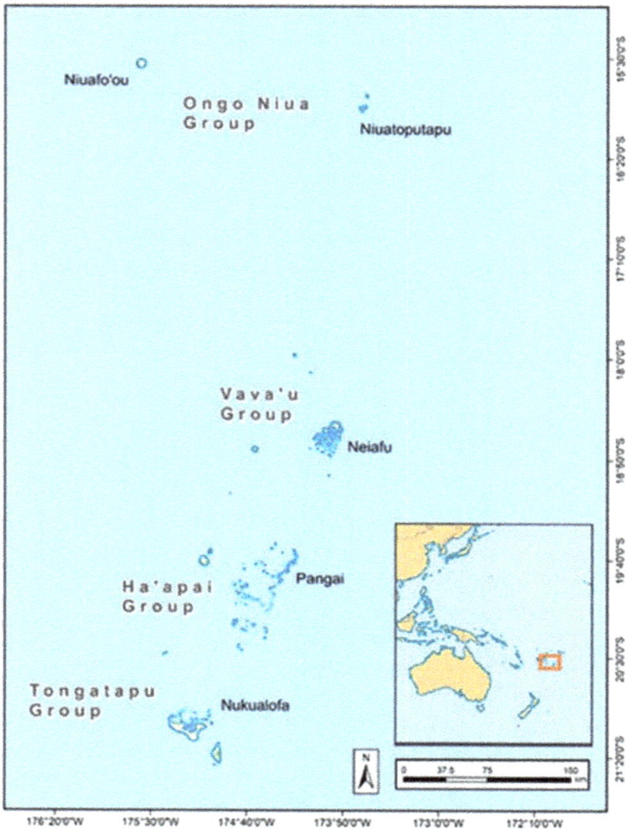

Figure 1. Map of the South Pacific Kingdom of Tonga, main Island Divisions and population centers [27].

Water supply in Tonga is mainly sourced from household rainwater harvesting systems and groundwater. Water is supplied from groundwater lenses or springs in the carbonate and sand islands. Many of the volcanic islands do not appear to have useable groundwater [27]. Because seawater underlies groundwater lenses, groundwater salinity in Tonga varies from fresh to brackish and even saline, depending on island geology, size, pumping method and rate, and ENSO and PDO conditions. Springs and groundwater used in 'Eua and most of the groundwater in Tongatapu have lower salinities than groundwater in Vava'u and Ha'apai [27]. Tongans prefer rainwater for drinking, and groundwater is used for all other purposes including toilet flushing. The main sanitation systems are cistern flush toilets or pour flush toilets discharging into septic tanks and pit toilets. Both septic tanks and pit toilets have the potential to contaminate groundwater.

In 2016, about three quarters of Tongans lived on the main island Tongatapu with Greater Nuku'alofa, the capital, having over a third of the country's population. About a quarter of the population is spread over the Kingdom's other four IDs (Figure 1). About 55% of the population is under the age of 25 with youth, aged 14 to 24, making up nearly 19% of the total population [25]. Gross domestic product (GDP) in 2017 was estimated to be US$5900 per person with an annual growth rate around 2.5% [28]. GDP varies widely between IDs with Tongatapu being 15% above the national average in 2013, while Ha'apai was 40% below the national average. Natural disasters have been estimated to cost 4.4% of GDP and TC Ian in 2014, which affected 70% of the population of Ha'apai, severely damaged 75% of the housing stock and cost 11% of GDP [16].

Piped water is supplied from groundwater or spring sources [27]. In Nuku'alofa, 'Eua, and population centers on Vava'u and Ha'apai, piped water is supplied by the Tonga Water Board. Village Water Committees are responsible for piped water supply in villages throughout Tonga. Village piped water supply is mostly intermittent unlike in Greater Nuku'alofa where supply is continuous.

In PICs, traditionally, sanitation was either a personal or household responsibility. In Tonga, sanitation is overseen by the Ministry of Health and, in villages, responsibility is delegated to Village Water Committees (VWCs). There are no public reticulated sewerage systems in Tonga and sanitation is mainly a household responsibility.

2.2. Demographics, Water Sources, and Sanitation Systems

The censuses in 2006 and 2016 [16,17] provide details of the national, ID, district, and village level population distribution as well as urban/rural data and can be used to identify changes. This work concentrates on the ID level. Census results are also used here to compare the use of different water sources for all non-drinking purposes, which includes use for sanitation, and to compare sanitation systems used by households across IDs and in Greater Nuku'alofa.

The very limited data on water use from different sources indicates that piped water systems in Tonga supply households (HH) with almost an order of magnitude more water than rain tanks [27]. Here, we make use of the water source ratio (WSR) as a measure of the ID availability of water for all non-drinking purposes [20]:

$$\text{WSR} = (\text{HH supplied by rain tanks})/(\text{HH with piped water supply}) \quad (1)$$

Sanitation systems in Tonga fall into two groups—those that require little or no water such as pit toilets and pour flush toilets, and cistern flush toilets that require more water. We introduce the sanitation type ratio (STR) as a crude, relative measure of improved sanitation systems:

$$\text{STR} = (\text{HH with pit} + \text{HH with pour flush toilets})/(\text{HH with cistern flush toilets}) \quad (2)$$

The use of these ratios is a result of the limited data available on water and sanitation systems and their use. We use WSR and STR to compare IDs and the capital Nuku'alofa. If availability of water is a significant factor in the choice of sanitation system [23], then the hypothesis is that the two ratios should be related.

2.3. Tonga Strategic Development Framework 2011–2014, TSDFI

TSDF1 [17] was developed after the first full parliamentary elections in Tonga. It built on the consultative process in the previous National Strategic Development Framework in October 2009. Its aim was to provide principles and directions to guide the work of the government administration over its four-year term, 2011 to 2014. It involved consultations over a nine-month period carried out in all five IDs in Tonga [29].

The government's vision for TSDFI was, "To develop and promote a just, equitable, and progressive society in which the people of Tonga enjoy good health, peace, harmony, and prosperity, in meeting their aspirations in life." The government's objectives are expressed at two levels: At the top level are the nine primary Outcome Objectives that contribute to the vision, and at the second level are four enabling themes that support the achievement of these outcomes. The enabling themes are more efficient and effective government, improving macro-economic and fiscal management, sustainable and accountable public enterprises, and coordinated and integrated approach to development partners. The Outcome Objectives have 27 Associated Strategies (ASs) and the enabling themes have a further 15 ASs. The ministries responsible for each Outcome Objective were identified and each were linked to Millennium Development Goals [17].

The Ministry of Finance and National Planning (MFNP) led preparation of TSDFI which was supported by the Australian Agency for International Development (AusAID) and the planning process was facilitated by the Asian Development Bank (ADB).

TDSFI was scanned for references to water, freshwater, rainwater, groundwater, water supply, WASH, sanitation and hygiene, and the Outcome Objectives and ASs were examined for their relevance to sanitation. The relative weight of sanitation to infrastructure and other services was determined.

2.4. Tonga Strategic Development Framework 2025–2015, TSDFII

The TSDFII [16] is structured differently to and is more complex than the preceding TSDFI. TSDFII is " ... the overarching framework of the planning system in Tonga. It provides an integrated vision of the direction that Tonga seeks to pursue." [16]. TSDFII is the top of a cascading system of planning and budgeting which is intended to guide:

- Medium term sector and district/island master plans.
- Three year rolling corporate plans and budgets for all ministries, departments, and agencies.
- Annual divisional and staff plans and job descriptions.
- Consultation, monitoring, and evaluation.

TSDFII identifies government priorities, assigns ministerial responsibilities, and aims to focus resources. TSDFII is arranged in a hierarchy where 29 Organizational Outcomes (OOs), grouped under three institutional pillars and two input pillars, feed into seven desired National Outcomes (NOs) that in turn feed into the single Outcome: "More inclusive sustainable growth and development" which supports the single planned National Impact of TSDFII: "A more progressive Tonga supporting a higher quality of life for all" which supports the motto of TSDFII, given by the reformer monarch Tupou I: "God and Tonga are my inheritance". TSDFII also lists 153 Strategic Concepts (SCs) which were issues raised during the consultation process but lie outside TSDFII but are intended as aids to sector, district and ministry, department, and agency planning and budgeting [16].

MFNP led development of TSDFII, supported by the ADB. It used wide but fairly rapid three month high-level consultations with "key sectors of the economy including the community (district and town officers), church leader forums, non-government organizations, and private business forums, including all the main sectors: agriculture, fisheries, tourism, commerce, manufacturing, and construction" [16]. Consultation meetings were held throughout Tongatapu and the IDs of 'Eua, Ha'apai, and Vava'u in the period between October 2014 and December 2014. The northern Ongo Niua ID was not covered. These consultations, together with lessons learnt from TSDFI, were incorporated into drafts of TSDFII which were circulated to government ministries, departments, and agencies for further review in early 2015 [16].

The TSDFII was scanned for references to water, freshwater, rainwater, groundwater, water supply, sanitation, hygiene, and WASH and the planned NOs, OOs, and SCs were examined for their relevance and applicability to sanitation. The weight given to these relative to other infrastructure and other services was determined.

2.5. Community Development Plans, 2016

Development of village CDPs [18] began in 2007, under the Local Government Division of the MIA. The CDPs were a response to the then National Vision in TSDFI "A Progressive Tonga Supporting Higher Life for All". Consultations throughout rural villages in Tonga's five IDs were implemented by the NGO Mainstreaming of Rural Development Innovation Trust Tonga (MORDI TT). The CDP process was supported by the International Fund for Agricultural Development (IFAD), UNDP, AusAID, and the Tonga Government. One of the requirements of the project was participation of 80% of the population of each rural village in the development, ranking of priorities, and endorsement of the village CDP. Priorities were ranked separately by women, youth (aged 14 to 24 years), and men. Consultations involved a lengthy process which culminated in District Officers and Town Officers of 136 village communities presenting their CDPs to the then Prime Minister on 4 October 2016 [18]. The CDPs are really ranked lists of local village concerns.

During the planning process the Department for Local Government was transferred from the Prime Minister's Office to the Ministry of Training, Employment, Youth, and Sports, and then to the MIA [18]. Analysis of and response to CDPs appear to have been deferred by these moves. We have not found any analysis of the valuable information on village development priorities in CDPs apart from our analysis of water supply [20].

Of the 136 CDPs presented, 117 are available on-line [19]. These represent 77.5% of all rural villages in Tonga. CDPs were downloaded and the priority rankings of each village that mentioned sanitation were recorded (Table S1). Particular note was made when sanitation ranked in the top three priorities for women, men, and youth, which also provided a village average. Village level results were aggregated to percentage of villages in each ID. Also recorded was the percentage of villages within each ID that ranked sanitation anywhere in their list of priorities.

3. Results

3.1. Changes in Demographics

Table 1 shows the demographic data from the 2006 [24] and 2011 [25] censuses for Tonga, for the five IDs, and for Greater Nuku'alofa (GN).

Table 1. Demographic data for Tonga, Island Divisions, and Greater Nuku'alofa from the 2006 [24] and 2016 [25] censuses and the percent change that has occurred since 2006.

Item	Year	Tonga	'Eua	Tongatapu	Vava'u	Ha'apai	Ongo Niua [1]	Greater [2] Nuku'alofa
Total Population	2006	101,991	5206	72,045	15,505	7570	1665	34,311
	2016	100,651	4945	74,611	13,738	6125	1232	35,184
Change (%)	*2016–2006*	*−1.3*	*−5.0*	*3.6*	*−11*	*−19*	*−26*	*2.5*
Population Density (pers/km^2)	2006	157	60	277	129	69	23	985
	2016	155	57	286	114	56	17	1010
Change (%)	*2016–2006*	*−1.2*	*−5.0*	*3.2*	*−11*	*−19*	*−26*	*2.5*
Number of Households (HH)	2006	17,529	905	12,012	2885	1377	350	5753
	2016	18,198	889	13,096	2745	1193	275	6240
Change (%)	*2016–2006*	*3.8*	*−1.8*	*9.0*	*−4.9*	*−13*	*−21*	*8.5*
Number People per HH	2006	5.8	5.8	6.0	5.4	5.5	4.8	6.0
	2016	5.5	5.6	5.7	5.0	5.2	4.5	5.6
Change (%)	*2016–2006*	*−5.5*	*−2.7*	*−5.0*	*−7.0*	*−5*	*−5*	*−6.1*
Number of Villages	2006	157	14	66	38	27	12	14
	2016	165	15	67	44	27	12	14
Change (%)	*2016–2006*	*5.1*	*7.1*	*1.5*	*16*	*0*	*0*	*0*
Area (km^2)		650	87	260 [3]	121	109	72	35

[1] Ongo Niua includes Niuafo'ou and Niuatoputapu. [2] Greater Nuku'alofa is made up of the districts of Kolofo'ou and Kolomotu'a in Tongatapu. [3] It appears that the area given for Tongatapu of 260 km^2 is the area occupied by villages.

Table 1 shows that in 2016 almost 75% of Tongans live on the main island Tongatapu, with Greater Nuku'alofa having 35% of the total population. About 25% of the population are spread over the Kingdom's other four IDs. The 2016 census defined 77% of the population as rural, although this includes villages within Greater Nuku'alofa [25].

Net migration from Tonga between 2006 and 2016 is evident in Table 1 but with significant inwards migration from the outer IDs to the main island Tongatapu and to Greater Nuku'alofa (Figure 2). The overall population density has decreased correspondingly in Tonga, with significant decreases in population density in Ha'apai and Ongo Niua partially offset by increases in Tongatapu. The population decrease in Ha'apai may be partly due to impacts of TC Ian in 2014 [16].

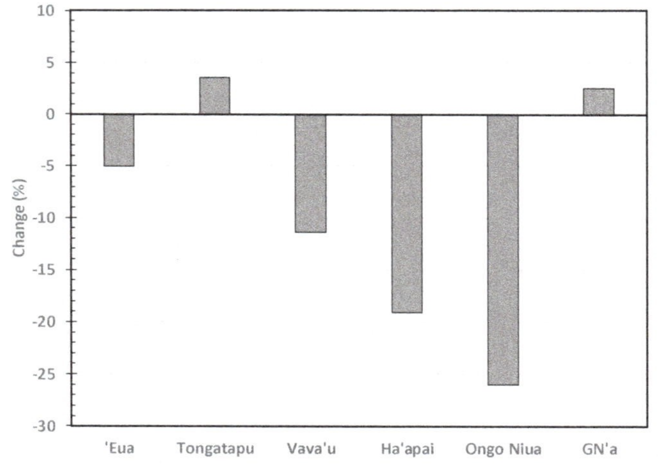

Figure 2. Percentage change between 2006 and 2016 in the population of Tonga's five Island Divisions and Greater Nuku'alofa (GN'a) relative to 2006.

Table 1 also reveals an overall increase in the number of households driven by large increases in Tongatapu and Greater Nuku'alofa but offset by decreases in the outer IDs, especially Ha'apai and Ongo Niua. Partly as a result of the shift in population (Figure 2) and the change in the number of households, the number of people per household decreased in Tonga by 5.5% between 2006 and 2016 with similar decreases across all IDs and Greater Nuku'alofa. Paradoxically, there has been an apparent increase in the number of villages in Tonga counted in the censuses, driven mainly by a large apparent increase in Vava'u.

3.2. Changes in Water Sources

Because of the intimate connection between water and sanitation, the use of different water sources is first examined. There is limited data on the amount of water used for different purposes in Tonga [20,27,30–32]. It is clear, however, that the greatest volume of household water use is for non-drinking purposes such as toilet flushing, washing, and bathing [33]. Table 2 lists the percentage of households accessing water from different sources for non-drinking purposes and the percentage changes in each between the 2006 and 2016 censuses for Tonga as a whole, the five IDs, and Greater Nuku'alofa.

For Tonga as a whole, Table 2 reveals a 6% increase in access to piped water supply, a 26% decrease in the use of household rank tanks, a large 62% decrease in the use of household water wells, and 40% decrease in other sources (not specified in the censuses) for non-drinking purposes between 2006 and 2016. For the IDs, 'Eua, and Ongo Niua had significant increases in access to piped water but, in Vava'u and Ha'apai, there was very little change in piped water supply. The national decrease in using household rainwater tanks was driven by very large decreases in 'Eua and Tongatapu and a smaller decrease in Ongo Niua. These decreases could be the result of improvements in piped water supply between 2006 and 2016 in 'Eua, Tongatapu, and especially Nuku'alofa, but also in Ongo Niua [27,34]

Although there was limited overall use of household wells in 2006, there were dramatic decreases in their use in 2016 in all IDs. In 'Eua and Vava'u household wells were abandoned as a source of water for non-drinking purposes in the 2016 data and use substantially reduced in Tongatapu, Ha'apai, and Greater Nuku'alofa. Household groundwater wells are easily contaminated by household sanitation systems such as pit toilets and leaking septic tanks [27].

Table 2. The percentage of households in Tonga, each Island Division, and Greater Nuku'alofa accessing water for non-drinking purposes from different water sources from the 2006 [24] and 2016 [25] censuses and the percentage change in sources relative to 2006.

Water Source	Year	Percentage of Households (%)						
		Tonga	'Eua	Tongatapu	Vava'u	Ha'apai	Ongo Niua	Greater Nuku'alofa
Piped Supply	2006 2016	83.2 88.3	80.1 95.6	88.5 93.3	80.8 80.7	51.7 52	52.1 59	87.5 92
Change (%)	2016–2006	*6.1*	*19*	*5.4*	*−0.2*	*0.5*	*13*	*5.1*
Rain Tank	2006 2016	14.7 10.9	19.1 4	10.0 6	16.8 18.9	40.1 44.6	47.3 40.3	10.4 7.1
Change (%)	2016–2006	*−26*	*−79*	*−40*	*13*	*11*	*−15*	*−32*
Own Well	2006 2016	1.6 0.6	0.3 0	0.9 0.5	2.2 0	7.8 3.1	0 0.4	1.2 0.6
Change (%)	2016–2006	*−62*	*−100*	*−41*	*−100*	*−60*	-	*−52*
Other	2006 2016	0.5 0.3	0.4 0.5	0.6 0.3	0.2 0.3	0.4 0.3	0.6 0.4	0.9 0.3
Change (%)	2016–2006	*−40*	*12*	*−49*	*44*	*−18*	*−30*	*−65*

Because non-drinking water use is the largest volumetric demand for water [33], the WSR, defined in Equation (1), provides a measure of the difficulty of meeting non-drinking water demand in villages and islands. This is because roof areas used to harvest rainwater are usually small and are often not very efficient, and average household size (Table 1) is relatively large [15]. The available data indicates that in outer islands, rain tanks supply about 20 L/person/day [30–32]. Figure 3 compares the WSP for Tonga's IDs as well as Greater Nuku'alofa for the 2006 and 2016 censuses.

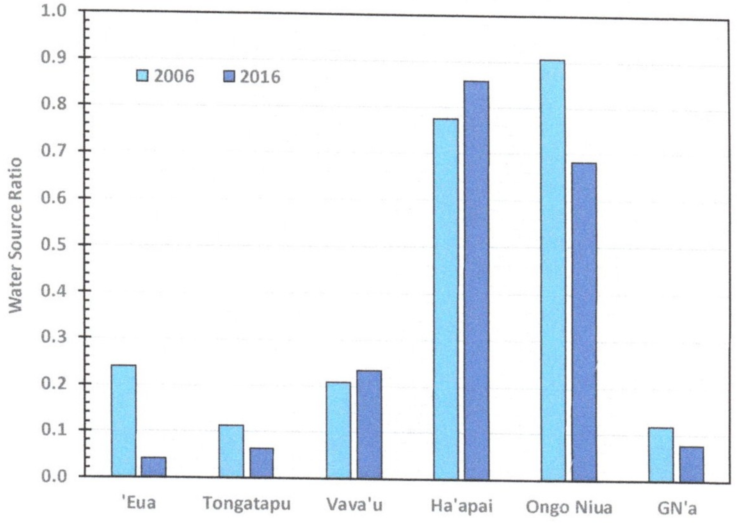

Figure 3. Water source ratio, defined in Equation (1), for non-drinking water uses in Tonga's Island Divisions and Greater Nuku'alofa (GN'a) using data from Table 2 for 2006 and 2016 censuses.

Ha'apai and Ongo Niua IDs stand out in Figure 3, with a much larger proportion of their non-drinking water use being supplied from household rain tanks. The increased access to piped water between 2006 and 2016 in 'Eua, Tongatapu, Ongo Niua, and Greater Nuku'alofa is also evident in Figure 3. In Vava'u and Ha'apai, WSR actually increased between 2006 and 2016 indicating a greater dependency on rain tanks. In these Island Divisions and in Ongo Niua, toilet flushing, washing, and bathing are more dependent on rainwater harvesting than elsewhere in Tonga.

3.3. Changes in Sanitation Systems

Table 3 lists the percentage of households using different types of sanitation systems in Tonga, in IDs, and in Greater Nuku'alofa from the 2006 and 2016 censuses.

Table 3. The percentage of households in Tonga, Island Divisions, and Greater Nuku'alofa using different types of toilets and the percentage changes between 2006 [24] and 2016 [25] relative to 2006. Flush is a cistern flush toilet; manual is a pour flush toilet and pit is a pit toilet.

Toilet Type	Year	Percentage of Households (%)						
		Tonga	'Eua	Tongatapu	Vava'u	Ha'apai	Ongo Niua	Greater Nuku'alofa
Flush	2006	70.2	60.4	80.2	53.9	38.1	33.8	87.5
	2016	82.3	77.4	88.2	71.2	52.6	62.3	92.5
Change (%)	*2016–2006*	*17*	*28*	*10*	*32*	*38*	*84*	*6*
Manual	2006	11.4	11.0	14.2	3.4	6.0	2.6	10.6
	2016	8.5	7.3	9.5	4.3	7.2	8.4	6.6
Change (%)	*2016–2006*	*−26*	*−33*	*−33*	*25*	*21*	*227*	*−38*
Pit	2006	18.1	27.9	5.9	42.3	55.8	63.0	1.7
	2016	9.0	15.0	2.2	24.2	39.2	29.3	0.8
Change (%)	*2016–2006*	*−51*	*−46*	*−64*	*−43*	*−30*	*−54*	*−53*
Other	2006	0.02	0.00	0.02	0.07	0.00	0.00	0.03
	2016	0.2	0.2	0.1	0.3	1.0	0.0	0.1
Change (%)	*2016–2006*	*918*	-	*778*	*376*	-	-	*180*

Between 2006 and 2016, dramatic changes are evident in Table 3 in the household use of different toilet systems in Tonga. Nationally, there was a 17% increase in the use of cistern flush toilets with a corresponding significant decrease in the use of manual pour flush toilets and a dramatic drop in the percentage of houses using pit toilets. Although the other toilet category is nationally very small, 1% or less, there was a major increase in use of this unspecified category.

All IDs increased access to household cistern flush toilets over the period 2006 to 2016 with the largest percentage increase in Ongo Niua and the smallest increase in Greater Nuku'alofa due to the already large percentage of households there with cistern flush toilets in 2006. All IDs also showed a large reduction in the use of pit toilets, with the largest reduction being in the main island of Tongatapu and the smallest, although still an appreciable reduction of 30%, in Ha'apai. 'Eua, Tongatapu, and Greater Nuku'alofa showed significant decreases in the use of manual pour flush toilets which, nationally, were offset by increases in their use in Vava'u and Ha'apai and a major increase of almost 230% in Ongo Niua. The "other" toilet category is not specified in either census, so it is uncertain why use of this category increased across all IDs. The 2006 census provided statistics on households which had no access to toilet facilities. This category is missing from the 2016 census and it may be that the increase in the other toilet facilities between 2006 and 2016 in Table 3 is due to the absence of the no toilet category in the 2016 census [24,25].

A measure of improvements in sanitation technology is the STR, defined in Equation (2). Figure 4 plots the STR for Tonga's IDs as well as Greater Nuku'alofa for both censuses.

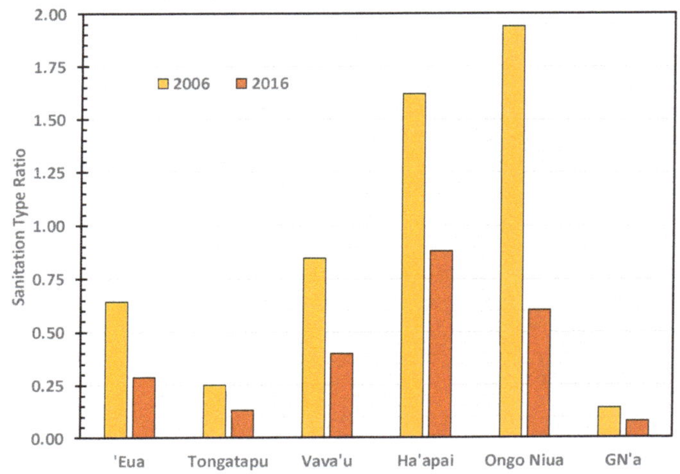

Figure 4. Sanitation type ratio, defined in Equation (2), for Tonga's Island Divisions and Greater Nuku'alofa (GN'a) using data from Table 3 for 2006 and 2016.

Figure 4 demonstrates that STR halved between 2006 and 2016 for Tonga and for all IDs and also for Greater Nuku'alofa. In 2006, Ongo Niua had the highest STR with Tongatapu the lowest. STR in Greater Nuku'alofa is half that of Tongatapu in both census years. In 2016, STR is highest in Ha'apai but STR remained lowest in Tongatapu. The results reflect the increases in household use of cistern flush toilets across all IDs in Table 3.

Of the many factors that determine choice of sanitation systems [23], one hypothesis is that the availability of adequate household water supply is a strong determinant. Figure 5 shows the relation between WSR and STR for Tonga's IDs and Greater Nuku'alofa using the 2006 and 2016 census results.

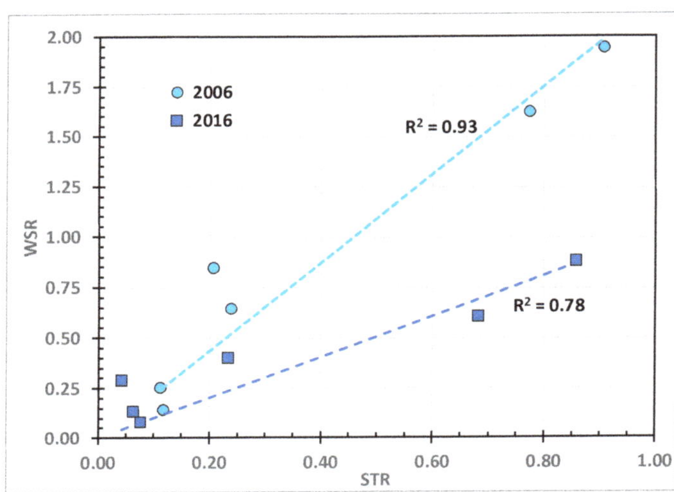

Figure 5. Relationship between the sanitation type ratio (STR), defined in Equation (2), and the water source ratio (WSR), defined in Equation (1), for non-drinking water uses for Tonga's Island Divisions and Greater Nuku'alofa (GN'a) from the 2006 and 2016 censuses.

For the 2006 data the relation between STR and WSR is given by:

$$STR = (2.2 \pm 0.1).\ WSR \tag{3}$$

The relation in Equation (3) has coefficient of determination $R^2 = 0.93$ and is highly significant ($p < 5.10^{-6}$). The relation for the 2016 data is:

$$STR = (1.0 \pm 0.1).\ WSR \tag{4}$$

The relation in Equation (4) has $R^2 = 0.78$ and is very significant ($p < 5.10^{-4}$). Equations (3) and (4) reveal that the census data demonstrate a strong correlation between the sanitation type used by households in Tonga and the source of water available to meet non-drinking water demands. They also show the marked improvement in STR relative to WSR between 2006 and 2016.

3.4. Sanitation Priorities in TSDFI

To support the Government's vision for Tonga, given in Section 2.3, the TSDFI identified nine Outcome Objectives which are listed in Table 4 together with the number of Associated Strategies (ASs) for each Outcome Objective.

Table 4. The Outcome Objectives and Associated Strategies of TSDFI. Shown in parentheses are the number of sub-Associated Strategies [17].

Outcome Objectives	Number of Associated Strategies
1. Strong inclusive communities	4
2. Dynamic public and private sector partnership as the engine of growth	2 (12)
3. Appropriate, well planned, and maintained infrastructure that improves the everyday lives of the people and lowers the cost of business	8
4. Sound education standards	2
5. Appropriately skilled workforce to meet the available opportunities in Tonga and overseas	2
6. Improved health of the people	2
7. Cultural awareness, environmental sustainability, disaster risk management, and climate change adaptation, integrated into all planning and implementation of programs	3
8. Better governance	2
9. Safe, secure, and stable society	2
Total	27(12)

Almost 30% of the ASs in Table 4 are linked to the Infrastructure Outcome Objective 3.

The word "sanitation" occurs 10 times throughout TSDFI. In 9 out of the 10 occurrences it occurs in the phrase "water and sanitation", emphasizing the intimate link between them. Table 5 compares the emphasis given to various services linked to infrastructure with that given to water and sanitation. Because of cross references in the ASs, several of the services are mentioned in more than one AS so the total of mentions is more than the number of ASs.

In Table 5, nearly 17% of the total references to services in ASs in TSDFI concern water and sanitation. This includes a mention of solid and liquid wastes under the wastes AS, part of which involves the disposal of septage from septic tanks. The emphasis on water and sanitation is equal to that for energy and for information and communications and half that given to transport.

Table 5. The number of references to different services in Associated Strategies under the Infrastructure Outcome 3 in TSDFI.

Service	References in Associated Strategies
Energy	2
Transport	4
Information and Communications	2
Wastes	1
Water	1
Water and Sanitation	2
Total	12

One of the 27 ASs (nearly 4%) solely concentrates on water and sanitation under Infrastructure Outcome Objective 3. TSDFI lists AS 13: "Maintaining and expanding access to safe water and sanitation for all communities." TSDFI provided performance indicators for this Strategy:

- Proportion of population using an improved drinking water source;
- Proportion of population using an improved sanitation facility;
- Installation of sewerage treatment plant (a septage waste disposal facility on Tongatapu); and
- Proportion of total water resources used.

TSDFI clearly identified water and sanitation as linked national priorities to be included by relevant ministries and agencies in their corporate plans and annual reports. In many PICs, sanitation falls under the Ministry of Health. TSDFI did not, however, include sanitation or hygiene under the Improved Health Outcome Objective 6 (Table 4). Hygiene is only mentioned once in TSDFI in a discussion of the necessity for water and sanitation legislation to ensure "improvement of the hygiene of sanitation facilities". There is no reference to WASH in TSDFI.

3.5. Sanitation Priorities in TSDFII

The seven desired NOs in TSDFII [16] are:

A. A more inclusive, sustainable, and dynamic knowledge-based economy.
B. A more inclusive, sustainable, and balanced urban and rural development across Island Divisions.
C. A more inclusive, sustainable, and empowering human development with gender equality.
D. A more inclusive, sustainable, and responsive good governance with law and order.
E. A more inclusive, sustainable, and successful provision and maintenance of infrastructure and technology.
F. A more inclusive, sustainable, and effective land administration, environment management, and resilience to climate and risk.
G. A more inclusive, sustainable, and consistent advancement of our external interests, security, and sovereignty.

The five pillars consist of three institutional pillars and the two input pillars:

Institutional Pillars:

1. Economic Institutions
2. Social Institutions
3. Political Institutions

Input Pillars:

4. Infrastructure and Technology Inputs
5. Natural Resource and Environment Inputs

These five pillars have 29 linked OOs.

Table 6 shows the number of OOs and SCs focused on services. All are listed under the Infrastructure NO E, except wastes, which was listed under the Environment NO F.

Table 6. The number of Organizational Outcomes and Strategic Concepts assigned to infrastructure services in TSDFII [16]. The total number of each are in parentheses.

Service	Organizational Outcome (29)	Strategic Concepts (153)
Energy	1	4
Transport	1	9
Information and Communications	1	9
Building and Structures	1	5
Wastes	1	4
Sanitation	0	0
Water Supply	0	1

Over 17% of the total OOs in TSDFII are the services listed in Table 6. Energy, transport, information and communications, building and structures, and wastes made up almost 3.5% each of the total OOs listed. Not one OO mentions water and/or sanitation in contrast to TSDFI which identified water and sanitation as a linked national priority under Infrastructure Outcome Objective 3.

Accompanying the OO are 153 SCs. Almost 21% of SCs involved the services in Table 6 but only one references water supply. Transport and information and communications each scored 6% of the SCs in contrast to water, 0.7%, and there is no mention of sanitation in any SC.

Under the natural resources pillar, within OO 5.2, "Improved use of natural resources for long term flow of benefits, SC b) records: "improve the management and delivery of safe water supply for business and households." There are no water supply key performance indicators (KPIs) associated with OO 5.2 SC b), so there is no way to measure performance. Since SCs are only considered an aid to planning there is no pressure for any ministry to address this SC.

Under the Health component of the Social Institutions Pillar, "Percentage of population with safe water supply" is listed as a KPI but has no associated SC or OO. Without this KPI being tied to an OO or an SC there is no guarantee that it would be addressed in ministry corporate plans or reports.

Sanitation is not mentioned in any of the 29 OOs or 153 SCs and it does not appear in any of the KPIs. The word "sanitation" only appears twice in TSDFII. In Annex 1 of TSDFII, "water and sanitation management" is mentioned as one of a long list of action topics of the Small Developing Island Countries Action Agenda in the SAMOA Pathway [3]. In Annex 2, TSDFII lists how NOs contribute to the United Nations SDGs. TSDFII claims that NOs F, E, and B, listed above, all contribute to fulfilling Tonga's commitments to SDG6: "Ensure availability and sustainable management of water and sanitation for all." There are no OOs or SCs under NO F, E, or B involving sanitation and water supply is only included as a single SC under NO F. TSDFII does not include any mention of hygiene or WASH.

The absence of water and sanitation from TSDFII after their inclusion in TSDFI is surprising and leaves the impression that water and sanitation are no longer sustainable development issues in Tonga despite the disparities between IDs revealed in the censuses.

3.6. Village Community Development Plans (CDPs)

Village CDPs are designated as plans. In practice, however, the available documents consist of ranked priorities of challenges to village development from women's, youths', and men's perspectives. There are no identified strategies for addressing challenges and no assignment of responsibilities or estimates of resources needed to address priorities. Table 7 provides details of the village CDPs that were available for analysis in each ID [19].

Table 7. The number of rural villages in 2016 [26], number of accessible village community development plans for Island Divisions in Tonga, and the medium number of identified priorities identified by women, youth, and men within each Island Division [19].

Island Division	Number of Rural Villages	Number of CDPs	Median No. of Priorities in Island Division		
			Women	Youth	Men
'Eua	12	13	8	7.5	7.5
Tongatapu	53	48	6	5	6
Vava'u	44	39	7	6	7
Ha'apai	27	5	9	7	10
Ongo Niua	12	12	5	5	5.5
Total	151	117			
	Country Median		7	6	7

In total, 117 CDPs of the original 136 that were presented in 2016 were available on-line [19]. These available CDPs are 77.5% of the 151 rural villages in Tonga (Table 7). CDPs for the main island Tongatapu, excluded the districts of Kolofo'ou and Kolomotu'a that make up Greater Nuku'alofa, so CDPs for Tongatapu in Table 7 represent rural areas of the main island. Table 7 reveals that only five of the 27 villages in Ha'apai had accessible CDPs. This may be due to TC Ian which severely damaged 75% of the housing stock on Ha'apai in 2014 [16].

Table 7 shows for Tonga as a whole, the median number of priority issues identified by women and men were the same while youth identified slightly less. Ongo Niua Island Division villages identified the least number of priority issues while Ha'apai Island Division villages identified the most. Youth in 'Eua identified slightly more priority issues than elsewhere. The maximum number of priorities identified for women, youth, and men in any village was 12, 11, and 12, respectively, while the minimum number was 4, 3, and 3.

3.7. Sanitation Priorities in Village CDPs

None of the women, youth, or men in any of the villages throughout Tonga's IDs identified sanitation as their highest priority concern. Only 4% of villages identified sanitation as their second priority with women (8%) nationally having a higher concern than men (3%). None of the youth in any ID rated sanitation as their second priority (0%). The highest average second priority concerns over sanitation were in Ha'apai (7%) and the lowest were in Tongatapu (1%). In Ha'apai women in 20% of villages rated sanitation as their second priority but none of the youth and men in Ha'apai rated sanitation as their second priority (0%).

Table 8 lists the percentage of villages that ranked sanitation within their top three priorities.

Table 8. The percentage of villages throughout Tonga's Island Divisions, as well as Tonga collectively, which listed sanitation within their top three priorities in terms of women's, youths', and men's perspectives and the average of all three.

Island Division	Percentage of Villages with Sanitation within the Top Three Priorities			
	Women	Youth	Men	Average
'Eua	8%	8%	0%	5%
Tongatapu	2%	0%	0%	1%
Vava'u	23%	3%	10%	12%
Ha'apai	40%	0%	40%	27%
Ongo Niua	0%	0%	8%	3%
Tonga	**15%**	**2%**	**12%**	**9%**

Overall, 9% of villages across Tonga in Table 5 placed sanitation within their top three priorities with women again showing more concern about sanitation than men. In contrast, youth in Tonga had little priority concerns about sanitation. The most concern about sanitation within the top three ranked issues was in the Ha'apai ID with both women and men showing equal concern (40%), in contrast to the absence of concern of youth. Tongatapu showed the least concern over sanitation (1%). Only women in 2% of the villages in Tongatapu ranked sanitation within the top three priorities.

Figure 6 plots the percentage of villages in IDs which identified sanitation as a concern in their list of all identified village priorities.

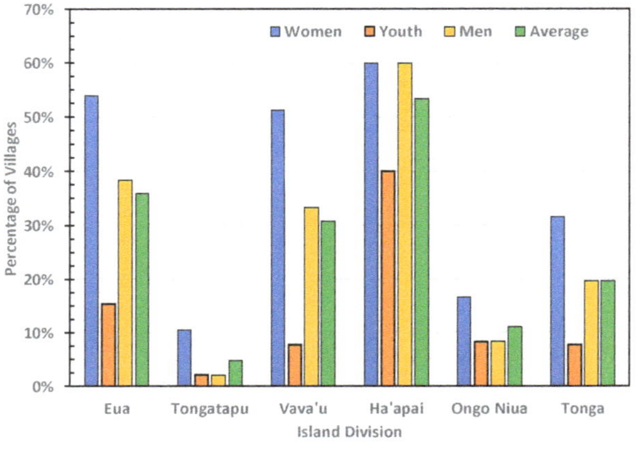

Figure 6. Percentage of villages in Island Divisions and in Tonga who ranked sanitation as a priority issue within their list of challenges to village development. Women's, youths', and men's priorities are given together with the average.

For Tonga as a whole, 20% of villages identified sanitation as a priority for village development. Women in Tonga expressed more concern about sanitation (32%) than men (20%) who were more concerned than youth (8%). The ID with the highest overall concerns about sanitation was Ha'apai (53%) with women and men having equal concerns (60%) and youth less (40%). The ID with the lowest number of villages identifying sanitation as a priority was Tongatapu (5%). There, women also expressed more concern (10%) than men or youth (both 2%). Women consistently identified sanitation as a concern in their village, more than men and youth, except in Ha'apai where women and men's concerns were equal.

The CDPs analyzed here were developed between 2007 and 2016, a period covered by both the 2006 and 2016 censuses. To determine if the distribution of the percentage of villages in IDs with percentage of priority sanitation concerns (PSC) is related to the Island distributions of STR (Figure 4), Figure 7 shows the relationship between PSC and STR.

The data in Figure 7 showed no significant relationships between PSC and STR for either census. The results, however, and the impact of a major tsunami on the ID suggest that the data for Ongo Niua may be outliers. If those results are omitted, a strong relationship is found for the other four IDs for the 2006 STR data:

$$\mathrm{PSC}(\%) = (36 \pm 4).\ \mathrm{STR} \quad (5)$$

Equation (5) has $R^2 = 0.96$ and is very significant ($p < 0.005$). The relationship for the 2016 STR data is:

$$\mathrm{PSC}\ (\%) = (68 \pm 10).\ \mathrm{STR} \quad (6)$$

Equation (6) has $R^2 = 0.93$ and is significant ($p < 0.01$). The difference in the coefficient between Equations (5) and (6) is due to the decrease in STR between 2006 and 2016 caused mainly by the large decrease in pit toilets.

Because of the strong relationship between STR and WSR, defined in Equations (3) and (4), there are also significant relationships between PSC and WSR for the 2006 Census results:

$$\text{PSC (\%)} = (80 \pm 16).\ \text{WSR} \tag{7}$$

Equation (5) has $R^2 = 0.89$ and is significant ($p < 0.02$). The relationship for the 2016 WSR data is:

$$\text{PSC (\%)} = (69 \pm 24).\ \text{WSR} \tag{8}$$

Equation (6) has $R^2 = 0.74$ and is weakly significant ($p < 0.07$). It is noted that within the standard errors of the coefficients in Equations (7) and (8), the two coefficients are not statistically different ($p > 0.1$). Combining the 2006 and 2016 data for all IDs except Ongo Niua results in the relationship:

$$\text{PSC (\%)} = (73 \pm 9).\ \text{WSR} \tag{9}$$

Equation (9) has $R^2 = 0.90$ and is highly significant ($p < 0.0001$).

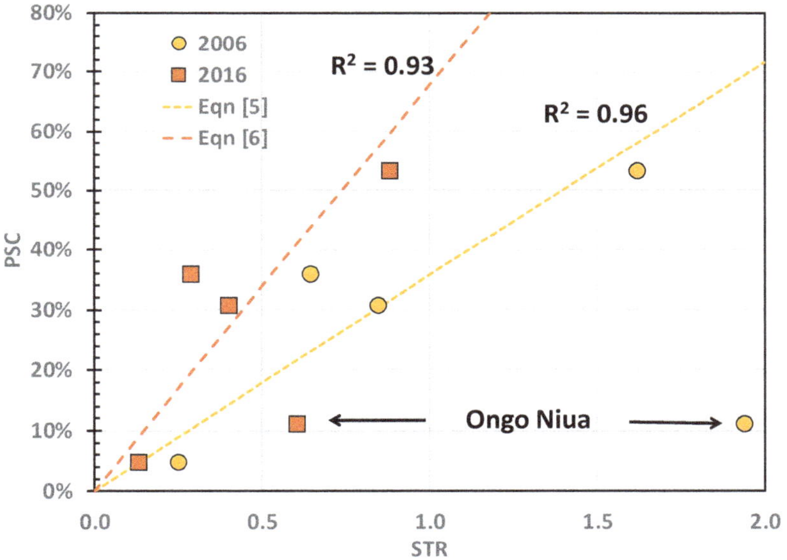

Figure 7. The relationship between the percentage of villages in each Island Division with priority sanitation concerns and the sanitation type ratio, defined in Equation (2), from the 2006 and 2016 censuses. The results for Ongo Niua are highlighted.

Equations (5)–(9) suggest that for four of Tonga's IDs, village priority concerns over sanitation are related to the type of sanitation being used which is related to whether water is being sourced from rain tanks or piped water supply, irrespective of which census results are used. The exception is Ongo Niua. It is noted in Figure 7 that Ongo Niua had the largest reduction in STR of any ID between 2006 and 2016, so that sanitation may have been of lesser relative concern there. These results suggest that, with the exception of Ongo Niua, the census results for sanitation type or water source used for non-drinking purposes provide pointers to village priority concerns over sanitation at the ID level. The special case of Ongo Niua will be discussed in the following.

4. Discussion

4.1. Information from Census Results

Data on water consumed by different uses, particularly sanitation, is sparse in most SIDS, which results in donors relying on average values when planning interventions or emergency relief [33]. Here we have explored the sort of information that can be gleaned about water sanitation from available census results in the specific case of Tonga with its multi-island IDs. The census results (Table 1) show population is concentrated in the main ID Tongatapu, with over a third of the country's population in the capital area, Greater Nuku'alofa. Slightly less than a quarter of the population is scattered across Tonga's other four IDs with population densities lower in northern islands.

Between 2006 and 2016, Tonga's overall population declined slightly but with significant internal shifts in population from the northern and southern-most IDs to Tongatapu and to Greater Nuku'alofa (Figure 2). This increased pressure on water supply and sanitation in the main island. This shift from outer islands to population centers is widespread across PICs [35], and is driven largely by the availability of services, including improved water supply and sanitation services in population centers. In Tonga, natural disasters such as TC Ian [16] and earthquakes and tsunamis [36] also contribute to the population movement.

The census data shows a national shift in water sources for non-drinking purposes between 2006 and 2016. The minimal and declining use of household wells is positive because of the potential for contamination of groundwater by local sanitation systems [10]. There has been increased access to piped water supply and decreased rainwater use, which is valued for drinking, but the shift is uneven (Table 2, Figure 3). In Vava'u and Ha'apai, household access to piped water has hardly changed over the decade but household access to rainwater supply increased in those Divisions. Since this water is used for washing, bathing, and toilet flushing it represents a considerable strain on rainwater harvesting systems, especially during the May to October dry season [10,27]. Even well-managed rainwater systems in Ha'apai are expected to run dry about one year in nine, compared with one year in 28 in Tongatapu and Vava'u and one year in 45 in Ongo Niua [20].

The main types of household toilets in use in Tonga are cistern flush, pour flush, and pit toilets (Table 3). There are no statistics on the use of compost toilets despite an early trial in Ha'apai [37]. Nationally, there has been an increase in household use of cistern flush toilets between 2006 and 2016, a corresponding decrease in pour flush toilets and a large 50% decrease in pit toilets, which is fairly uniform across all IDs and Greater Nuku'alofa. The decrease in pit toilets is particularly important for reducing local groundwater contamination and improving hygiene.

The STR, defined in Equation (2), was introduced as an approximate relative measure of sanitation systems which use less water compared to those requiring more. STR halved in Tonga between 2006 and 2016 with decreases occurring in all IDs, and particularly Ongo Niua. This demonstrates a national shift to sanitation systems with larger water use. Despite this shift, the STR for Ha'apai and Ongo Niua in 2016 remain high compared to other IDs and particularly to Greater Nuku'alofa.

One of the questions this work sought to answer was: "Do census data show a link between island sanitation types and island water sources?" While there are many factors that influence the choice by households of the type of sanitation system [23], particularly in PICs, an adequate supply of water appears crucial. We found very strong and highly significant ($p < 5.10^{-4}$) correlations between STR and WSR for all Tonga's IDs for both 2006 and 2016 census results (Equations (3) and (4)). This implies that, even in Tonga's outer islands, the move to improved sanitation systems is governed by access to adequate water supply. The change in STR revealed in the census data also implies that cistern flush toilets are the desired system by householders. The analysis of census data also showed continuing large differences in reliable water sources and improved sanitation systems available to households between the main island, Tongatapu including the capital region, Greater Nuku'alofa, and the other IDs.

One of the questions this work aimed to address was: "In the absence of detailed information, can census data be used to identify both improvements in sanitation and the diversity of sanitation needs in PICs?" Comparison of the 2006 and 2016 censuses has shown marked changes in sanitation types used and revealed IDs which continue to be disadvantaged relative to the main island. We will now consider if these changes and differences are reflected in TSDFI and TSDFII.

4.2. National Versus Local Sustainable Development Plans and Sanitation

TSDFI, spanning the period 2011 to 2014, recognized explicitly the intimate link between water and sanitation evident in the census results. Of its nine national Outcome Objectives, two are directly relevant to water and sanitation. About 30% of the 27 Associated Strategies in TSDFI concerned infrastructure services, and 17% of them are focused on improving access to safe water and sanitation under the Infrastructure Outcome Objective. There are no water or sanitation Associated Strategies or related performance indicators under the Improved Health Outcome Objective. This is surprising in Tonga. All village water supply systems as well as national sanitation are the responsibility of the Ministry of Health.

TSDFII, covering the period 2015 to 2025, was built on lessons learnt through TSDPI but its structure is more complex and its objectives and strategies more numerous. TSDFII abandoned the linkage between water and sanitation in TSDFI. It contains no mention of water and/or sanitation in any of its 29 Organizational Outcomes. This is despite the fact that TSDFII claims it contributes to Tonga's commitments to SDG6. Of the 153 Strategic Concepts raised in the three-month consultations leading to TSDFII, one concerned water supply but, again, there was no mention of sanitation. This means that ministries have no national obligation to include sanitation in corporate plans or reports. This appears to imply that while water and sanitation were a priority in TSDFI, with the improvements in water supply and sanitation evidenced in the 2006 and 2016 censuses, water supply and sanitation are no longer national priorities in TSDFII, despite the continuing disparity between IDs evident in the censuses.

The available local village CDPs, presented in 2016, were developed over a nine-year period. Ha'apai had the highest median number of priority development concerns, perhaps partly reflecting the impacts of TC Ian in 2014 [16], while Ongo Niua had the lowest (Table 7). Sanitation was not ranked highest priority concern by any village in Tonga, and only 9% of villages ranked sanitation within their top three priorities, with women more concerned than men and both much more concerned than youth. Ha'apai villages had a much higher concern over sanitation than in Tongatapu which had the lowest concerns (Table 8). Nationally, 20% of villages identified sanitation within their list of development priorities (Figure 6). The percentage of villages expressing concern in Ha'apai (53%) about sanitation was an order of magnitude higher that in Tongatapu (5%). Nationally, again women had a higher concern over sanitation as a development priority than men or youth.

The results for sanitation are in contrast to the priority rankings for water supply in the CDPs where 55% of villages identified water supply as highest priority [20]. The lower priority for sanitation appears to stem from the fact that traditionally in PICs, sanitation was either a personal or household responsibility. There is no reticulated sewage system in Tonga. Piped water supply, however, is a community responsibility through Village Water Committees or, in population centers, the Tonga Water Board.

Sanitation priority development concerns expressed by IDs in their CDPs were strongly correlated with the sanitation type and water supply ratios from the census data, Equations (5)–(8), provided the results for Ongo Niua are ignored. This implies that the census results may be useful in identifying sanitation priorities in NSDS processes, certainly at the ID level. The apparent anomalous results in Ongo Niua appear to be due to major improvements in sanitation systems there between 2006 and 2016. Ongo Niua will be further discussed below.

Another of the questions this work sought to answer was: "Do top-down NSDS plans give the same weight to sanitation priorities as that identified in nation-wide bottom-up village development plans and is there an evolution of priorities in NSDS?" It has been shown here that while TSDFI did recognize water and sanitation as national priorities in line with the village CDPs, TSDFII did not. Suggested reasons for this discrepancy follow.

4.3. Why is Ongo Niua an Outlier?

Unlike other IDs in Tonga, the apparent lack of village concerns over sanitation in the CDPs of the remote northern Ongo Niua did not match the water and sanitation metrics in the 2006 and 2016 censuses. Ongo Niua ID is made up of three islands Niuatoputapu, Niuafo'ou, and Tafahi. Niuatoputapu has 57% of the population of Ongo Niua (1232 in 2016) while Tafahi has 2.5%. In Niuafo'ou, household water supplies are from rainwater harvesting [32] and sanitation is mainly provided by pit toilets and pour flush toilets. Niuatoputapu has a piped groundwater supply as well as household rain tanks and households access cistern flush toilets as well.

On 30 September 2009, Niuatoputapu island was struck by an 8.3 magnitude earthquake whose epicenter was 190 km northeast of the island. Three tsunami waves up to 17 m high struck the island, penetrating over one kilometer inland and inundated about 46 percent of the island. This resulted in the death of nine people, the complete destruction of one third of the houses and all government buildings, water supply and sanitation infrastructure [36].

A World Bank post-tsunami reconstruction project [36], completed in 2014, moved houses inland, constructed 73 new houses and a new piped water system, reconstructed roads, improved household rainwater harvesting systems, and installed pedestal cistern flush toilets and septic tanks. The project noted that there were still some pit and compost toilets on the island.

It appears that this reconstruction project is the reason that the STR in Ongo Niua decreased dramatically between 2006 and 2016 (Figure 4) and is probably the reason for the low ranking of sanitation in village CDPs in Ongo Niua. The 2009 tsunami, 2014 TC Ian, 2017 TC Gita, and 2020 TC Harold re-emphasize the vulnerability of sanitation and water supply infrastructure in SIDS to frequent natural disasters.

4.4. The Mismatch between National and Local Development Planning for Sanitation

Four possible explanations are advanced for the absence of priority given to water and sanitation in TSDFII in 2015 compared with the emphasis in TSDFI. The first is that the significant national improvements shown in sanitation and water supply in the 2016 census were recognized prior to the census being conducted in 2016 so that water and sanitation were no longer considered a priority for national planning. The second is that, traditionally, sanitation has been a household/individual responsibility, as in all PICs, and with no reticulated sewerage system in Tonga it is not considered government business, even though the government has to dispose of septage. This, however, does not explain that water and sanitation was a linked priority in TSDFI. The third is that all ministries and most of the government agencies with responsibility for TDSFII are located in Tongatapu and mostly in Greater Nuku'alofa where water supply is continuous and relatively plentiful and, as a consequence, improved cistern flush sanitation systems are widely in use. The fourth is that TDSFI was based on nine months of consultations in all IDs whereas TDSFII relied on consultations over three months in four of the five IDs.

4.5. Sanitation in Islands Reliant on Rainwater Harvesting

We found IDs with higher dependence on rainwater harvesting have generally more village-level concerns about the type of household sanitation system than in islands with greater access to piped water. The exception was Ongo Niua, where, in some islands, their volcanic geology means that groundwater appears absent or is very limited. Restricted rainwater storage, small roof areas, lack of maintenance, six month dry seasons, ENSO- and PDO-related droughts, and significant household demand mean most rainwater harvesting systems are not able to supply reliably sufficient water for flush toilets.

Several solutions have been trialed in small remote island countries and a range of zero or minimal water use systems are available [23]. Pit toilets do not use water and are relatively cheap, but they are regarded as less convenient and do pollute groundwater. Compost toilets were trialed in Lifuka in the Ha'apai ID [37], but they are not included in census surveys. CDPs coupled with census results show a general preference for flush toilets which are more convenient and require less upkeep than compost toilets. The use of seawater or brackish water for flushing has been used in high population density atolls either from reticulated supply systems or local wells. Leakage from saline reticulation systems can salinize fresh groundwater [10]. Desalinated water can also be used to provide water for flushing. However, desalination systems are expensive to operate and difficult to operate and maintain in remote outer islands [10].

While a number of economic, social, cultural, environmental, technical, and cultural factors influence the choice of household sanitation systems [23], we have shown here that an adequate supply of water is a major factor. Remote island communities look to the population centers and see flush toilets as the standard. Indeed, in Tonga, cistern flush toilets discharging into septic tanks are the desired goal. This is despite the fact the normal construction of septic tanks with concrete blocks inevitably leads to leakages with the potential to pollute soil and groundwater. Molded plastic septic tanks appear a safer but more expensive option [10].

4.6. Sanitation, a Dilemma for Island Governments

The dramatic change in emphasis on water and sanitation between TSDFI and TSDFII is evidence of a dilemma many PIC governments face. Water supply traditionally was the responsibility of the extended family, while sanitation was largely a personal matter which has evolved to a household responsibility. The transition to modernity in PICs has resulted in high density urban centers where water and sanitation are often government responsibilities but with lower density rural areas where tradition is often still strong. Addressing the diversity of needs in disparate situations equitably and responding to international programs and agreements such as SDG6, which focus on government responsibilities in sanitation, counter long-held traditional thinking.

In Tonga the situation is more complex. Most households prefer rainwater for drinking, even in the capital area, with government agencies or Village Water Committees supplying piped groundwater for other uses, where possible. Households, therefore, still play a role in water supply, which is more prominent in rural communities. Even in the capital and population centers, sanitation is a household responsibility with government responsible for collecting and disposing of septage in some areas. So, there is uncertainty about the appropriate role of government in sanitation as evidenced by its absence as a national priority from TSDFII. Clearly, implementing improved health, building, and environmental standards nationally are key government functions in sanitation and the facilitation of improved sanitation facilities through micro-financing schemes is a possibility [23]. Here we have shown that rural women are more concerned about sanitation than men or youth. Perhaps the election of more women to parliament would sharpen the national focus on sanitation?

4.7. Limitations of this Work

This work only examined aggregated ID level data for the census and CDP results. Both contain a wealth of information at the district and village level which could better inform both planning and intervention. We have not used here more sophisticated sanitation or WASH indices [21–23] to identify differences between IDs. This is due to the limited information available at the ID, district, and village level. Even per capita water consumption is not available except for the capital and population centers [27], and a handful of villages [30–32]. The recent passage of Tonga's Water Resources Act 2020 provides stimulus for improved data collection across the country.

The crude measures we have used here, WSR and STR, have inbuilt assumptions. The WSR assumes that all piped water systems in IDs are comparable. They are not. Greater Nuku'alofa and major centers in three other islands have access to a continuous supply of water, but in most rural villages piped water systems operate intermittently requiring household storage for continuous use. The STR assumes that flush toilets are what island communities desire irrespective of the adequacy of water supply. In the case of Ongo Niua, the large decrease in STR documented reconstruction efforts following a major natural disaster.

5. Conclusions

WASH has been identified as a major continuing national challenge in PICs [9] and improved governance is seen as a key step in addressing that challenge [14]. National sustainable development strategies and plans have been advocated as efficient governance instruments to identify overarching national priorities, select appropriate solutions, allocate responsibilities and resources, and fulfil international and regional commitments, especially in sustainable development. Inevitably, they are largely top-down processes. Transplanted governance institutions from one country may not suit another and standardized processes or formulaic approaches may not be apposite in PICs [6,20], particularly those with village communities dispersed over islands spread across large ocean and with a wide range of development challenges and traditional practices.

PICs have ideal strengths for bottom-up processes [7]. One of the inherent drawbacks of bottom-up processes is the inevitable time and cost penalties they incur. The Kingdom of Tonga provided the opportunity to examine how both top-down and bottom-up planning processes handle sanitation and water at the Island Division level. Consultations leading to TSDFII took three months, TSDFI took nine months, while the community consensuses building leading to CDPs evolved over nine years.

The previous two censuses in Tonga provide a clear picture of the household use of sanitation types and water sources, the changes in usage over time, and show large variations between the main island Tongatapu, especially in the capital area, and the other Island Divisions. The census data revealed that the type of sanitation system chosen is strongly correlated with access to piped water supply at the Island Division level. Censuses provide valuable data about differences in sanitation type and water sources which correlated well with aggregated village sanitation priorities in most Island Divisions. These relationships may also be evident in other PICs. The changes in sanitation metrics also revealed responses to natural disasters, so frequent in SIDS.

The first Tonga Strategic Development Framework, TDSFI, from 2011 to 2014 acknowledged the linkage between sanitation and water and their importance under an Infrastructure Outcome Objective. There is no recognition in the current TDSFII of that linkage and no mention of improved water and sanitation services under the Infrastructure Pillar or within the Health Outcome Objectives. The available bottom-up village CDPs show a markedly different set of priorities for water [20] and sanitation across Island Divisions, with the main Island Division being largely satisfied with the sanitation available while outer islands, and particularly Ha'apai, had significant development concerns over sanitation. One of the strengths of the CDP process was that it provided women and youth with the opportunity to voice their concerns as well as men who are normally the only group consulted. Overall, village CDPs showed higher concerns over sanitation by women than men with youth the least concerned.

A question raised here is: Can NSDS in multi-island countries be improved for sanitation and water outcomes? We have found here a major difference in the national priority given to sanitation and water in TSDFI and the diverse sanitation priorities expressed in village CDPs, especially by women, and their total absence in TSDFII. TSDFI reflects the sanitation data in the 2006 census. TSDFII does not acknowledge the large differences in sanitation systems evident in either the 2006 or 2016 censuses. Instead, TSDFII appears to reflect the lack of village priority concerns over sanitation found in village CDPs from the main island, Tongatapu.

Refining census questions to better target sanitation types and sources and reliability of water supply would assist NSDS processes. Synchronizing NSDS processes to follow and draw on the census data would allow the census results to be integrated into NSDS national priorities. Community development planning processes at the village level should be continuing processes, rather than a one-off process, drawing on local strengths and run by the villages so that their results can be fed directly into NSDS. The advantage of TSDFI was that it had a less complex structure with objectives identifiable at the village level. A simple structure concentrating on island-relevant issues such as:

- Health
- Food supply, fisheries, and agriculture
- Land and marine environments, climate, and extreme events
- Infrastructure services
- Economic opportunities and employment
- Education and training
- Culture and security
- Governance and international relations

could be used as a template for village CDPs to feed directly into NSDS, with water and sanitation being covered under infrastructure, health, environment, and governance.

In the four years since their presentation in 2016, none of the priorities in CDPs had been analyzed until now. There is a pressing need for other identified priorities in the CDPs to be analyzed and incorporated into a revised TSDFII and updated ministry operational plans and reports. Without that, villagers will lose confidence in the relevance of both national and local planning processes.

We have had here the opportunity to analyze the priorities given to water and sanitation in top-down NSDSs and nation-wide, bottom-up village level Community Development Plans. The differences found here raise the question of whether NSDSs in other Pacific Island countries have the same mismatches and have to cope with similar differences between urban, rural and outer island development priorities.

Supplementary Materials: The following are available online at http://www.mdpi.com/2071-1050/12/22/9379/s1, Table S1: Sanitation Priorities in Community Development Plans, Tonga.

Author Contributions: Conceptualization, I.W., T.F. and T.K.; methodology, I.W., T.F. and T.K.; formal analysis, T.F. and I.W.; writing—original draft preparation, I.W.; writing—review and editing, I.W., T.F. and T.K.; project administration, T.K. All authors have read and agreed to the published version of the manuscript.

Funding: This research received no external funding.

Acknowledgments: This work arose as a result of a strategic planning project supported by the World Meteorology Organization under its Climate Risk and Early Warning Systems Initiative for Pacific Island Countries focusing on improved governance. WMO Pacific are thanked for that support. We are grateful to Karin Nagorcka for editorial assistance. The authors thank an unnamed reviewer whose comments improved this paper.

Conflicts of Interest: The authors declare no conflict of interest.

References

1. UN. *Report of the Global Conference on the Sustainable Development of Small Island Developing States (SIDS), Bridgetown, Barbados, 25 April–6 May 1994*; United Nations: New York, NY, USA, 1994; p. 77. Available online: https://undocs.org/pdf?symbol=en/A/CONF.167/9 (accessed on 9 September 2020).
2. UN. *Report of the International Meeting to Review the Implementation of the Programme of Action for the Sustainable Development of Small Island Developing States. Port Louis, Mauritius, 10–14 January 2005*; United Nations: New York, NY, USA, 2005; p. 112. Available online: https://www.un.org/ga/search/view_doc.asp?symbol=A/CONF.207/11&Lang=E (accessed on 24 July 2020).
3. UNGA. *Report of the Third International Conference on Small Island Developing States. Apia, Samoa, 1–4 September 2014*; UN General Assembly: New York, NY, USA, 2014; p. 70. Available online: http://www.un.org/ga/search/view_doc.asp?symbol=A/CONF.223/10&Lang=E (accessed on 30 September 2020).
4. WHO; UNICEF. *Progress on Drinking Water, Sanitation and Hygiene: 2017 Update and SDG Baselines*; World Health Organization and the United Nations Children's Fund: Geneva, Switzerland, 2017; p. 114. Available online: https://www.unicef.org/publications/files/Progress_on_Drinking_Water_Sanitation_and_Hygiene_2017.pdf (accessed on 15 October 2020).

5. UNDESA. *Report: Workshop on National Sustainable Development Strategies in Pacific Island Developing States. New York, NY, USA, 4–5 May 2006*; Division for Sustainable Development, United Nations Department of Economic and Social Affairs: New York, NY, USA, 2006; p. 48. Available online: https://www.un.org/esa/sustdev/natlinfo/nsds/workshop/final_report.pdf (accessed on 21 September 2020).
6. Lamour, P. *Foreign Flowers: Institutional Transfer and Good Governance in the Pacific Islands*; University of Hawaii Press: Honolulu, HI, USA, 2005; p. 220.
7. Barnett, J. Titanic states? Impacts and responses to climate change in the Pacific Islands. *J. Int. Aff.* **2005**, *59*, 203–219. Available online: https://www.jstor.org/stable/24358240 (accessed on 20 October 2020).
8. Daniell, K.A. *Co-Engineering and Participatory Water Management, Organisational Challenges for Water Governance*; UNESCO IHP International Hydrology Series; Cambridge University Press: Cambridge, UK, 2012; p. 215.
9. WHO; SOPAC. *Sanitation, Hygiene and Drinking-Water in the Pacific Island Countries—Converting Commitment into Action*; World Health Organization and Pacific Islands Applied Geoscience Commission: Suva, Fiji, 2008; p. 72. Available online: https://www.pacificwater.org/userfiles/file/water%20publication/PacificReport_Final%20Version%206.pdf (accessed on 20 October 2020).
10. Falkland, T.; White, I. Freshwater availability under climate change. In *Climate Change and Impacts in the Pacific*; Kumar, L., Ed.; Springer Nature: Cham, Switzerland, 2020; Chapter 11, pp. 403–448. [CrossRef]
11. UNICEF; WHO. *Progress on Sanitation and Drinking-Water—2015 Update and MDG Assessment*; United Nations International Children's Emergency Fund/World Health Organization: New York, NY, USA, 2016; p. 90. Available online: https://apps.who.int/iris/bitstream/handle/10665/177752/9789241509145_eng.pdf;jsessionid=677BFB8F2BA902C267BE57C04203E1E9?sequence=1 (accessed on 30 September 2020).
12. ABoM; CSIRO. Climate Variability, Extremes and Change in the Western Tropical Pacific: New Science and Updated Country Reports. In *Pacific-Australia Climate Change Science and Adaptation Planning Program Technical Report*; Australian Bureau of Meteorology and Commonwealth Scientific and Industrial Research Organisation: Melbourne, Australia, 2014; p. 372. Available online: https://www.pacificclimatechangescience.org/wp-content/uploads/2014/07/PACCSAP_CountryReports2014_WEB_140710.pdf (accessed on 20 October 2020).
13. Diamond, H.J.; Lorrey, A.M.; Renwick, J.A. A Southwest Pacific tropical cyclone climatology and linkages to the El Niño–Southern Oscillation. *J. Clim.* **2013**, *26*, 3–25. [CrossRef]
14. UNDP Water Governance Facility/UNICEF. WASH and Accountability: Explaining the Concept. In *Accountability for Sustainability Partnership*; UNDP Water Governance Facility at SIWI and UNICEF: New York, NY, USA, 2015; p. 30. Available online: http://www.watergovernance.org/ (accessed on 15 October 2020).
15. National Sustainable Development Strategies. Available online: https://www.un.org/esa/sustdev/natlinfo/nsds/pacific_sids/pacific_sids.htm (accessed on 21 July 2020).
16. Government of Tonga. *Tonga Strategic Development Framework (2015–2025)*; Ministry of Finance and National Planning: Nuku'alofa, Tonga, 2015; p. 130. Available online: http://www.finance.gov.to/tonga-strategic-development-framework (accessed on 4 September 2020).
17. Government of Tonga. *Tonga Strategic Development Framework, TSDF, 2011–2014*; Ministry of Finance and National Planning: Nuku'alofa, Tonga, 2011; p. 42. Available online: http://www.adb\T1\textgreater{}files\T1\textgreater{}cobp-ton-2014-2016-oth-02 (accessed on 30 September 2020).
18. Pacific Islands Report. *Tonga Launches First Community Development Plans*; East-West Center: Honolulu, HI, USA, 2016. Available online: http://www.pireport.org/articles/2016/10/13/tonga-launches-first-community-development-plans (accessed on 25 July 2020).
19. Tonga Local Government; Ministry of Internal Affairs. Tonga Community Development Plans. Available online: https://www.tongalocal.gov.to/community-development-plans (accessed on 2 September 2020).
20. White, I.; Falkland, A.F.; Kula, T. Meeting SDG6 in the kingdom of Tonga: The mismatch between national and local sustainable development planning for water supply. *Hydrology* **2020**, *7*, 81. [CrossRef]
21. Schwemlein, S.; Cronk, R.; Bartram, J. Indicators for monitoring water, sanitation, and hygiene: A systematic review of indicator selection methods. *Int. J. Environ. Res. Public Health* **2016**, *13*, 333. [CrossRef] [PubMed]
22. Cronk, R.; Luh, J.; Meier, B.M.; Bartram, J. *The Water, Sanitation, and Hygiene Performance Index: A Comparison of Country Performance in Realizing Universal WASH*; University of North Carolina: Chapel Hill, NC, USA, 2015; Available online: http://waterinstitute.unc.edu/wash-performance-index-report/ (accessed on 24 October 2020).

23. Hashemi, S. Sanitation sustainability index: A pilot approach to develop a community-based indicator for evaluating sustainability of sanitation systems. *Sustainability* **2020**, *12*, 6937. [CrossRef]
24. TSD. *Census of Population and Housing: Administrative Report and Basic Tables, January 2009*; Tonga Statistics Department: Nuku'alofa, Tonga, 2009; Volume 1, p. 254. Available online: https://tongastats.gov.to/census/population-statistics/#22-25-wpfd-2011-1498194829 (accessed on 30 September 2020).
25. TSD. *Census of Population and Housing: Basic Tables and Administrative Report, October 2017*; Tonga Statistics Department: Nuku'alofa, Tonga, 2017; Volume 1, p. 259. Available online: https://tongastats.gov.to/census/population-statistics/ (accessed on 9 September 2020).
26. Thompson, C.S. *The Climate and Weather of Tonga*; New Zealand Meteorological Service, Te Ratonga Tirorangi: Wellington, New Zealand, 1986; p. 60.
27. WRMS. *National Water Resources Report (Draft)*; Water Resources Management Section, Natural Resources Division; Prepared for Climate Resilience Sector Project, Tonga, ADB Grant (03-78 TON); WRMS: Nuku'alofa, Tonga, 2019; p. 97.
28. CIA. The World Factbook. In *Australia-Oceania: Tonga*; US Central Intelligence Agency: Washington, DC, USA, 2020. Available online: https://www.cia.gov/library/publications/the-world-factbook/geos/tn.html (accessed on 15 September 2020).
29. ADB. Technical Assistance Consultant's Report. In *Tonga: Integrated Strategic Planning, Medium-Term Fiscal Framework and Budgeting*; Asian Development Bank: Mandaluyong, Philippines, 2006; p. 204. Available online: https://www.adb.org/sites/default/files/project-document/68408/38159-ton-tacr.pdf (accessed on 15 September 2020).
30. WRMS. *Integrated Water Resources Management Plan for Koloa, Vava'u*; Water Resources Section, Natural Resources Division, Ministry of Lands and Natural Resources; Prepared for Climate Resilience Sector Project; ADB Grant (03-78 TON); WRMS: Nuku'alofa, Tonga, 2018; p. 43.
31. WRMS. *Integrated Water Resources Management Plan for Nomuka, Ha'apai*; Water Resources Section, Natural Resources Division, Ministry of Lands and Natural Resources; Prepared for Climate Resilience Sector Project, Tonga, ADB Grant (03-78 TON); WRMS: Nuku'alofa, Tonga, 2018; p. 68.
32. WRMS. *Integrated Water Resources Management Plan for Niuafo'ou, Ongo Niua*; Water Resources Section, Natural Resources Division, Ministry of Lands and Natural Resources; Prepared for Climate Resilience Sector Project, Tonga, ADB Grant (03-78 TON); WRMS: Nuku'alofa, Tonga, 2018; p. 77.
33. World Health Organization. *Chapter 9, Technical Notes on Drinking Water, Sanitation and Hygiene in Emergencies*, 2nd ed.; Prepared by Water, Engineering and Development Centre; Loughborough University: Leicestershire, UK, 2013; p. 4.
34. Fielea, Q.V.; (Tonga Water Board, Nuku'alofa, Kingdom of Tonga). Personal communication, 2020.
35. Ward, R.G. *Widening Worlds, Shrinking Worlds? The Reshaping of Oceania*; Pacific Distinguished Lecture edition (1 January 1999); Centre for the Contemporary Pacific, Australian National University: Canberra, Australia, 1999; p. 38.
36. TWB. *Implementation Completion and Results Report (Ida-H6280) on an Ida Grant in the Amount of Sdr 3.32 million (US$ 5.0 Million Equivalent) to the Kingdom of Tonga for a Tonga Post Tsunami Reconstruction Project*; The World Bank, Sustainable Development Department, East Asia and Pacific Region: Washington, DC, USA, 2014; p. 50. Available online: http://documents1.worldbank.org/curated/en/684061468111888947/pdf/ICR30360P120590IC0disclosed07010140.pdf (accessed on 20 October 2020).
37. Crennan, L. *Composting Toilet Trial, Lifuka, Tonga, Final Report*; Prepared for Tonga Water Board Institutional Project, April 1999; Institute for Sustainable Futures, University of Technology: Sydney, Australia, 1999; p. 29.

Publisher's Note: MDPI stays neutral with regard to jurisdictional claims in published maps and institutional affiliations.

sustainability

MDPI

Article

The Synergy of Ecosystems of Blue and Green Infrastructure and Its Services in the Metropolitan Area—Chances and Dangers

Rafał Blazy [1,*], Hanna Hrehorowicz-Gaber [1], Alicja Hrehorowicz-Nowak [1] and Arkadiusz Płachta [2]

1 Department of Spatial Planning, Urban and Rural Design, Faculty of Architecture, Cracow University of Technology, 31-155 Kraków, Poland; hanna.hrehorowicz-gaber@pk.edu.pl (H.H.-G.); h.alicja@o2.pl (A.H.-N.)
2 Department of Climatology, Institute of Geography and Spatial Management, Jagiellonian University, 30-387 Kraków, Poland; arkadiusz.plachta@doctoral.uj.edu.pl
* Correspondence: rblazy@pk.edu.pl

Abstract: Ecosystems have become synanthropized, and the degree of their transformation depends on their susceptibility to anthropopressure, but they are necessary for the functioning of the anthropogenic environment. They provide many ecosystem services, yet they are often not protected in any way, and their value is not taken into account at all in the process of creating local development plans. The analysis of the blue and green infrastructure covered three municipalities: Łapanów, Gdów, and Dobczyce. To calculate the benefits of ecosystem services, the method of calculating the Ve coefficient was adopted, which would enable a more accurate financial evaluation of the local development plan and make the previously synthetic economic coefficient of net present value (NPV) real. Besides, the impact of water bodies on the financial benefits of ecosystem services was analyzed. Only the protection of ecosystems introduced by including it in the local development plan enables full ecosystem synergy. Next to anthropological ecosystems, there are also natural ecosystems, which are necessary for the proper functioning of the commune. The network of those includes green (in the case of vegetated areas) and blue (in the case of surface waters) infrastructure, and their synergy is the key to the sustainable development of the commune.

Keywords: blue infrastructure; spatial planning; urban planning; ecosystem services

Citation: Blazy, R.; Hrehorowicz-Gaber, H.; Hrehorowicz-Nowak, A.; Płachta, A. The Synergy of Ecosystems of Blue and Green Infrastructure and Its Services in the Metropolitan Area—Chances and Dangers. *Sustainability* **2021**, *13*, 2103. https://doi.org/10.3390/su13042103

Academic Editor: Shervin Hashemi

Received: 7 January 2021
Accepted: 11 February 2021
Published: 16 February 2021

1. Introduction

A modern city in the conditions of Atlantic-European culture is usually a product of the planning process. Therefore, when considering the problem of synergism in the perspective of urban issues, one should, first of all, refer to the relationship between the structure of the environment and planning, especially in such unique ecosystems [1] as the Carpathians. In this regard, the structural plan should perform the function of aligning individual functional components of the city, determining the mutual relations of elements, and setting the directions of their development and the principles of cooperation between individual units, areas, and strategic points. In this case, for cities with a significant cultural and environmental potential [2], systemic solutions allowing for the cohesion of elements of structures created by humans with green and blue infrastructure are important. Preventing fragmentation and isolation of individual elements seems to be crucial. From the point of view of synergy, the greatest integration should take place in key places where all elements are layered and where additive integration should be sought at all levels of spatial and functional connections.

1.1. Synergy as a Word Concept for Urban Ecosystems

Synergy is a special form of coexistence and interdependence, characterized by the openness of individual elements and the exchange between them. This openness should be primarily functional [3] but also spatial of composed forms, elements, or urban assumptions. Absence of unnecessary barriers and spatial boundaries between individual objects of the created structure characterizes the easy accessibility of individual functions.

Thinking about synergy, we focus on acceleration and activation synergy, in which dynamic actions play an important role, about constitutive and harmonizing combat. Cooperation on a given team increases the level of its performance. The concept of synergy is often encountered in various revitalization programs, where it is treated as the so-called "Multiplier effect" [4] or "leverage effect" [5]. Sometimes we have the "apollo syndrome" [6], which is an effect of negative synergy in which one of the elements, due to its position or a joint attempt, subordinates other elements of the system.

In Polish cities, despite the efforts of urban planners, the investment pressure disrupts the synergy by dominating the other elements, including the natural environment. The most appropriate seems to be the synergy of equivalent components or micro components, i.e., insignificant, minor, and average elements. Additive synergism occurs when an effect results from the summation of the actions of individual ingredients. We can also talk about hyperadditional synergism [7] when there is a significant intensification of the interaction of components. It is worth transferring these general principles of synergism and considering each with individual adjacent functions and urban spaces. However, it is difficult to assess the degree of influence or intensification of the action of one element on the other. Taking into account the common urban planning theories and a certain canon of planning rules, activities of a synergistic nature should include joining (also in the sense of preventing fragmentation), coordination, activities leading to an increase in the effectiveness of individual elements (environment, human, and structure), and the aforementioned integration leading to the complementarity of the selected structure.

1.2. Ecosystems and Their Importance for Urban Areas

Urbanization is progressing and expanding around the world, but human survival depends on nature. Cities and their prosperity depend on the surrounding ecosystems, but they are also based on the ecosystems in the city itself. The definition of blue-green infrastructure includes a network of natural and semi-natural solutions with multiple functions. It takes into account many forms of retention, including ponds, basins, depressions of the land, reservoirs, and rain gardens. On the one hand this serves the purpose of rainwater management, on the other it serves purification, green and wetland areas, and other ecosystem services. These are the services provided by the environment for human use and which we can distinguish and define economically. Bolund and Hunhammar [8] identify seven different urban ecosystems in their research: street trees, lawns and parks, urban forests, farmland, wetlands, lakes/sea, and streams, and they identify six ecosystem services that are relevant and interact locally and directly to Stockholm and its population. Among these services, they list air filtration, microclimate, noise reduction, rainwater drainage, wastewater treatment, and recreational and cultural values.

The essence and importance of internal ecosystems in the city were discussed in the strategic documents of the European Union, which indicates the need to implement green-blue infrastructure to improve the continuity of green structures and rainwater management due to economic, social, and environmental benefits. In May 2016, the Urban Agenda for the EU was created with the Amsterdam Pact. The EU Urban Agenda approaches policy and legislation in EU cities in an integrated and coordinated way. It aims to boost economic growth, make life more attractive, and innovate European cities. It also deals with the effective identification and solving of social problems in their area. The EU Urban Agenda is an important framework in preventing urban fragmentation and urban sprawl. It is a tool for achieving the goals of sustainable development [9].

Internal relations between elements of the urban structure and ecologically active areas should lead to harmony and synergy, i.e., understanding and cooperation on a given level, which in the case of individual entities may take place through interdependence which in this case is not a negative feature. Schematically, the stratification of the individual components of the synergistic structure of the city is presented in the figure below.

2. Materials and Methods

The strongest changes in phytocoenosis and zoocenosis and their interrelationships are observed in anthropogenic urban ecosystems [10]. It seems particularly important to maintain correct interconnections in areas of natural value, such as the Carpathians, where the environment is a value of a transboundary nature. There are 200 municipalities in the Polish Carpathians, including about 10 cities and over 50 towns. Many of them, especially those in the area of influence of larger cities, are subject to structural transformations, most often at the expense of green areas, waters, and forests (i.e., open areas). For research purposes, three locations in the Carpathian foothills in the area of influence of a large city of Krakow—Gdów, Dobczyce, and Łapanów (Figure 1)—were compiled. These three sites were also selected because of the connecting feature in an ecological aspect: they are the catchment areas of the Raba River. This is important since the river basin can be a reference field for mapping ecosystem services. Sub-catchments are one of the seven basic reference fields according to R.U. Syrbe and U. Walz (2012) [11] which are

- single patches, spatial landscape elements;
- "least common geometric unit" is generated automatically in GIS by superimposing maps of different components;
- administrative units;
- sub-catchments of rivers (or higher-order);
- natural units (soil, vegetation, etc.);
- landscape units;
- regular artificial geometric units (e.g., raster mesh).

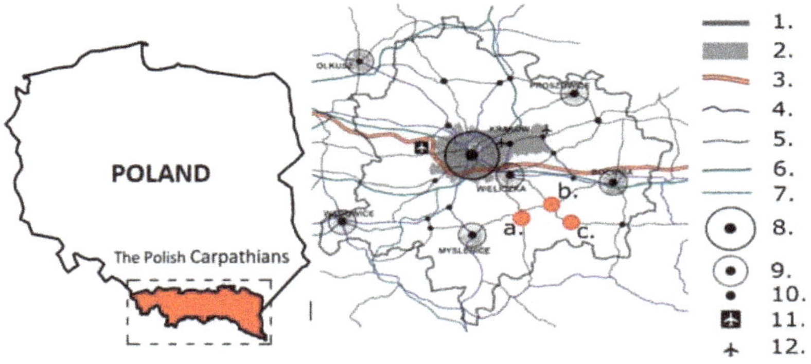

Figure 1. Locations considered in the study. a. Dobczyce; b. Gdów; c. Łapanów; 1. border of the Krakow metropolitan area; 2. the main urban centers; 3. highway; 4. national and provincial roads; 5. railway lines of national importance; 6. and 7. railway lines of regional importance; 8. a transportation hub of national importance; 9. a transportation hub of regional importance; 10. a transportation hub of local importance; 11. passenger airports; 12. sport airports. Source: www.Dobczyce.pl/planowanieprzestrzenne (accessed on 27 January 2021).

The study was made based on acts of local law MPZP (Local Spatial Development Plan) and municipal data regarding financial forecasts for the presentation of the Local Spatial Development Plan. The forecast of financial effects [12–14] did not take into account the revenues from ecosystem services; therefore, the thesis was put forward that the neglected financial revenues from ecosystem services should be included in the process of creating the financial forecast [15].

2.1. Ecosystem Services and Their Importance in Terms of Blue Infrastructure

The basis for economic water management , capture, and not quick discharge—as is the case in urbanized and urbanizing areas—which transfers the effects of floods and inundations to the areas below. The blue infrastructure aims to improve local retention by managing the rainwater where it is produced, in the area where rainfall has occurred supporting the traditional drainage system, e.g., in the event of sudden rainfall.

Gray infrastructure should complement the blue-green infrastructure and include hydro-technical infrastructure aimed at collecting and draining water, consisting of a sewage system that collects water from roads, squares and buildings, storm collectors and a sewage treatment system. Dobczyckie Lake (Figure 2) is a dam reservoir, located about 30 km from Krakow and in the town of Dobczyce, a town with a population of several thousand, in the Małopolskie Voivodeship, in the area affected by the Kraków agglomeration. The reservoir was created in 1986 by damming the Raba's waters with 30 m high and 617 m long dam. The reservoir has an area of approx. 10.7 km^2 and a tota capacity of 127 million m^3.

Figure 2. A view of the Dobczyce Lake. Source: author Krzysztof Cabak.

The green ecosystem and the water environment perform social and health functio Outdoor recreation places are created or made more attractive, and health conditio are improved. Even a small number of plants can produce a relatively large amou of oxygen [16,17]: 155 m^2 of green area produces enough oxygen for a person's dai oxygen demand. Greenery absorbs CO2 and dust, even as it cleans the air of harmf chemicals. The advantages of this are especially noticeable in cities on hot days becau they experience the so-called "Islands of Heat" [18]. The more urbanized the structu with a smaller percentage of the biologically active area, the clearer the effects. They a created by replacing natural green areas with buildings, pavements, and concrete surfac that absorb and retain heat. The local temperature drops by 2–3 °C in the place of green water areas.

Greenery and water in the city are also natural, landscape, and spatial functio Benefits for nature include increasing biodiversity through plant development and creati habitat conditions for insects and other species of small animals, including insects and a phibians, as well as water management in containers, ponds, and basins. Caring for spat values must be reflected in planning records. They must contain prejudices concerni both the shaping of the structure of buildings and the development of building plots a open areas. What is important for the Carpathian landscape is the multifaceted synergy the internal elements of the urban structure with the surrounding Carpathian landscap

Greenery in the city plays an important decorative role; ponds and basins w wetlands store and purify water from urban pollution, as well as improve the aesthetic the local landscape. Green roofs that capture rainwater in the limited space of the city act as corridors necessary to ensure landscape continuity; they can also have an aesth function because they are a distinct green decorative element that can be used to m unattractive places. The green and blue infrastructure are fully correlated with activities the development of tourism in the Raba Valley. Investments implemented in the Dobcz Commune as part of the Raba Open River project (Figure 3) will serve to improve tourist infrastructure along the Raba River by eliminating the existing barriers to transport infrastructure and increasing the accessibility of the river embankment for tourists. The preparation of a diverse offer of tourist products will extend the stay of tourists.

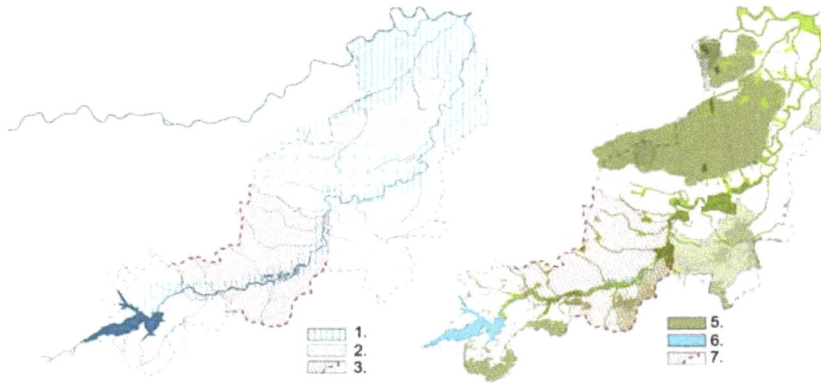

Figure 3. Schema of the green and blue infrastructure in the upper reaches of the Raba River. 1., 2.—blue infrastructure; 3., 7.—zone of influence of recreational ecosystem services, 5.—green infrastructure, 6.—blue infrastructure. Source: authors' materials.

Monetary valuation of ecosystem services assumes the determination of Total Economic Value [19], but in reality, it is only a fraction of the total valued environmental resource, including the ecosystem, because we are imperfectly able to inventory resources and assess non-market utility. A 2013 survey of European citizens revealed how people value ecosystem services that serve humanity. With a choice between different services, Europeans have found that maintaining biodiversity is a priority they are willing to pay for. The issue of ES valuation is of interest to various research disciplines: natural, economic, and social. The concept of environmental services appeared in 1981 [20], but already in the 1970s attempts were made to evaluate the so-called "Non-productive functions" [21]. In scientific literature, the valuation of ecosystem services has been discussed for many years [22–24]. The comparison of the valuation of ecosystems, including ecosystem services, can be carried out using several selected environmental valuation methods, such as conditional valuation [25], by defining Willingness to Pay (WTP), Willingness to Accept (WTA), hedonic pricing, travel cost, and cost methods.

In the case study, the valuation of the entire green-blue infrastructure of the urban ecosystem as a whole was estimated since ecosystems provide many services simultaneously (Table 1), the "availability of which depends on the functioning of the ecosystem as a whole, consisting of many mutually synergistic elements" [26,27].

Table 1. Levels and measures of biodiversity reduced by Feledyn-Szewczyk B., Sienkiewicz 2010.

Bio-Diversity	Quantity	Quality
ecosystem brand	range and range of species of communities	The variety of ecosystems and communities
species brand	species richness	Species differentiation, equality
genetic brand	mutations	Allele variability in the gene pool and gene exchange

The problem of biodiversity loss was noticed relatively long ago; in 1992, during the Convention on Biological Diversity signed in Rio de Janeiro [28], the necessity to reduce the loss of biodiversity by 2010 was recorded. This goal was not met back then, so similar actions were taken at the Rio + 20 Conference in 2012, which resulted in the creation of the EU Biodiversity Strategy 2020 [29].

Protection of water reservoirs should be an indispensable planning document (spatial planning structure in Poland is shown in Figure 4), while these provisions have a real and significant impact on the condition of water reservoirs, which in turn also translates into the provision of ecosystem services by them. The introduction of provisions for the protection of water bodies into spatial planning directly impacts the quality of life of their environment. The current hierarchical spatial planning layout is shown in the chart below. It is similar to the planning systems in the entire Carpathians.

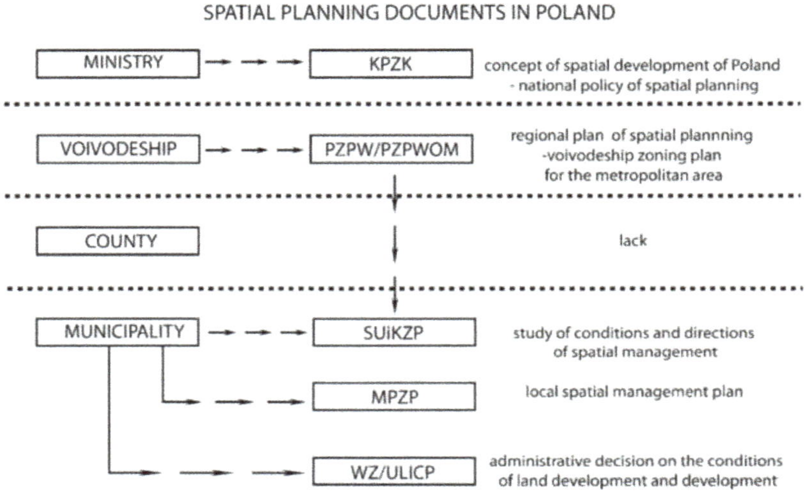

Figure 4. The spatial planning system in force in Poland. Source: authors' study.

Due to the scale, it is only the commune level that makes it possible to define real rules for shaping the space. The directional document of communes in the study of the conditions and directions of spatial development (SUiKZP) is the basic document creating general spatial policy and local rules of municipal development. In the study prepared on a scale of 1:10,000 (with possible exceptions), the commune authorities determine the basic scope of spatial policy and indicate areas intended for development, the initial location of technical infrastructure, scenic and landscape protection, as well as protected areas in terms of natural and cultural resources. The study also designates areas for various reasons excluded from the possibility of their development and areas for which local spatial development plans should be prepared. In their absence, the basis for spatial management is administrative decisions on development conditions, which do not have to comply with SUIKZP arrangements and results in the lack of protection of space and the environment. As a result, in areas not covered by the local spatial development plan, it is possible to implement investments contrary to the assumptions of the local policy and the spatial structure of the commune. Local spatial development plans are the direct basis for issuing building permits. The arrangements of the local development plan define the conditions of development and regulate the way of developing specific properties located in the area covered by the plan, which is to ensure transparency and stability of the policy of local authorities and to preserve the spatial order.

The Act of 27 March 2003, on spatial planning and development introduced the obligation to prepare an environmental impact assessment plan and financial impact forecast as an appendix to the spatial development plan. The scope of the forecasts is defined in the Regulation of the Council of Ministers of 26 August 2003, on the require scope of the local spatial development plan project. The forecast of the impact of the plan's findings on the environment includes guidelines for shaping green-blue infrastructure to prevent its fragmentation, forms of nature protection, and other elements significant for the environment.

Ecosystem services are mainly public goods that are not a market product; they do not have a price and, as such, become worthless for the inhabitants. The lack of valuation is a major cause of ecosystem degradation and loss of biodiversity. If we want to guide our ecological safety, we must "measure" ecosystems and biodiversity and record it in local planning documents. Most of the services are classified as indirect benefits resulting from various natural processes, the effects of which are often delayed in time and the resulting changes are non-linear [30,31]. The table below (Table 2) shows the commonly used methods of ecosystem services valuation.

Table 2. Methods of ecosystem valuation.

Assessment Method	Assessed Ecosystem Services	Advantages of the Method	Method Limitations	Comments
Market prices	Only valid for the market	Market data readily available and reliable	Limited to market-related services	They set the lower limit of values, do not include subsidies, do not take into account useless values
Replacement costs	They depend on the existence of the real estate market placeholders	Market data readily available and reliable	It is possible to overestimate the value	For example, the costs of building flood embankments
Costs of hedonistic behavior	Services that affect air quality and visual-aesthetic values	Market data readily available and reliable	A huge amount of data is required, often missing and limited to land ownership	The price of the property is influenced by the natural features of the place, its quality, and environmental hazards
Travel costs	Services that affect the recreational value of a place	It is based on observations and surveys	It is just for recreational benefits. Trips with multiple destinations and destinations are not included.	The cost incurred by a tourist to arrive at a given place (travel, overnight) is a measure of the recreational value.

Ecosystem services can only be valued in scope, and possibly never fully valued. Economic quantification may only be possible for that part of the services that are relatively well understood and for which there are sufficient data. The economic valuation of ES is difficult and error-prone.

An example of calculations for the preparation of a forecast of financial effects with ecosystem services of Dobczyce Lake is presented below. The data were ranked down by the FEGS Ecosystem Ranking System, Final Ecosystem Goods and Services Classification System (FEGS-CS) [32], followed by a specific type of ecosystem providing calculation of potential, resource, and supply numerical indicators. The calculations of the indicators do not uniquely calculate the scheme for each ecosystem, and their calculation depends on the nature of the ecosystem for the user of the year. The analysis of the results was based on the method of calculating the evaluation and valuation indicators developed by Piotr Mikołajczyk from the Environment Information Center UNEP/GRID-Warsaw [33].

In the case of procurement and cultural services, the calculation of the real benefits is easier due to the possibility of using the already collected data [34]. Nevertheless, regulatory benefits to define and convert are also possible. The Dobczyce Lake, if necessary, regulates water circulation [35,36] and, as a result, changes in insolation. In his research, Matuszko (2005) cites data showing an increase in the number of sunny days a year.

"The average annual sunshine duration before the creation of the reservoir (1971–1980) is 1451 hours, and after its formation (1991–2000) it increased and amounts to 1544 hours." In a whole year, cultivation is more efficient and economical; the number of hours of light in greenhouse increases and crops and this is affecting directly on the cost of products (and has an impact on the lives of the city's inhabitants).

The table below (Table 3) shows that the lake, as an ecosystem structure, provides services that bring measurable financial benefits and which should be included in the forecast of financial consequences for the commune.

Table 3. Sample conversion of ecosystem services into financial benefits calculated for Dobczyce based on data from the Commune Office.

Service According to FEGS	Description of the Benefit	Type of Ecosystem Providing the Service	Potential Indicators	Pool/Resource Indicators	Supply Indicators	Assessment Possible Based on Existing Data
00.0601	Cultural: Tourist site	Lagoon	25.27 € [2]	107,589 (number of accomodations [1])	27,187.74 €	Low error threshold
00.0801	Cultural: Educational cruise on the reservoir	Lagoon	32.97 € [4]	28,800 persons [3]	949,536.00 €	Low error threshold
00.0205	Supply: Electricity production	Lagoon	0.12 € [5]	9.6 GWh [6]	1,160,401.44 €	Low error threshold
31	Regulating: Influence on water circulation and insolation	Lagoon	0.12 € [5]	11 kWh [7]	(+224.83 €) [8]	Low error threshold

[1] Data averaged based on the number of offers sold in the city in the commune's strategy for 2016–2022. [2] Average price of a night in the city. [3] Number of people flowing into the city related to services. The data for 2019 show that 4800 used the weekend cruises within 2 months. [4] Cost of communication both ways, the price of accommodation and meals. [5] Average price (Tauron data) for households per 1 kWh. [6] Annual electricity production at the hydroelectric power plant in Dobczyce. [7] The amount of kWh is equivalent to the time of receiving solar irradiation within 1 hour for a greenhouse with 100 lamps. [8] Cost saved after the tank—The product of the hourly difference in insolation before and after connecting the tanks (93 h) and the cost of electricity consumption.

However, the forecast of financial effects does not refer to the valuation of ecosyste services that the environment provides to man, and thanks to the method of valuation these services, a clear economic effect can be indicated. Including the indicator of ecosyste services in the forecast of financial effects could effectively protect the environmental val enshrined in local law, which is green and blue infrastructure. In such a case, the lo spatial development plan would become a real (legally valid) protection tool.

2.2. *The Method of Analysis in the Aspect of Environmental Value—Obtaining and Compiling Municipality Data*

Considering the values in Section 2 in favor of the significant role of green and bl infrastructure and its benefits (including financial), it seems reasonable to include issue of ecosystem services in the planning process, and in particular when creatin forecast of financial consequences. The net present value method (NPV method for sh is commonly used in creating financial forecasts, belongs to the category of dynai methods, and is based on the discounted cash flow analysis at a given discount rate [$NPV^n{}_{t=1} = \frac{\Sigma CF_t}{(1+r)^t} - CF^0$, where NPV is net present value; CF_t is free cash flows each year of the plan, as the difference between effects and costs; costs (outlays) relate the transformation of the area in year 0 (before the plan enters into force); r is discount 0.13% (in Poland on 27 January 2021) [38] (as a decimal); t is duration of the plan (10 ye in this case); $\frac{1}{(1+r)^t}$ is discounting factor.

To be able to determine the forecast of financial effects with the NPV method and to enrich it with the coefficient resulting from the benefits of ecosystem services, the table below (Table 4) lists the data necessary to convert the net present value.

Table 4. Communal data necessary to prepare a forecast of the financial effects of local development plan.

Wartość	Łapanów	Gdów	Dobczyce
Price [1] per 1 m^2 of a plot of land or land in a commune in the Małopolskie Voivodeship	8.57 €/m^2	12.74 €/m^2	16.70 €/m^2
Change in prices of plots of land or land in the commune in the Małopolskie Voivodeship in the last 12 months	11%	−8%	124%
Average plot or land value [1] in a commune in the Małopolskie Voivodeship	23,020.34 €	41,129.32 €	71,232.25 €
Number of plots or land for sale in the commune in the Małopolskie Voivodeship	62	64	10
Average area plot or land in the commune in the Lesser Poland Voivodeship	2942 m^2	3703 m^2	5189 m^2
Average price [1] of an agricultural plot per 1 ar	3.95 €	3.29 €	8.79 €
Average house value [1]	109,847.45 €	109,847.45 €	118,635.25 €

[1] Value calculated based on the exchange rate on 27/01/2020 (1 PLN = 0.219773 EUR).

The calculation of the forecast financial impact for the plan is based on the area of the plan which is converted from agricultural to building areas. Moreover, to calculate the coefficient enriching the previous formula with financial benefits from ecosystem services, it is necessary to measure ecosystem areas. For this reason, the necessary areas are listed in Table 5.

Table 5. Levels and measures of biodiversity reduced by Feledyn-Szewczyk B., Sienkiewicz 2010.

	Łapanów	Gdów	Dobczyce
MPZP/act of a local law change	Resolution amending: Local Plan No. XIX/140/2012—3.49 km^2 [1]	Resolution No. XXXVI/249/2017 of the Gdów Commune Council of 8 June 2017—1.013 km^2	Resolution No. XXVI/164/16 of the City Council of Dobczyce —12.940 km^2
Agricultural plots	35 ha	12.10 ha	502.08 ha
building plots	69.64 ha	165.01 ha	233.73 ha
Water ecosystem blue infrastructure	3.5 ha	16.55 ha	140.00 ha
Alluvial green infrastructure	68.76 ha	19.11 ha	65.54 ha
recreation green infrastructure	5 ha	53.73 ha	18.45 ha
forest green infrastructure	23.48 ha	4.5 ha	20.35 ha

[1] Area of the fragment in the vicinity of the reservoir to which we refer (because the local development plan was prepared for the entire commune).

In the analysis of ecosystem services, it is important to pay attention to the surface of the water ecosystem in Dobczyce. Lake Dobczyckie is the third largest lake in the Małopolskie Voivodeship, and its condition is improving compared to previous years. Therefore, it should be suspected that after updating the NPV formula and extending it with the ecosystem services factor, its value will increase significantly concerning Gdów (no large ecosystem areas) and Łapanów (much smaller water ecosystem). Table 6 presents artificially created water reservoirs in Małopolska. The reservoirs have been classified in terms of their general condition based on data on the status of waters in the Małopolskie Voivodeship obtained from the website of the Regional Water Management Authority in Kraków. The condition and tendency of changes in the condition of the water reservoir were determined by placing the sign "↓" next to them when the condition of the reservoir is in degradation, the sign " = " when the condition of the reservoir does not change, and the sign "↑" when the condition of the reservoir improves. Then, the types of protection that the tanks are subject to have been defined. SUIKZP (study of conditions and directions of spatial management) and MPZP documents were analyzed for the areas where the reservoirs are located (to help the reader understand the topic, Figure 4 shows the spatial planning system in Poland), and checked whether provisions are protecting the reservoir or ecosystem services. The summary also includes other types of protection including water reservoirs, such as sanitary protection or ecological protection.

Table 6. Artificial reservoirs in Małopolska. Prepared based on summaries prepared by the Regional Water Management Authority in Kraków. https://krakow.wody.gov.pl/data (accessed on 27 January 2021) for years 2008–2016.

Name of the Water Reservoir	Area (ha)	Max. Depth. (m)	Impact [1]	Genesis	Text in MPZP [2]	Text in SUIKZP [2]	Condition	Ratio Area (%) [3]
Bagry (Kraków)	31.4	10.0	yes (Kraków)	Post-industrial excavation flooded	no protection	protection	=	1.3 [8]
Balaton (Trzebinia)	3.0	9.5	yes	flooded quarry-drinking water reservoir	no protection	protection	↓	0.02
Chechelskie Lake	54	No data	no	A dam reservoir	no protection	protection	No data	0.5
Czorsztyńskie Lake	1100.0	50.0	no	A dam reservoir	no protection	protection	↓	0.05
Dąbski Pond	2.54	No data	yes (Kraków)	Post-industrial excavation flooded	ecological area	protection	=	1.3 [8]
Dobczyckie Lake	1065.0	28.0	yes	A dam reservoir	no protection	protection	↑	16.0 [7]
Gdow-Nieznanowice	16.55	No data	yes	Post-industrial excavation flooded	Protection	protection	=	No data
Klimkowskie Lake	310.0	25.0	no	A dam reservoir	no protection	protection	=	1.0
Kryspinów Lagoon	1500.0	8.0	Yes [5]	No data	no protection	protection	↑	Not specified [6]
Łapanów Lagoon	3.5	No data	yes	A dam reservoir	no protection	protection	↓	No data
Mucharskie Lake	1035.0	50.0 [4]	no	A dam reservoir	no protection	protection	No data	13.0
Nowohucki Lagoon	7.0	2.5	yes (Kraków)	Recreational reservoir	no protection	protection	↓	1.3 [8]
Płaszowski Pond	7.9	No data	yes (Kraków)	Post-industrial excavation flooded	no protection	protection	=	13.0 [8]
Przylasek Rusiecki (lake reservoirs)	0.0086	No data	yes (Kraków)	Natural oxbow of the Vistula	no protection	protection	↓	1.3 [8]
Rożnowskie Lake	1600.0	35.0	no	A dam reservoir	Protection	protection	↓	4.5
Zakrzówek Lagoon	23.0	32.0	yes (Kraków)	Flooded quarry	ecological area	protection	↑	1.3 [8]
Zesławicki Lagoon	No data	No data	yes [5]	Post-industrial excavation flooded	no protection	protection	↓	Not specified [6]

[1] In the impact of the Krakow agglomeration. [2] Regarding reservoir or ecosystem services. [3] Ratio of water to the area of the commune. [4] Height of the dam. [5] Partially in the city of Krakow. [6] Not specified due to merger across multiple communes. [7] The lake covers 82% of the area of the city of Dobczyce. [8] Percentage of the area of the City of Krakow calculated in total for Krakow.

3. Results

Adoption of NPV method and ecosystem services factor

From the collected data for the local development plan of each commune the values of CFt and CF0 were calculated to apply the NPV method and then analyze the application of the ecosystem services ratio. Averaged price of an agricultural plot, the average price of an undeveloped building plot, and average prices of buildings (Table 7) were used for calculations. As a result of the above actions, the NPV economic coefficient was obtained

Table 7. Average prices of agricultural plots, undeveloped building plots, and buildings, based on municipals statistics. Values calculated based on the exchange rate on 27 January 2020 (1 PLN = 0.219773 EUR).

	The Average Price of an Agricultural Plot [€] per m^2	The Average Price of an Undeveloped Building Plot [€] Per m^2	The Price of Building [€]
Łapanów	3.95	8.57	109,906.20
Gdów	3.29	12.74	109,906.20
Dobczyce	8.79	16.70	118,698.70

To take into account the additional benefits of ecosystem services, the percentage share of ecosystem area (from Table 5) concerning the built-up area was calculated. As a result of the analysis, it was assumed that the benefit from ecosystem services will not exceed 1% of the obtained value. The coefficient of ecosystem services marked as V_e is, therefore, a promil of the ratio of ecosystems area to building area.

$V_e = \frac{\Sigma e_s \times 100\%}{\Sigma b_s} \times 1‰$, where V_e is the coefficient of ecosystem services, e_s is ecosystem surfaces, and b_s is areas intended for development (built-up areas).

The table below presents the NPV values for local development plans for municipalities. The values of V_e are calculated from the data contained in Table 8, and NPVe is updated by the amount being the product of NPV and V_e (see Table 9).

Table 8. NPF calculated for Łapanów, Gdów, and Dobczyce municipalities. Values calculated based on the exchange rate on 27/01/2020 (1 PLN = 0.219773 EUR).

	CF_t [1]	CF_0 [2]	NPV [3]
Łapanów	5,130,228,433.03 €	277,056,043.31 €	3,489,118,707.36 €
Gdów	15,105,062,302.82 €	543,828,778.80 €	10,545,547,414.64 €
Dobczyce	277,634,186,789.86 €	2,056,663,606.55 €	201,795,942,848.67 €

[1] Free cash flows for each year of the plan, as the difference between effects and costs. [2] Costs (outlays) related to the transformation of the area in year 0 (before the plan enters into force). [3] Net present value.

Table 9. NPV_e updated with Ve coefficient NPV value calculated for Łapanów, Gdów, and Dobczyce municipalities. Values calculated based on the exchange rate on 27/01/2020 (1 PLN = 0.219773 EUR).

	NPV	V_e	NPV_e	Financial Difference Taking into Account the Coefficient V_e
Łapanów	3,489,118,707.36 €	0.19	4,152,051,261.76 €	662,932,554.40 €
Gdów	10,545,547,414.64 €	0.03	10,861,913,837.08 €	316,366,422.44 €
Dobczyce	201,795,942,848.67 €	0.32	266,370,644,560.24 €	64,574,701,711.57 €

The table shows a directly proportional increase in monetary profits to the increase in the area of ecosystems. The presence of water reservoirs as ecosystems in Dobczyce and Łapanów is also clearly visible here.

4. Discussion

The implementation of solutions in the field of green and blue infrastructure in urban space requires a change in the approach of many professional environments to solving problems. Including ecosystem services in climate change adaptation plans gives hope for the implementation of systemic solutions supporting the development of legal, financial, and educational (including training) tools in the field of green and blue infrastructure. Its use in smaller cities located in agglomeration zones of large cities also increases the ability to adapt to changing climatic conditions. The subject of the valuation of ecosystem services was undertaken in 1997 by R. Constanza [39], who noted that "The issue of valuation is inseparable from the choices and decisions we have to make about ecological system Some argue that the valuation of ecosystems is either impossible or unwise, that we canno place a value on such intangibles as human life, environmental aesthetics, or long-tern ecological benefits. But, in fact, we do so every day." Constanza points that the searcl for choices entails the necessity of valuation. Accordingly, this also includes decision regarding spatial development and ecosystems for valuation. The difference can onl lie in the accuracy and whether the valuation of production with an understanding o the topic in the reduction of ecological knowledge. In the considerations of De Groo (2002), he shows that when assessing ecosystem services available, it is necessary to tak into account their totality for a given area due to the synergy that should characteriz the relationship between these services. "Since most functions and related ecosyster processes are inter-linked, sustainable use levels should be determined under comple system conditions".

To achieve the desired effect of "synergy" of ecosystems and to protect Green Urban I frastructure together with the ecosystem services it provides, efficient and well-thought-ou spatial planning is necessary. A tool that can significantly improve the level of protection c ecosystems is the inclusion of ecosystem services in the forecast of financial effects for loc development plans based on NPV method [40]. Such action will not only allow for a mo precise determination of the financial benefits resulting from the preservation of ecosyste areas but also, as a result, help protect these areas against excessive use for developmen

In particular, forecasts of financial effects should include the following:

(1) a forecast of the impact of the local zoning plan on the commune's income an expenses, including real estate tax and other income related to the commune's real esta turnover as well as the fees and compensation referred to in Art. 36 of the act;

(2) a forecast of the impact of the provisions of the local spatial development pla on expenditure related to the implementation of investments in the field of technic infrastructure, which are part of the commune's tasks;

(3) conclusions and recommendations regarding the adoption of the proposed solutio to the draft local plan, resulting from taking into account their financial multiplier effects;

(4) financial impact of ecosystem services expressed by adding the value of V_e to t economic factor NPV.

The first three elements listed above are included in most financial impact foreca for the local spatial development plan performed in Polish municipalities. They do n yet include point 4, which has an important protective function for ecosystems and th services and facilitates the translation of ecosystem services into measurable benefits. Su a procedure will also significantly improve the attitude of the local community to t protection and leaving of ecosystem areas in a condition that enables the continuity service provision. Therefore, taking into account the factor of income from these servic may become an important advantage during public consultations that take place wh making local plans.

The research presented in the article may also refer to the smaller and larger sc of the problem (national, European, or the scale of a single local project) and constit a universal tool for calculating and comparing the benefits of leaving ecosystems a protecting them. This will help to maintain a balance between the urbanized space and area of green and blue infrastructure.

5. Conclusions

Documents of local law relating to spatial planning in the area of a commune or lar administrative unit are the basic tool for determining the directions of spatial developm Often, these documents do not take into account additional conditions for the prese of ecosystems, and ecosystem services, which are necessary for the appropriate syne of anthropogenic and natural ecosystems, and also are a source of measurable, finan benefits. As a result, ecosystem services are not anticipated in spatial developmen protected in any way. Similar proposals, e.g., the InVEST method of ecosystem servi valuation, have already been made for more specialized areas such as marine areas, "Integrated Valuation of Ecosystem Services and Tradeoffs (InVEST) evaluates how alternative management or climate scenarios yield changes in the flow of ecosystem services. outputs are expressed in biophysical units (e.g., landed biomass) or socioeconomic units (e.g., net present value (NPV) of finfish)" [41]. This illustrates the need to obtain economic data about ecosystem services.

In connection with the above study, it should be concluded that it is possible to include ecosystem services in financial forecasts in the field of spatial planning. The method commonly used to forecast the effects of investments, and thus the effects of introducing a new act of local law, both in Poland and abroad, is the NPV economic indicator method [42]. The factor added to NPV can be a way to protect ecosystems as it represents the legitimacy of leaving ecosystems undeveloped and in a state where they can provide ecosystem services. These studies can also be considered in a broader scope and are applicable not only in Poland but also in planning in other countries.

Author Contributions: Conceptualization, R.B., H.H.-G.; methodology R.B., H.H.-G.; validation, R.B., H.H.-G.; data analysis, H.H.-G., A.H.-N., A.P.; writing—Original draft preparation R.B., H.H.-G., A.H.-N., A.P.; writing—Review and editing, R.B., A.H.-N., visualization, H.H.-G., A.H.-N.; supervision, R.B.; All authors have read and agreed to the published version of the manuscript.

Funding: This research received no external funding.

Institutional Review Board Statement: Not applicable.

Informed Consent Statement: Informed consent was obtained from all subjects involved in the study.

Data Availability Statement: The data presented in this study are available on request from the corresponding author. The data are not publicly available due to municipalities policy.

Conflicts of Interest: The authors declare no conflict of interest.

References

1. Grilli, G.; Jonkisz, J.; Ciolli, M.; Lesinski, J. Mixed Forests and Ecosystem Services: Investigating Stakeholders' Perceptions in a Case Study in the Polish Carpathians. *For. Policy Econ.* **2016**, *66*, 11–17. [CrossRef]
2. Zhang, J.; Yin, N.; Wang, S.; Yu, J.; Zhao, W.; Fu, B. A Multiple Importance–Satisfaction Analysis Framework for the Sustainable Management of Protected Areas: Integrating Ecosystem Services and Basic Needs. *Ecosyst. Serv.* **2020**, *46*, 101219. [CrossRef]
3. Karimi, A.; Yazdandad, H.; Fagerholm, N. Evaluating Social Perceptions of Ecosystem Services, Biodiversity, and Land Management: Trade-Offs, Synergies and Implications for Landscape Planning and Management. *Ecosyst. Serv.* **2020**, *45*, 101188. [CrossRef]
4. Damania, R.; Fredriksson, P.G.; List, J.A. The Multiplier Effect of Globalization. *Econ. Lett.* **2004**, *83*, 285–292. [CrossRef]
5. Derlatka, A. Synergia Jako Kryterium Ocenyprojektów Rewitalizacji Śródmieść. *Bud. Archit.* **2017**, *16*, 19–28. [CrossRef]
6. Belbin, R.M. Chapter 2—The Apollo syndrome. In *Management Teams*, 3rd ed.; Belbin, R.M., Ed.; Butterworth-Heinemann: Oxford, UK, 2010; pp. 13–23. ISBN 978-1-85617-807-5.
7. Myers, N.D.; Prilleltensky, I.; Hill, C.R.; Feltz, D.L. Well-being self-efficacy and complier average causal effect estimation: A substantive-methodological synergy. *Psychol. Sport Exerc.* **2017**, *30*, 135–144. [CrossRef]
8. Bolund, P.; Hunhammar, S. Ecosystem Services in Urban Areas. *Ecol. Econ.* **1999**, *29*, 293–301. [CrossRef]
9. Urban Agenda for the EU. Available online: https://ec.europa.eu/futurium/en/urban-agenda (accessed on 27 January 2021).
10. Feledyn-Szewczyk, B. Bioróżnorodność Jako Wskaźnik Monitorowania Stanu Środowiska. *Studia Rap. IUNG PIB* **2016**, *47*, 105–124. [CrossRef]
11. Solon, J.; Roo-Zielinska, E.; Affek, A.; Kowalska, A.; Kruczkowska, B.; Wolski, J.; Degórski, M.; Grabińska, B.; Kołaczkowska, E.; Regulska, E.; et al. Świadczenia Ekosystemowe w Krajobrazie Młodoglacjalnym. In *Ocena Potencjału i Wykorzystania [Ecosystem Services in a Postglacial Landscape. Assessment of Potential and Utilisation]*; Wydawnictwo Akademickie SEDNO: Warsaw, Poland, 2017; ISBN 978-83-7963-062-2.
12. Document. Available online: https://bip.malopolska.pl/uglapanow,a,197387,uchwala-nr-vii4403-rady-gminy-laparow-z-dnia4lipca-2003-r-w-sprawie-uchwaleniamiejscowego-planu-zago.html (accessed on 23 January 2021).
13. Document. Available online: https://www.dobczyce.pl/sites/default/files/upload/uchwala_xviii.110.16_bienkowice_dz.urz_.pdf (accessed on 20 January 2021).
14. Document. Available online: http://www.gdow.e-mpzp.pl/tekst/XXXVI_249_2017_ror.pdf (accessed on 20 January 2021).
15. Hrehorowicz-Gaber, H. Contemporary Changes, and Development of Spatial Structures in the Carpathians. Conditions and Development Models. Cracow University of Technology: Cracow, Poland, 2019.

16. Coutts, C.; Hahn, M. Green Infrastructure, Ecosystem Services, and Human Health. *Int. J. Environ. Res. Public Health* **2015**, *12*, 9768–9798. [CrossRef]
17. Blazy, R. Living Environment Quality Determinants, Including PM2.5 and PM10 Dust Pollution in the Context of Spatial Issues—The Case of Radzionków. *Buildings* **2020**, *10*, 58. [CrossRef]
18. Haashemi, S.; Weng, Q.; Darvishi, A.; Alavipanah, S.K. Seasonal Variations of the Surface Urban Heat Island in a Semi-Arid City. *Remote Sens.* **2016**, *8*, 352. [CrossRef]
19. Fromm, O. Ecological Structure and Functions of Biodiversity as Elements of Its Total Economic Value. *Environ. Resour. Econ.* **2000**, *16*, 303–328. [CrossRef]
20. Riechers, M.; Strack, M.; Barkmann, J.; Tscharntke, T. Cultural Ecosystem Services Provided by Urban Green Change along an Urban-Periurban Gradient. *Sustainability* **2019**, *11*, 645. [CrossRef]
21. Solon, J. Koncepcja "Ecosystem Services" i jej zastosowania w badaniach ekologiczno-krajobrazowych. *Probl. Ekol. Krajobrazu* **2008**, *21*, 28–30.
22. McCauley, D.J. Selling out on nature. *Nature* **2006**, *443*, 27–28. [CrossRef] [PubMed]
23. Costanza, R.; de Groot, R.; Sutton, P.; van der Ploeg, S.; Anderson, S.J.; Kubiszewski, I.; Farber, S.; Turner, R.K. Changes in the global value of ecosystem services. *Glob. Environ. Chang.* **2014**, *26*, 152–158. [CrossRef]
24. Kronenberg, J. Betting against Human Ingenuity: The Perils of the Economic Valuation of Nature's Services. *BioScience* **2015**, *65*, 1096–1099. [CrossRef]
25. Bjornstad, D.J.; Kahn, J.R. (Eds.) *The Contingent Valuation of Environmental Resources*; Edward Elgar: Cheltenham, UK, 1996.
26. Kronenberg, J. *Usługi Ekosystemów—Nowe Spojrzenie na Wartość Środowiska Przyrodniczego*; Wydawnictwo Uniwersytetu Łódzkiego: Łódź, Poland, 2016; ISBN 978-83-7969-576-8.
27. De Groot, R.S.; Wilson, M.A.; Boumans, R.M.J. A typology for the classification, description and valuation of ecosystem functions, goods and services. *Ecol. Econ.* **2002**, *41*, 393–408. [CrossRef]
28. Convention on Biological Diversity (CBD). 1992. Available online: https://www.cbd.int/ (accessed on 20 January 2021).
29. The EU Biodiversity Strategy to 2020. Available online: https://ec.europa.eu/environment/nature/biodiversity/strategy_2020/index_en.html (accessed on 27 January 2021).
30. Dubel, A. Comparative analysis of methods of environmental vaulation. case study of błonia krakowskie. *Prace Naukowe Uniw. Ekon. Wroc.* **2017**, 122–131. [CrossRef]
31. Atkinson, G. Environmental valuation and benefits transfer. In *Cost-Benefit Analysis and Incentives in Evaluation. The Structural Funds of the European Union*; Florio, M., Ed.; Edward Elgar Publishing: Cheltenham, UK, 2006.
32. Landers, D.H.; Nahlik, A.M. *Final Ecosystem Goods and Services Classification System (FEGS-CS)(EPA/600/R-13/ORD-004914)*; US Environmental Protection Agency, Office of Research and Development: Washington, DC, USA, 2013.
33. Mikołajczyk, P. Mapowanie i Ocenaekosystemów i Ich Usług w Polsce. *UNEP GRID—Warszawa*, 7 October 2015.
34. Pueyo-Ros, J. The Role of Tourism in the Ecosystem Services Framework. *Land* **2018**, *7*, 111. [CrossRef]
35. Dancewicz, A. Pole Opadów Atmosferycznych w Otoczeniu Sztucznych Zbiorników Wodnych Otmuchów i Nysa Przed i Po Ich Powstaniu. *Wiad. Inst. Meteorol. Gospod. Wodnej* **1993**, *XVI*, 70–85.
36. Hall, D.K.; Frei, A.; DiGirolamo, N.E. On the frequency of lake-effect snowfall in the Catskill Mountains. *Phys. Geogr.* **2018**, *39*, 389–405. [CrossRef] [PubMed]
37. Matuszko, D. Próba określenia wpływu zbiornika wodnego na zachmurzenie i usłonecznienie (na przykładzie Zbiornika Dobczyckiego). In *Rola Stacji Terenowych w Badaniach Geograficznych*; Instytut Geografii i Gospodarki Przestrzennej Uniwersytetu Jagiellońskiego: Kraków, Poland, 2005; pp. 79–91.
38. Available online: https://www.nbp.pl/home.aspx?f=/dzienne/stopy.html (accessed on 5 January 2021).
39. Costanza, R.; d'Arge, R.; De Groot, R.; Farber, S.; Grasso, M.; Hannon, B.; Limburg, K.; Naeem, S.; O'neill, R.V.; Paruelo, J. The Value of the World's Ecosystem Services and Natural Capital. *Nature* **1997**, *387*, 253–260. [CrossRef]
40. Knoke, T.; Gosling, E.; Paul, C. Use and Misuse of the Net Present Value in Environmental Studies. *Ecol. Econ.* **2020**, *174*, 106664. [CrossRef]
41. Guerry, A.D.; Ruckelshaus, M.H.; Arkema, K.K.; Bernhardt, J.R.; Guannel, G.; Kim, C.-K.; Marsik, M.; Papenfus, M.; Toft, J.E.; Verutes, G.; et al. Modeling benefits from nature: Using ecosystem services to inform coastal and marine spatial planning. *Int. J. Biodivers. Sci. Ecosyst. Serv. Manag.* **2012**, *8*, 107–121. [CrossRef]
42. Yemshanov, D.; McCarney, G.R.; Hauer, G.; Luckert, M.M.; Unterschultz, J.; McKenney, D.W. A real options-net present value approach to assessing land use change: A case study of afforestation in Canada. *For. Policy Econ.* **2015**, *50*, 327–336. [CrossRef]

sustainability

MDPI

Article

Environmental Urban Plan for Failaka Island, Kuwait: A Study in Urban Geomorphology

Ahmed Hassan [1,*], Muhammad G Almatar [2], Magdy Torab [3] and Casey D Allen [4,5]

1 Kuwait Ministry of Education, Kuwait 50001, Kuwait
2 Department of Geography, College of Social Sciences, Kuwait University, Kuwait 5969, Kuwait; muhammad.almatar@ku.edu.kw
3 Department of Geography, Damanhur University, Damanhur 22511, Egypt; magdytorab@art.dmu.edu.eg
4 Department of Biological and Chemical Sciences, Faculty of Science and Technology, The University of the West Indies, Bridgetown TA22, Barbados; casey.allen@cavehill.uwi.edu
5 Stone Heritage Research Alliance, LLC., Kaysville, UT 84037, USA
* Correspondence: ameh812000@gmail.com

Received: 18 August 2020; Accepted: 27 August 2020; Published: 1 September 2020

Abstract: Failaka Island, located in the far east of Kuwait Bay about 20 km from the State of Kuwait's coast, represents a focal point for regional geography and history, including natural wonders and archaeological sites dating to the Bronze, Iron, Hellenistic, Christian and Islamic periods. According to environmental data and in coordination with local authorities to develop an urban plan, the island is set to become the first tourist destination for the State of Kuwait. To achieve the Vision of Kuwait 2035, one of the planning objectives centers on Urban Planning for the Establishment of Environmental Cities that Achieve (UPEECA) environmental sustainability criteria. The article then, aims to propose the environmental urban plan for Failaka Island. Based around Environmental Analytical Hierarchical Processes (EAHP) and using the Field Calculator and ModelBuilder functions in ArcGIS, this research centers on the feasibility of carrying out an urban plan using suitability modeling that incorporates 4 factors and 13 criteria covering the island's ecological and human composition. This study utilizes both remote sensing (Unmanned aerial vehicles **UAVs** for 3D imaging) and field study (ground truthing) to identify changes in land use and land cover—such as using sample analysis of the historical sites and soils for tracing evidence and creating/updating a soil map—and create the first geographic information systems (GIS) database for the island that can lead capable of generating a suitability model.

Keywords: urban geomorphology; sustainable environment; suitability model-GIS; environmental analytical hierarchical process (EAHP); Failaka Island

1. Introduction

Global urban population exceeded the rural population for the first time in human history in 2007. Since then, the proportion of people living in urban areas has continued growing and it is expected that by 2050 nearly two-thirds of the global population will be urban [1] Accordingly, researchers in all fields related to sustainable development—including geoheritage and urban geomorphology—should strive for urban planning based on assessment of environmental potentials that include on-site assessment and land-use suitability analyses, with the aim of identifying the most appropriate spatial pattern for the future land use in order to achieve a sense of sustainable development. In recent years, land-use suitability analysis has been applied to the assessment of agricultural land [2], for the study and evaluation of urban services planning [3], to determine a solar farm location in the Legionowo District, Poland [4], landscape evaluation and planning [5], regional planning and environmental impact assessment [6,7], and help with deciding on optimal solar power plant locations in Kuwait [8].

Additionally, these types of analyses have been used to help determine land habitats for animal and plant species [9], as well as identifying potential future land use conflicts in North Central Florida [10].

While suitability analysis focuses on the process and procedures used to establish the appropriateness of a system according to the needs of a stakeholder [3], in most lesser-developed countries, people often construct residential buildings without considering resources for the new residential areas. Therefore, it becomes the government's problem to provide the required resources for these areas. Occasionally, some countries adopt urban planning strategies that are incompatible with the environment, increasing environmental problems that make life difficult for the population.

Site suitability analysis is the process of determining the fitness of a given tract of land for a defined use [11]. To find a suitable site for construction of an amenity or urban area, however, sophisticated analyses that consider critical components, such as technical, environmental, physical, social, and economic factors, are required. Tools for identification, comparison, and analyzing multi-criterion decisions for urban site development require proper planning and management, and often incorporate vital geospatial tools, such as remote sensing, geographic information systems (GIS), global positioning systems (GPS), and analytical hierarchical process (AHP) modeling [12,13]. Based on these facets, this study suggests building Environmental Analytical Hierarchical process (EAHP), centered around an integrated technique utilizing the AHP and GIS, to support the assessment and the selection of suitable areas for urban development on Failaka Island in the state of Kuwait.

2. Site Setting and Goals

The second-largest Kuwaiti island, after Boubyan island, Failaka Island is located 20 km off the Kuwait coast between 29°23′ and 29°28′16″ North and 48°16′ and 48°24′10″ East, with a size approximately 13.8 km long and varying from 1.8 to 6.5 km in width, creating an area of approximately 46.36 square km (Figure 1; [14]). Triangular shaped with its base in the west and its head in the southeast, Failaka Island remains relatively flat apart from a small hill approximately nine meters high on its western side [15,16].

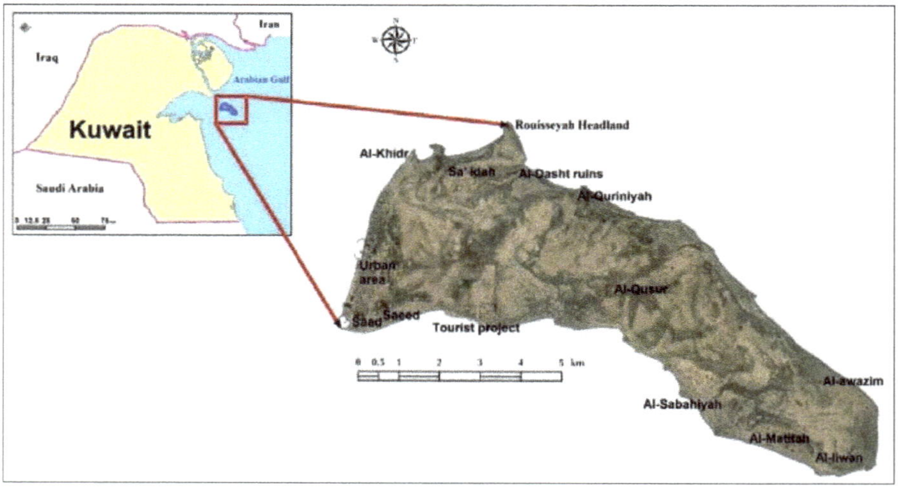

Figure 1. Location of the study area, Source: [14].

A focal point of geography and history for Kuwait, Failaka Island offers unspoiled beaches and archaeological sites dating back to the Bronze, Hellenistic, Christian, and Islamic periods. Although the first population figure recorded for Failaka was in 1839 (150 inhabitants), it represents the only island off Kuwait's coast that has been known to host a population for thousands of years, even increasing during the twentieth century, from 1500 people in 1930 to over 5800 people in 1985 [17]. Failaka Island is considered a unique urban case study as it has had no occupation, and therefore no urbanization, since 1990 when the Iraqi army displaced its residents by forcing them to leave the island. After Kuwait's liberation, the population did not return to the island, and it has been abandoned since that time [17–19]. Given these circumstances, the state and other decision-makers have a unique opportunity to implement potential mitigation and planning strategies for the island—taking into account analyses from specialists in environmental planning, as well as incorporating both human and physical characteristics (i.e., geomorphology, geoarchaeology, soils, coastline types, land use, and geology)—before attempting reconstruction.

To achieve such goals, the government of Kuwait has previously embarked on implementation of various development plans, each of which aims to turn the island into a major tourist destination based on Failaka Island's regional significance. Previous plans concentrated on various development issues, such as a tourist attraction center, residential development projects, and urban infrastructure development. In this context, state plans with differing foci appeared over the past several decades [14,17,19–21], as well as plans conducted by researchers with individual efforts and different scientific backgrounds e.g., [19,22]. While each of these previous studies included specific criteria, none integrate them in a significant way, leading to an incomplete assessment of the island. To remedy these deficiencies, this study proposes a method for environmental planning of Failaka Island by using the Environmental Analytical Hierarchical Process (EAHP) method, incorporated into a geographic information system (GIS), for assessment of multiple factors, including scoring-based scientific literature and local knowledge applied to Multi-Criteria Decision Analysis (MCDA).

3. Materials and Methods

Computer modeling has become one of the most important tools in the field of land use and land cover, as GIS and remote sensing are utilized with increasing frequency to investigate landscape changes such as urban expansion [23]. The availability of advanced computer software and tremendous amounts of spatial data encourages GIScientists and remote sensing specialists to implement such technology in modeling land use and land cover classes, and can be a valuable tool for sustainable planning and evaluating social and environmental consequences of landscape fragmentation [24,25]. Given their flexibility, models can also be utilized for studying relationships and interactions amongst system elements and for creating testable hypotheses and predictions regarding patterns and mechanisms [26]. The analyses presented in this article utilize components from various model types and incorporate multiple factors to determine optimal site locations. It bears noting that the plan outlined below takes it cue from often-used criteria-based suitability models.

Cartographic models, for example, are usually analyzed to identify suitable locations based on a set of criteria determined by analysts, such as suitability models—initially specified in qualitative conditions, such as high, medium, and low—but the actual set of criteria is generated according to quantitative measurements required for the ranking process, where the overall suitability ranking process is generated based on the importance of different classes of criteria [27]. For example, Carr and Zwick [10] ranked suitability using values ranging from 1 to 9 for each single suitability layer, where 1 represents the lowest suitability and 9 identifies the highest suitability. These layers were then combined based on land use categories into three main suitability criteria: high, moderate, and low.

Another example of a cartographic model is what Blostad [27] calls weighting among criteria. This process's final step allows for combining all suitability layers into one single layer that determines the best locations. Blostad [27] stated that assigning suitability which depends on non-quantifiable measures is usually difficult, and the process is easier when the importance of the diverse criteria is expressed on a common scale (e.g., 1 to 9 or 1 to 5). Still, there are different kinds of weighting processes that can be used in suitability models to generate the final output. For example, Carr and Zwick [10] used a pairwise comparison methodology called Analytic Hierarchy Process (AHP) to determine the weight of the layers to produce the final suitability. Similarly, Blostad [27] utilized an importance ranking to assign the weights for slope effects, soil types, and noises cost in his use of a suitability model to determine the best location for home constriction. In the context of Failaka Island, to perform a multi-criteria suitability analysis for the potential environmental urban plan, four main steps must be included: identifying site criteria, data preparation, developing a composite suitability index, and selecting a suitable site (Figure 2).

3.1. Identifying Site Criteria

Criteria of land-use suitability for urban development are derived from multi-disciplinary scientific theories related to physical, socio-economical, and ecological attributes [28]. Often, planners and decision makers use only socio-economic parameters in urban planning, and rarely consider environmental aspects. The criteria in Figure 2 can be classified into two main categories: Boolean (exclusionary) and favorable criteria (inclusionary). While Boolean criteria assist in identifying suitability for a particular purpose in a binary manner, favorable criteria reveal this suitability in gradual levels [4,8].

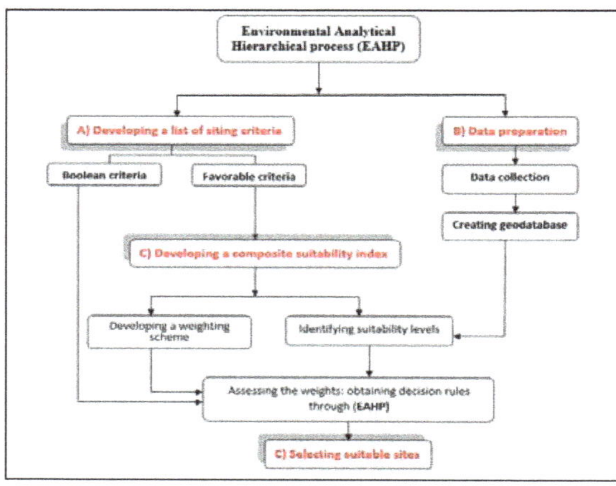

Figure 2. Environmental Analytical Hierarchical process (EAHP) Framework. The four steps involved in calculating the (EAHP) are denoted by letters A, B, C, and D.

3.2. Data Preparation

Data preparation represents an important step in land suitability analysis [4,6]. Upon identifying analysis criteria, datasets of various indicators were collected. These datasets included:

- A digital elevation model (DEM) from an aerial survey with a spatial resolution of 25 cm, including several topographical analyses such as slope, aspect, and creation of a contour map [14].
- A geomorphological map, derived from various sources (i.e. [14], (ground truthing), and aerial photography by a Mavic2 pro drone UVA.
- Soil analysis—including pH, cations, and anions—for 34 samples covering most parts of the island, leading to an updated soil map for the island, and analyses in the Faculty of Science Laboratory, at Alexandria University (Egypt).
- Coastline types (as a shapefile, from [15].
- Land use/land cover. This represents perhaps the most important dataset for urban and environmental planning, and the most fundamental data of the evaluation model for urban development suitability. It can be derived from multiple sources [14]. To obtain data on the history of urbanization on the island in the past century, the study relied on several historical maps from Mohamed [16,21].

- A 1:50,000 scale geological map [29].
- Cost-effectiveness criteria. One of the most important criteria for saving expenses and maintaining the environment with the lowest potential so the island does not lose its distinctive environmental character because of human-caused environmental changes. The island was divided into four sectors from west to east, with sector 1 being closest to the port, and sector 4 being the farthest, the aim being to focus urbanization on the western half of the island.
- Land availability criteria. In order to protect archaeological areas and urban areas from human and natural hazards, criteria are selected that provide an urban plan to help preserve historical monuments of the island, especially since the island has several important archaeological sites dating back more than 5000 years [30,31]. For archaeological areas, the island was divided into four archaeological categories (see Land Availability, number 9, Table 1). With regard to the natural hazards, Almatar et al. [15] provided a map of the areas exposed to the danger of sea level rise and sabkhas. This also divided the island into four sectors (Land Availability, number 10, Table 1).
- Socio-economic criteria are divided into two overarching layers: distance from the main town and the productive capacity of the land. Firstly, the island was again divided into four sectors. Secondly, 34 soil samples were analyzed and compared with the results of a study conducted by Abbadi and El-Sheikh [32]. Furthermore, for understanding soil and land use planning in terms of sustainable development, it was also necessary to define the interactions and competitions which exist between the different uses of soil and land [33].

These collected datasets were then integrated into a geodatabase, including many vector feature classes, such as land use highlighting various human activities, as well as the geomorphological map, soil type, and archaeological/historical sites. The vector layers were subsequently employed to create several raster surfaces for further analyses.

Table 1. List of criteria and relevant indicators.

Factors	Weight %	Criteria	Weight %	Rank			
				9	6	3	0
				High	Moderate	Low	Unsuitable
Environmental	60	1. Elevation (M)	8	10–7.5	7.4–5	4.9–2.5	2.4–0
		2. Geomorphology	15	Old urban	Flat dry land	Sabkhas	Archaeology site
		3. Soil PH	8	8.43–8.87	8.19–8.42	8–8.18	7–7.9
		4. Coastline type	15	Sandy	Sandy rocky	Muddy and gravelly	Rocky
		5. Geology	10	Cemented coastal deposits	Strand line deposit	Dipdipah formation	Sabkhas deposits
		6. Slope	4	0.93–2.57	0.46–0.92	0.2–0.45	0–0.19
Cost-effectiveness	15	7. Distance to port	8	Very near (west)	Near	Middle	Far (East)
		8. Distance to water and energy line	7	Very near (west)	Near	Middle	Far (East)
Land Availability	10	9. Archaeological area	6	Rest of land	–	buffer zone 400 m	Sites
		10. Natural hazard	4	Zone 1 (6–10 m)	Zone 2 (3–6 m)	–	Zone 3 (0–3 m)
Socio-economic	15	11. Land use	10	Urban area	Non- urban	Salt deposits	Archaeology site
		12. Distance from main town	2	Zone 1	Zone 2	Zone 3	–
		13. The productive capacity of the land	3	Zone 1	Zone 2	Zone 3	–

3.3. *Developing a Composite Suitability Index*

For the purpose of this study, a composite suitability index was developed that considered four factors reflecting the four favorable analysis criteria (Table 1). To establish the weighting matrix, 13 criteria were included, with higher values given to overarching landform features, such as geomorphology, coastline type, and geology. The relative weighted values of each factor and criteria were then discussed and reviewed by both the authors and experts from different disciplines in the geosciences (see Table 1 for full discussion).

3.4. *Selecting Suitable Sites*

This step involved calculating a composite suitability index through aggregating various favorable indicators according to their weights. Based on these results, and with little modification (Figure 3), the 13 suitability criteria for the proposed urban plan were further divided into four types: unsuitable, low, medium, and high.

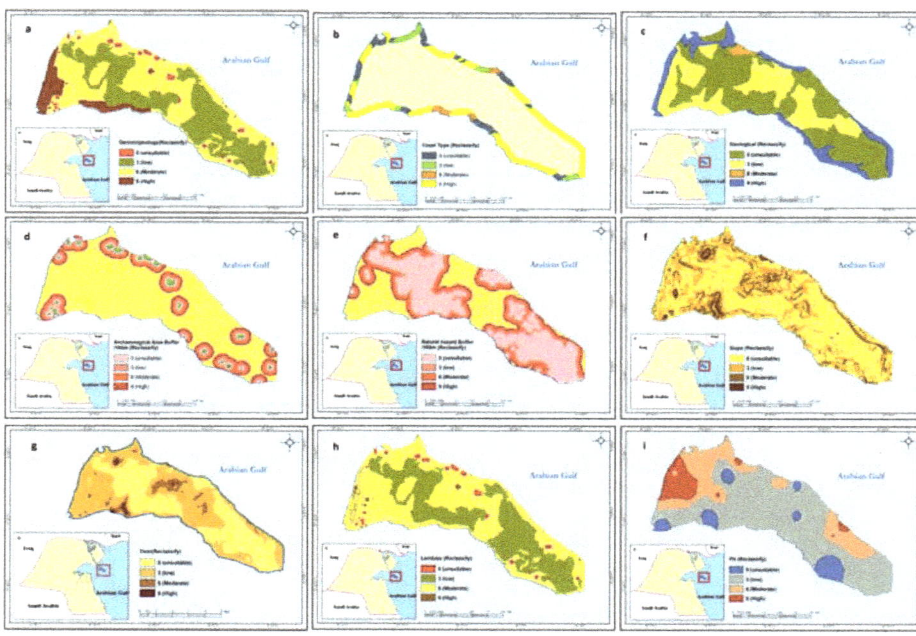

Figure 3. Basic layouts of criteria reclassification: (**a**) geomorphology units, (**b**) types of beaches, (**c**) geological formations, (**d**) archaeological sites, (**e**) natural hazard potential, (**f**) slopes, (**g**) digital elevation model (DEM), (**h**) land-use, (**i**) pH.

4. Physical and Human Geography of Failaka Island

Examining the physical and human properties of any location is crucial in understanding ecosystems [34]. In this context, the study of climate, geology, soil, geomorphology, history, and archaeological sites for Failaka Island reveals more of the environmental aspects that characterize the island, which in turn contributes to the development of an urban plan based on environmental potential. This potential further contributes to achieving the sustainable development goals (SDGs). Officially known as "Transforming Our World"—the 17 goals set by the United Nations Organization 2015-2016—the 17 Sustainable Development Goals are included in the 2030 Agenda for Sustainable Development, especially goals 11, 14, and 15 [35].

4.1. Climate

The island's climate is hot and arid with moderate winds, and the temperature reaches a maximum of 50 °C in summer and a minimum of 4 °C in winter. The average monthly wind speeds on the island are 3.3–7.3 m/s, while the approximate annual rainfall is 125.1 [22,36]. Table 2 displays some climatic characteristics of the study area.

Table 2. Climatic characteristics of Failaka Island 2007–2019

Month	Temp. Ave. (c)	Temp. Min. (c)	Temp. Max. (c)	Relative Humidity Ave (%)	Rainfall Amount (mm)	Wind Speed Ave (m/s)	Wind Speed Max. (m/s)	Evaporation Ave Airport. (mm)
JAN	13.5	4.6	21.7	68.4	14.9	4.6	19.1	5.4
FEB	15.9	6.5	25.4	63.3	11.9	4.8	19.9	7.2
MAR	20.4	11.1	31.7	54.7	14.3	4.8	20.8	9.8
APR	25.3	15.7	37.5	49.7	6.2	4.6	24	12
MAY	31.4	21.5	44.2	42	5.6	4.6	20.5	15.3
JUN	35.5	25.8	48.1	30.2	0	5.7	19.1	17.8
JUL	36.7	27.4	49.2	34.5	0	5.2	17.5	17.5
AUG	36.3	27.5	47.8	41.9	0	4.6	15.7	16.4
SEP	33.7	23.5	45.6	42.7	0.2	4.4	16.2	14.7
OCT	28.5	18.5	41.1	52.6	10.8	4.3	19.7	11
NOV	20.9	11.7	31.1	60.6	45	4.6	18.9	6.7
DEC	15.5	7	24.4	66.2	16.2	4.6	17	5.2

Source: Kuwait Metrological Department [36], 2020. All data from Failaka Island Station except evaporation from Kuwait Airport Station.

4.2. Geology

The geological map represents a very important tool for urban planning and, accordingly, understanding geological formations represents an important component for any urban development project. Fortunately, Failaka Island's geological formations are straightforward [18,37] and divided into four categories:

- Cemented coastal deposits or oolitic complexes. An oolitic complex is a series of part marine and part aeolian ridges. These ridges formed alongside Failaka's barrier beaches and coastal dunes, which are separated from the Gulf by coastal lagoons and sabkhas. Likely deposited during high water stands, they consist of oolitic sand, sandstone, and limestone. The area of this formation is 6.9 km^2, which constitutes 14.88% of the island's total area.
- Dibdibah formations. These formations, created during the Pleistocene Epoch, appear as coarse-grained, pebbly sand with thin intercalation of clayey sand and clay. These sediments are sometimes combined with calcium carbonate and gypsum, producing conglomerates. They cover 18.81 km^2 of the island, or 40.57% of the island's total area.
- Sabkha deposits. Most of these coastal mudflats are widespread on the island, with surface levels varying between 0 and 3 m in height, but being extremely flat and usually barren of any vegetation [29,31].This type of sediment is usually under the constant influence of saline groundwater and susceptible to floods at certain times, often leaving deposits of different evaporite minerals. The sabkha deposits cover an area of 20.1 km^2 constituting 43.35% of the total area of the island [14].
- Strand line deposit: These shell-sand and shingle sites appear on the coast in the form of beach gravel and minor unconsolidated beach sand. This geological formation covers the lowest area

and percentage of the island (0.55 km^2 and 1.17%, respectively), perhaps due to the formation of these sediments on the northern coast, where strong sedimentation streams that formed during rainy periods of the Pleistocene Epoch were washed away by the currents from the Shat al-Arab region when it collided with the Headland of Rouisseyah [29].

4.3. Soil

Soils represent a potential source of considerable information in urban planning studies across scales, from specific archaeological sites to broad regional assessments. Applications of soils in the study of the human past include soil geomorphology, the geoarchaeological principles of site formation processes, and the identification of human impacts [38]. Soil surveys are a useful tool for interpreting landscapes, but with the caveat that they were designed for modern planning and land-use studies. Soils are also an integral component of stratigraphic studies at archaeological sites. Owing to these complexities, it was necessary to compare the soil types of the island with the designated plan sites [19]. This endeavor represented one of the most important criteria of our proposed urban plan, as relying on an updated soil map was key to establishing optimum land use planning efforts.

The soil map was prepared based on 34 samples collected in the winter of 2018 covering most parts of the island (Figure 4). Soil study relied on analyses of such criteria as pH, SAR, electrical conductivity, cation and anion exchanges and ratios, and $CaCO_3$ (calcium carbonate), as well as comparing its results with an earlier study by Abbadi and El-Sheikh [39]. Soil production capacity was calculated based on soil pH and cation-anion exchange capacity, and these results, displayed as green areas in Figure 4, were excluded from being urban areas, and instead proposed as agricultural areas within the island plan.

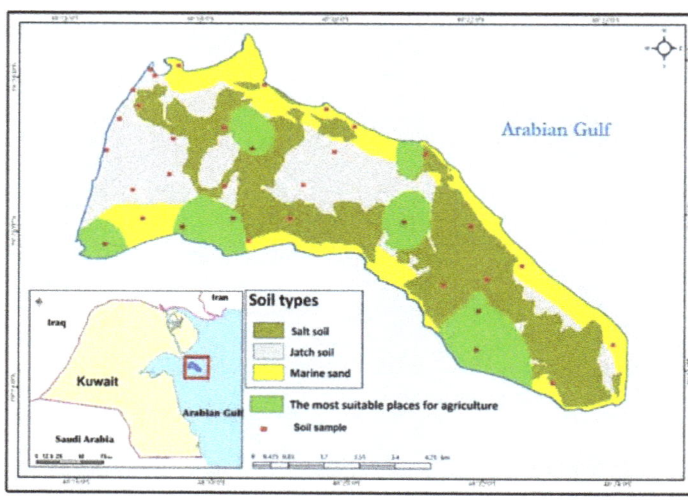

Figure 4. Soil types on Failaka Island and the proposed areas for agriculture. Source: [19,39] modified for soil analysis samples procured during this study.

Chemical soil analyses of Failaka Island represent a very important indication of the best arable areas and potential archaeological sites. For example, the detailed role of soil calcium represents an important component of identifying food preparation areas of the past because it is present in the ash of burning charcoal and found in high amounts in teeth and bones. Similarly, middens found in situ represent the source for some of the site's bone and ash disposal areas and contain significant amounts of calcium [32]. These factors make soil calcium a crucial soil research component in archaeology.

Our soil analyses confirmed previous findings, indicating a rise in calcium in sample points 5, 10, 22, 28, and 29, with value amounts ranging from 75–106 (mEq L^{-1}).

4.4. Geomorphology

During the planning of an urban environment, usually only economic and social parameters are considered [40]. Yet, geomorphology remains an important and effective tool to aid planners and decision-makers in low-cost environmental planning and environmental sustainability [41]. For example, the sabkhas on Failaka Island constitute about 50% of the landmass [15], and therefore building a city or extending a road on the sabkhas areas would cost a significant amount. In this case, understanding the island's geomorphology (i.e., landforms) becomes an effective tool to choose the most appropriate and cost-effective route through the sabkhas, as well as the best location for the city.

Further, the geomorphic processes of erosion and sedimentation processes can be related to the rates of change in the beaches. For example, 68% of the coastline of New England and the mid-Atlantic region of the United States has undergone erosion in recent decades for a variety of reasons, but often related to anthropogenic influences [42]. In contrast, Failaka Island's beaches remain untouched for the past few decades because people have not been allowed to inhabit or visit the island since 1990.

Based on such geomorphic factors then cf., [40,42] the island can be divided into two types of geomorphological forms: those related to the coastline and those on the landmass proper (Figure 5). Indeed, use of the island for touristic purposes cannot be planned without taking the coastline type into account, and thus this study also took into consideration shoreline pattern classification along the island's 38.75 km total coastline length.

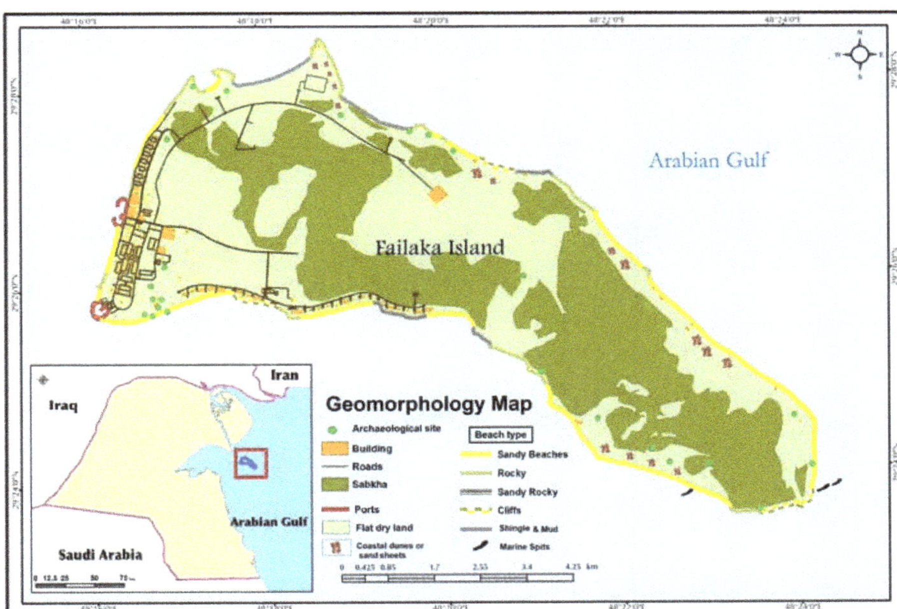

Figure 5. Geomorphology map of Failaka Island. Source: Field study and [14].

Coastal landforms can be further classified into two types: erosional landforms and sedimentation landforms (Figure 6). Analysis of satellite images, drone imagery, and field study/ground truthing, reveals five beach types (Figure 5 and Table 3). The classification of beach types remains an essential pillar of the proposed urban development plan, because the Kuwait government wishes the island to function as a tourist destination. That means limiting the use of sandy beaches for tourism development areas remains an important consideration. As Table 3 demonstrates, half of the island's coasts are sandy beaches, making these areas suitable for urban development and creation of public beaches.

Figure 6. Basic geomorphological landforms on Failaka island. (**A**) Marine headlands and bays. (**B**) Sandy beaches and urban area. (**C**) Some of the erosion landforms: Marine cliffs and terraces. (**D**) Inland sabkhas, nebkhas (e.g., coppice dunes), and chemical erosion. (**E**) Marine spits and headlands. (**F**) Sand sheets and coastal sand dunes. Photos captured with Mavic2 pro UAV. A, B, and E from an altitude of 300–456 m and photos C and D from 52 m. Photo E was taken from ground level.

Table 3. Lengths of beach types on Failaka Island.

Type	Length (km)	%
Sandy beaches	19.91	51.38
Rocky beaches	9.6	24.75
Shingle and Mud beaches	3.06	7.89
Sandy rocky beaches	2.3	5.96
Cliffs	2.7	6.97
Ports	1.18	3.05
Total	38.75	100

Source: (Field study/ground truthing, [14]).

Geomorphological processes play an important role in balancing the sediment accumulation/loss as well, given that, between 1976 and 2016, the northern beaches accumulated sand at 77 cm/year, while southern beaches eroded by 78 cm/year [43]. Developing and urbanizing in/near sand dunes or in areas where some other geomorphic danger, such as a foundation being weakened by unstable bedrock, represents potential catastrophe, including loss of life. Due to this, the study sought to understand the geomorphological characteristics of the island, characterized by four anthropogeomorphological units: depressions (sabkhas), flat drylands, previously-built urban areas, and archaeological sites.

4.5. Historic and Archaeological Sites

Failaka Island hosted several factors that made it ideal for the establishment of ancient civilizations, including being a strategic location along trade routes, suitability as a natural port, and mediation of other ancient civilizations [30]. According to Bibby [44], the island represents one of the oldest settlements (3000 BC) within marine range of the Arabian Gulf. Bibby [44] also asserts that the island had a distinctive pattern of civilization due to its close contact with neighboring civilizations. Nevertheless, the human settlement of Failaka was not stable throughout history, as it witnessed influxes and exoduses depending on trade activities in the Arabian Gulf [18], as well as periodic epidemics, each of which adversely affected continual habitation on the island. Based on available documents, Failaka civilization can be categorized into three ages: (1) ancient civilization, including the archaeological sites of the Bronze and Iron Age, (2) middle civilizations, represented by the Hellenistic civilization in the southwest of the island, and (3) modern civilizations, including Islamic and Christian civilizations [18,44].

The archaeological sites on Failaka Island extend along the shoreline and through the middle of the island (Figure 7). Most archaeological missions have focused on the southwestern and northwestern parts of the island, and those records support the island's occupation from the late Islamic to the Modern Eras. The Danish archaeological excavations have been operating since 1958, making them the first archaeological excavation in the history of Kuwait [15,45]. Yet, several archaeological sites—such as Al-Dasht, Al-Sabahiyah, Al-Sa'id, Al-Ali, Al-Awazim, and Matitah—have been only surveyed and not studied in any detail. The following sites represent the most important archaeological sites located on the island to date, and would therefore make the most interesting tourism sites.

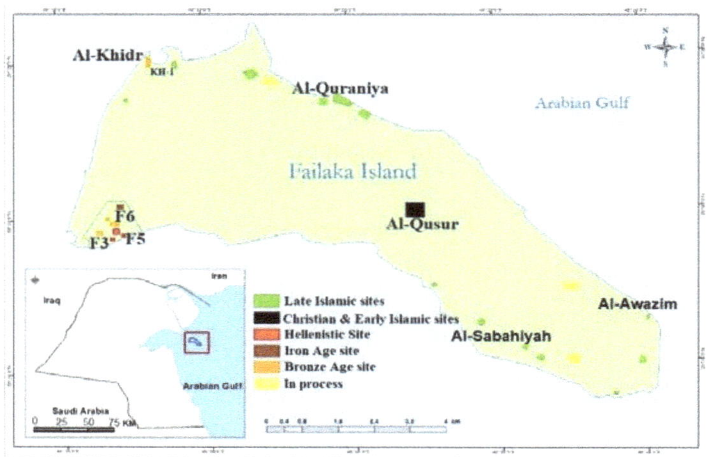

Figure 7. Archaeological sites in Failaka island, Source: Field study and [14].

Al-Khidr: A Bronze Age site, known as the Dilmun-Culture port, located in the northwest of Failaka Island. It was fully excavated in 2004. The survey indicated the presence of numerous sherds, stone structures, metal objects, faunal and floral remains, pearls and fishing hooks, and over 600 Dilmun stamp seals. The site is located directly below a modern Islamic cemetery that stretches along the western shore of the shallow Al-Khidr bay (Figure 8c), which served as a port in the past [46,47]. Today, the Al-Khidr site consists of three visible and two less visible mounds.

Figure 8. Several archaeological sites on Failaka island. (**a**) Oblique photo by UAV, Mavic 2 pro, altitude 41 M, 19 March 2019, (**b**) Mosaic for Hellenistic castle by UAV, Mavic 2 pro, altitude 143 M, 19 November 2019. (**c**) Oblique photo by UAV to Al-khidr, KH-1 and old port.

The KH-1 mound is a low mound roughly 3 to 3.5 m above the contemporary seabed of Al-Khidr bay, stretching 150 m in a north-south direction. It is a sandy dune with shrubby vegetation. Due to high tides, water erosion (2–4 m) has influenced/damaged the growth of its vegetation [46].

F3 mound, a Bronze Age mound referenced by the Danish mission in 1958, and sometimes also known as Tell Sa'ad. It is roughly 9 m above sea level and contains a residential settlement that likely hosted domestic activities, including archaeological evidence of an elite or temple-like structure, the skeletons of gazelles and goats, 170 round stamp seals, and some kilns for an unknown purpose [45]. On the top of this mound, the summerhouse of Ahmad Al-Jaber Al-Sabah, the sheikh of Kuwait, was built during the 1920s [47,48].

F5 mound, also known as Tell Sa'id, is an Iron Age mound that represents the Seleucid culture, post-Alexander the Great. The site was a subject of the first archaeological mission by the Danes (Figures 7 and 8b). It is now understood to be a Hellenistic fortress that consisted of multiple residential units, two temples, multiple storage areas, two gates, and a trench. A French archaeological mission has been working at the site since the 1980s [45].

F6 mound, also known as "the palace", represents a Bronze age mound located near Tell Sa'ad and Tell Sa'id. It is the oldest archaeological site on the island associated with the third Ur dynasty and the Dilmun occupation (~2200 BC). The site consists of a palace-like structure and a Dilmun temple at 4.20 m above sea level [45].

Al-Qusur: Situated in the middle of the island with elevations ranging from roughly two- to five-meters above sea level and surrounded by sabkhas, Al-Qusur represents the largest archaeological site on Failaka Island, covering roughly 2 × 2.6 km [15,45,47]. Over one hundred structures have been found by Italian, Slovakian, and French archaeological missions, and include structural foundations of courtyard houses, church buildings, a central building, and an oval building of unknown purpose. The archaeological records support the presence of a Christian community and a monarchy sometime between the 7th to 9th century AD [15,47].

The Al-Quraniya mound, located in the west-central part of Failaka Island along the north shore, is roughly 550 m long and 250 m wide. With an elevation of roughly 6–8 m above sea level, Al-Quranyia is one of the highest architectural sites on the island, and it contains an old village with 30 structures, mixed grasses and shrubs, and a deserted modern farm [49]. The old village dates to roughly 200–440 years ago, according to the features of the Arabic corner tower, the square courtyard with buildings, the plain and incised pottery, and the glazed ware.

5. Results and Discussions

Failaka Island represents one of the most important heritage areas in the State of Kuwait, and there are plans from the state to construct a nature reserve, a residential area, and a tourist marina [15]. Failaka's good environmental condition, freshwater, and fertile soil have contributed to previous initiation of agricultural activities and human settlement as well. Lack of coral reefs on the coastal areas coupled with deep, wide gaps along the coastline have provided for the establishment of several ports on the island in the past ([19] and Figure 8c). Many researchers and archaeological missions concluded that Failaka Island was one of the important cultural centers in the Arabian Gulf region, as it has a distinctive cultural character among other places in the region [15,18,47]. The old general plan of Kuwait [20] mentioned the possibility of exploiting the island for tourism development, with the aim to make it like the Princess Islands (Turkey), Capri (Italy), or the Canary Islands (Spain).

5.1. Old Urban Plans

On top of socioeconomic characteristics, physical and environmental factors should also play a fundamental role in the site selection of a development plan. Failaka Island, with no exception, has environmental inputs that cannot be ignored in the site selection processes [19]. While there are multiple other urban plans for Failaka island formulated between 1964 and 2005 (see Figure 9), many reasons exist that make these plans invalid for implementation. Several examples of why previous urban plans for Failaka Island remain unacceptable include:

- Environmental changes on the island. For instance, an increase in the area of sabkhas and changes in land cover by the length of the island's population abandonment and change in State policy regarding settlement of the island.
- Disregard for important archaeological areas that should to be left empty and protected or developed as part of the tourist or urban areas to generate extra income.
- Spreading of sabkha on the island, which now constitutes 43.3% of the land area, and which should be left empty because of the difficulty in road construction. An intelligent environmental plan will reduce costs to the State rather than resorting to continual soil treatments and raising the level of sabkha, especially considering the effects of sea level rise since 1990, which increased the areal extent of sabkha during that timeframe.
- Ignoring beach type. Even going so far as to put a public beach in the area of rocky beaches, and camping areas in sabkhas ranging in depths from 0–3 m, creating hazardous areas for tourists.
- Lack of intensive fieldwork to survey the island, especially the most recent study, issued by the Municipality of Kuwait, showing that they did not consider the archaeological and environmental components of the island, or several other scientific standards (e.g., soils, geomorphology, prior land use).
- Overlooking the many environmental changes that took place in Failaka in recent years. For example, deserting the island reduced the agricultural areas used, leading to desertified areas where sabkhas have grown to cover 43.3% of the total land area [14,15].
- Being focused solely on the land layout of the island without considering the link between the existing land use patterns and those proposed. For example, it is necessary to compare the soil types (and condition) of the island with the designated site plans using a new soil map of Failaka Island based on recent studies [19,32], and creating a new soil map analysis via soil samples to determine soil production capacity based on total nutrients in the soil.

With regards to inclusion of soils, it should be noted that, although a few of the old plans (cf., [14,19,20,22](suggested site selection decision was heavily affected by soil fertility, these plans utilized a self-contained city model with agricultural areas that would cover a small population's food needs. The preservation of the agricultural land and allocation of a high proportion of land (50%) of these plans' areas for agricultural utilization is strong evidence of soil fertility's impact on site selection decisions. Even so, the studies focus remained on the most suitable places for agriculture only.

Urban planning based on engineering without considering environmental implications/biodiversity, geoarchaeological sites, and spatial modeling are no longer as suitable as they were in the past. Through spatial modeling based on multiple criteria, it is possible to produce urban plans that achieve the goal of sustainable development, inclusive of any factors influences that ensure preservation of environmental diversity. Additionally, numerous discoveries have occurred at Failaka's archaeological sites over the last two decades, and these should be taken into account when considering any type of plan moving forward.

Given these important reasons, the old urban plans are no longer suitable. Indeed, if those older plans were to be implemented, not only would any prior human footprint be erased by damaging/destroying archaeological areas, but the potential for environmental destruction and geomorphological hazards would also increase. Based on these criteria, there is an urgent need for a new urban plan.

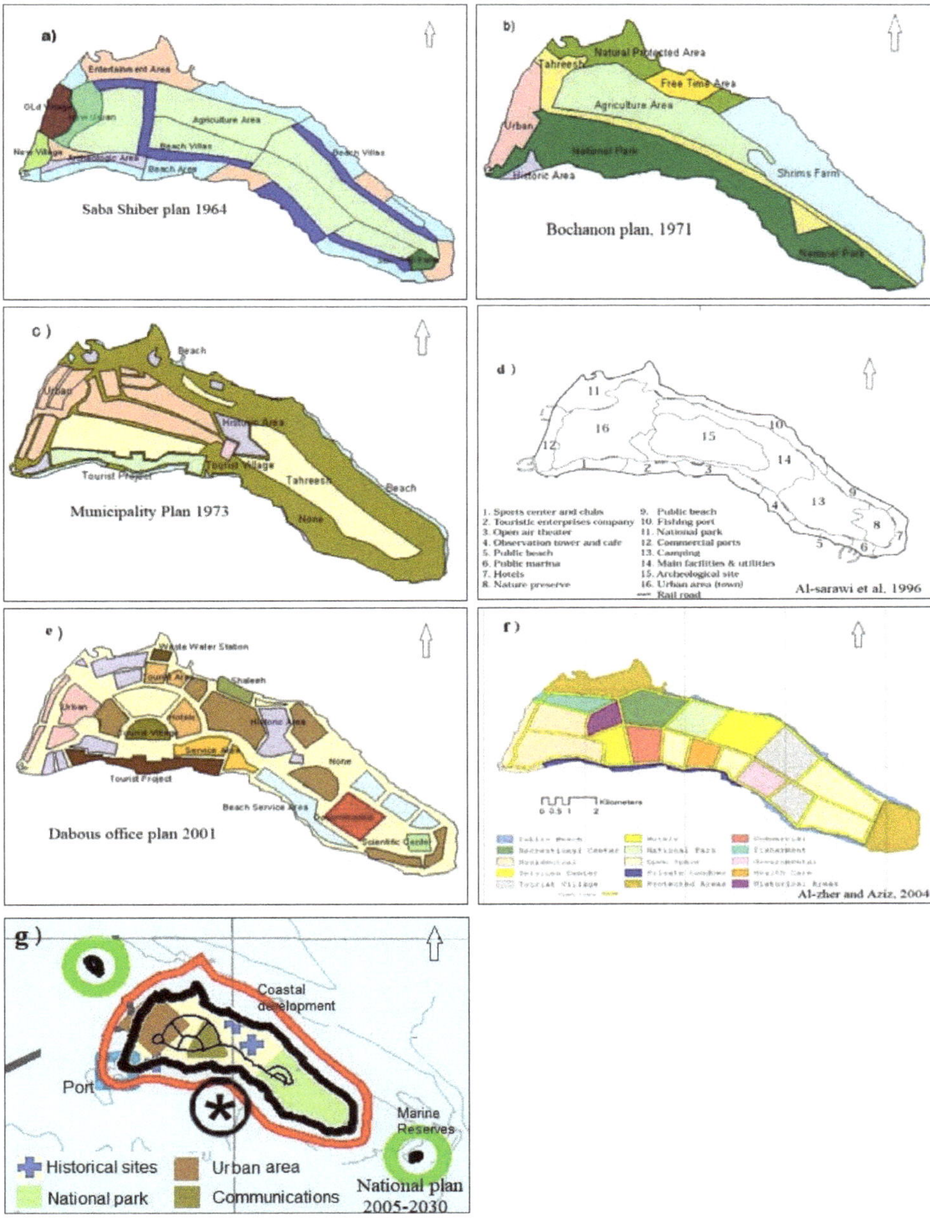

Figure 9. Several old urban plans of Failaka island. Source [14,19,20,22].

5.2. New Urban Plan for Failaka Island

The proposed urban plan for Failaka Island presented herein is based on environmental factors and natural characteristics, as well as socioeconomic factors. The selection and analysis of sites was dependent on different criteria, as noted in Table 4, and the proposed urban plan utilizes both high-precision data and high-resolution topographic surveys that relies on three main data sources that produced 13 individual data layers (Table 4): intensive field work, aerial survey by UAV, and high-resolution satellite imagery—all components neglected by previous plans.

Table 4. List of data sets used in this study.

Factors	Criteria	Type	Data Source and Setup
Environmental	Geomorphology	Favorable	Field work, drone image, World Veiw2, 2010, 2018
	Coastline type		Field work, drone image, World Veiw2, 2010, 2018
	Geology		KOC,1980
	Elevation (M)		Aerial photography, 2004, Kuwait Municipality, 25 cm
	Slope		From DEM
	Soil PH		Al-zaher and Aziz, 2004; Abbadi and El-Sheikh, 2002; Analysis of 34 soil samples
Cost-effectiveness	Distance to port Distance to water and energy line	Favorable	The island divided into four sectors from west to east, where priority was given to urbanization areas in the west than east
Land Availability	Archaeological area	Boolean	Archaeological areas have been added buffer zone to protect it from urbanism
	Natural hazard		Sabkha area
Socio-economic	Land use	Favorable	The island is divided into four zones, according to the type of land use
	Distance from main town		The island divided into four sectors from west to east, where priority was given to urbanization areas in the west than east
	The productive capacity of the land		Analysis of 34 soil samples

After correlating the multiple layers, environmental, cost-effectiveness, and socio-economic factors can be considered favorable criteria, while land availability can be considered a Boolean criterion. Upon further analysis of the layers, a set of indicators were identified to reflect criterion in Table 4, and the relationship between each indicator and suitability for a specific urban style that fits each proposed location.

Table 5 and Figure 10 specify the area of each anthropogeomorphological landform on the island and embody an essential pillar for understanding the Earth's surface on Failaka Island. For example, the already-existing urban area represents 5% of the island's total area, which means building on it is very important if the island's heritage sites are to remain intact (and perhaps subsequently used as an income-generating tourist attraction). In the same context, though the current road network is concentrated in the west of the island with a length of about 15 km, it is old, worn-out, and in need of maintenance. However, with a new plan, all new, proposed road networks could be moved away from archaeological areas and geomorphological hazards.

Table 5. Area of landforms and percentage of Failaka Island.

Landforms or Land Use	km^2	%
Sabkhas	20.12	43.3
Archaeology sites	1.79	3.9
Urban area	2.34	5
Non	22.19	47.8
Total	46.44	100

Source: Field study and [14].

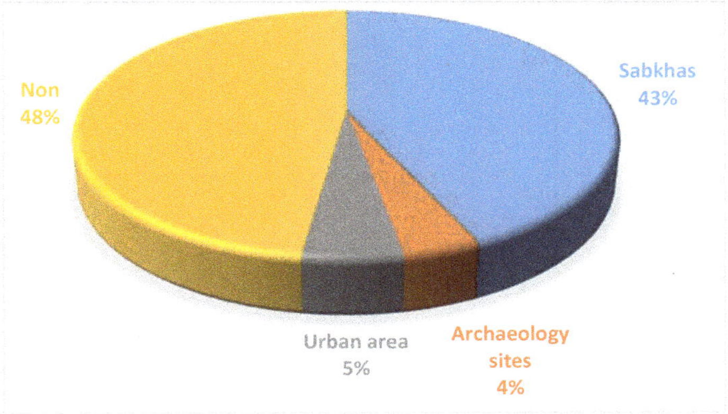

Figure 10. Landform percentage of Failaka Island.

Using the EAHP resulted in four suitability ranges: unsuitable, low, moderate, and high. From these data, a suitability map for Failaka island was created (Figure 11). As noted in Table 6, the total area of high suitability for urban use is 6.87 km^2 (14.81% of total land area), whereas the total area for moderate urban use is 5.63 km^2 (12.1% of total land area). Thus, just over one quarter of the island is potentially suitable for urban development based on the 13 criteria proposed by this study ("High" or "Moderate"). The rest of the island remains unsuitable for development because of the increasing sabkhas and/or rocky/muddy beaches.

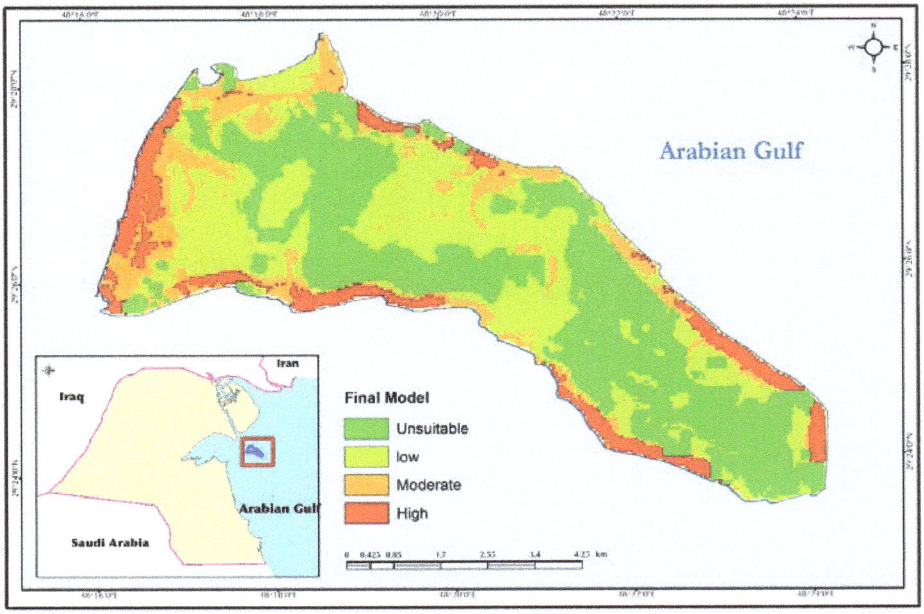

Figure 11. The most suitable areas for the proposed urban plan.

Table 6. Area and percentage of each category suitable for urbanization.

Category	Area km^2	%
Unsuitable	18.38	39.64625
Low	15.48	33.39085
Moderate	5.63	12.14409
High	6.87	14.81881
Total	46.36	100

6. Conclusions and Recommendations

Urban planning based on environmental criteria represents an essential pillar for achieving sustainable development for future cities. This study suggested the Environmental Analytical Hierarchical Process (EAHP) method and incorporated multiple accompanying factors into a geographic information system (GIS) for further analysis. Out of the 46.36 km^2 total land area of Failaka Island, only 12.5 km^2, or 27.9%, was found as "suitable" area for development ("High" and "Moderate", see Table 6 and Figure 11). Analysis also showed that sandy beaches of the island represent approximately 51% of the island's total beaches (and about 20 km in length) and, alongside archaeological sites, these remain an essential component for attracting weekend tourists.

Based on years of intense fieldwork by the current research team—as well as previous expeditions and up-to-date modeling and analysis of spatial data—this study suggests several recommendations for moving forward with the planning process of Failaka island as a tourist destination and potential UNESCO World Heritage Site. These include:

- Using transportation that does not pollute the environment, such as a solar-powered electric metro line, circumnavigating the island to avoid the range of Sabkhas and depressions located in the center of the island.
- Preventing vehicles from creating random passages in desert paths, which affects vegetation and living organisms and contributes to non-natural erosion patterns.
- Creating two areas for biodiversity on the island: "Al-Khidr Bay" in the northwest of the island, and "Sabkhas" located in the east of the island between Al-Awazim and Al-Sabahiya.
- Building a road along the coastline that encircles the island (perhaps adjacent to the solar-powered metro line) and two roads that cross the island from north to south, specifically avoiding desert depressions and sabkhas with an elevation of less than 3 m, to avoid the roads turning into salt lakes during the rainy season.
- Enhancing the natural areas of Failaka Island through increased protection and preservation efforts. These natural areas include archaeological sites, coastal dunes, and the coral reef within the intertidal flat.
- Allowing the inland sabkhas to remain in their natural state, using the coastal dunes as a coastal defense in some locations.

Given all the evidence from ground-truthing and GIS analysis, it is clear Failaka Island has environmental components and diversity in its archaeological sites which require Kuwaiti officials and decision-makers to plan in a way that ensures preservation of these factors for future generations—including Kuwait's largest heritage site. Failaka Island's unique physical and cultural diversity has made it a magnet for governmental officials, who want it to become a World Heritage Site, even hosting a recent UNESCO delegation visit. The plan presented herein supports the efforts to designate Failaka island as a World Heritage Site, build a residential city in the west of the island for occasional visitors and a small on-island population, and offer geoheritage and natural tourist attractions, such as the rich archaeological sites and calm, sandy beaches that comprise half of the island's coastline.

Author Contributions: Project creation, A.H. and M.G.A.; Funding acquisition, M.G.A.; Methodology, A.H., M.T., and C.D.A.; Project administration, M.G.A.; Writing original draft, A.H. and M.G.A.; Writing—review and editing, A.H., M.T. and C.D.A. All authors have read and agreed to the published version of the manuscript.

Funding: This research was funded by [Kuwait University, research sector] grant number [OG01\16].

Conflicts of Interest: The authors declare no conflict of interest.

References

1. UNDESA. *World Urbanization Prospects: The 2014 Revision*; United Nations, Department of Economic and Social Affairs: New York, NY, USA, 2014.
2. Feizizadeh, B.; Blaschke, T. Land suitability analysis for Tabriz County, Iran: A multi-criteria evaluation approach using GIS. *J. Environ. Plan. Manag.* **2013**, *56*, 1–23. [CrossRef]
3. Parry, J.; Showkat, G.; Sultan, B. GIS based land suitability analysis using AHP model for urban services planning in Srinagar and Jammu urban centers of JandK, India. *J. Urban Manag.* **2018**, *7*, 46–56. [CrossRef]
4. Mierzwiak, M.; Calka, B. Multi-Criteria Analysis for Solar Farm Location Suitability. *Rep. Geod. GeoInform.* **2017**, *104*, 20–32. [CrossRef]
5. Girvetz, E.; Thorne, J.; Berry, A.; Jaeger, J. Integration of landscape fragmentation analysis into regional planning: A statewide multi-scale case study from California, USA. *Landsc. Urban Plan.* **2008**, *86*, 205–218. [CrossRef]
6. Marull, J.; Pino, J.; Mallarach, J.; Cordobilla, M. A land suitability index for strategic environmental assessment in metropolitan areas. *Landsc. Urban Plan.* **2007**, *81*, 200–212. [CrossRef]
7. Rojas, C.; Pino, J.; Jaque, E. Strategic environmental assessment in Latin America: A methodological proposal for urban planning in the Metropolitan Area of Concepcion (Chile). *Land Use Policy* **2013**, *30*, 519–527. [CrossRef]
8. Hassaan, M.; Hassan, A.; Al-Dashti, H. GIS-based Suitability Analysis for Siting Solar Power Plants in Kuwait. *Egypt. J. Remote Sens. Space Sci.* **2020**, in press.
9. Store, R.; Kangas, J. Integrating spatial multi-criteria evaluation and expert knowledge for GIS-based habitat suitability modeling. *Landsc. Urban Plan.* **2001**, *55*, 79–93. [CrossRef]
10. Carr, M.; Zwick, P. Using GIS suitability analysis to identify potential future land use conflicts in North Central Florida. *J. Conserv. Plan.* **2005**, *1*, 89–105.
11. Steiner, F.; McSherry, L.; Cohen, J. Land suitability analysis for the upper Gila River watershed. *Landsc. Urban Plan.* **2000**, *50*, 199–214. [CrossRef]
12. Shukla, A.; Kumar, V.; Jain, K. Site suitability evaluation for urban development using remote sensing, GIS and analytic hierarchy process (AHP). In *Proceedings of the International Conference on Computer Vision and Image Processing*; Springer Science+Business Media: Singapore, 2017. [CrossRef]
13. Sandipan, D.; Anirban, B.; Sagar, M. Study on urban land suitability assessment using remote sensing and GIS: A case study of Khairagarh, in Chhattisgarh. *Int. J. Comput. Appl.* **2013**, *74*, 20–26.
14. Satellite Images: Worldview 2-50 cm -2010 and 2018; Google Earth 2018 and 2020; Esri-World Imagery, 2020; Aerial photography-2004 from Kuwait Municipality. Available online: https://apollomapping.com/worldview-2-satellite-imagery (accessed on 26 May 2020).
15. Al-Matar, M.; Hassan, A.; Hassan, A. Investigating Land-Use and Land-Cover Changes of Failaka Island: A Study in Geography and Geoarchaeology. *J. Soc. Sci.* **2020**, *48*. in press.
16. Mohamed, K. *Failaka Island Historical Map*; Kuwait Research and Studies Center: Kuwait City, Kuwait, 1997.
17. Ministry of Planning. *Census 1985*; Ministry of Planning: Kuwait City, Kuwait, 1985.
18. Mostafa, O. *Failaka Island Regional Study*; Kuwait Foundation of Advancement Science (KFAS): Kuwait City, Kuwait, 1988.
19. Al-Zaher, A.; Aziz, M. Optimal Urban Land Use in Failaka Island, Kuwait: Applied Study Using GIS and Remote Sensing. *Bull. Egypt. Geogr. Soc.* **2004**, *77*, 105–134.
20. Shiber, S. *The Kuwait Urbanization: Documentation, Analysis, Critique*; Kuwait Gov't Printing Press: Kuwait City, Kuwait, 1964; Available online: https://aukt.wordpress.com/2013/09/30/the-kuwait-urbanization-by-saba-george-shiber/ (accessed on 24 February 2020).
21. Mohamed, K. *Failaka Island: The Most Famous Kuwaiti Islands*; Kuwait Research and Studies Center: Kuwait City, Kuwait, 2006.

22. Al-Sarawi, M.; Marmoush, Y.; Lo, J.; Al-Salem, K. Coastal Management of Failaka Island. *J. Environ. Manag.* **1996**, *47*, 299–310. [CrossRef]
23. Herold, M.; Couclelis, H.; Clarke, K.C. The role of spatial metrics in the analysis and modeling of urban land use change. *Comput. Environ. Urban Syst.* **2005**, *29*, 369–399. [CrossRef]
24. Cabral, P.; Zamyatin, A. Three land change models for urban dynamics analysis in Sintra-Cascais area. In Proceedings of the first Workshop of the EARSEL SIG on Urban Remote Sensing: Challenges and Solutions. Humboldt-Universität zu Berlin, Berlin, Germany, 2–3 March 2006.
25. Henríquez, C.; Azócar, G.; Romero, H. Monitoring and modeling the urban growth of two mid-sized Chilean cities. *Habitat Int.* **2006**, *30*, 945–964. [CrossRef]
26. Fragkias, M.; Seto, K.C. Modeling urban growth in data-sparse environments: A new approach. *Environ. Plan. B Plan. Des.* **2007**, *34*, 858–883. [CrossRef]
27. Bolstad, P. *GIS Fundamentals: A First Text on Geographic Information Systems*; Eider Press: White Bear Lake, MN, USA, 2016.
28. Liu, R.; Ke, Z.; Zhijiao, Z.; Alistair, B. Land-use suitability analysis for urban development in Beijing. *J. Environ. Manag.* **2014**, *145*, 170–179. [CrossRef]
29. KOC–Kuwait Oil Company. *Geology Map of the State of Kuwait, Kuwait, Scale 1:50000*; KOC–Kuwait Oil Company: Al Ahmadi, Kuwait, 1980.
30. Al-Duwaysh, S. *Cultural Sites on the Western Coast of the Persian Gulf until the Third Century BC*; Center for Research and Studies of Kuwait: Kuwait City, Kuwait, 2015; Available online: http://www.crsk.edu.kw/ (accessed on 16 April 2018).
31. El-Baz, F.; Al-Sarawi, M. *Atlas of the State of Kuwait from Satellite Images*; Kuwait Foundation for the Advancement of Sciences: Kuwait City, Kuwait, 2000.
32. Vranova, V.; Danso, M.; Rejšek, K. Soil Scientific Research Methods Used in Archaeology–Promising Soil Biochemistry: A Mini-review. *Acta Univ.* **2015**, *63*, 1417–1426. [CrossRef]
33. Blum, W. Land Use Planning and Policy Implication: Bridging Between Science, Politics and Decision Making. In *Developments in Soil Classification, Land Use Planning and Policy Implications*; Shabbir, S., Faisal, T., Mahmoud, A., Eds.; Springer: Berlin, Germany, 2013.
34. Freedman, B. *Environmental Science: A Canadian Perspective*; Dalhousie University Libraries Digital Editions: Halifax, NS, Canada, 2018; Available online: https://ecampusontario.pressbooks.pub/environmentalscience/ (accessed on 12 June 2019).
35. United Nations (UN). The Sustainable Development Goals. Available online: https://www.un.org/sustainabledevelopment/ (accessed on 6 July 2020).
36. KMD–Kuwait Metrological Department. 2020. Available online: https://www.met.gov.kw/Climate/climate_hist.php?lang=arb (accessed on 21 May 2020).
37. Geologic Map of the State of Kuwait. Compiled by the Geological Survey of Austria, Vienna. 1968. Available online: http://nla.gov.au/nla.obj-2132686417 (accessed on 25 April 2020).
38. Holliday, V. *Soils in Archaeological Research*; Oxford University Press: New York, NY, USA, 2004.
39. Abbadi, G.; El-Sheikh, M. Vegetation analysis of Failaka Island (Kuwait). *J. Arid Environ.* **2002**, *50*, 153–165. [CrossRef]
40. Bathrellos, G.; Kalliopi, P.; Hariklia, S.; Dimitrios, P.; Konstantinos, C. Potential suitability for urban planning and industry development using natural hazard maps and geological–geomorphological parameters. *Environ. Earth Sci.* **2012**, *66*, 537–548. [CrossRef]
41. Thornbush, M.J.; Allen, C.D. *Urban Geomorphology: Landforms and Processes in Cities*; Elsevier: Amsterdam, The Netherlands, 2018.
42. Goudie, A.; Viles, H. *Geomorphology in the Anthropocene*; Cambridge University Press: Cambridge, UK, 2016. [CrossRef]
43. AlSahli, M.; Almutairi, F. Change of Northern Kuwait Shoreline and Its Related Physical Factors. *J. Soc. Sci.* **2019**, *47*, 1.
44. Bibby, G. *Looking for Dilumn*; Alfred A. Knopf. Inc.: New York, NY, USA, 1970.
45. Ashkanani, H. *Interregional Interaction and Dilmun Power in the Bronze Age: A Characterization Study of Ceramics from Bronze Age Sites in Kuwait and Its Neighbors*; NCCAL: Kuwait City, Kuwait, 2016.
46. Benediková, L.; Barta, P. A Bronze Age Settlement at Al-Khidr, Failaka Island, Kuwait. In Proceedings of the Seminar for Arabian Studies, London, UK, 24–26 July 2008; Volume 39, pp. 43–56.

47. Al-Mutairi, H. *Islamic Archaeology on the Northwest Coast of Failaka Island-A Comparative Analytical Archaeological Study*; National Council for Culture, Arts and Letters: Kuwait City, Kuwait, 2017.
48. Ashkanani, H. *Failaka Island in Postcards*; National Council for Culture, Arts, and Letters: Kuwait City, Kuwait, 2018.
49. Shehab, S.; Bartík, M.; Tirpák, J.; Ďuriš, J.; Barta, P.; Benediková, L. Bielich. Survey and Mapping of Al-Quraniya, Failaka Island, State of Kuwait. *Službách Archeol.* **2009**, *2*, 17–22.

sustainability

MDPI

Article

Academic Policy Regarding Sustainability and Artificial Intelligence (AI)

Muhammad Tanveer [1], Shafiqul Hassan [2,*] and Amiya Bhaumik [1]

[1] Lincoln University College, Kota Bharu 15050, Malaysia; cans_tanveer@hotmail.com (M.T.); amiya@lincoln.edu.my (A.B.)

[2] Prince Sultan University, Riyadh 66833, Saudi Arabia

* Correspondence: sHassan@psu.edu.sa

Received: 10 September 2020; Accepted: 27 October 2020; Published: 12 November 2020

Abstract: Artificial intelligence (AI) has grown, and technologies have intensified across all fields of life, particularly in education. AI has been applied to resources to improve skills giving teachers the time and freedom to provide understanding and adaptability and drive performance. This paper, written for policymakers in the field of education, highlights the impact of AI and advancements in academic policy. These academic policymakers generate ideas and strategies for applying AI across various disciplines. There is also discussion around AI implementation in education throughout developing nations for moving towards and ensuring affordable, high-quality education for every individual. Education for sustainable development (ESD) aims to promote the development of knowledge, skills, understanding, values and actions necessary to build a sustainable world, to protect and preserve the environment, and promote social equity and economic sustainability. This paper analyses how AI can be used to update learning probabilities by providing examples of how it can be integrated with existing educational systems, using data to improve educational capital and quality in developing countries. It goes on to discuss whether policymakers and institutions can reinvent and rework educational programs to polish graduates' skills for the growing presence of AI across all disciplines. There are four main parts to this work: (1) different dimensions regarding the complexities and potential implications, (2) the pros and cons of educational sustainability policy related to AI, (3) carving out AI and its outstanding execution, and finally (4) the linkage of AI with higher education within the context of educational expansions. In conclusion, the paper focuses on AI's applications, benefits and sustainable development education challenges.

Keywords: university education; academic policy; educational system; educational sustainability; artificial intelligence and higher education

---

1. Introduction

Artificial intelligence (AI) [1] refers to the simulation of human intelligence in machines that are programmed to think like humans and mimic their actions [2]. AI is an increasingly technical field that can even transform the way we socialise. AI is everywhere—it exists in innovative teaching and learning methodologies and it is studied in various contexts. AI requires complex technology and a vibrant, innovative community. Yet, what about the urgencies for AI in developing nations? Could AI resolve the issues caused by physical and social gaps in these nations? These are among the questions explored in this paper.

Historically, AI was developed as an academic field in 1955, and there was excitement at first, followed by a lack of funding and disappointment. AI research began in 1956 in a workshop at Dartmouth College, where John McCarthy coined the word "artificial intelligence" to differentiate from cybernetics, and to evade Norbert Wiener's influence [3].

From its commencement at the Dartmouth discussion in 1956, AI has drawn attention in business and technology [4]. Few technological advances have been as steadily unattended as AI. Although AI has existed for almost 60 years, it has remained a niche technology until recent fundamental changes known as "the great leap". The great leap brought computer abundance (big data), computational resources, economic exposure and developments in machine learning. This report focuses on AI and analytics, the latest decade's major technology buzzwords, and other topics in which algorithms and learning analytics are relevant, being innovations that are fit for each other.

While there is no clear and accepted definition of AI, numerous works of literature offer variations on the definition, including [5].

The study of artificial intelligence is to proceed based on the conjecture that every aspect of learning or any other feature of intelligence can in principle be so precisely described that a machine can be made to simulate it [6,7].

AI is progressing rapidly in the current age and having a great impact on the higher education sector. Universities are already using and benefiting from the AI of the IBM Watson supercomputer. IBM Watson offers assistance to students at Deakin University in Australia at any time of day, 365 days of the year [8]. AI is integral for university "outsourcing" using intelligent machines [9].

2. Literature Review

AI is a branch of computer science concerned with building smart machines that are capable of performing and even outperforming human jobs. The development of AI is shaping a growing assortment of sectors. AI is predicted to affect global productivity [10], equivalence and enclosure [11], eco-friendly outcomes [12] and numerous other areas, in both the short-and long-term [13,14]. AI has proven its position in a variety of fields as a game-changing force, triggering unprecedented transformations. Expert systems using AI can be programmed to communicate with the world through capacities such as visual intuition, speech recognition and smart actions that we consider to be inherently human.

There is comprehensive literature on the theory and practice of reasoning in AI. Researchers have investigated the growth, logical reasoning and qualitative temporal theoretical properties of AI [15]. For example, AI growth in Pakistan is attracting significantly however, contemporary national policy has also produced conditions in which private sector enterprises have gained considerable influence and agency within regional development networks [16]. Private education institutions have taken advantage of these favorable political conditions to develop, expand and apply AI technologies to their specific areas of extracurricular provision [17]. AI can also be applied in countries according to their cultural values and wealth. Advanced AI technology, research and resources are available through large computing centers, which have very high-energy requirements [18].

2.1. Dimensions of AI

The dimensions of AI are discussed in the following sections.

2.1.1. Judging Humanly

In the complete and literal sense, this is the new and exciting attempt to make machines "feel" through AI as with brains [19]. AI serves and inserts factors that influence our thoughts and how our actions can be implemented. This includes the automation of activities that we associate with human thinking, activities such as decision-making, problem solving and learning [20].

2.1.2. Behaving Humanly

The action of developing machines to perform tasks once performed by humans requires creativity [21]. The emergence of AI and its growing impact on many sectors necessitates an assessment of its effect on the attainment of sustainable development [22]. AI includes the study of how machines do activities that humans can do better [23].

2.1.3. Thinking Logically

To build AI that thinks logically, computer scientists must investigate cognitive skills via computer simulations [24]. Specific calculations can enable perceiving, reasoning and acting. Logical thinking and AI efforts to influence human wisdom. AI is not only capable of processing numerical data, but also learning human wisdom-based operations.

2.1.4. Acting Logically

Acting logically is the process of acting systematically to achieve given goals and beliefs. Computational intellectual capacity is the analysis of the intelligent operative design [25]. Acting logically is difficult to achieve, as stated by Vinuesa and colleagues:

> "On the other hand, it may also lead to additional qualification requirements for any job, consequently increasing the inherent inequalities, and acting as an inhibitor towards the achievement of this target [26]".

AI is influenced by a knowledge artefact attitude [27]. Figure 1 illustrates judging humanly in an attempt to make machines feel and behave like humans thinking logically and acting logically.

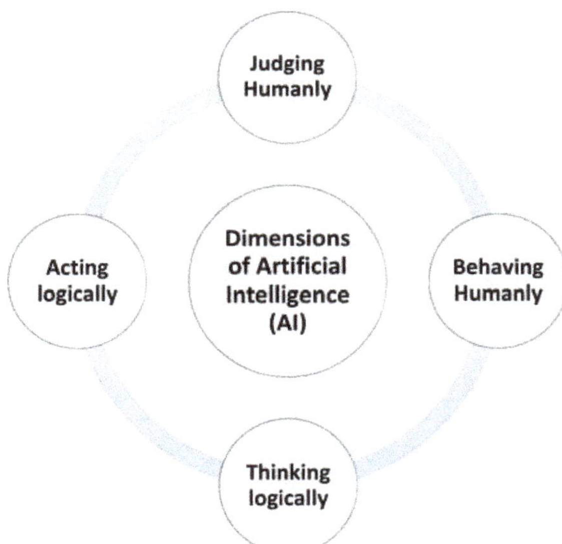

Figure 1. Developed by author through literature.

2.2. The Concept of Machine Learning and Analytics

The concept of AI is to address how we should create information systems that systematically evolve with practice. So, the question is: what are the simple laws concerning all learning methods? The term AI is commonly used in reference to any kind of programmed machine learning [28]. It is starting to replace "big data" and its "advanced analysis" and "predictive analytics". In general, machine-learning algorithms may decide how and where to handle important functions [29]. Deep learning is software development that builds a computer's scientific knowledge (and art) so that it learns from digital records and data to make decisions independently of humans [30]. It endows applications with just the power to read and respond to computer authoring tools. Machine learning is the process of achieving AI. It is also worth noting that without the use of machine learning, AI is still possible, but it would involve the construction of millions of coding and decoding structures.

2.3. Clarification

Figure 1 illustrates that judging humanly means attempting to make machines feel, think logically and act logically like humans. Work in AI has concentrated primarily on both the aforementioned elements of competence (understanding, thinking, data analysis, comprehension) and use of words. There are two forms of AI: statistics-driven AI and information-dependent AI. Both are centred on an abstract description of technical knowledge that a system is affecting. Most of AI's recent success can be attributed to advancements in statistics-driven AI. In 1959, just a few years after AI's emergence, Arthur Samuel coined the term "machine learning", describing the idea as "the ability to understand without specific programming" [31].

Deep learning is another commonly used term, being one of several machine-learning methods. Some alternative methods of deep learning include learning from the algorithm, inductive logic coding, clustering, confirming analysis and Bayesian systems. Deep learning is a machine-learning subfield. It is a new advancement to data learning representations that emphasises the learning of continuous levels of progressively significant depictions.

AI flourishes with information. With more information, the outcomes of AI systems become more reliable. Big data allows AI to stand on its own feet; without big data there can be no databased AI. A contemporary illustration of the term big data is a group of data sets with size far beyond the capability to collect, store, handle and analyse typical database software techniques [31].

Academic data analytics and curriculum management are two specific areas where machine learning is used in academic courses. Data mining is a subject in computer science that defines the process of finding interesting and informative trends and interactions in vast amounts of information. Educational data mining (EDM) creates approaches and employs mathematical, machine learning and data analysis strategies to analyse data obtained while teaching. Learning analytics (LA) is the assessment, compilation, review and reporting of learner data and their contexts to understand and improve the learning process effectively. The objective is to scale up the real-time use of LA by students, teachers/scientists and virtual school systems to improve the contributions of students at both the course and personal levels.

3. Using AI to Improve Performance and Productivity

Though AI has been facing many issues in the education sector, part of the problem is the hesitation of this sector to transform conventional methods into technological systems. AI became part of the narrative of improving schooling through the development of tutoring programs that could customise education. This is beginning to change as technology advances, with new models around the globe questioning AI in the field of education. The next section outlines examples of government policies, entrepreneurial commitments and corporate sector projects in mounting economies, as a peek into the initial phases of AI-integrated educational measures.

Promoting Personalisation and Optimising Learning Outcomes

AI can promote interactive learning. Among the most revolutionary elements of collective education supported by technology is that seen in situations when participants are not in the same location practically. This gives vector options for students to choose when and where they want to learn. Online discussion groups also play an important role in machine-supported interactive learning.

AI frameworks used to track distributed group discussions have focused on AI technologies such as machine learning and seamless document analysis while supplying educators with knowledge about student conversations or facilitating learner interaction and understanding. Teaching staff spend too much time on regular and logistical activities, including creating homework and addressing frequently asked questions (FAQs). Using a dual-teacher design featuring an instructor and a remote substitute teacher who can perform most of the teacher's daily role, teachers have more time to concentrate on student instruction and one-on-one contact.

Computer-assisted learning (CAL) provides options to virtual and AI learning approaches for students [32]. Using technologies to support students across various content pathways with the aid of their schools and teachers, AI can tailor education and maximise learning opportunities. As reported recently, smart, continuing education structures are part of current technical opportunities for expanding education in underdeveloped countries [33].

A great deal of time is spent on exams and assignments, so AI could be used as an analysis tool to rate instructors and save time. AI not only works for automating repetitive classification tasks, but also for examining papers in data analytics in education management information systems (EMIS) and for evolving learning management systems (LMS).

Overall, EMIS is an ideological movement of information and data systems that collaborate, process, archive, interpret and transmit strategic planning knowledge in the education sector. These data systems are commonly used at state, municipal and academic levels for political officials, decision-makers and administrators, and for the collection of official statistics. Applied to student performance testing data, results-driven decision making (DDDM) is a focus of many college and community initiatives, partly due to state- and federal-based transparency strategies. A structured and conceptual EMIS helps participants throughout all school-sector grades to obtain helpful information to organise and apply a system of education more effectively, create realistic and value-effective strategies, devise appropriate initiatives and track and analyse academic performance. In nations where information is comprehensive, trustworthy, consistently collated, analysed and stratified, EMIS will have tougher skills and the capacity to analyse facts and figures instantly and produce data analytics from school, state and federal levels. EMIS even opens up the prospect of creating lookup tables for policymaking.

4. Youth to Grow with AI in the Coming Years

Overall, businesses are quick to adopt solutions based on AI, suggesting an investment in new job types and abilities connected to a sector's use of AI. As such, the education sector has a clear need to react to rework teaching methods and revise regulations. No country on earth is prepared for smart automation. Even professionals in the system, as a logical answer to smart automation, are only beginning to emerge.

4.1. A Modern Framework for a Future Driven by the Internet and AI

Training plays a key role in making future, intelligence-ready workplace initiatives. Striking a balance in AI capabilities goes beyond introducing evermore effective learning-friendly innovations. In many universities, AI has penetrated and has led the growing concern that intelligent machines will soon replace many human beings. This article explores the complementarity of humans and AI, to offer a more constructive and realistic viewpoint on how everyone can contribute their power in context. This includes evaluating the curriculum and approaches that are used to provide training at all educational institutions. Hybrid learning is an educational model in which some students attend classes directly, while others attend remotely. Educators simultaneously instruct remote and onsite students using hardware and software for video conferencing [34].

4.1.1. Frameworks of Digital Skills

A core concept of data literacy is:

The capacity to securely and properly navigate, handle, recognize, incorporate, interact, analyse and generate content across mobile devices and enabling technology for social and economic activity. It involves skills that have been known as computer literacy, information communication technology literacy, knowledge literacy and intercultural communication [34].

Overall, the information below reveals the skills needed, characterised as a function of this structure. In collaboration with big private players, for example Cisco, the International Society for Technology in Education, Microsoft and Intel, the Communication and Information Technology

Proficiency Formula for Teachers (ICT CFT) was created by the United Nations Educational, Scientific and Cultural Organization (UNESCO) in 2011, and it was revised and updated in 2018. This system defines the skills that instructors now have to incorporate through their educational policies to improve essential awareness in participants.

Transmedia connects directly to changing educational needs in this transformative age. It not only helps students learn to use AI-led resources, but also to live and interact in a world of hybrid societies of human and digital agents. Furthermore, the system underlines the role of computer technology in serving the following six primary information areas:

1. Curriculum and evaluation;
2. Comprehension of information and communication technology (ICT);
3. Pedagogy;
4. Practiced education for professionals;
5. Organisation and establishment.

The program identifies three levels of gaining knowledge:

The European Digital Competence Framework or Dig Comp *(Joint Research Centre, 2018)* is another framework developed by the European Union (EU) to contribute to the professional development of an individual's technical capabilities. The framework maintains that the language is critical, and requires new technology in an integrated, effective, interactive and useful way to encourage people to achieve their objectives of service, research, leisure, inclusion and digital community engagement. The system is organised into five areas of expertise that define the essential components of technological skills, namely knowledge and software literacy, connectivity and interaction, digital media creation, security and critical thinking. It considers these aspects and traces them across four skill levels, including basic, advanced, qualified and highly skilled [35].

4.1.2. Computational Thinking (CT)

CT came to light as one of the key skills to boost teachers to prosper in a community operated by AI. The Computer Science Teachers Association in the United States of America (USA) describes CT as a major issue-resolving method for the activities indicated below:

1. Helping solve difficulties using a computer and other equipment;
2. Rationally arranging and collecting data;
3. Using templates and models to depict data;
4. Streamlining answers using the algorithms (a series of ordered steps);
5. Determining, examining and applying solutions to accomplish the most efficient and productive variety of paths and infrastructure; and
6. The major issue-solving method extended and generalised to a broad range of things *(International Association of Engineering Teachers of Learning and Computer Science, 2011)*.

Therefore, although it clearly belongs to the computer-engineering realm, CT is an ability that has roles in other areas. Due to the growing use of AI in working environments, CT has become a crucial skill for educators to come to terms with adjusting learning needs to the employment market. Therefore, many states have started to incorporate CT into their core curriculum.

Computer thinking uses heuristic data to find solutions in the face of uncertainty to learn and prepare. It is our collated searches that lead to a list of webpages—a winning strategy in computational thinking. Large volumes of data in time and space, processing power and storage capacity all contribute to computing thinking.

A State Commission survey revealed that while national governments of the EU are at different stages of incorporating CT into their national curricula, they have all started to work on it. Regarding the levels of CT integration in curricula around the world, the study [35] found three groupings:

1. Curriculum reform and modernisation began (during the last three to five years) in the United Kingdom France, Portugal, Italy and Finland;
2. Planning for integration of CT into the core curriculum in Denmark, Finland, Norway and the Czech Republic; and
3. Long records of practical education in Austria, Poland and Lithuania.

4.2. Building AI Capacity through Post-Basic Training and Education

There is a rising number of countries that have already created a comprehensive plan for AI. China, France, and as of recently, the USA have comprehensive strategies that cover initiatives conducted by businesses to innovate and incorporate new products and services and conduct research and development (R&D). The development process is often the first step. It shows the inordinate impact in the training of workers competent in AI. Overall, R&D encourages engagement in tertiary studies. By contrast, Finland has made choices through the development of a larger platform to easily achieve these goals of AI literacy, reaching one percent of their total workforce.

4.3. Higher Education and AI

AI is hypothetically, and early in practice, officially entering the higher education realm. One of the key strategies used by policymakers to resolve specific skill gaps is to develop AI skills across science and research. Most countries are looking to make AI science and practical careers more appealing to increase their skills in AI and become innovators.

4.4. Technical and Vocational Education Training (VET) and AI

VET organisations must also be able to offer systems that integrate AI-associated skills, particularly if they are planning and aiming to develop students with expertise that is adapted to the workforce. Long-term literacy in some countries is mentioned in practical skills training and services for adult learning. Thus, lifelong learning encompasses these and other policies, and it includes a broad spectrum of measures. Germany and Singapore have specific training plan structures. In Germany, eligible people looking for work or already employed may collect teaching gift cards from their corresponding engineering firm or workplace to use for appropriate training and professional development. Such grants are available at universities that are approved for further education *(Federal Ministry of Employment and Foreign Affairs, 2012)*.

5. AI induction in Education—Challenges and Implications for Policy

AI-powered providers have become widespread in many parts of the world, including the poorest countries. For example, in Kenya, robots are used for secure, discreet and anonymous responses to questions concerning family planning and for connecting people to medical professionals. These robots depend on AI technology. AI programs have also surfaced across many countries in the African farming sector.

Monetary and public freight sectors are further examples of areas where AI-powered technologies are changing people's lives. Tala is the number-one competition for financing in Kenya. It provides low-interest mortgages and clear arrangements on reimbursement. There seems to be no installing loan with potential customers. This demonstrates that even under harsh conditions, innovation is possible. Some scholars have built a structure for analysing health-conscious developments as developed through necessity under challenging conditions.

Questions being asked both by foreign aid and government policies include: How might developing countries with crippling social health issues cope with the intricate ecosystem in literacy that is necessary for AI solutions to grow? Why would government policy inspire instructors as being are central performers and not just bystanders? In developing countries, AI in education is often ignored as a component of a technologically advanced dialogue that guides a securely educated workforce and knowledge ecosystem service.

The final section of this report discusses the six key challenges of integrating AI throughout schooling, as both a way of improving wealth and academic value and encouraging knowledge. It covers the second main point of this article: AI as a new learning experience, and how schooling can support students and potential employees in an AI-powered environment. Figure 2 brings together the areas that AI have infiltrated in the education sector.

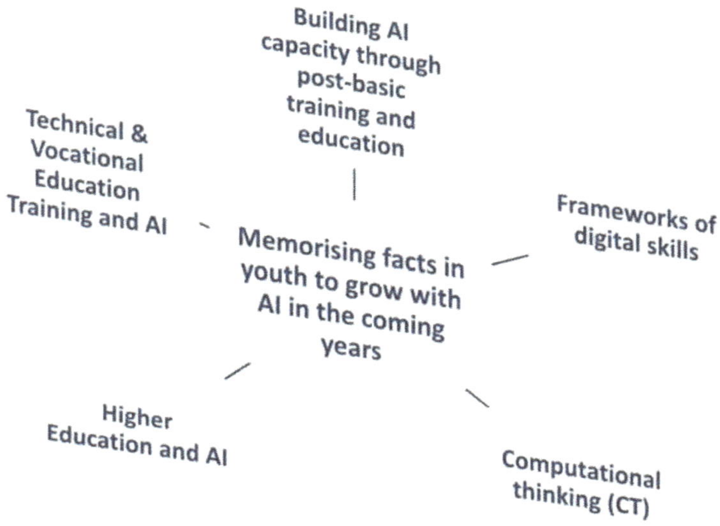

Figure 2. Developed by the author, deduced through literature.

AI is the perfect combination of something artificial with intelligence that makes it capable of learning. In addition, so many different views have been expressed about AI, and shown below are some of society's most important assessments against AI. Figure 2 shows the aspects of AI in the coming years, which are derived from the discussion made earlier. Figure 2 also presents the framework of AI induction in education.

5.1. An Extensive AI Policy for Sustainability

In the face of rapid advances in AI-powered technologies, the education industry is both a consumer and a user. As seen in France, China, Australia, Denmark, South Korea, Singapore and the USA, the educational dimension is crucial as companies grow international AI initiatives. AI has fantastic potential to increase educational systems; yet, the question is: how might AI help students, professors, administrative staff and lawmakers? Educational standards must provide students with the necessary skills to function in an AI-rounded society. Organisations including IBM, DreamBox, Pearson, Smart Sparrow, McGraw Hill, LightSide, Knewton, Coursera and Cerego are also making good progress in incorporating personalised research across smart technologies that customise studying using analytics.

AI deals with information, creativity, industry and new restrictions inside diverse systems. Analysts have predicted that the machine learning market will grow by 50 percent between 2017 and 2021. What were different countries' position, in a sense? Can these countries deal with the stress of the speed of private markets driving changes in technology? For developing countries, the best way to answer these burning questions is via more research.

To introduce the benefits of AI into education, governments must solve problems concurrently to produce ideas and policies and build or promote environments for creativity. Education policy implementation for AI is still in its adolescence, but it is likely to increase rapidly over the next decade. This is challenging in these early stages, as there are several holes in policy that need to be filled. Several policymakers are beginning to expand people's understanding and political debate on these topics to build a diverse environment for AI and for collective good. Throughout this process, it is vital to seek advice from experts and form teams to design plans, guidelines and policies that cover the following:

- Laws, policies and government institutions capable of dealing with AI in the field of innovation;
- Policies to protect private companies' use of AI in terms of security uses, anonymity and accountability of how architectures are developed;
- Policy for the educational and ethical implications of AI.

New national sector organisations and institutions are essential for maintaining economic development in a conceptual AI sense. The development of academic and research labs for student training, as well as training of AI professionals, are important to build new work opportunities. Research into AI is also underway in many countries:

- Regions are building labs and fertility clinics with state funds as part of plans to establish AI programs supporting government services;
- Regions are building labs and research labs with public money as part of plans to establish AI programs supporting essential services.

Collaborations between academic institutions can speed up AI competence growth. An important component of improving the AI curriculum and testing is public–private collaboration. Countries listed in this document have formed alliances between industry and government to exchange content and political power, as well as to ensure that training programs are well-matched with the needs of the workforce. Nevertheless, collaborations should never be confined to specific disciplines; intra-sector alliances are just as relevant as inter-sector alliances.

5.2. Ensuring Enclosure in Employment and Equality in AI

Equity and inclusion should be core values in the design of AI policies for education. Legislators must ask a set of questions regarding inclusion and equity when developing their reforms. For example, what environments and infrastructure are urgently needed in emerging countries to make AI possible in literacy? What have we learned from past observations, in terms of internet connectivity, to create secure and equal requirements for internet privacy? Numerous studies have plotted the barriers to introducing AI in academic achievement in underdeveloped countries. The major ones are the affordability and availability of:

1. Electricity;
2. ICT hardware;
3. Student basic ICT capabilities;
4. Data;
5. Culturally sensitive subject matter;
6. Language;
7. Internet robustness [33].

5.3. Training of AI-Enabled Educators and Students to Grasp Learning

There is no evidence of process-wide adoption of AI-based learning and teaching, including for systems management, even if new developments have yet to be halted by the educational software industry. Their fundamental problem is that they endorse new avenues for coordinating education that conflict with conventional approaches, instead of discussing the current problems and challenges facing students. There needs to be more detailed tests to reflect the possible benefits of new approaches.

Faculty training is a vital parameter for motivating them to seek guidance from instructional data to trace pedagogy and encourage pervasive use of AI in the classroom. That said, AI-enabled systems already offer incentives to eliminate some of the repetitive and logistical duties educators perform, including marking and maintaining records. Optimising these activities will free up instructors' resources, motivating teachers to dedicate greater attention to their organisation's artistic, empathic and stimulating facets.

5.4. Value Production and Functional Data Systems

Since data fuels AI, trustworthy and expeditious data are a mandatory requirement to installing increased big data systems for AI. A functional data analysis program with thorough and up-to-date data provides opportunities for probabilistic and computational modelling algorithms that have been enabled by AI. No algorithm, no matter how complex, will function correctly with incorrect data. A statistics-rich environment is, therefore, a necessity for AI-enabled programs. Nonetheless, the quality of data is a requirement that is essential but inadequate. AI-enabled programs are only as strong as the information they comprise. It is important to note that the education system is not the only database for information on the quality of education. The UNESCO Institute for Statistics (UIS) is the official and trusted source of internationally comparable data on education, science, culture and communication. Thus, residential results from UIS can also provide clues into extenuating factors that account for school teaching hardships.

5.5. The Relevance of AI Concepts and Research-Based Education

There is now a tremendous need to endorse AI technologies in learning in terms that serve to make society better adapted to AI-driven economic needs and practices. Ensuring efficient learning of AI is not only a question of resources, but also of measuring and assessing what works in training—distributing it in formats that are accessible to instructors and ideal for optimisation. The current issue is to understand AI usage and how it can contribute to educational achievement. This can be discussed using parallels and organised gradually with systematic studies and tests to create a reliable body of knowledge.

Instead of saying more research is required, the implication is that research needs to be directed with specific goals in mind. In learning, the so-called "prevalent associations" have had an omnipotent problem—the total number of associations that expand the uncertainty of separating impacts or merging the effects of published research papers. Since the impacts of instruction approaches are very difficult to eliminate, any explanation of strategies for optimising performance will reflect student skills and abilities, social class, encouragement and the connections between these. Because education anomalies are complicated and nuanced, the relevant questions are not whether to include AI in learning; the question is which AI technologies will help address the changing learning needs that instructors are tackling in the classroom, while taking into account circumstances and prospects [36]. If universities or organisations plan so that they spend less and get more, they will be successful and able to maintain and sustain success [37]. Nonetheless, in using AI, the real disruption of education is yet to come.

5.6. Transparency in Collecting Data

AI has many important implications; however, there are social and ethical issues that need attention. Many people read about arbitrarily biased AI programs [38] that are trying to take existence-impact life choices in a semi-transparent manner, anticipating to take all of our vacancies [31] and take control of the population *(The Register, 2018)*. In discussions on ethical data standards, data protection comes up almost immediately. The challenge is the willingness to use personal details while preserving the safety of personally identifiable information and privacy preferences. In addition, it is critical to install safeguards to discourage data theft. This becomes even more difficult in a school setting with young learners who, legally speaking, are unable to give written consent to the processing and use of their personal data [39].

6. Conclusions

Until now, the private industry has taken AI as a guiding light in most regions of the world. Technology giants of the USA and China dominate in the race to advance AI-enabled growth. The rapid growth of start-up companies has also played a key role in increasing AI acceptance. Particularly significant is the continued expansion of the EdTech sector, with AI-enabled learning tools in use throughout classrooms. Moreover, given the enormous impact of AI on all aspects of life, increasing numbers of policymakers can systematically introduce practical strategies to AI. Several nations have even published national AI plans, such as Denmark, China, France, Australia, the USA and South Korea. Research is a comprehensive element in any of these responses.

Unavoidably, AI is an industry that drives innovation, thereby enhancing the profitability of jurisdictions. Education and AI have penetrated the business sector, creating both socio-economic and economic opportunities. As robots become cheaper and more advanced, they move to new industries from their traditional roles in operations. The future of almost every industry and every human being is exaggerated by AI. The main drivers of emerging technologies, such as big data, robots and the Internet of Things are all part of AI and will continue to be technological innovators.

However, there is also space left for information sharing, the premise of which is experience sharing. It is now time to gather more information about how countries can move forward with this unsure and continuously evolving territory to promote collaborations and adopt detailed AI interpretations in the education sector. This can be a forum for community engagement and peer networking to build an AI education "aurora" to look at AI projects in communication and advise on national and international AI policy proposals. This observation tower (with a strong emphasis on the developing world) will help boost the dialogue between decision-makers, based on empirical evidence. The report concludes that the proposals of AI put forward are important starting steps that can direct the development of a real policy system. Certainly, engaging sectors in developing such a paradigm is critical, as AI's impact spreads across areas of the economy.

Moreover, as shown by the examples listed in this paper, multiple responses can be seen to the revisions that accompany the emergence of AI in respective disciplines of human movement. As evidence to the hypothesis, recently, a conference was scheduled to highlight the impact of studying AI on the progression of Pakistan [40,41].

Author Contributions: Main contribution to writing—original draft, M.T.; writing—review, S.H.; editing, M.T.; supervision, A.B. All authors have read and agreed to the published version of the manuscript.

Funding: All authors of this article would like to thank the Governance and Policy Design Research Lab of Prince Sultan University for their financial and academic support to conduct this research and publish it in Sustainability.

Conflicts of Interest: The authors declare that there is no conflict of interest.

References

1. Turing, A.M. *Computing Machinery and Intelligence*; Oxford University Press: Oxford, UK, 1950. [CrossRef]
2. Simmons, A.B.; Chappell, S.G. Artificial Intelligence-Definition and Practice. *IEEE J. Oceanic Eng.* **1988**, *13*, 14–42. [CrossRef]
3. Buchanan, B.G. A (very) brief history of artificial intelligence. *Ai Mag.* **2015**, *26*, 53.
4. Cummings, M.L. *International Security Department and US and the Americas Programme. Artificial Intelligence and the Future of Warfare*; Chatham House: London, UK, 2017.
5. McCarthy, J.; Minsky, M. A proposal for the dartmouth summer research project on artificial intelligence. *Al Mag.* **2006**, *27*, 12.
6. Russell, S.J.; Norvig, P. *Artificial Intelligence: A Modern Approach*, 3rd ed.; Prentice-Hall: Upper Saddle River, NJ, USA, 2010.

7. Gao, J.; Huang, X.; Zhang, L. Comparative Analysis between International Research Hotspots and National-Level Policy Keywords on Artificial Intelligence in China from 2009 to 2018. *Sustainability* **2019**, *11*, 6574. [CrossRef]
8. Deakin University. IBM Watson Now Powering Deakin. A New Partnership That Aims to Exceed Students' Needs. 2014. Available online: http://archive.li/kEnXm (accessed on 18 April 2020).
9. Grove, J. TeachHigher 'Disbanded' Ahead of Campus Protest. Times Higher Education. 2 June 2015. Available online: https://www.timeshighereducation.com/news/teachhigher-disbanded-ahead-campus-protest (accessed on 18 April 2020).
10. Acemoglu, D.; Restrepo, P. *Artificial Intelligence, Automation, and Work*; National Bereau of Economic Research: Cambridge, MA, USA, 2018.
11. Bolukbasi, T.; Chang, K.-W.; Zou, J.; Saligrama, V.; Kalai, A. Man is to computer programmer as woman is to homemaker? Debiasing word embeddings. *Adv. Neural Inf. Process. Syst.* **2016**, *29*, 4349–4357.
12. Norouzzadeh, M.S.; Nguyen, A.; Kosmala, M.; Swanson, A.; Palmer, M.S.; Packer, C.; Clune, J. Automatically identifying, counting, and describing wild animals in camera-trap images with deep learning. *Proc. Natl Acad. Sci. USA* **2018**, *115*, E5716–E5725. [CrossRef]
13. Tegmark, M. *Life 3.0: Being Human in the Age of Artificial Intelligence*; Random House Audio Publishing Group: New York, NY, USA, 2017.
14. Courtland, R. Bias detectives: The researchers striving to make algorithms fair. *Nature* **2018**, *558*, 357–360. [CrossRef]
15. Oke, S.A. A literature review on artificial intelligence. *Int. J. Inform. Manag. Sci.* **2008**, *19*, 535–570.
16. Tanveer, M.; Hassan, S. The role of new and creative ideas in developing industries of education, software and manufacturing in Pakistan. *J. Entrep. Educ.* **2020**, *23*.
17. Knox, J. Artificial intelligence and education in China. *Learn. Media Technol.* **2020**, *45*, 298–311. [CrossRef]
18. Jones, N. How to stop data centres from gobbling up the world's electricity. *Nature* **2018**, *561*, 163–166. [CrossRef] [PubMed]
19. Hageland, J. *Artificial Intelligence the Very Idea*; Mit Press: Cambridge, MA, USA, 1985.
20. Bellman, R.E.; Zadeh, L.A. Decision-Making in a Fuzzy Environment. *Manage. Sci.* **1970**, *17*. [CrossRef]
21. Kurzweil, R. *The Age of Intelligent Machines*; Mit Press: Cambridge, MA, USA, 1990.
22. Vinuesa, R.; Azizpour, H.; Leite, I.; Balaam, M.; Dignum, V.; Domisch, S.; Felländer, A.; Langhans, S.D.; Tegmark, M.; Nerini, F.F. The role of artificial intelligence in achieving the Sustainable Development Goals. *Nat. Commun.* **2020**, *11*, 233. [CrossRef]
23. Rich, E.; Knight, K. *Artificial Intelligence*; Tata Mcgraw-Hill: New Delhi, India, 1991.
24. Charniak, E.; Riesbeck, C.K.; McDermott, D.V.; Meehan, J.R. *Artificial Intelligence Programming*; Psychology Press: New York, NY, USA, 2013.
25. Poole, D.; Mackworth, A.; Goebel, R. *Computational Intelligence: A Logical Approach*; Oxford University Press: New York, NY, USA, 1998.
26. Nagano, A. Economic growth and automation risks in developing countries due to the transition toward digital modernity. In Proceedings of the 11th International Conference on Theory and Practice of Electronic Governance—ICEGOV', Galway, Ireland, 18 April 2018.
27. Nilsson, N. *Artificial Intelligence: A New Synthesis*; Elsevier: Amsterdam, The Netherlands, 1997.
28. Chou, J.S.; Ngo, N.T.; Chong, W.K. The use of artificial intelligence combiners for modeling steel pitting risk and corrosion rate. *Eng. Appl. Artif. Intell.* **2017**, *65*, 471–483. [CrossRef]
29. Domingos, P. A Few Useful Things to Know about Machine Learning. *Commun. ACM* **2012**, *55*, 78. [CrossRef]
30. Géron, A. *Hands-on Machine Learning with Scikit-Learn and Tensor Flow: Concepts, Tools, and Techniques to Build Intelligent Systems*; O'Reilly Media, Inc.: Sebastopol, CA, USA, 2017.
31. Manyika, J.; Chui, M.; Brown, B.; Bughin, J.; Dobbs, R.; Roxburgh, C.; Byers, A.H. Big Data: The Next Frontier for Innovation, Competition, and Productivity. Available online: https://www.mckinsey.com/business-functions/mckinsey-digital/our-insights/big-data-the-next-frontier-for-innovation# (accessed on 16 April 2020).
32. Schittek, M.; Mattheos, N.; Lyon, H.C.; Attström, R. Computer Assisted Learning. A Review. *Eur. J. Dental Educ.* **2001**, *5*, 93–100. [CrossRef]
33. Nye, B.D. Intelligent Tutoring Systems by and for the Developing World: A Review of Trends and Approaches for Educational Technology in a Global Context. *Int. J. Artif. Intell. Educ.* **2015**, *25*, 177–203. [CrossRef]

34. Tanveer, M.; Karim, D.; Mahbub, A. Higher Education Institutions and the Performance Management. Library Philosophy and Practice (e-journal). 2018, p. 2183. Available online: https://digitalcommons.unl.edu/libphilprac/2183 (accessed on 16 April 2020).
35. Hilbert, M. Big Data for Development: A Review of Promises and Challenges. *Dev. Policy Rev.* **2015**, *34*, 135–174. [CrossRef]
36. Hoodbhoy, Z.; Hasan, B.; Siddiqui, K. Does Artificial Intelligence Have Any Role in Healthcare in Low Resource Settings? *J. Med. Artif. Intell.* **2019**, *2*, 13. [CrossRef]
37. Tanveer, M.; Karim, D.; Mahbub, A. The Use of Performance Measurement in Universities of Pakistan. Library Philosophy and Practice (e-journal). 2019, p. 3010. Available online: https://digitalcommons.unl.edu/libphilprac/3010 (accessed on 16 April 2020).
38. ProPublica. Machine Bias. 2016. Available online: https://www.propublica.org/article/machine-biasrisk-assessments-in-criminal-sentencing (accessed on 18 April 2020).
39. Raissi, M.; Perdikaris, P.; Karniadakis, G.E. Physics informed deep learning (part I): Data-driven solutions of nonlinear partial differential equations. *arXiv* **2017**, arXiv:1711.10561.
40. Komal, A.S. Artificial Intelligence (AI), Machine Learning (ML) and Implications for Pakistan. Available online: https://thesvi.org/svi-in-house-seminar-panel-discussion-report-artificial-intelligence-ai-machine-learning-ml-and-implications-for-pakistan-august-20-2019/ (accessed on 21 April 2020).
41. Tanveer, M.; Bhaumik, A.; Haq, I.U. Pakistani Universities and Leadership Reconnoitering the Prospects of Furtherance. *J. Seybold Rep.* **2020**, *15*, 10.

sustainability

MDPI

Article

Long Range Transport of Southeast Asian PM2.5 Pollution to Northern Thailand during High Biomass Burning Episodes

Teerachai Amnuaylojaroen [1,2,*], Jirarat Inkom [1], Radshadaporn Janta [3] and Vanisa Surapipith [3]

1 School of Energy and Environment, University of Phayao, Phayao 56000, Thailand; 55086396@up.ac.th
2 Atmospheric Pollution and Climate Change Research Units, School of Energy and Environment, University of Phayao, Phayao 56000, Thailand
3 National Astronomical Research Institute of Thailand, Chiang Mai 53000, Thailand; radshadaporn@narit.or.th (R.J.); vanisa@narit.or.th (V.S.)
* Correspondence: teerachai.am@up.ac.th or teerachai4@gmail.com

Received: 14 October 2020; Accepted: 10 November 2020; Published: 2 December 2020

Abstract: This paper aims to investigate the potential contribution of biomass burning in PM2.5 pollution in Northern Thailand. We applied the coupled atmospheric and air pollution model which is based on the Weather Research and Forecasting Model (WRF) and a Hybrid Single-Particle Lagrangian Integrated Trajectory Model (HYSPLIT). The model output was compared to the ground-based measurements from the Pollution Control Department (PCD) to examine the model performance. As a result of the model evaluation, the meteorological variables agreed well with observations using the Index of Agreement (IOA) with ranges of 0.57 to 0.79 for temperature and 0.32 to 0.54 for wind speed, while the fractional biases of temperature and wind speed were 1.3 to 2.5 °C and 1.2 to 2.1 m/s. Analysis of the model and hotspots from the Moderate Imaging Spectroradiometer (MODIS) found that biomass burning from neighboring countries has greater potential to contribute to air pollution in northern Thailand than national emissions, which is indicated by the number of hotspot locations in Burma being greater than those in Thailand by two times under the influence of two major channels of Asian Monsoons, including easterly and northwesterly winds that bring pollutants from neighboring counties towards northern Thailand.

Keywords: PM2.5; biomass burning; long-range transport of PM2.5; source of PM2.5

1. Introduction

Air pollution is a widespread problem that affects human health and other atmospheric aspects, i.e., it can contribute to the warming of the atmosphere and can affect rain and cloud patterns. Air pollution is released from a number of man-made and natural sources including fossil fuels burning in electricity production, transportation, industry and households, agriculture, and waste processing. According to a 2014 World Health Organization (WHO) report the premature deaths of about 7 million people worldwide were caused by air pollution [1]. It is estimated that in developing countries approximately 300,000 to 700,000 people can be saved from premature death if aerosol levels are reduced to a safe level (an Air Quality Index (AQI) number under 100 signifies good or acceptable air quality) [2].

Southeast Asia is a region with frequent air pollution problems every year, particularly at the beginning of the year, from January to April. Biomass burning strongly dominates air pollution from the regional to local scale in Southeast Asia [3]. Moreover, severe haze events in this region caused by particulate pollution have become more intense and frequent in recent years. Widespread biomass burning occurrences and particulate pollutants from human activities other than biomass burning play important roles in degrading the air quality in Southeast Asia [4,5]. In addition, appropriate meteorological and topographic effects are also favorable conditions that contribute to the air pollution problem in Southeast Asia. In northern Thailand, all these factors are combined. Most cities in northern Thailand are located in the mountainous area and surrounded by paddy fields. Larger villages like Chiang Mai face increasing problems due to traffic jams, but farmers also burn stubble in preparation for the coming rain and rice planting at this time of the year, and these narrow valleys provide perfect bowls for this smog and smoke.

The contributing pollutants in Northern Thailand, however, are not only from national sources but also from long-distance air pollutants introduced by meteorological factors, i.e., wind, temperature, and humidity [6–9]. The climate of northern Thailand is characterized by monsoons. The period from mid-February until the end of May is the transition period from Northeast monsoon (most prevalent in December–January) to Southwest monsoon (most prevalent in July–August) climates. The hottest weather observed during March–April, coinciding with the presence of intensive thermal lows in the area [10]. During the transition period, winds can transport air pollutants from the surrounding area into northern Thailand [11]. The northeastern monsoon transfers air pollutants from China to the southern Chinese sea from mid-December to mid-April and can then flow continuously to the mainland Southeast Asia.

Exposure to high levels of air pollution can cause a variety of adverse health outcomes. PM2.5 is the most important air pollutant and strongly affects human health. Recent calculations of global premature mortality rates on the basis of high-resolution global O3 and PM2.5 models show that Southeast Asia and the Western Pacific account for around 25% and 45% of world mortality [12,13]. PM2.5 can penetrate deeply into the respiratory tract and enter the lungs. Exposure to small particles can also impair the function of the lungs and exacerbate medical conditions such as asthma and heart disease [14–16]. At the beginning of 2020, the PM2.5 measurements of air pollution in Chiang Mai, one of the largest cities in northern Thailand, reached a staggering level of 330 on the PM2.5 concentration over the weekend, making the northern city the most polluted city in the world, while the airborne pollution levels across Thailand's northern region vary between 100 and 390 of the air pollutants concentrations, according to AirVisual data. There have been a few studies exploring the source contributions of PM2.5 in this region. For example, Chueinta et al. [17] reported the characterization and source identification of fine and coarse particles collected in urban and suburban residential areas in Thailand and later performed an extended study on the Bangkok metropolitan curbside [18]. Leenanupan et al. [19] carried out similar work on the characterization of fine particulate pollution in the Mae Hong Son province in the north of Thailand. A few collaborative studies on fine and coarse particulate air pollution at the Asia Pacific regional scale were also reported, e.g., Oanh et al. [6,7]; Ebihara et al. [20,21]; Hopke et al. [22]. Moreover, only a few long-term PM2.5 and $PM_{10-2.5}$ monitoring data are available for elsewhere in this region. However, it is unclear how much neighboring countries contribute to the air pollution problem in northern Thailand. In addition, Lee et al. [5] found that nitrate aerosol was the major component of PM2.5 particles in Southeast Asia.

This work applied atmospheric model coupling with the air pollution model to investigate the potential contributions of biomass burning in air pollution in Northern Thailand. We used the Weather Research and Forecasting (WRF) model version 3.8.1 to simulate the meteorological conditions on March 2016. The model's initial and boundary conditions were taken from the Final Analysis Data (FNL) [23], and the modeled temperature and wind speed were compared to a dataset from the Pollution Control Department (PCD). To clarify the model's capabilities, statistical analyses such as Index of Agreement (IOA) and Fractional Bias FB were used for the model evaluation. The output from the WRF model was used as the meteorological conditions to be inserted into the Hybrid Single-Particle Lagrangian Integrated Trajectory (HYSPLIT) model to determine the long-range transport of regional air pollutants from the neighboring countries of Southeast Asia toward northern Thailand.

2. Materials and Methods

We use a coupled atmospheric model called the WRF model version 3.8 [24] to generate the meteorological input for HYSPLIT, which is an air quality model. The results from HYSPLIT were then used to identify the PM2.5 pathway in Southeast Asia.

The WRF model was developed to study several atmospheric factors and also to be used for operational weather forecasting. It is a non-hydrostatic mesoscale model consisting of several physical schemes, including radiation, cumulus, and microphysics. In this study, we designed 1 WRF domain with a horizontal resolution of 20 km grid spacing. In addition, the model sets 30 vertical levels up to 50 hPa. The outer domain entirely covers the upper mainland of Southeast Asia and some areas of East and South Asia, such as the South of China and East of India, as shown in Figure 1. The model configuration was listed in Table 1. Southeast Asia is influenced by East Asian monsoons, which carry the air mass from high latitudes to this region. The transboundary transport from the western border, such as Myanmar and India, also affect the quality of the air in northern Thailand, while the inner domain is located in northern Thailand. To solve the water vapor, cloud, and precipitation process, the model was configured using the WRF Single-Moment 3-class scheme according to Hong et al. [25]; Hong and Lim [26]. This method predicts a simple-ice system with three types of hydrometers: vapor, cloud water, and rain. The calculation of these processes is based on the mass content of the diagnostic relationship. The Kain–Fritsch scheme [27] is the sub-grid scale process used for the convective resolution. It has the potential to use a cloud model with updrafts and downdrafts, as well as to consider the effects of detrainment and training on cloud formation. The similarity theory scheme is used to emulate the thermal gradient over the surface responsible for friction velocity and wind over the surface [28–32]. The model spin-up was conducted from 15 to 28 February 2012 to reduce the effects from the initial conditions. From 1 March 2012 to 1 April 2012, the WRF model was designed to simulate the weather conditions for the HYSPLIT model. The main meteorological variables of the WRF model, i.e., wind (U, V, W), temperature (T), surface pressure (Psfc), and relative humidity (RH), were used as input data for the HYSPLIT model.

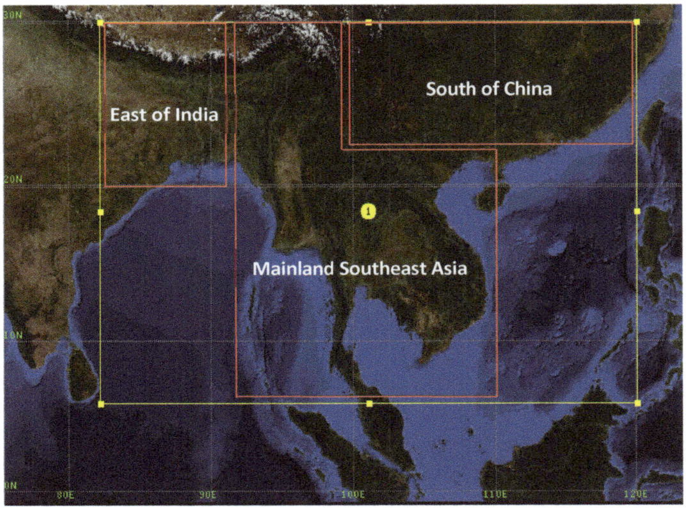

Figure 1. Domain configuration.

Table 1. Physics option in model configuration.

Scheme	Option
Microphysics	WRF Single Moment (WSM) 3-class simple ice scheme
Radiation	rrtm scheme
Surface Phyaics	Duhia
Land Surface Model	Noah Land Surface Model
Convective	Kain–Fritsch (new Eta) scheme

The HYSPLIT model [33] is designed to calculate both simple air plot pathways and complex dispersal and deposition simulations. This model is used to determine air pollutants in source–receptor relationships through trajectory analyses and predict dispersion for a number of events, such as volcanic eruptions, the transportation of wildfire smoke, and episodes of dust storms (http:/ready.arl.noaa.gov/index.php). HYSPLIT is a Lagrangian model, meaning that its scatter calculations follow the vector of transport, and only the computing point meteorological fields are required. HYSPLIT has the same horizontal coordination and map projections as the meteorological input. The meteorological profiles for the vertical grid are interpolated linearly onto an internal model following the field coordination. Particles released from a source are advected by a mid-wind field in a particle dispersion simulation and dispersed according to the random components caused by atmospheric turbulence. The meteorological data fields of each meteorological input source are then converted from (Gridded Binary (GRIB1 and GRIB2), netCDF, MM5, etc.) to a common format required for the running of HYSPLIT. Typically, 1 or 3 h weather data are the most commonly used for the calculation of transport and dispersion and are then temporarily interpolated linearly between the input data times to the HYSPLIT integration time.

The HYSPLIT model was driven to produce back trajectories of parcels originating from an observational site via the above-described WRF simulation atmospheric fields. The backward trajectory provides the Lagrangian paths of the air parcels within the timeframe chosen (24 h in the current study). This path is useful for identifying the source locations of the pollutants that fall within the backward trajectory. Based on the PM2.5 concentrations in 8 observation sites in northern Thailand in March, the maximum PM2.5 concentration for the back-trajectory analysis was found on 27 March 2012. Once the sources were identified, the HYSPLIT model was applied to each of the identified sites for a 24-h period. The computer domain in HYSPLIT was designed to have a horizontal grid of 300 to 300 cells, each with a resolution of 0.01° and eight vertical levels (500, 1000, 1500, 2000, 2500, and 3000 m above ground level). A full 3D dispersion particulate model with a total of 5000 particles released for each emission cycle was used. The speed of the horizontal and vertical turbulence was calculated using the Kantha and Clayson method [34]. The stability of the boundary layer was estimated by heat and momentum, and the mixed depth of the layer was derived from the meteorological model.

3. Results and Discussion

3.1. Meteorological and Trajectory Modelling

To analyze the potential contribution of biomass burning in air pollution in Northern Thailand, the meteorological conditions and tracing of the dispersed emission sources due to biomass burning from many locations are required. In this paper, we use the Lagrange particle dispersion model known as Hybrid Single-particle Lagrangian Integrated Trajectory (HYSPLIT) with meteorological data from the WRF simulation in March 2016 to determine the transport of particle air parcels. A Lagrangian approach was chosen here as this approach is suitable to study the transboundary transport of particles with backward trajectories of air parcels across the Southeast Asian region due to biomass burning during the dry season. In addition, we used the output from the WRF model to analyze the meteorological patterns in March 2016.

3.1.1. Meteorological Model Evaluation

Since we used the WRF simulation output to analyze the meteorological patterns and input the data into HYSPLIT, the model output needs to be evaluated by comparing the output with actual data from ground-based measurements. We compared the model results from the WRF model to the ground-based observations from the Pollution Control Department (PCD), as shown in Table 2. In general, the Thai PCD measures the hourly concentrations of six pollutants: PM_{10}, PM2.5, CO, NO_2, SO_2, and O_3. The PCD measurement sites are located near urban areas, so the results are likely to be influenced by the emissions of motor vehicles. Since the PCD recently measured PM2.5 in Thailand, some station sites offer a PM2.5 dataset. Here, we compared the modeled results to the PCD dataset at 3 locations, based on the complete PM2.5 dataset shown in Table 3. We used the statistical indicators, the Index of Agreement (IOA) and Fractional Bias (FB), to examine the efficiency of the model.

Table 2. Location of the Pollution Control Department (PCD) measurements in northern Thailand.

Location	Latitude	Longitude
Nan	18.78	100.77
Lampang	18.25	99.76
Chiang Mai	18.78	98.98

Table 3. Statistical analysis between the Weather Research and Forecasting (WRF) model and ground-based measurements.

Station	Temperature		Wind Speed	
	The Index of Agreement (IOA)	Fractional Bias (FB)	IOA	FB
Chiang Mai	0.79	2.5	0.54	2.1
Lampang	0.66	1.3	0.46	1.5
Nan	0.57	2.4	0.32	1.2

A comparison between the hourly output of the model and the observational data is shown in Figure 2. In general, the model results accurately reflected the real observations. The modeled temperature at 2 m was slightly higher than the ground-based measurements by 2–3 °C, while the wind speed was overestimated by 1 m/s. In addition, the statistical analysis shown in Table 3 indicates that the WRF and PCD datasets agree well in the hourly temperature and wind speed for most sites, with an acceptable IOA in the range of 0.57 to 0.79 for temperature and 0.32 to 0.54 for wind speed. However, the model generally overestimated the temperature by 1.3–2.5 °C and the temperature by 1.2–2.1 m/s, as indicated by Fractional Bias (FB). These errors are most likely to be caused by the rough resolution of the 20 km grid spacing model simulation, which has difficulty simulating sub-grid scale processes, such as convection and wind. In particular, the simulation of wind requires correction of the Large Eddy Simulation (LES). As discussed in Wurps et al. [35], a good LES simulation corresponds to a very fine resolution, such as 2.5 m, 10 m, and 20 m.

3.1.2. General Meteorology

The general meteorological conditions for the month of March 2016 in Southeast Asia can be better understood by analyzing the wind flow provided by the WRF modelling results. The Asian Winter Monsoon, which occurs from November to March, circulates air from mainland Asia towards the oceanic regions. During March 2016, the winds entered the north of Thailand through two major channels (Figure 3). The first channel is characterized by winds blowing from eastern Asia (e.g., eastern China and Taiwan) towards Laos and northern Thailand, while the second channel is characterized by northwesterly winds blowing from Burma and entering into northern Thailand. These flow patterns indicate that northern Thailand tends to be influenced by emissions from a wide range of biomass burning sources located in Laos and Burma.

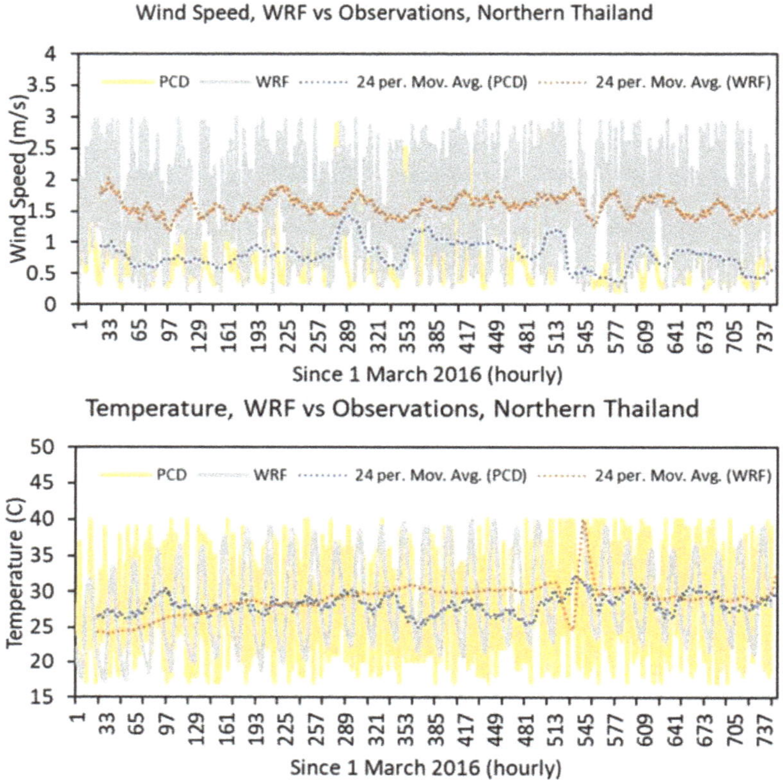

Figure 2. Average wind speed and temperature between the 1-h WRF model output (gray) and the hourly observations from PCD (yellow), as well as the 24-h WRF model output (orange-dots) and the 24-h observation from PCD (blue-dots).

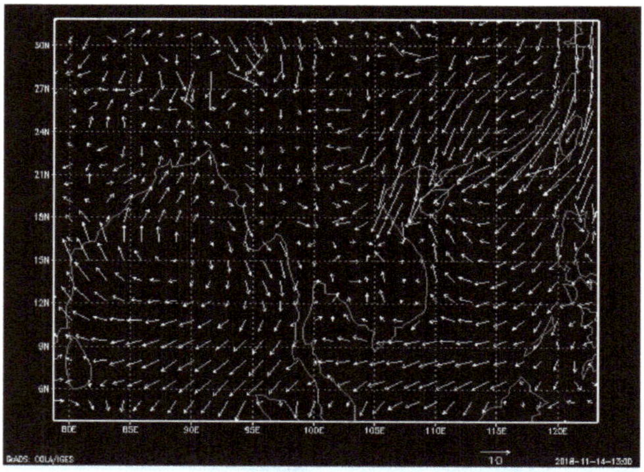

Figure 3. Average wind speed and direction from the WRF model in March 2016.

In addition, the influence of weather on air pollution in northern Thailand was analyzed using a linear and non-linear analysis. Data from PM2.5 and meteorological sources, such as temperature and wind, were used by the Pollution Control Department (PCD) to analyze the relationship between the weather factor and PM2.5 during 20–27 March 2016, indicating a high concentration of PM2.5 in northern Thailand. In this section, we use both linear and non-linear correlation analyses to determine this relationship. The linear analysis of the correlations was based on the Pearson correlation (1), while the non-linear correlation was based on the Spearman Equation (2).

Pearson's correlation coefficient is as follows:

$$r_{x,y} = \frac{\sum(x_i y_i - n\overline{xy})}{(n-1)s_x s_y} \tag{1}$$

where

$$s_x = \sqrt{\frac{1}{n-1}\sum_{i=1}^{n}(x_i - \overline{x})^2} \tag{2}$$

$$s_y = \sqrt{\frac{1}{n-1}\sum_{i=1}^{n}(y_i - \overline{y})^2} \tag{3}$$

$$\begin{aligned} \overline{x} &= \tfrac{1}{n}\textstyle\sum_{i=1}^{n} x_i \\ \overline{y} &= \tfrac{1}{n}\textstyle\sum_{i=1}^{n} y_i \end{aligned} \tag{4}$$

where n is the number of samples, and x_i, y_i are the individual samples indexed.

The Spearman Equation is as follows:

$$r_s = 1 - \frac{6\sum_{i=1}^{n} {d_i}^2}{n(n^2-1)} \tag{5}$$

where d_i is the difference between the two ranks of the dataset, and n is the number of datasets.

In general, the linear correlation analysis shows a slightly positive wind speed coefficient and a negative temperature coefficient of 0.082 and (−0.46) (Table 4). The non-linear correlation analysis shows a negative wind speed correlation with (−0.53), but a slightly positive temperature correlation with 0.18. Figure 4 shows a certain relationship between temperature and wind and a high concentration of PM2.5 during 23–26 March 2016. When the concentration of PM2.5 gradually increased from 60 μg/m^3 on 23 March to 90 μg/m^3 on 25 March, the temperature also decreased slowly from 28.5 °C on 23 March to below 27 °C on 25 March. At the same time, the wind speed was reduced from 3 m/s to 2.5 m/s. These results make sense in the context of meteorological factors affecting the quality of the air because low temperatures can suppress near-surface air pollution, while low wind speeds deteriorate air quality relative to near-ground pollutants due to limited air ventilation [36].

Table 4. Relationship between wind and temperature and PM2.5 in northern Thailand in March 2006.

	R	R_s
Wind vs. PM2.5	0.082	−0.53
Temp vs. PM2.5	−0.46	0.18

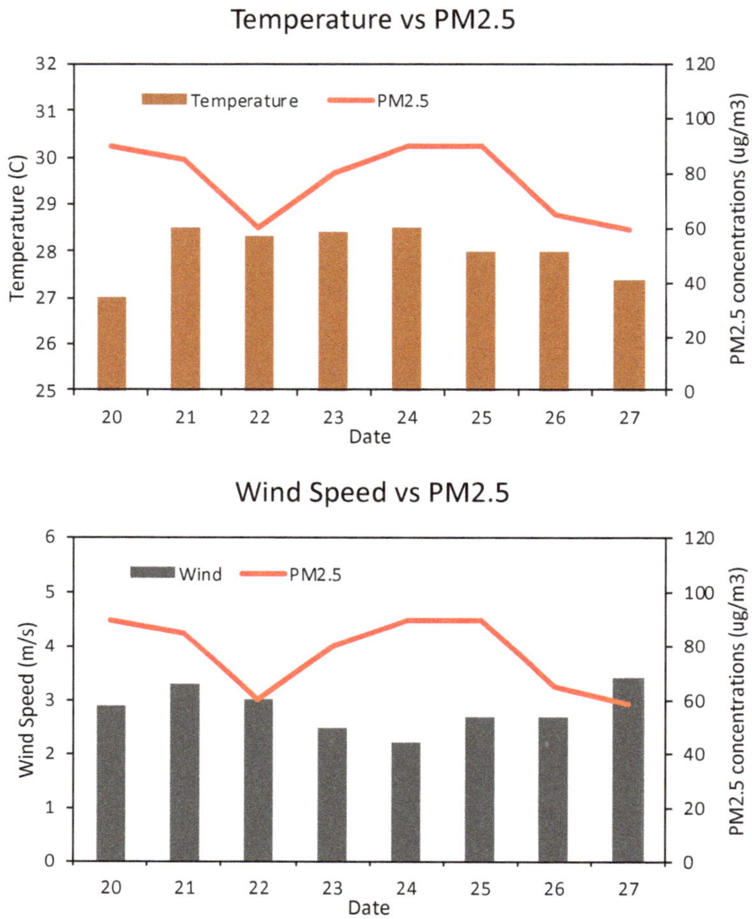

Figure 4. Relationship between temperature and PM2.5 (**up**), wind speed, and PM2.5 (**low**).

3.1.3. Backward Trajectory

To analyze the sources of PM2.5, we used the Lagrange particle dispersion model Hybrid Single-particle Lagrangian Integrated Trajectory (HYSPLIT) in backward mode with meteorological data from the WRF simulation from 20 to 27 March 2016 to determine the transport of particulate air parcels. HYSPLIT trajectory plots with eight trajectories were generated for the period from 0000 UTC 20 to 0000 UTC 27 March 2016, as shown in Figures 5 and 6. The backward trajectories started above eight locations and were based on the 8 provinces of northern Thailand at altitudes ranging from 500 m to 3000 m. In general, the north-eastern monsoon affected the north of Thailand. The backward trajectory map showed that the air mass that reached northern Thailand originated in the northeast. The winds were of continental origin and picked up during the day. Another channel was from the northwest airflow, with the backward trajectory map showing that the air mass reaching northern Thailand originated in Burma and some parts of India. At 500 m, the airflow from the north-east was located over northern Thailand, which is likely to bring some pollutants from eastern China, northern Vietnam, and Laos to the north of Thailand. However, at 3000 m, transport from Burma and India strongly dominate the emission sources of pollutants to the north of Thailand.

3.2. Hot Spot and Biomass Burning in Southeast Asia

To analyze the biomass burning situation in this region, we used Near Real Time (NRT) Moderate Imaging Spectroradiometer (MODIS) hotspot position data to evaluate Southeast Asian fires. A 1 km pixel center with the MODIS Fire and Thermal Anomaly Algorithm [37] was used to represent one or more fires that were thermal/active fires. This is the simplest method used to detect active fires and other thermal phenomena, including volcanoes. The fire hotspot data for January to April 2016 were obtained from https:/firms.modaps.eosdis.nasa.gov/fireinformation for the resource management system (FIRMS).

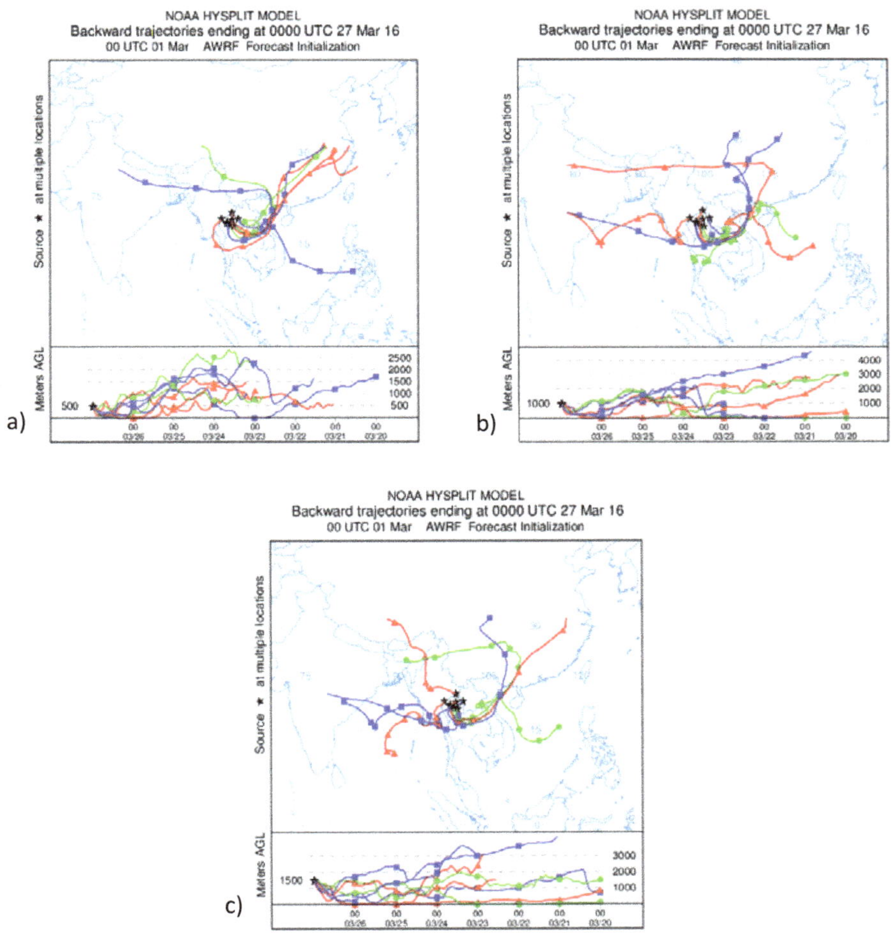

Figure 5. Hybrid Single-Particle Lagrangian Integrated Trajectory (HYSPLIT) backward ensemble Table (**a**) 500. m, (**b**) 1000 m, and (**c**) 1500 m, beginning on 0000 UTC 20 March 2016 and ending on 0000 UTC 27 March 2016 in northern Thailand.

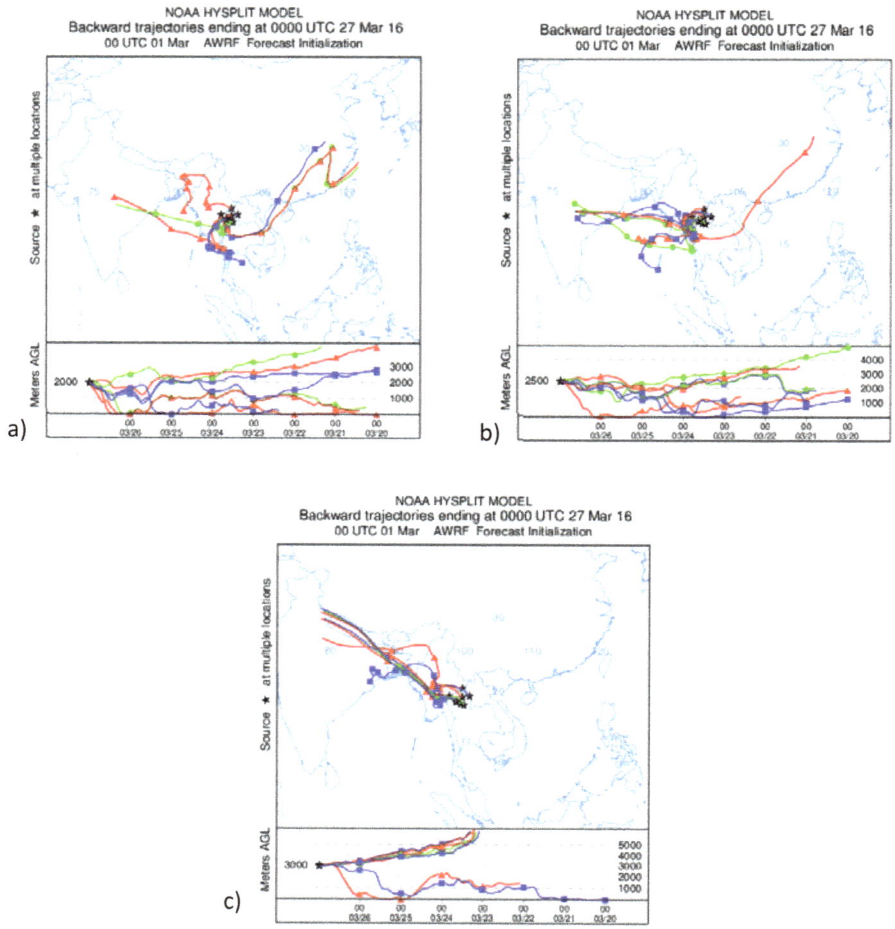

Figure 6. HYSPLIT backward ensemble trajectories initiated at (**a**) 2000 m, (**b**) 2500 m, and (**c**) 3000 m, beginning on 0000 UTC 20 March 2016 and ending on 0000 UTC 27 March 2016 in northern Thailand.

Table 5 shows the number of fire hotspot locations in Southeast Asia between January and April 2016. From January to April, the number of hotspot sites steadily increased three times. Cambodia contributed a significant number of hotspot locations (>10,000) in January and March, representing >40% of the total hotspots in south-east Asia, while Burma and Laos were the major contributors of hotspot locations (23,132 and 24,570, respectively) in March and April, representing approximately 37% and 28% of the hotspot locations in Southeast Asia, as mentioned in Figure 7. Indeed, the hotspot position in Thailand increased by 5 times between January and March. Focusing on hotspot locations in northern Thailand in March 2016 showed that this area contributed about 30% of all hotspots in Thailand, and Mae Hong Son province was the greatest contributor of hotspot locations, as listed in Table 6. The hotspots in Thailand, however, were about 16% lower than those in Burma by 2 times, while Laos still contributed around 18%. Hotspot locations in neighboring countries and in Thailand grew the average monthly PM2.5 concentrations in northern Thailand, which also continually increased to 70 μg/m^3 from October 2015 to March 2016 (Figure 8). The highest concentration of PM2.5 was approximately 96 μg/m^3 on 20 and 24 March 2016 in northern Thailand (Figure 8).

Table 5. Number of hotspot locations in Southeast Asia between January and April 2016. (https://firms.modaps.eosdis.nasa.gov/).

Country	January	February	March	April
Indonesia	3295	933	1899	1524
Burma	1507	4643	23,132	19,013
Cambodia	10,149	11,408	6763	6841
Laos	497	1855	11,284	24,570
Thailand	2580	5730	10,275	7819
Vietnam	1520	2612	5357	5094
Philippines	756	736	2344	2318
Malaysia	288	306	1195	1535
Total	20,609	28,221	62,260	68,730

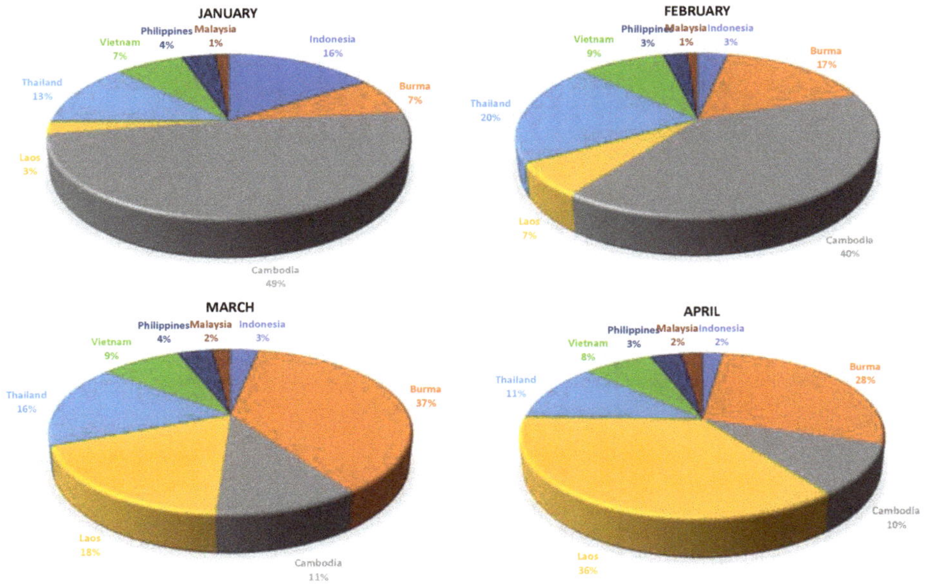

Figure 7. The proportion of hotspot locations for each country in Southeast Asia.

Table 6. Number of hotspot locations in northern Thailand in March 2016. (https://firms.modaps.eosdis.nasa.gov/).

Province	No. of Hotspot
Chiang rai	196
Chiang Mai	566
Nan	326
Phayao	147
Phrae	471
Mae Hong Son	795
Lampang	461
Lamphune	97
Total	3059

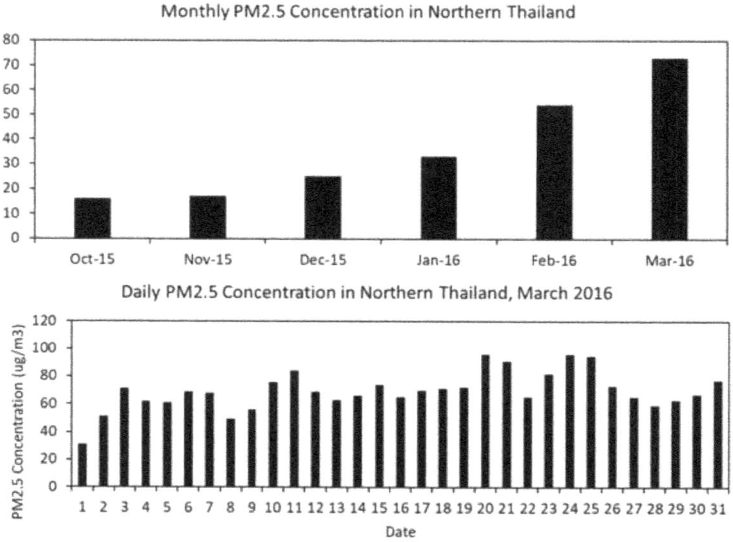

Figure 8. Monthly and daily PM2.5 concentrations in March 2016 in northern Thailand.

4. Conclusions

This paper aimed to understand the potential contributions of biomass burning to PM2.5 pollution in Northern Thailand. We developed a coupled atmospheric and air pollution modeling system based on the Weather Research and Forecasting Model and a Hybrid Single-Particle Lagrangian Integrated Trajectory Model (HYSPLIT). The WRF model was used with 1 domain of 20 km grid spacing covering Southeast Asia, some parts of India, and China. The final analysis data were used as a meteorological initial and boundary condition. The model results, including temperature and wind speed, were compared with the ground-based measurements from the Pollution Control Department (PCD) in northern Thailand. The model's capability was shown to be acceptable when compared to observations, as indicated by the Index of Agreement (IOA) in ranges of 0.57 to 0.79 for temperature and 0.32 to 0.54 for wind speed, while the fractional bias of temperature and wind speed were 1.3 to 2.5 °C and 1.2 to 2.1 m/s. The meteorological and trajectory analysis found that the influence of the Asian Winter Monsoon can carry air pollutants to northern Thailand through two major channels. The first channel is characterized by winds blowing from eastern Asia towards Laos into northern Thailand, while the second channel is characterized by northwesterly winds blowing from Burma and entering northern Thailand. Additionally, the low temperature and wind speed during this time in northern Thailand provide favorable conditions that contribute to the air pollution problem. The analysis of Near Real Time (NRT) Moderate Imaging Spectroradiometer (MODIS) hotspot data indicated that the biomass burning from Burma has greater potential to contribute to the air pollution problem in Thailand compared to national emissions.

Author Contributions: Conceptualization, T.A.; methodology, T.A. and J.I.; software, T.A.; validation, T.A. and J.I.; formal analysis, T.A.; investigation, T.A. and J.I., resources, T.A.; data curation, T.A. and J.I.; writing—original draft preparation, T.A.; R.J. writing—review and editing, T.A. and V.S.; visualization, T.A.; supervision, T.A. All authors have read and agreed to the published version of the manuscript.

Funding: This research was funded by National Astronomical Research Institute of Thailand (NARIT).

Acknowledgments: The author would like to thank Pollution Control Department (PCD) from Thailand for ground-based measurement dataset. We also acknowledge the use of data and imagery from LANCE FIRMS operated by NASA's Earth Science Data and Information System (ESDIS) with funding provided by NASA Headquarters.

Conflicts of Interest: The authors declare no conflict of interest.

References

1. WHO. 7 Million Premature Deaths Annually Linked to Air Pollution. Available online: http://www.who.int/mediacentre/news/releases/2014/air-pollution/en/ (accessed on 11 November 2020).
2. Seinfeld, J.H.; Pandis, S.N. *Atmospheric Chemistry and Physics: From Air Pollution to Climate Change*; John Wiley & Sons: Hoboken, NJ, USA, 2016.
3. Yin, S.; Wang, X.; Zhang, X.; Guo, M.; Miura, M.; Xiao, Y. Influence of biomass burning on local air pollution in mainland Southeast Asia from 2001 to 2016. *Environ. Pollut.* **2019**, *254*, 112949. [CrossRef] [PubMed]
4. Lee, H.H.; Iraqui, O.; Gu, Y.; Yim, S.H.L.; Chulakadabba, A.; Tonks, A.Y.M.; Yang, Z.; Wang, C. Impacts of air pollutants from fire and non-fire emissions on the regional air quality in Southeast Asia. *Atmos. Chem. Phys.* **2018**, *18*, 6141–6156. [CrossRef]
5. Lee, H.H.; Iraqui, O.; Wang, C. The Impact of Future Fuel Consumption on Regional Air Quality in Southeast Asia. *Sci. Rep.* **2019**, *9*, 2648. [CrossRef] [PubMed]
6. Oanh, N.K.; Upadhyay, N.; Zhuang, Y.-H.; Hao, Z.-P.; Murthy, D.; Lestari, P.; Villarin, J.; Chengchua, K.; Co, H.; Dung, N. Particulate air pollution in six Asian cities: Spatial and temporal distributions, and associated sources. *Atmos. Environ.* **2006**, *40*, 3367–3380. [CrossRef]
7. Oanh, N.T.K.; Leelasakultum, K. Analysis of meteorology and emission in haze episode prevalence over mountain-bounded region for early warning. *Sci. Total Environ.* **2011**, *409*, 2261–2271. [CrossRef]
8. Amnuaylojaroen, T.; Kreasuwun, J.; Towta, S.; Siriwitayakorn, K. Dispersion of particulate matter (PM10) from forest fires in Chiang Mai province, Thailand. *Chiang Mai J. Sci.* **2010**, *37*, 39–47.
9. Amnuaylojaroen, T.; Kreasuwun, J. Investigation of fine and coarse particulate matter from burning areas in Chiang Mai, Thailand using the WRF/CALPUFF. *Chiang Mai J. Sci.* **2012**, *39*, 311–326.
10. Thai Meteorological Department. The Climate of Thailand. Available online: https://www.tmd.go.th/en/archive/thailand_climate.pdf (accessed on 11 November 2020).
11. Cheng, I.; Zhang, L.; Blanchard, P.; Dalziel, J.; Tordon, R. Concentration-weighted trajectory approach to identifying potential sources of speciated atmospheric mercury at an urban coastal site in Nova Scotia, Canada. *Atmos. Chem. Phys.* **2013**, *13*, 6031–6048. [CrossRef]
12. Amnuaylojaroen, T.; Macantangay, R.C.; Khodmanee, S. Modeling the effect of VOCs from biomass burning emissions on ozone pollution in upper Southeast Asia. *Heliyon* **2019**, *5*, e02661. [CrossRef]
13. Lelieveld, J.; Barlas, C.; Giannadaki, D.; Pozzer, A. Model calculated global, regional and megacity premature mortality due to air pollution. *Atmos. Chem. Phys.* **2013**, *13*, 7023–7037. [CrossRef]
14. Vichit-Vadakan, N.; Ostro, B.D.; Chestnut, L.G.; Mills, D.M.; Aekplakorn, W.; Wangwongwatana, S.; Panich, N. Air pollution and respiratory symptoms: Results from three panel studies in Bangkok, Thailand. *Environ. Health Perspect.* **2001**, *109*, 381–387. [PubMed]
15. Tsai, F.C.; Smith, K.R.; Vichit-Vadakan, N.; Ostro, B.D.; Chestnut, L.G.; Kungskulniti, N. Indoor/outdoor PM10 and PM2.5 in Bangkok, Thailand. *J. Expo. Anal. Environ. Epidemiol.* **2000**, *10*, 15–26. [CrossRef] [PubMed]
16. Jinsart, W.; Tamura, K.; Loetkamonwit, S.; Thepanondh, S.; Karita, K.; Yano, E. Roadside particulate air pollution in Bangkok. *J. Air Waste Manag. Assoc.* **2002**, *52*, 1102–1110. [CrossRef] [PubMed]
17. Chueinta, W.; Hopke, P.K.; Paatero, P. Investigation of sources of atmospheric aerosol at urban and suburban residential areas in Thailand by positive matrix factorization. *Atmos. Environ.* **2000**, *34*, 3319–3329. [CrossRef]
18. Chueinta, W.; Bunprapob, S. *Elemental Quantification and Source Identification of Airborne Particulate Matter in Pathumwan District*; Chemistry and Material Science Research Program, Office of Atoms for Peace: Bangkok, Thailand, 2003.
19. Leenanupan, V.; Harnvong, T.; Sritusnee, U.; Bovornkitti, S. Elemental composition of atmospheric particulates in Mae Hong Son province. *J. Health Sci.* **2002**, *11*, 525–533.
20. Ebihara, M.; Chung, Y.; Chueinta, W.; Ni, B.-F.; Otoshi, T.; Oura, Y.; Santos, F.; Sasajima, F.; Wood, A. Collaborative monitoring study of airborne particulate matters among seven Asian countries. *J. Radioanal. Nucl. Chem.* **2006**, *269*, 259–266. [CrossRef]

21. Ebihara, M.; Chung, Y.; Dung, H.; Moon, J.; Ni, B.-F.; Otoshi, T.; Oura, Y.; Santos, F.; Sasajima, F.; Wee, B. Application of NAA to air particulate matter collected at thirteen sampling sites in eight Asian countries: A collaborative study. *J. Radioanal. Nucl. Chem.* **2008**, *278*, 463–467. [CrossRef]
22. Hopke, P.K.; Cohen, D.D.; Begum, B.A.; Biswas, S.K.; Ni, B.; Pandit, G.G.; Santoso, M.; Chung, Y.-S.; Davy, P.; Markwitz, A. Urban air quality in the Asian region. *Sci. Total Environ.* **2008**, *404*, 103–112. [CrossRef]
23. NCAR. NCEP FNL Operational Model Global Tropospheric Analyses, Continuing from July 1999. Available online: https://doi.org/10.5065/D6M043C6 (accessed on 11 November 2020).
24. Skamarock, W.; Klemp, J.B.; Dudhia, J.; Gill, D.O.; Barker, D.; Duda, M.G.; Huang, X.-Y.; Wang, W. *A Description of the Advanced Research WRF Version 3. NCAR Technical Note NCAR/TN-475+STR*; University Corporation for Atmospheric Research: Boulder, CO, USA, 2008. [CrossRef]
25. Hong, S.-Y.; Dudhia, J.; Chen, S.-H. A revised approach to ice microphysical processes for the bulk parameterization of clouds and precipitation. *Mon. Weather Rev.* **2004**, *132*, 103–120. [CrossRef]
26. Hong, S.-Y.; Lim, J.-O.J. The WRF single-moment 6-class microphysics scheme (WSM6). *Asia Pac. J. Atmos. Sci.* **2006**, *42*, 129–151.
27. Kain, J.S. The Kain–Fritsch convective parameterization: An update. *J. Appl. Meteorol.* **2004**, *43*, 170–181. [CrossRef]
28. Paulson, C.A. The mathematical representation of wind speed and temperature profiles in the unstable atmospheric surface layer. *J. Appl. Meteorol.* **1970**, *9*, 857–861. [CrossRef]
29. Dyer, A.; Hicks, B. Flux-gradient relationships in the constant flux layer. *Q. J. R. Meteorol. Soc.* **1970**, *96*, 715–721. [CrossRef]
30. Webb, E.K. Profile relationships: The log-linear range, and extension to strong stability. *Q. J. R. Meteorol. Soc.* **1970**, *96*, 67–90. [CrossRef]
31. Beljaars, A.C. The parametrization of surface fluxes in large-scale models under free convection. *Q. J. R. Meteorol. Soc.* **1995**, *121*, 255–270. [CrossRef]
32. Zhang, D.; Anthes, R.A. A high-resolution model of the planetary boundary layer—Sensitivity tests and comparisons with SESAME-79 data. *J. Appl. Meteorol.* **1982**, *21*, 1594–1609. [CrossRef]
33. Draxler, R.R.; Hess, G.D. An overview of the HYSPLIT_4 modelling system for trajectories. *Aust. Meteorol. Mag.* **1998**, *47*, 295–308.
34. Kantha, L.H.; Clayson, C.A. Small scale processes in geophysical fluid flows. *Int. Geophys. Ser.* **2000**, *68*, 883.
35. Wurps, H.; Steinfeld, G.; Heinz, S. Grid-Resolution Requirements for Large-Eddy Simulations of the Atmospheric Boundary Layer. *Bound. Layer Meteorol.* **2020**, *175*, 1–23. [CrossRef]
36. Jones, A.M.; Harrison, R.M.; Baker, J. The wind speed dependence of the concentrations of airborne particulate matter and NOx. *Atmos. Environ.* **2010**, *44*, 1682–1690. [CrossRef]
37. Giglio, L.; Schroeder, W.; Justice, C.O. The collection 6 MODIS active fire detection algorithm and fire products. *Remote Sens. Environ.* **2016**, *178*, 31–41. [CrossRef] [PubMed]

Publisher's Note: MDPI stays neutral with regard to jurisdictional claims in published maps and institutional affiliations.

sustainability

MDPI

Article

Characteristics of Non-Smokers' Exposure Using Indirect Smoking Indicators and Time Activity Patterns

Byung Lyul Woo [1,2], Min Kyung Lim [3], Eun Young Park [4], Jinhyeon Park [5], Hyeonsu Ryu [1], Dayoung Jung [1], Marcus J. Ramirez [2] and Wonho Yang [1,5,*]

[1] Department of Occupational Health, Daegu Catholic University, Gyeongsan-si, Gyeongbuk 38430, Korea; yissoyi@gmail.com (B.L.W.); clover92@cu.ac.kr (H.R.); ekdud37@naver.com (D.J.)
[2] Industrial Hygiene, Preventive Medicine, Force Health Protection, U. S. Army Medical Department Activity-Korea/65th Medical Brigade, Unit # 15281, APO AP 96271-5281, USA; marcusram2003@gmail.com
[3] Department of Cancer Control and Population Health, National Cancer Center, Graduate School of Cancer Science and Policy, Goyang-si, Gyeonggi-do 10408, Korea; mickey@ncc.re.kr
[4] Division of Cancer Prevention & Early Detection, National Cancer Control Institute, National Cancer Center, Goyang-si, Gyeonggi-do 10408, Korea; goajoa@ncc.re.kr
[5] Center of Environmental Health Monitoring, Daegu Catholic University, Gyeongsan-si, Gyeongbuk 38430, Korea; venza11@naver.com
* Correspondence: whyang@cu.ac.kr

Received: 12 October 2020; Accepted: 30 October 2020; Published: 1 November 2020

Abstract: Since the global enforcement of smoke-free policies, indoor smoking has decreased significantly, and the characteristics of non-smokers' exposure to secondhand smoke (SHS) has changed. The purpose of this study was to assess the temporal and spatial characteristics of SHS exposure in non-smokers by combining questionnaires and biomarkers with time activity patterns. To assess SHS exposure, biomarkers such as cotinine and 4-(methylnitrosamino)-1-3-(pyridyl)-1-butanol (NNAL) in urine and nicotine in hair were collected from 100 non-smokers in Seoul. Questionnaires about SHS exposure and time activity patterns were also obtained from the participants. The analysis of biomarker samples indicated that about 10% of participants were exposed to SHS when compared with the criteria from previous studies. However, 97% of the participants reported that they were exposed to SHS at least once weekly. The participants were most exposed to SHS in the outdoor microenvironment, where they spent approximately 1.2 h daily. There was a significant correlation between the participants' time spent outdoors and self-reported SHS exposure time (r^2 = 0.935). In this study, a methodology using time activity patterns to assess temporal and spatial characteristics of SHS exposure was suggested. The results of this study may help develop policies for managing SHS exposure, considering the time activity patterns.

Keywords: secondhand smoke; time activity; smoke-free; exposure assessment

1. Introduction

Secondhand smoke (SHS) is defined as a mixture of the smoke from the burning of tobacco products and smoke exhaled by smokers [1]. SHS exposure can cause adverse health effects such as respiratory disease, cardiovascular disease, and lung cancer [2,3]. In particular, exposure to SHS has become an important health issue due to its association with sudden infant death syndrome (SIDS) [4,5]. The World Health Organization (WHO) has estimated that, globally, more than 6 million people die from smoking, and the mortality rate from SHS exposure has reached 900,000 [6,7].

The WHO has organized the Framework Convention on Tobacco Control (FCTC) and encourages countries to implement education and awareness programs regarding the risks of SHS with the aim of a "Smoke Free World" [8]. According to the FCTC, fifteen developed and developing countries have successfully implemented smoke-free policies in indoor environments such as homes, workplaces, and public places during the last decade [9,10].

In Europe, under the theme of "lifting the smokescreen", the importance of SHS monitoring was recognized, and the results are being utilized to assess smoke-free policies [11]. In Canada, the Smoking Regulatory Index (a new way to measure public health performance) was developed and utilized to assess as an indicator of how well people are protected from SHS and the effectiveness of the policies to minimize SHS exposure [12]. Since the enactment of the National Health Promotion Act in Korea in 1995, indoor non-smoking areas have been gradually expanded, while a few outdoor areas have been designated as non-smoking areas through revisions of the corresponding laws in 2010 and 2012 [13,14]. As a result, the exposure rate of non-smokers to indoor SHS has significantly decreased [15].

However, as most indoor environments were designated as non-smoking areas, smokers had to shift their designated smoking areas from indoors to outdoors [16]. Consequently, non-smokers are more likely to be exposed to SHS at entrances or terraces of buildings with a smoking ban [17]. Hence, several countries are expanding non-smoking areas to outdoor locations of hospitals, kindergartens, restaurants, sports facilities, and public transportation [18,19]. However, smoking is still prevalent at the entrances of buildings such as restaurants, bars, shops, and universities in Korea [13]. Sureda et al. investigated the concentration of particulate matter with diameter less than 2.5 μm ($PM_{2.5}$) as an indicator of SHS at entrances and corridors of buildings with a smoking ban and compared the results, which showed that there was no significant difference in concentrations at either of the two places [20].

Previous studies have mainly used questionnaires and biological sampling for SHS indicators [21,22]. However, these methods provide limited information on specific SHS exposure duration and microenvironments. Therefore, this study attempted to understand the temporal and spatial characteristics of SHS exposure using time activity patterns, and evaluated the relationship between these characteristics and the biological sampling results.

2. Materials and Methods

2.1. Selection of Study Participants

Among the participants recruited for the basic SHS exposure survey conducted by the National Cancer Center, 100 non-smokers aged 19 to 74 and residing in Seoul were randomly selected. Questionnaires, time activity diaries, and biomarker samples were collected from April to May 2017. Past smokers who had quit smoking for more than 6 months and those who had not dyed their hair within 3 months were included. On the first day, the survey consent form and questionnaire were filled out along with collection of the first urine and hair samples. The participants were required to record their time activity diary for 7 days. On the seventh day, the time activity diary and the second urine samples were collected. This study was approved by the National Cancer Center for IRB vide approval number NCC2017-0074.

2.2. Questionnaire and Time Activity Diary

The questionnaire included demographic characteristics such as the subject's gender, age, occupation, housing type, and income, exposure characteristics to SHS such as places where the participants were mainly exposed to SHS, time spent in different microenvironments, the number of smokers within their family, and their viewpoint on the smoking policy. The used questionnaire was adapted from the previous studies [15,23–27]. In addition, the records of the participants' time activity patterns were collected to assess the exposure. The participants' activities and microenvironments were recorded every 30 min [28]. The number of smokers observed, the number of cigar butts found, and the number of self-reported exposure to SHS in the microenvironments were also recorded.

2.3. Cotinine

Cotinine has a half-life of 18–24 h, which means that the cumulative exposure to SHS can last 2–3 days [29–31]. In addition, cotinine is often used as an index of SHS exposure as internal cotinine concentration is rarely affected by other factors [32]. High-performance liquid chromatography tandem mass spectrometry (HPLC-MS/MS, Agilent 1100 series, and API 4000, AB Sciex, Santa Clara, USA) was used for analysis [33].

2.4. NNAL

NNAL (4-(methylnitrosamino)-1-3-(pyridyl)-1-butanol, a metabolite of 3-pyridyl-1-butanone) can be found in the urine of non-smokers exposed to SHS, and the half-life of NNAL is approximately 1 to 2 weeks. [21]. As NNAL and NNAL-Gluc are biomarkers of lung carcinogens released from cigarettes, they are not only extremely useful indicators, but have also been reported to be superior to other biomarkers in the comprehensive assessment of various biomarkers related to SHS [34]. Liquid chromatography tandem mass spectrometry (LC-MS/MS; Agilent 1260 series & Triple Quadrupole 5500: Turbo Ion Spray TM source, AB Sciex, Santa Clara, USA) was used for the NNAL analysis [33].

2.5. Nicotine in Hair

Hair is a stable specimen that is easy to collect and is hardly metabolized by drugs. As its growth rate is slow, hair reflects relatively long-term exposure to chemicals, and approximately 1 cm of hair reflects SHS exposure of approximately a month [35]. LC-MS/MS was used to analyze nicotine in hair samples [36].

2.6. Criteria for Biomarkers to Assess SHS Exposure

The criteria for classifying exposure to SHS of non-smokers were established by reviewing literature and articles suggested by several organizations [21,37,38]. The non-smokers were classified to be exposed SHS when the concentration above the limit of quantitation (LOQ) was detected as the results of biomarker analysis, or according to cut-off concentration (1 ng/mL, 3.77 pg/mL, and 2 ng/mg, cotinine in urine, NNAL in urine and nicotine in hair respectively). The limits of quantitation (LOQ) of cotinine, NNAL, and nicotine were obtained according to the Clinical & Laboratory Standards Institute (CLSI) guidelines.

2.7. Statistical Analysis

Correlation was used to compare the time activity patterns, concentrations of biomarkers, number of cigarette butts found, number of smokers observed, and the number of exposures to SHS. The results of correlation analysis were presented by coefficient of correlation (R^2), and statistical analysis was performed using IBM SPSS (Version 19).

3. Results

3.1. Characteristics of Participants and SHS Exposure at Microenvironments

Table 1 shows the demographic characteristics of the participants. All subjects were distributed uniformly by gender, age, and occupation. Of these, 16% were unemployed, 87% had an education level of college graduates or higher, 69% were married, and 70% resided in apartments or multi-family houses. Additionally, 27% were past smokers who had quit smoking, and 22% of the participants drank more than twice a week. A total of 97% participants reported that they were exposed to SHS at least once a week and that they experienced SHS exposure outdoors most of the time (92%). The SHS exposure at public places was 62%, which was higher than at home (36%) and workplaces (39%).

Table 1. General characteristics and major microenvironments where secondhand smoke (SHS) exposure occurred.

Characteristics	N	Home (%)	Workplace (%)	Public Place (%)	Boundary [1] (%)	Outdoor (%)
Gender						
Male	50	15 (30)	23 (46)	32 (64)	40 (80)	45 (90)
Female	50	21 (42)	16 (32)	30 (60)	39 (78)	47 (94)
Age						
20–29	17	6 (35)	6 (35)	13 (76)	15 (88)	16 (94)
30–39	25	8 (32)	14 (56)	18 (72)	21 (84)	24 (96)
40–49	23	7 (30)	10 (43)	12 (52)	18 (78)	21 (91)
50–59	16	7 (44)	4 (25)	11 (69)	12 (75)	15 (94)
60 ≤	19	8 (42)	5 (26)	8 (42)	13 (68)	16 (84)
Job						
Supervisory/profession	24	11 (46)	11 (46)	15 (63)	19 (78)	24 (100)
Office worker	23	8 (35)	9 (39)	15 (65)	19 (83)	20 (87)
Service/retail	21	6 (29)	13 (62)	14 (67)	19 (90)	21 (100)
Technician/labor	16	5 (31)	6 (38)	8 (50)	11 (69)	13 (81)
Unemployed	16	6 (38)	0 (0)	10 (63)	11(69)	14 (88)
Education						
High school or lower	13	5 (38)	5 (38)	6 (46)	9 (69)	11 (85)
College or higher	87	31 (36)	34 (39)	56 (64)	70 (80)	81 (93)
Household income [2]						
<USD 30,000	7	3 (43)	3 (43)	5 (71)	5 (71)	7 (100)
USD 30,000–60,000	31	10 (32)	11 (35)	20 (65)	22 (71)	27 (87)
USD 60,000–80,000	22	10 (45)	8 (36)	15 (68)	19 (86)	20 (91)
>USD 80,000	23	7 (30	11 (48)	14 (61)	18 (78)	22 (96)
Marital status						
Married	69	25 (36)	25 (36)	40 (58)	53 (77)	64 (93)
Not married	31	11 (35)	14 (45)	22 (71)	26 (84)	28 (90)
Residential type						
Single-family home	5	1 (20)	1 (20)	2 (40)	3 (60)	5 (100)
Apartment	70	25 (36)	25 (36)	45 (64)	55 (79)	62 (89)
Townhouse	20	8 (40)	10 (50)	12 (60)	18 (90)	20 (100)
Multi-purpose building [3]	5	2 (40)	3 (60)	3 (60)	3 (60)	5 (100)
Past smoker						
Yes	27	12 (44)	12 (44)	19 (70)	23 (85)	25 (93)
No	73	24 (33)	27 (37)	43 (59)	56 (77)	67 (92)
Alcohol consumption						
Never	22	7 (32)	4 (18)	13 (59)	17 (77)	19 (86)
once/month	28	12 (43)	11(39)	19 (68)	21 (75)	26 (93)
once/week	28	9 (32)	12 (43)	18 (64)	22 (78)	26 (93)
≥twice/week	22	8 (36)	12 (55)	12 (55)	19 (86)	21 (95)
Total	100	36 (36)	39 (39)	62 (62)	79 (79)	92 (92)

[1] Boundary: entrance, porch, terrace, top of building, parking lot, and so on. [2] Thirteen unemployed and one respondent did not answer to this question. [3] Mixed use buildings, e.g., residential and commercial apartment, office, and hotel.

3.2. SHS Exposure Duration at Microenvironments

Figure 1 shows the duration of SHS exposure for each microenvironment. The participants reported that they were exposed to SHS for an average of 4.5 h during a week, and the exposure duration was longer indoors than it was outdoors. The SHS exposure duration was the longest in indoor workplaces and shortest in indoor public places. There was no statistically significant difference in exposure time by microenvironment. The SHS exposure duration was longer in restaurants, followed by bars and transportation-related facilities.

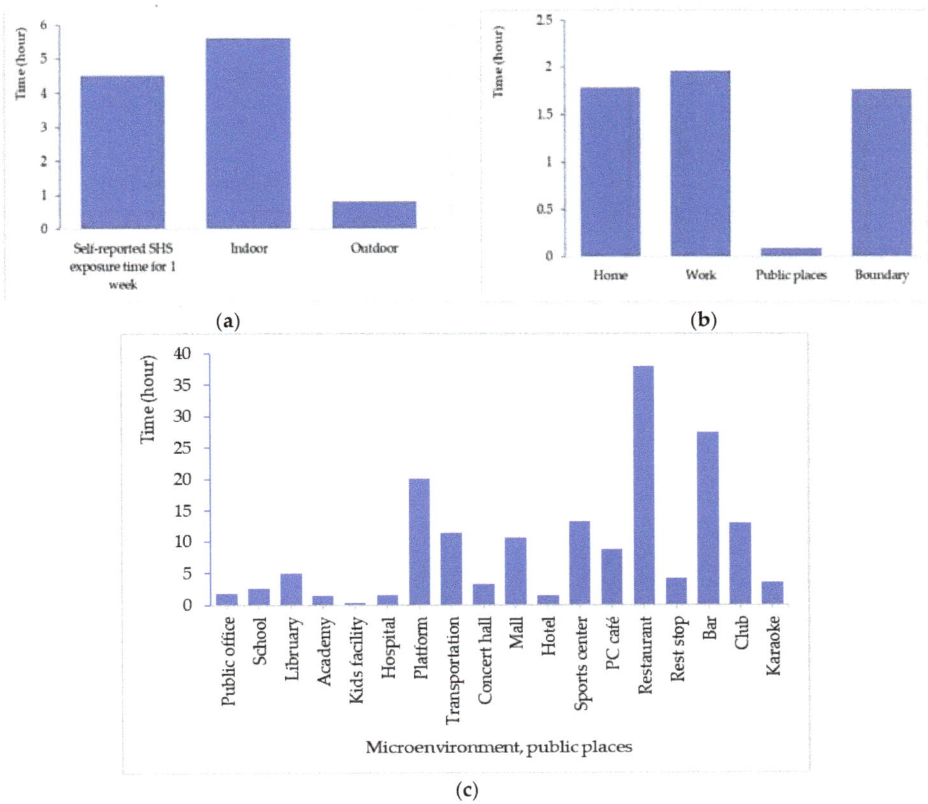

Figure 1. SHS exposure duration: (**a**) average for a week, indoor and outdoor; (**b**) home, workplace, public place, and boundary; and (**c**) public places.

3.3. Time Activity Pattern

The microenvironments where the participants stayed and the time spent in each microenvironment during a week are shown in Table 2. The participants spent most of their time indoors and spent approximately 1.2 h outdoors per day. The participants' time spent at the workplace and public transportation for weekdays was relatively higher than on weekends. In contrast, the participants spent more time at home, and the duration of private car use increased during the weekend.

Frequency distributions of the time spent in each microenvironment for weekdays and weekends are shown in Figure 2. The participants stayed at home most of time, and their working hours were approximately from 7 am to 6 pm during weekdays. The rate of visiting and using restaurants peaked at noon, and time spent outdoors increased at 8 a.m., 12 p.m., and 6 p.m. In addition, the rate of transportation usage peaked at 8 a.m. and 7 p.m. during weekdays. The participants' time spent at home increased, and working hours significantly decreased on weekends. The time spent on leisure, other indoor events, outdoor occasions, and transportation was centralized at around 3 p.m.

Table 2. Time spent (h) indoors, outdoors, and in transportation.

Microenvironment	Weekday			Weekend			Total		
	Mean ± SD	Rate (%)	Sum (%)	Mean ± SD	Rate (%)	Sum (%)	Mean ± SD	Rate (%)	Sum (%)
Indoor			84.08			84.72			84.26
Home	13.18 ± 3.20	54.92		16.71± 4.27	69.61		14.19 ± 2.87	59.12	
Workplace	5.04 ± 3.37	20.98		1.11 ± 2.51	4.60		3.91 ± 2.60	16.30	
Leisure	0.33 ± 0.68	1.38		0.28 ± 0.89	1.15		0.31 ± 0.56	1.31	
Restaurant/bar	0.61 ± 0.56	2.54		0.59 ± 0.74	2.46		0.60 ± 0.52	2.52	
Others	1.02 ± 2.11	4.26		1.66 ± 2.74	6.90		1.25 ± 2.15	5.01	
Outdoor			7.60			9.15			8.04
Home	0.01 ± 0.06	0.05		0.09 ± 0.44	0.38		0.03 ± 0.14	0.14	
Workplace	0.26 ± 0.83	1.09		0.05 ± 0.27	0.22		0.20 ± 0.63	0.84	
Leisure	0.26 ± 0.62	1.09		0.47 ± 0.84	1.95		0.32 ± 0.58	1.34	
Restaurant/bar	0.15 ± 0.30	0.60		0.19 ± 0.65	0.79		0.15 ± 0.36	0.66	
Others	0.15 ± 1.00	4.77		1.40 ± 1.37	5.81		1.22 ± 0.94	5.07	
Transportation			8.32			6.14			7.7
Bus	0.51 ± 5.89	2.24		0.28 ± 0.72	1.15		0.46 ± 0.51	1.93	
Subway	0.75 ± 0.80	3.14		0.31 ± 0.59	1.28		0.63 ± 0.60	2.61	
Taxi	0.03 ± 0.09	0.13		0.02 ± 0.09	0.09		0.03 ± 0.07	0.12	
Car	0.59 ± 0.99	2.45		0.82 ± 1.04	3.40		0.65 ± 0.89	2.72	
Others	0.09 ± 0.25	0.36		0.05 ± 0.25	0.22		0.08 ± 0.24	0.32	

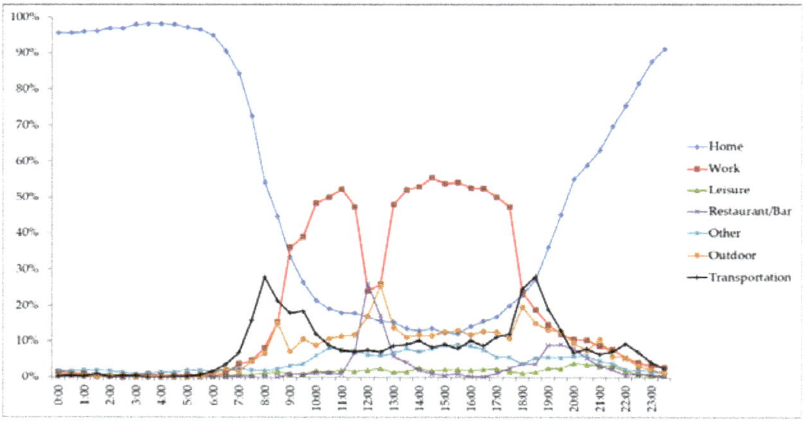

(**a**) Weekday

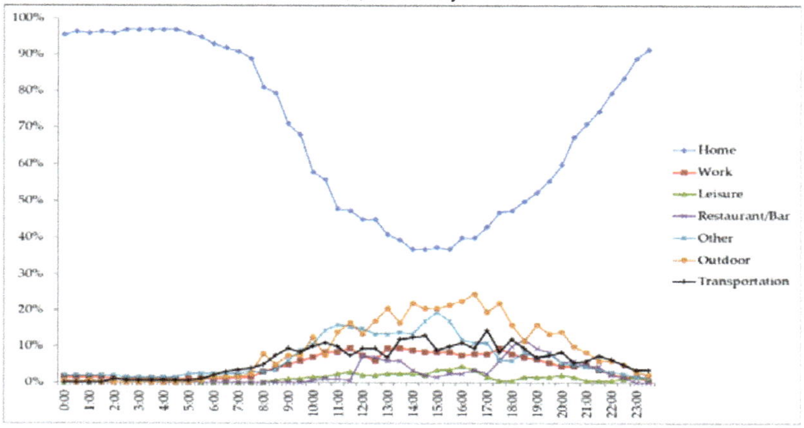

(**b**) Weekend

Figure 2. Percentage of participants at each microenvironment by the time of day for 24 h in weekdays and weekends.

3.4. SHS Exposure Pattern

Figure 3 shows the number of smokers observed, the number of cigarette butts found, and the number of self-reported exposures to SHS by time for 7 days. The number of cigarette butts, smokers, and exposure to SHS increased during commuting (8 a.m. and 6 p.m.) and lunch hours (12 p.m. to 1 p.m.) during weekdays. During the weekend, the number of cigarette butts found was highest between 10 a.m. and 11 a.m., and the second highest number was observed at 2 p.m. The number of exposures to SHS was consistently distributed between 8 a.m. and 11 p.m., and the number of smokers observed was divided into two periods (3 a.m.–12 p.m. and 4 p.m.–11 p.m.).

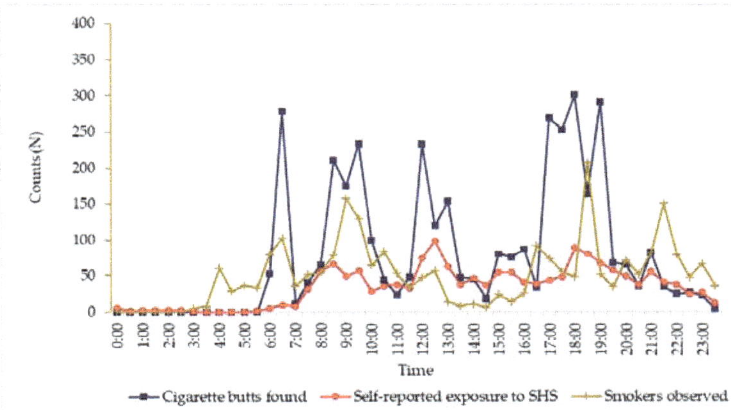

(a) Weekday

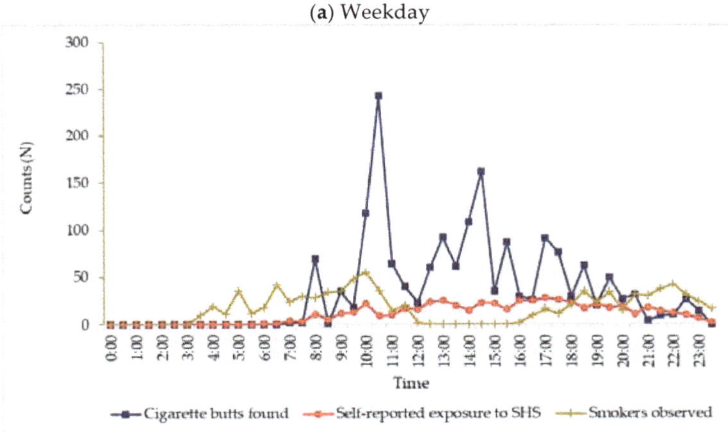

(b) Weekend

Figure 3. The number of cigarette butts found, smokers observed, and self-reported exposure to SHS during weekdays and weekends.

3.5. Time Activity Pattern and SHS Exposure Pattern

The results of the correlation analysis between time activity patterns and SHS exposure are presented in Table 3. The number of cigarette butts found and the number of self-reported SHS exposures significantly decreased as the time spent indoors increased. However, there was a statistically significant positive relationship between the time spent at work, leisure, restaurant, other indoor spaces, and the number of cigarette butts found and the number of self-reported SHS exposures. In particular, the number of cigarette butts found and self-reported SHS exposures were significantly correlated with the time spent outdoors (Figure 4).

Table 3. Correlation between the time spent at each microenvironment and number of cigarette butts found, smokers observed, and self-reported SHS exposure by time.

Time Spent	Weekday			Weekend		
	Cigarette Butts Found	Self-Reported Exposure to SHS	Smokers Observed	Cigarette Butts Found	Self-Reported Exposure to SHS	Smokers Observed
Indoor						
Home	−0.519 **	−0.783 **	−0.212	−0.637 **	−0.916 **	0.126
Workplace	0.321 *	0.494 **	0.060	0.684 **	0.864 **	−0.100
Leisure	0.252	0.638 **	0.210	0.450 **	0.720 **	−0.232
Restaurant/bar	0.371 **	0.590 **	0.047	0.212	0.656 **	−0.002
Others	0.248	0.572 **	0.040	0.678 **	0.719 **	−0.217
Outdoor	0.605 **	0.935 **	0.261	0.561 **	0.930 **	−0.191
Transportation	0.621 **	0.766 **	0.527 **	0.672 **	0.867 **	0.108

* $p < 0.05$; ** $p < 0.01$.

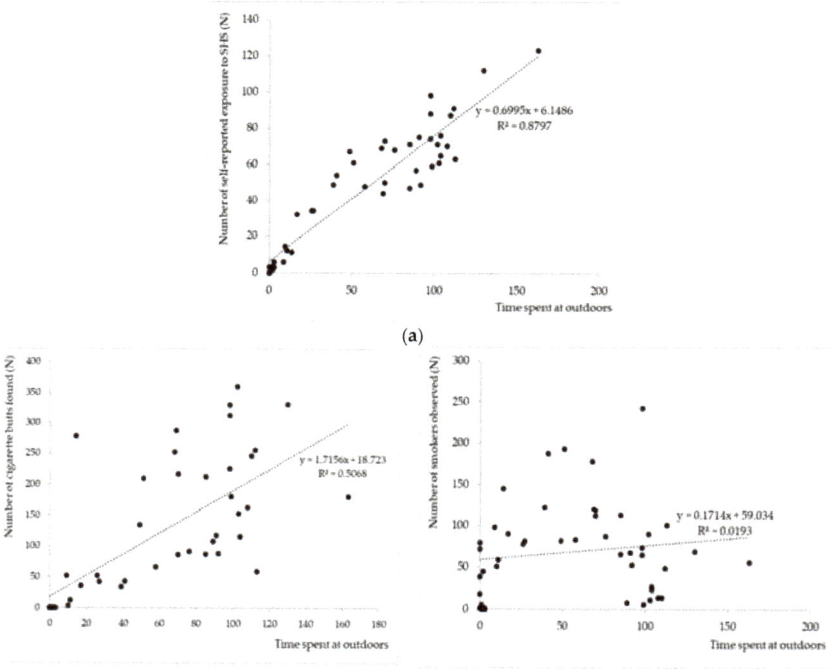

Figure 4. Correlation between time spent outdoors and (**a**) self-reported SHS exposure; (**b**) cigarette butts found; and (**c**) smokers observed for 7 days.

3.6. Biological Levels

The concentrations of cotinine, NNAL, and nicotine are presented in Table 4. Urine samples for cotinine and NNAL were corrected by measuring concentrations of creatinine in urine to minimize physiological reactions caused by kidney diseases. Urine samples for cotinine and NNAL were collected twice with weekly intervals to determine the half-life. However, there was no significant difference between the pre- and post-sampling results. Additionally, 64–69% of cotinine, 75% of NNAL, and 78% of hair nicotine samples were higher than the LOQ and were considered to be exposed to SHS.

However, the samples that indicated the participants' exposure to SHS were less than 10% when using cutoffs from published literature [21,37,38]. The results showed a statistically significant relationship between the analytes (Table 5).

Table 4. Concentrations of cotinine and 4-(methylnitrosamino)-1-3-(pyridyl)-1-butanol (NNAL) in urine and nicotine in hair.

N = 100	GM [1] (GSD) [2]	95% CI	*p*-Value [3]	SHS Exposure Criteria	
				>LOQ [4] (%)	>Cut-Off [5] (%)
Cotinine1 (ng/mL)	0.58 (4.72)	0.43–0.79	0.508	64	6
Cotinine2 (ng/mL)	0.53 (3.86)	0.40–0.69		69	7
NNAL1 (pg/mL)	1.12 (2.93)	0.91–1.39	0.397	75	9
NNAL 2(pg/mL)	1.04 (2.62)	0.86–1.26		75	8
Nicotine, hair (ng/mg)	0.09 (2.63)	0.07–0.10		79	0

[1] Geometric mean, [2] Geometric standard deviation, [3] Paired t-test, [4] Limit of quantification, [5] cut-off for cotinine, 5 ng/mL; NNAL, 3.77 pg/mL; and hair nicotine, 2 ng/mg.

Table 5. Correlations between cotinine, NNAL and nicotine levels.

	Cotinine 1	NNAL 1	Cotinine 2	NNAL 2	Nicotine in Hair
Cotinine 1	1				
NNAL 1	0.412 **	1			
Cotinine 2	0.537 **	0.443 **	1		
NNAL 2	0.381 **	0.646 **	0.519 **	1	
Nicotine in hair	0.335 **	0.339 **	0.341 **	0.455 **	1

** $p < 0.01$.

3.7. Biological Levels and Time Activity Pattern

Correlation analysis between the participants' time spent in each microenvironment for 7 days and their biological sampling results are presented in Table 6. NNAL concentrations showed a marginal negative correlation with the time spent indoors at home, whereas the NNAL2 concentration had a positive relationship with the time spent at other transportations. The nicotine concentration in hair showed a statistically significant negative correlation with the time spent at indoor leisure facilities. However, a statistically significant positive correlation was observed between the hair nicotine concentration and the time spent at outdoor leisure facilities. Cotinine concentrations did not show any significant correlation with the time spent in each microenvironment.

Table 6. Correlations between biological sampling results and time spent in each microenvironment.

Variables	Cotinine 1	NNAL 1	Cotinine 2	NNAL 2	Nicotine in Hair
Indoor					
Home	−0.070	−0.281 **	−0.025	−0.228 *	−0.073
Work	0.129	0.142	0.117	0.070	−0.050
Leisure	0.047	−0.063	−0.117	0.015	−0.199 *
Restaurant/bar	0.123	0.022	0.109	0.072	0.005
Others	−0.094	0.094	−0.185	0.109	0.121
Outdoor					
Home	−0.054	−0.124	−0.051	−0.021	0.021
Work	−0.039	−0.013	0.087	0.001	−0.037
Leisure	0.152	0.094	0.075	0.162	0.204 *
Restaurant/bar	−0.033	0.013	0.043	0.117	0.079
Others	0.021	0.022	−0.005	−0.117	−0.003
Transportation					
Bus	0.086	0.089	0.013	0.152	0.019
Subway	−0.072	0.099	−0.055	0.122	0.107
Taxi	−0.027	0.001	0.022	0.135	−0.080
Car	−0.104	0.087	0.075	−0.066	−0.036
Others	0.036	0.078	0.136	0.264 **	0.100

* $p < 0.05$; ** $p < 0.01$.

3.8. Biological Levels and SHS Exposure Indices

The results of the correlation analysis between biological levels and SHS exposure indices are presented in Table 7. There was a marginal positive relationship between the concentration of cotinine and the number of smokers observed and self-reported SHS exposure. In addition, NNAL had a marginal positive relationship with the number of smokers observed. There was no significant relationship between the number of cigarette butts found with any biological sample and the concentration of hair nicotine and any of the SHS exposure indices.

Table 7. Correlations between self-reported SHS exposure indices and biological exposure indices.

Variables	Cotinine 1	NNAL 1	Cotinine 2	NNAL 2	Nicotine in Hair
Number of cigarette butts found	0.049	0.042	0.103	0.081	0.065
Number of smokers observed	0.207 *	0.255 *	0.233 *	0.331 **	0.157
Number of self-reported SHS exposure	0.220 *	0.147	0.306 **	0.19	0.091

* $p < 0.05$; ** $p < 0.01$.

4. Discussion

Considering the health effects of SHS, various policies, including the prohibition of indoor smoking, are being implemented worldwide. The pattern of SHS exposure is changing from indoor to outdoor environments. Therefore, this study assessed the degree of SHS exposure in outdoor environments, taking into account the pattern of people's time activities. The methodology used in this study and the results derived might be used in non-smoking policy.

The results of SHS exposures were compared with previous study conducted several countries. A total of 97% of the participants responded that they had been exposed to SHS at least once during the last week. The SHS exposure rates were 92% for outdoors, 62% at public places, 39% at workplaces, and 36% at home. In comparison with a study by Eriksen et al., SHS exposure rates were the highest after China, Bangladesh, Egypt, Vietnam, Greece, and Indonesia [9]. These SHS exposure rates were higher than the Korea National Health and Nutrition Examination Survey (home: 4.7%, workplaces: 12.7%, and public places: 21.1%) [22]. The results indicate that SHS exposure rates obtained by a simple questionnaire may be subjective and overestimated depending on the assessment method. Non-smokers can perceive not only the smell of cigarettes directly, but also smokers' breathing and body odor as SHS exposure. Therefore, it is necessary to clarify the route of SHS exposure.

Based on the questionnaire, the microenvironment where the participants were most exposed to SHS was outdoors. Kaufman et al. and Sureda et al. have reported that smoking areas are shifting from indoors to outdoors [16,20]. In the results of the time activity survey conducted in this study, participants spent approximately 8% of their daily time outdoors. Time activity studies previously conducted by Klepis et al. and Yang et al. indicated that people spent 5% of their time outdoors, which is approximately 1 to 2 h per day [39–41]. As the time spent outdoors was relatively short, the effectiveness of the outdoor smoking ban has been widely debated [42,43]. However, Lopez et al. reported that the concentration of nicotine and other substances from SHS in the air can be higher in terraces or corridors of buildings with a smoking ban, as compared to their concentrations outdoors. In addition, building types and ventilation conditions could also affect the concentration of airborne SHS indicators [20,44]. Therefore, building users may be intermittently exposed to high concentrations of SHS when they enter or exit buildings, which may have a negative psychological effect.

The number of cigarette butts found, smokers observed, and self-reported SHS exposure showed similar patterns over time, which were also similar to the participants' time activity patterns for outdoors. This indicates that the participants were frequently exposed to SHS outdoors while moving from one space to another. Several studies have reported that exposure at entrances and exits of buildings has been an issue, and some countries are expanding non-smoking areas to these outdoor locations [16,18,20,44].

The classification of non-smokers exposed to SHS, based on existing literature, showed that most of their exposures to SHS were low. However, according to the classification by CLSI, 64–78% of non-smokers were exposed to SHS. Therefore, further studies are required to set the criteria for biological samples for SHS exposure. In addition, Hecht et al. reported that concentrations of NNAL and NNAL-Gluc in some human subjects could be detected even after 281 days since smoking cessation [45]. Therefore, these factors should be considered when selecting non-smokers in further SHS exposure assessment studies.

The number of smokers observed during a week showed a marginal correlation with the concentrations of cotinine and NNAL, and the number of self-reported exposures to SHS had a significant positive correlation with the concentrations of cotinine. Although the correlations between time spent at each location and concentrations of biomarkers were marginal, the results showed that the time activity and SHS exposure patterns could be individually used to assess non-smokers' exposure to SHS. As most residential areas prohibit smoking, non-smokers would have a lower chance of SHS exposure as they stay at home for longer durations. Likewise, time spent at other indoor areas would also decrease SHS exposure and biomarker concentrations. However, non-smokers' SHS exposure and biomarker concentrations would be higher if they frequently move within different microenvironments, which would increase the time spent outdoors. In addition, their exposure to SHS would increase as smokers would smoke at entranceways, exits, or terraces of buildings with a smoking ban [46]. There was no significant correlation between the concentration of cotinine and time activities. This can be explained by the half-life of cotinine being 18–24 h, which cannot be compared with weekly activities [47].

In this study, the main microenvironment for SHS exposure was outdoors, especially entrances to restaurants, bars, clubs, and places related to transportation. In addition, it was also found that people were mainly exposed to SHS during peak movement hours, such as rush hour while commuting and lunch hours. In Korea, smoking is prohibited indoors; however, smokers can be seen in areas near entrances and exits of buildings such as restaurants, bars, malls, and clubs. Even though the duration of SHS exposure in such a microenvironment was relatively low, the facility users could be exposed to high concentrations of SHS, causing discomfort. Therefore, management and countermeasures are required to control the exposure to SHS in such microenvironments. The number of participants in this study was relatively low, and the types of public places visited by each participant were limited. In addition, their time spent in some of the microenvironments was very short. Hence, there were limitations in analyzing the relationship between the time spent in every microenvironment and the respective concentrations of biomarkers. Therefore, in future studies, a longer-term study period and larger-scale study of the population would be required to obtain a detailed exposure assessment and the relationship between the times spent in microenvironments and the concentrations of biomarkers for SHS exposure. The sample size and of 100 non-smokers may not be representative to assess the exposure to SHS. However, this study suggested the methodology by combining questionnaires and biomarkers with time activity patterns. The results of this study could be used as a basis for establishing an expansion to the non-smoking area policies, designation of outdoor smoking areas, non-smoking while walking, and outdoor smoking during rush hour, which would help reduce SHS exposure.

5. Conclusions

While indoor smoking is decreasing worldwide, outdoor smoking is increasing. Thus, non-smokers can still be exposed to SHS. The SHS exposure assessment using time activity patterns can be used to identify non-smokers' temporal and spatial characteristics of SHS exposure. This method can also be used to establish management priorities. In this study, the main microenvironments where non-smokers are exposed to SHS were outdoors, mainly related to public places such as restaurants, bars, clubs, entrances, and transportation. In addition, the participants experienced exposure to SHS during commuting time and lunch hours. Non-smokers may have more chances to be exposed to SHS if they visit any of these places frequently. Therefore, time activity surveys need to be an essential factor in future SHS exposure assessment studies. The data from these studies could be used to identify detailed information on time and the microenvironment of SHS exposure to assess long-term SHS exposures. The study can also help reinforce smoking bans at entrances and surrounding spaces around buildings such as restaurants, bars, and clubs.

Author Contributions: Data collection and writing—original draft, B.L.W.; supervision, conceptualization, review, and editing, M.K.L. and E.Y.P.; investigation, J.P., H.R. and D.J.; review and editing, M.J.R.; supervision, conceptualization, writing, review, and editing, W.Y. All authors have read and agreed to the published version of the manuscript.

Funding: This research was funded by the Korea Environmental Industry and Technology Institute (KEITI) through the Environmental Health Action Program, grant number 2018001350001.

Conflicts of Interest: The authors declare no conflict of interest.

References

1. U.S. Environmental Protection Agency. *Secondhand Smoke*; EPA-402-F-93-004; US EPA: Washington, DC, USA, 1993.
2. Brennan, P.; Buffler, P.A.; Reynolds, P.; Wu, A.H.; Wichmann, H.E.; Agudo, A.; Pershagen, G.; Jöckel, K.H.; Benhamou, S.; Greenberg, R.S.; et al. Secondhand smoke exposure in adulthood and risk of lung cancer among never smokers: A pooled analysis of two large studies. *Int. J. Cancer* **2004**, *109*, 125–131. [CrossRef] [PubMed]
3. Kawachi, I.; Colditz, G.A. Workplace exposure to passive smoking and risk of cardiovascular disease: Summary of epidemiologic studies. *Environ. Health Perspect.* **1999**, *107*, 847–851. [CrossRef] [PubMed]
4. Misra, D.P.; Nguyen, R.H.N. Environmental tobacco smoke and low birth weight: A hazard in the workplace? *Environ. Health Perspect.* **1999**, *107*, 897–904. [CrossRef] [PubMed]
5. Mohlman, M.K.; Levy, D.T. Disparities in maternal child and health outcomes attributable to prenatal tobacco use. *Matern. Child Health J.* **2016**, *20*, 701–709. [CrossRef]
6. Word Health Organization. *WHO Global Report: Mortality Attributable of Tobacco*; WHO Libr. Cat. Data; World Health Organization: Geneva, Switzerland, 2012; p. 4.
7. Öberg, M.; Jaakkola, M.S.; Woodward, A.; Peruga, A.; Prüss-Ustün, A. Worldwide burden of disease from exposure to second-hand smoke: A retrospective analysis of data from 192 countries. *Lancet* **2011**, *377*, 139–146. [CrossRef]
8. Word Health Organization. *WHO Framework Convention on Tobacco Control (WHO FCTC)*; World Health Organization: Geneve, Switzerland, 2003.
9. Eriksen, M.; Mackay, J.M.; Schluger, N.; Islami, F.; Drope, J. Secondhand. In *The Tobacco Atlas*, 6th ed.; American Cancer Society: Atlanta, GA, USA, 2015; pp. 22–23.
10. Christie, J. Legislative smoking bans for reducing secondhand smoke exposure, smoking prevalence and tobacco consumption. *Public Health Concern Smok. Alcohol Subst. Use* **2013**, 195–199. [CrossRef]
11. Diethelm, P.; McKee, M. *Lifting the Smokescreen. Tobacco industry Strategy to Defeat Smoke Free Policies and Legislation*; European Respiratory Society: Lausanne, Switzerland; Institut National du Cancer: Boulogne-Billancourt, France, 2006.
12. Rosenfield, D.; Manuel, D.G.; Alter, D.A. The Smoking Regulatory Index: A new way to measure public health performance. *CMAJ* **2006**, *174*, 1403–1404. [CrossRef] [PubMed]
13. Park, E.Y.; Lim, M.K.; Yang, W.; Yun, E.H.; Oh, J.K.; Jeong, B.Y.; Hong, S.Y.; Lee, D.H.; Tamplin, S. Policy effects of secondhand smoke exposure in public places in the Republic of Korea: Evidence from $PM_{2.5}$ levels and air nicotine concentrations. *Asian Pac. J. Cancer Prev.* **2013**, *14*, 7725–7730. [CrossRef]
14. Levy, D.T.; Cho, S.I.; Kim, Y.M.; Park, S.; Suh, M.K.; Kam, S. SimSmoke model evaluation of the effect of tobacco control policies in Korea: The unknown success story. *Am. J. Public Health* **2010**, *100*, 1267–1273. [CrossRef]

15. Korea Centers for Disease Control and Prevention. Korea National Health and Nutrition Examination Survey (KNHANES) 2016. 2017. Available online: http://www.mohw.go.kr/react/modules/download.jsp?BOARD_ID=140&CONT_SEQ=342756&FILE_SEQ=220287 (accessed on 22 September 2020).
16. Kaufman, P.; Griffin, K.; Cohen, J.; Perkins, N.; Ferrence, R. Smoking in urban outdoor public places: Behaviour, experiences, and implications for public health. *Health Place* **2010**, *16*, 961–968. [CrossRef]
17. Government of Canada. Canadian Tobacco Use Monitoring Survey (CTUMS): Smoking prevalence 1999 to 2012. Available online: https://www.canada.ca/en/health-canada/services/publications/healthy-living/canadian-tobacco-use-monitoring-survey-smoking-prevalence-1999-2012.html (accessed on 22 September 2020).
18. *Global Smokefree Partnership FCTC Article 8-Plus Series the Trend toward Smokefree Outdoor Areas*; University of Otago, Wellington Campus: Wellington, NZ, USA, 2009; pp. 1–14.
19. Repace, J.L. Benefits of smoke-free regulations in outdoor settings: Beaches, golf courses, parks, patios, and in motor vehicles. *William Mitchell Law Rev.* **2008**, *34*, 1621–1638.
20. Sureda, X.; Martínez-Sánchez, J.M.; López, M.J.; Fu, M.; Agüero, F.; Saltó, E.; Nebot, M.; Fernández, E. Secondhand smoke levels in public building main entrances: Outdoor and indoor $PM_{2.5}$ assessment. *Tob. Control* **2012**, *21*, 543–548. [CrossRef] [PubMed]
21. Avila-Tang, E.; Al-Delaimy, W.K.; Ashley, D.L.; Benowitz, N.; Bernert, J.T.; Kim, S.; Samet, J.M.; Hecht, S.S. Assessing secondhand smoke using biological markers. *Tob. Control* **2013**, *22*, 164–171. [CrossRef] [PubMed]
22. Korea Centers for Disease Control and Prevention. Korea National Health and Nutrition Examination Survey (KNHANES) 2017. 2018. Available online: http://www.mohw.go.kr/react/modules/download.jsp?BOARD_ID=140&CONT_SEQ=346600&FILE_SEQ=241156 (accessed on 22 September 2020).
23. Coleman, B.; Rostron, B.; Johnson, S.E.; Persoskie, A.; Pearson, J.; Stanton, C.A.; Choi, K.; Anic, G.; Goniewicz, M.L.; Michael, K.; et al. Transitions in electronic cigarette use among adults in the Population Assessment of Tobacco and Health (PATH) Study, Waves 1 and 2 (2013–2015). *Tob. Control* **2019**, *28*, 50–59. [CrossRef] [PubMed]
24. National Institutes on Drug Abuse. Monitoring the Future: National Results on Adolescent Drug Use: Overview of Key Findings 2016. 2017. Available online: http://www.monitoringthefuture.org//pubs/monographs/mtf-overview2016.pdf (accessed on 22 September 2020).
25. The Ontario Tobacco Research Unit. Ontario Tobacco Survey. 2010. Available online: https://otru.org/wp-content/uploads/2012/07/OTS_Technical_Report.pdf (accessed on 22 September 2020).
26. Statistics Canada. *Canadian Health Measures Survey (CHMS) Data User Guide: Cycle 2*; Available upon request (infostats@statcan.gc.ca); Statistics Canada: Ottawa, ON, Canada, 2013.
27. Centers for Disease Control and Prevention. National Health and Nutrition Examination Survey 2015–2016 Data Documentation, Codebook, and Frequencies. In *Smoking—Secondhand Smoking Exposure (SMQSHS_I).*; 2017. Available online: https://wwwn.cdc.gov/Nchs/Nhanes/2015-2016/SMQSHS_I.htm (accessed on 22 September 2020).
28. Freeman, N.C.G.; De Tejada, S.S. Methods for collecting time/activity pattern information related to exposure to combustion products. *Chemosphere* **2002**, *49*, 979–992. [CrossRef]
29. Jarvis, M.J. Children's exposure to passive smoking: Survey methodology and monitoring trends. In *International Consultation on ETS and Child Health Report*; World Health Organization: Geneva, Switzerland, 1999; Available online: https://www.who.int/tobacco/media/en/jarvis.pdf?ua=1 (accessed on 22 September 2020).
30. Knight, J.M.; Eliopoulos, C.; Klein, J.; Greenwald, M.; Koren, G. Passive smoking in children: Racial differences in systemic exposure to cotinine by hair and urine analysis. *Chest* **1996**, *109*, 446–450. [CrossRef]
31. Rickert, W.S. Environmental tobacco smoke: Properties, measurement techniques and applications. In *International Consultation on Environmental Smoke (ETS) and Child Health*; World Health Organization: Geneva, Switzerland, 1999; Available online: https://www.who.int/tobacco/resources/publications/ets/en/ (accessed on 22 September 2020).
32. Benowitz, N.L.; Jacob, P.; Denaro, C.; Jerkins, R. Stable isotope studies of nicotine kinetics and bioavailability. *Clin. Pharmacol. Ther.* **1991**, *49*, 270–277. [CrossRef]
33. Park, E.Y.; Yun, E.H.; Lim, M.K.; Lee, D.H.; Yang, W.; Jeong, B.Y.; Hwang, S.H. Consequences of incomplete smoke-free legislation in the Republic of Korea: Results from environmental and biochemical monitoring: Community based study. *Cancer Res. Treat.* **2016**, *48*, 376–383. [CrossRef]

34. Hecht, S.S. Human urinary carcinogen metabolites: Biomarkers for investigating tobacco and cancer. *Carcinogenesis* **2002**, *23*, 907–922. [CrossRef]
35. Benowitz, N.L.; Hukkanen, J.; Jacob, P. Nicotine chemistry, metabolism, kinetics and biomarkers. In *Handbook of Experimental Pharmacology*; Springer: Berlin/Heidelberg, Germany, 2009; Volume 192, pp. 29–60. [CrossRef]
36. Ryu, H.J.; Seong, M.W.; Nam, M.H.; Kong, S.Y.; Lee, D.H. Simultaneous and sensitive measurement of nicotine and cotinine in small amounts of human hair using liquid chromatography/tandem mass spectrometry. *Rapid Commun. Mass Spectrom.* **2006**, *20*, 2781–2782. [CrossRef]
37. Blue Care Network of Michigan. Cotinine testing. 2010. Available online: http://d2xk4h2me8pjt2.cloudfront.net/webjc/attachments/179/272dd1d-nicotine-cotinine-interpretive-guidelines-.pdf (accessed on 22 September 2020).
38. Moyer, T.P.; Charlson, J.R.; Enger, R.J.; Dale, L.C.; Ebbert, J.O.; Schroeder, D.R.; Hurt, R.D. Simultaneous analysis of nicotine, nicotine metabolites, and tobacco alkaloids in serum or urine by tandem mass spectrometry, with clinically relevant metabolic profiles. *Clin. Chem.* **2002**, *48*, 1460–1471. [CrossRef] [PubMed]
39. Klepeis, N.E.; Nelson, W.C.; Ott, W.R.; Robinson, J.P.; Tsang, A.M.; Switzer, P.; Behar, J.V.; Hern, S.C.; Engelmann, W.H. The National Human Activity Pattern Survey (NHAPS): A resource for assessing exposure to environmental pollutants. *J. Expo. Anal. Environ. Epidemiol.* **2001**, *11*, 231–252. [CrossRef]
40. Yang, W.; Lee, K.; Yoon, C.; Yu, S.; Park, K.; Choi, W. Determinants of residential indoor and transportation activity time in Korea. *J. Expo. Sci. Environ. Epidemiol.* **2011**, *21*, 310–316. [CrossRef] [PubMed]
41. Park, J.; Ryu, H.; Kim, E.; Choe, Y.; Heo, J.; Lee, J.; Cho, S.; Sung, K.; Cho, M.; Yang, W. Assessment of $PM_{2.5}$ population exposure of a community using sensor-based air monitoring instruments and similar time-activity groups. *Atmos. Pollut. Res.* **2020**. [CrossRef]
42. Chapman, S. Banning smoking outdoors is seldom ethically justifiable. *Tob. Control* **2000**, *9*, 95–97. [CrossRef]
43. Chapman, S. Should smoking in outside public spaces be banned? No. *BMJ* **2008**, *337*, 2008–2010. [CrossRef]
44. López, M.J.; Fernández, E.; Gorini, G.; Moshammer, H.; Polanska, K.; Clancy, L.; Dautzenberg, B.; Delrieu, A.; Invernizzi, G.; Muñoz, G.; et al. Exposure to secondhand smoke in terraces and other outdoor areas of hospitality venues in eight European Countries. *PLoS ONE* **2012**, *7*, e42130. [CrossRef]
45. Hecht, S.S.; Carmella, S.G.; Chen, M.; Dor Koch, J.F.; Miller, A.T.; Murphy, S.E.; Jensen, J.A.; Zimmerman, C.L.; Hatsukami, D.K. Quantitation of urinary metabolites of a tobacco-specific lung carcinogen after smoking cessation. *Cancer Res.* **1999**, *59*, 590–596. [PubMed]
46. Berman, B.A.; Wong, G.C.; Bastani, R.; Hoang, T.; Jones, C.; Goldstein, D.R.; Bernert, J.T.; Hammond, K.S.; Tashkin, D.; Lewis, M.A. Household smoking behavior and ETS exposure among children with asthma in low-income, minority households. *Addict. Behav.* **2003**, *28*, 111–128. [CrossRef]
47. Benowitz, N.L. Cotinine as a biomarker of environmental tobacco smoke exposure. *Epidemiol. Rev.* **1996**, *18*, 188–204. [CrossRef]

Publisher's Note: MDPI stays neutral with regard to jurisdictional claims in published maps and institutional affiliations.

sustainability

MDPI

Review

The Fourth Industrial Revolution and the Sustainability Practices: A Comparative Automated Content Analysis Approach of Theory and Practice

Vasja Roblek [1], Oshane Thorpe [2], Mirjana Pejic Bach [3,*], Andrej Jerman [4] and Maja Meško [5,6,*]

1 Faculty of Organisation Studies in Novo Mesto, 8000 Novo Mesto, Slovenia; vasja.roblek@gmx.com
2 College of Media and Mass Communication, American University in the Emirates, Dubai 503000, UAE; Oshane.thorpe@aue.ae
3 Faculty of Economics, University of Zagreb, 100000 Zagreb, Croatia
4 LPP, 1000 Ljubljana, Slovenia; andrejjerman1@gmail.com
5 Faculty of Management, University of Primorska, 6000 Koper, Slovenia
6 Faculty of Organizational Sciences, University of Maribor, 4000 Kranj, Slovenia
* Correspondence: mpejic@net.efzg.hr (M.P.B.); maja.mesko@fm-kp.si (M.M.)

Received: 23 September 2020; Accepted: 13 October 2020; Published: 15 October 2020

Abstract: (1) Background: The article provides a methodologically coherent analysis of technological development in the context of the fourth industrial revolution or Industry 4.0 and its impact on changes in sustainable development policy. (2) Methods: Using a Comparative Automated Content Analysis (ACA) approach, the article compares recent scientific work on sustainable development and the fourth industrial revolution with the discourse in the news media on sustainable development and industry 4.0. (3) Results: The scientific literature focuses more on changes in business models, production processes, and technologies that enable sustainable development. Newspaper and magazine articles write more about sustainable or green investments, sustainable standards, and sustainable reporting. The focus is on topics that are directly relevant to current sustainable business development and the promotion of research and development of clean and smart technologies and processes. (4) Conclusions: The ACA allows a more systematic comparison of different data sources. The article provides a starting point for sustainable development professionals to gain useful insights into a specific context with the help of the ACA.

Keywords: sustainability; industry 4.0; sustainable investment; corporate social responsibility; sustainable standards; sustainable reporting; smart manufacturing; renewable energy; cleaner production

---

1. Introduction

In the second decade of the 21st century, humanity is confronted with the emergence of the Fourth Industrial Revolution or Industry 4.0 (IR 4.0) and the demand to implement the 17 Sustainable Development Goals (SDGs) set out in Agenda 2030 for Sustainable Development. The Agenda balances and links the three dimensions of sustainable development—economic, social, and environmental—and stipulates that the SDGs must be adopted in all countries of the world by 2030 [1].

Most of the scientific literature described I4.0 primarily from a technical point of view [2], but there are less researched topics like organizational management [3–5], as well as the ecological and social aspects within I4.0. Researchers such as Birkel et al. [6] point out that there are still rare integrative researches of economic, ecological, and social aspects. It is tough to simultaneously maintain economic profitability whilst improving the environmental, as well as the social aspects of industrial value creation. Thus, the challenges and potential of I4.0 appear in stark contradiction to the three-dimensions mentioned in the Triple Bottom Line. The I4.0 concept raises fears of job losses and growing inequality. Therefore, an interdisciplinary, integrative study of I4.0 is required, which does more than merely balance the ecological and social potential, but also connects them to market success [7,8].

To this end, a comparative study between the current peer-reviewed academic articles and the less rigorous news media can serve to identify the themes that are of importance and aid in the understanding of I4.0. An enquiry of this nature, therefore, delves deep to identify changes in sustainable development policy, and with the gathered knowledge can provide guidelines for further research that will later further contribute to the theoretical and practical development of sustainable policies.

The method of content analysis in the media, whether it is done manually or through the aid of natural language processing artificial intelligence, is an incredibly exciting area, since it offers a wealth of sources to be analyzed. It is also essential as we cannot divorce ourselves from the impact of journalistic discourse in the present day. Therefore, in this paper, journalistic discourse will be extrapolated from newspaper reports (articles or "newspaper story").

According to van Dijk's [9] newspaper report refers to "the kind of text that provides information about recent events". Rea [10] also added to this definition, by stating that "Information about recent events that are of interest to large enough group or that can affect the lives of large enough groups of people".

The comparative research delves into the opportunities offered by I4.0, including, but not limited to, improvement of various production processes, which includes robotization, as well as how I4.0 has stimulated research on the possibilities and effects outside the smart factories themselves [11,12]. There are 10 major global trends in I4.0, most of which already exist but have been improved, in terms of the features introduced: Demographic shifts, urbanization, knowledge growth, deindustrialization, market globalization compared to protectionism, advanced business models, technology convergence, increase robotics, cybersecurity, climate changes, and global sustainability [13,14]. The immediate sustainability result of I4.0 is manufacturing-economic sustainability. The digitalization of the manufacturing industry influences manufacturing efficiency, supply chain mergers, energy efficiency, the emergence of business model innovation, cost-saving, financial sustainability, human resource skills development, and corporate profitability. It is vital to replace fossil fuels with renewable energy, which also aids in the decarbonization of the society. I4.0 is eventually crucial for promoting and enabling environmental protection and emission reduction [15]. There are a large number of articles, both empirically and theoretically based, which span cases of scientific research in the case of academic journal articles and others can be found in newspapers and magazines. In this research, an alternative approach was chosen that was capable of efficiently and successfully categorizing vast quantities of data and enabled the reader to obtain appropriate explanations of the research phenomenon understandably. For the topic under discussion, an automated content analysis method (ACA) was used to identify the key themes and the concepts of interest to researchers [16,17].

This paper consists of six chapters. First, the introduction, followed by the research method, which includes data collection and literature selection. The fourth chapter provides data analysis and the results of the ACA. The paper concludes with a discussion of results and conclusions, which include a comparative analysis of findings, research limitations, and propose research in the future.

2. Content Text Analyses Methods

2.1. Introduction of a Classical Content Analysis

Content analysis is traditionally located in the field of quantitative methods, though it has mixed methodology applications. In the early applications of this method, research results were usually expressed in quantitative form. When formulating the first more precise definition of the method, Berelson [18] had in mind what was most often represented in practice and also influenced the perception of content analysis as a quantitative method. According to Berelson [19] "content analysis is a research technique for the objective, systematic and quantitative description of the manifest content of communication". The understanding of the method as quantitative is also a consequence of the necessity to collect data through research, which are used to solve existing problems, i.e., they are socially conditioned. The content analysis "experiences its own experience" [18]. Content analysis flourished when it became instrumental in practice (during the Second World War and the so-called Cold War when it was used to study enemy propaganda).

Discussion of quantitative and qualitative approach in the content analysis began in the middle of the last century, after the appearance of the first methodological study on content analysis- Berelson's text about the content analysis [19]. Kracauer [20], the first proponent of qualitative content analysis, reacted to Berelson's view of the method, which leads Schreier [21] to the conclusion that the qualitative form of the procedure was developed from his quantitative form. The debate on quantitative and qualitative content analysis was one of the three main topics at the Allerton House Conference in 1955 [22], and at the end of the same decade, George [23] published a text advocating the application of the procedure in qualitative form.

Following the mentioned attempts to affirm the qualitative application of the analysis of the media content, a decades-long silence follows. However, the debate was revived at the beginning of the 21st century. Schreier [21] points out that silence was only present in the English-speaking world, especially in the United States and England. However, the qualitative content analysis continued to develop in the rest of Europe, primarily in Germany, by the turn of the 21st century, the discussion was revived in the English-speaking literature. Mayring [24] stood out in terms of influence, and his work became the standard literature on the qualitative application of the methodology [21].

The content analysis thus was developed to determine what can be investigated and how the inquiry can be undertaken using the approach. Based on the existing definitions, two phases can be observed in determining the qualitative content analysis. In the first years, when the application of the method in a qualitative form was advocated, it was defined by comparison with the quantitative form, whereas the more recent definitions deal exclusively with the specifics of qualitative content analysis, and in very brief terms [25]. The difference between older and newer definitions is not surprising, since the former arose at the time when the qualitative content analysis was confirmed as a form of research procedure and the latter when its status was no longer questioned. At the same time, we should not lose sight of the fact that a sharp dividing line between qualitative and quantitative content analysis cannot be drawn. Krippendorff [26] was questioned about the distinction between these two forms of procedure in terms of their usefulness and validity.

Newer determinations are, as already indicated, focused on the specifics of qualitative content analysis, without comparison with the quantitative form of the procedure. Besides, the definitions are concise, without a more thorough consideration of the peculiarities of the qualitative form of the procedure at the level of determination. For Hsieh and Shannon [27], qualitative content analysis is "a research method for subjectively interpreting the content of textual data through the systematic classification of coding processes and the identification of topics or patterns". Similarly, Schreier [21] defines qualitative content analysis as the process by which "systematic description of the meaning of qualitative data" is performed in a relevant context.

Classic content analysis is based on manual text content analysis; this means that the researcher manually (personally) checks different sources and identifies ideas and topics based on his perceptions and views. Manual analysis restricts the process of text content research itself, as the researcher is limited by the time and physical ability to analyze the content, so it leads to the inevitable limitation of the scope of a sample text. The limitation is perceived as insufficient sampling (biased), which reduces the efficiency and effectiveness of the results of the analysis [26].

2.2. Automated Content Analyses

Technological developments have led to the development of automatic text analysis, which is specifically designed to use information technology to extract statistically manipulative information about the presence, intensity, and/or frequency of thematic and/or stylistic features of texts [28,29]. From a methodological perspective, automatic text analysis has several very significant advantages. Since a computer performs the analysis based on a predefined algorithm, the data obtained in this way are objective, verifiable, and reproducible. Moreover, the use of this method minimizes the measurement error that is a consequence of the individual differences between assessors and allows maximum methodological equivalence of different studies using the same text analysis program. Furthermore, the data collected in this way do not show any methodological differences from explicit methods that are frequently used in management (e.g., expert assessments, assessments by stakeholders, questionnaires, among others) [30].

The authors used Leximancer 5.0 to analyze content comparisons of texts based on ACA, which is used in this research for the content analyses of the articles, based primarily on probabilistic models generated by algorithms [17,31]. ACA represents a text-mining tool [28], and it is used for the analysis of text using AI/machine learning techniques. This subsect of computer science focuses on identifying patterns, to also generate predictions and to identify and define thematic/themes in the selected text collection (terms) [32]. The ACA enables researchers to simultaneously analyze a large body of text for themes and similar concepts [33]. This ability makes ACA comparable to the classification model in remote sensing found in the feature-or object method [34]. The method finds terms that have a high probability of being linked based on their repetitive proximity in the literature. The concepts are gleaned from the repetition across multiple texts and are found to be closely related [35]. The concepts are subsequently used in the categorization of the literature. This process is far more complicated than word sense and wordnet because it does not merely only count frequency but also takes into account the semantics and natural language complexities found in English such as synonyms, co-decision frequencies, and sentence structure [36].

According to Smith and Humphreys [37], the process is aiming to replicate in software the method for "transforming lexical co-occurrence information from natural language into semantic patterns in an unsupervised manner". The program is based on semantic and relational co-occurrence information extraction. It uses different algorithms for each phase, which are statistical. A characteristic of the algorithms is that they use nonlinear dynamics and machine learning. Smith and Humphreys [37] also validated the Leximancer according to the typology presented by Krippendorff [26] (face validity, stability (including sampling validity of members, reproducibility (including sampling validity of representatives and predictive validity), which also covers structural validity in the case of concept network comparisons, correlative validity (also including semantic validity), and (5) functional validity).

In recent years, various researchers have established a rigorous procedure for characterizing various content (e.g., research literature, journal articles, corporate sustainability reports, government policy, etc.) with the topic of sustainable development, technological transformation, and need for sustainability processes. Such papers are:

- Amini, Bienstock, and Narcum [38]: Status of corporate sustainability: A content analysis of Fortune 500 companies;
- Cheng and Edwards [16]: A comparative automated content analysis approach on the review of the sharing economy discourse in tourism and hospitality;
- Kim and Kim [39]: Sustainable Supply Chain Based on News Article and Sustainability Reports: Text Mining with Leximancer and DICTION;
- Lock and Araujo [40]: Visualizing the triple bottom line: A large-scale automated visual content analysis of European corporations' website and social media images;
- Nunez-Mir et al. [28]: An automated content analysis of forestry research: are socioecological challenges being addressed?
- Paolone, F., Sardi, A., Sorano, E., and Ferraris, A. [41]: Integrated processing of sustainability accounting reports: a multi-utility company case study;

- Pucihar [17]: The digital transformation journey: content analysis of Electronic Markets articles and Bled eConference proceedings from 2012 to 2019;
- Roblek et al. [42]: The interaction between internet, sustainable development, and emergence of Society 5.0;
- Sullivan et al. [43]: Using industrial ecology and strategic management concepts to pursue the Sustainable Development Goals;
- Keller et al. [44]: News media coverage of climate change in India 1997–2016: Using automated content analysis to assess themes and topics.

The next chapter provides the research design.

3. Research Design

3.1. Data Collection

This research presents an insight into the studies about the relations between I4.0, emerging changes to the economy, production, sustainability, and the environment. The study analyzed academic journal articles published from 2010 to June 2020 as well as newspapers and magazines articles published between 2015 and 2020. I4.0 is influential on the development of new technology and business models as well as on social change. In order to know the problems and the current situation, which also affects society, not only corporations, it is necessary to observe and analyze news articles that reflect a broader spectrum of trending events, or "what is going on" (colloquially) [45]. In this way, "hot themes" from a particular area are to be designed and presented to the public and achieve an "agenda-setting" effect [46]. Newspaper articles bring to the attention of the public those news items that are of interest to them at the time, and that could influence changes in regulatory policy and practice [47,48]. The purpose of academic scientific articles is to present new scientific findings and thus to provide the scientific and professional public with access to further knowledge in a particular scientific field [49].

Literature Selection

The literature selection was prepared in the three-step screening process. First, the papers were searched by keywords on Industry 4.0, 4th Industrial revolution, sustainability, clean production, sustainable development, environment, green investment, smart factory, and sustainable corporate responsibility on the WOS database. In the second phase, only peer-review papers were selected. The third step includes manual review and selection of peer review papers titles, abstracts, and conclusions.

The research platform Web of Science was used for the search of Science Citation Index Expanded (SCI-EXPANDED), Social Sciences Citation Index (SSCI), and Arts abd Humanities Citation Index (A & HCI) to identify relevant papers. The Boolean keyword combination used for this research are as follows, (TS = (Industry 4.0 * and sustainability) and language: (english) and document types: (Article)Indexes = SCI-EXPANDED, SSCI, A & HCI). The research focused on collecting a decade of data. Therefore, the time frame of 1990 to 2020 was used. The results of the search were limited to the articles published in the refereed journals only. The peer review was limited to scientific journals written in English and was therefore not intended to provide a comprehensive assessment of the totality of the state of the subject.

The research paper uses ACA for comparison and analysis of the current knowledge of the particular topic and to identify research gaps for preparing future researches [50]. The literature review is prepared according to the Prisma 2009 technique. The process is presented in Figure 1 [51].

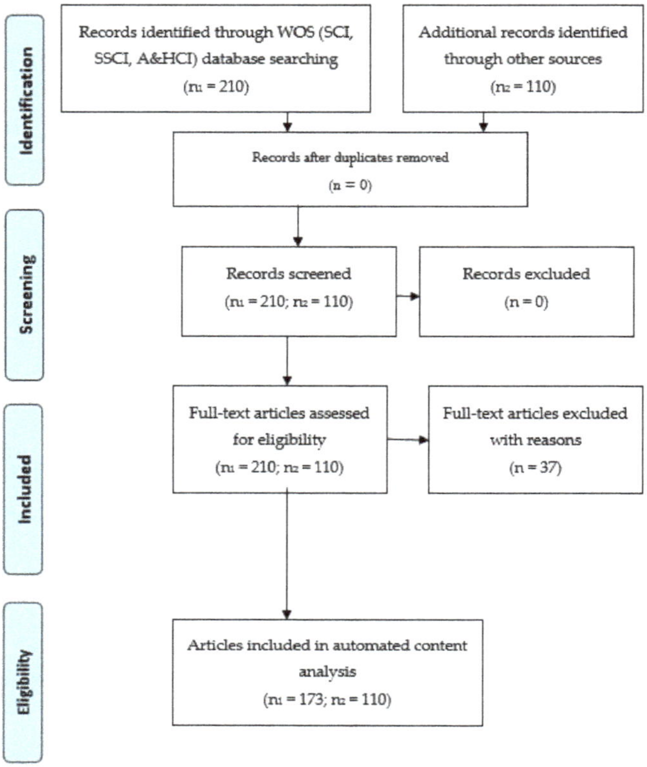

Figure 1. Prisma 2009 flow diagram.

A search of the Web of Science yielded 210 articles published in 92 peer-reviewed journals, the authors' analysis utilized data from Clarivate Analytics, in the Web of Science database. Review of the abstracts as well as the full texts was undertaken, unearthing 37 irrelevant articles, i.e., those where the main text was incongruent with the inquiry into the Industry 4.0 impact on sustainability were subtracted from the sample and analysis, which resulted in a final sample of 173 articles published across 77 journals.

Among the leading scientific journals that have published research with I4.0 and sustainability/sustainable development in the last 10 years (the first article was published in 2010), are Journal of Cleaner Production, International Journal of Production Research, Journal of Manufacturing Technology Management, Sustainability, Process Safety and Environmental Protection, Computers Industrial Engineering, Production Planning Control, International Journal of Production Economics, Technological Forecasting and Social Change, and Journal of Business Research. Most of the papers came from the WOS categories of the environmental sciences (35), green sustainable science technology (31), and environmental studies (24). These categories are followed by the engineering industrial (22), engineering manufacturing (15), engineering environmental (12), operations research management science (11), management (9), computer science interdisciplinary application (8), and energy fuels (6).

The newspapers and magazine articles were taken from the ProQuest recent newspapers database and global news stream. These two databases are considered to deliver complete news coverage. The ProQuest recent newspapers include 50+ current edition digitized newspapers, and global news stream contains indexed and entirely stored articles from more than 2500 national, regional, national, and international English newspapers [52]. The option to eliminate duplicate news has been added to the search. When saving, essential information about the source of the article (issue date, the title of the newspaper) was automatically captured. The search returned 110 articles from magazines and newspapers in the English language. All articles were published between 1 January 2015 and 16 June 2020. Among the leading newspapers and magazines selected were The Times, The Wall Street Journal, The New York Times, The Financial Times, The Economist, Reuters, and The Guardian.

3.2. Data Analysis

The text mining approach was used in the research, with the aid of ACA [53]. The data analysis was completed using the very proficient natural language processing tool Leximancer 5.0.

Unlike previously rudimentary ACA tools that merely count frequency, Leximancer analyzes the meanings in text extracts utilizing its algorithms, which firstly extrapolate the critical concepts. The resulting concepts are used to conduct qualitative analyses; it additionally uses the quantitative method with new sets of algorithms for the phrases [54].

As a high-level natural language processing software, Leximancer starts with no pre-conceptions about the data that is inputted. Therefore, the analysis is data-driven, reducing inherent bias as the analysis emerges from the data itself. The process is underpinned by Bayesian theory; without the need for human intervention in the analysis of the data, the iterative approach is entirely unsupervised [37]. As such, "fragmented pieces of evidence" in documents "can be used to predict what is happening in a system" [55]. Leximancer shows its results through the usage of heat maps for relevant themes. Themes are color-coded for convenience, where the most important themes are conveniently shown in hot colors, and the warmer colors denote less critical themes [56]. Through the Leximancer visualizer output, concepts that are nearby to each other on the heat map have a significant semantic relationship [37,57].

The analysis was carried out in four steps. The first step involved the selection of the document, the second step involved the generation of concept seeds, the third step involved the creation of a thesaurus, and the fourth step enabled the generation of the results. It should be noted that the parameters can be changed based on the needs of the research. In detailed research, all the words that were not meant for the content of the research were excluded. Leximancer already contains a stop word list for individual words (e.g., a, an, me, you, via). The remaining words were manually removed from further analysis (e.g., "paper", "article", "study", "research", "methodology", "author information", "acknowledgement", and "references in journal articles").

In comparing the text with Leximancer, two fundamental crossroads were faced. Both types of data were not similar in many ways. The first crossroad faced was the issue of a comparative equivalence [16], which is significant, as peer-reviewed academic journal articles, as well as newspapers, have a niche readership in mind (i.e., the general public vs. academics). To this end, the concepts generated by Leximancer are context-dependent with variations in their definitions and scopes from both sources (one often employs figurative language, while the other tends to be more literal). To ensure that the analysis shows consistency across the two different types of literature, the authors examined each concept and manually linked it to the original text to understand its context. Concepts were reworded where needed. The second issue is the different sizes of the samples [58].

Differences in sample size can affect the final comparison of the data. When one does an ACA, concepts are systematically chosen to characterize the entire dataset, therefore it is not unfounded that a source (journal article) that is lengthier will contribute more data than a shorter one (newspaper article). Similarly, when the Leximancer thesaurus algorithm discovers a reoccurring concept, primarily if that concept occurs in multiple sources, the concept will be contributed to mainly in part by the more comprehensive source, thus conflating the importance of specific concepts based on a small sample of large sources, making comparison difficult. There are two solutions to this problem. The first is to create a unique and separate project file for each data source and compare the results, as ACA tools (Leximancer) will avoid mixing two different mental models. The second solution is to increase the number of automatically discovered concepts, which would create a large net that feeds the software with a large enough set of concepts that would reduce the bias towards lengthy journal articles. The results are discussed in the following section [54].

In the next subchapter, the research results are presented.

4. Results

4.1. The Thematic Concerns in the Scientific Journal's Articles

Leximancer generated a total of 20 concepts and five themes from the titles, abstracts, and keywords of 173 articles published in 77 journals. The authors used the slider % visible concepts to 100% and changed the number of concepts visible on the map from 50% (automatically) to 100%. The theme size was moved from 33% (automatic) to 61%. A theme is defined as a group or cluster of concepts that are correlated based on certain commonalities or connections. The commonality is gleaned from the proximity of concepts (represented by colored circles) on the concept map. The names of the themes are taken from the concept that is the most prominent in the group of interrelated concepts [17]. Table 1 presents the themes, hits, and related concepts.

Table 1. Themes and concepts in the journals in the years from 2010 to June 2020.

Theme	Hits	Concepts
Sustainability	435	sustainability, industry, manufacturing, management, supply, data, industrial, waste
Technologies	252	technologies, energy, systems, efficiency
Model	239	model, process, production, business, development
Environmental	114	environmental, performance, economic
Emissions	38	emissions, carbon, consumption

Source: Authors' work.

Figure 2 represents the diagram generated by the software, visualizing the concepts of interrelated terms. Themes on a Leximancer concept map are visualized using the heat map concept. Therefore, like on a heat map, fiery colors (red, orange) denote the most important themes, while cool colors (blue, green) denote those less critical [54]. The four most important themes generated are "sustainability", "technologies", "model", and "environmental".

Figure 2 shows that the circles of specific themes overlapping with the circles of other themes, thus forming cross-sections that contain individual concepts, which thus fall into both overlapping themes. For example, there is an overlap among »sustainability« other themes »environmental«, »model«, and »technologies«, whilst »technologies« overlap with »emissions«, »model«, and »sustainability«. The »environmental« theme overlaps with »sustainability« and »model«. There is an overlap among the themes »model«, »environmental«, »sustainability«, and »technologies«. The theme »emissions« overlaps with the themes »technologies« and »model«. Additionally, Figure 2 shows that the concepts »waste«, »production«, »process«, and »efficiency« can be found in two themes, with an intersection of the themes »sustainability« and »model«. The concepts »social« and »performance« lies between the intersection of the themes »environmental« and »sustainability«. The concept "potential" lies between the intersection of the theme's "development" and "services". The concepts »manufacturing«, »internet of things«, »industrial«, and »digital« lie between the intersection of the themes »sustainability« and »technologies«. The concept »consumption« lies between the intersection of the themes »technologies« and »emissions« and the concepts »business«, »demand«, »efficiency«, »process«, »waste«, and »development« lie between the intersection of the themes »technologies« and »model«.

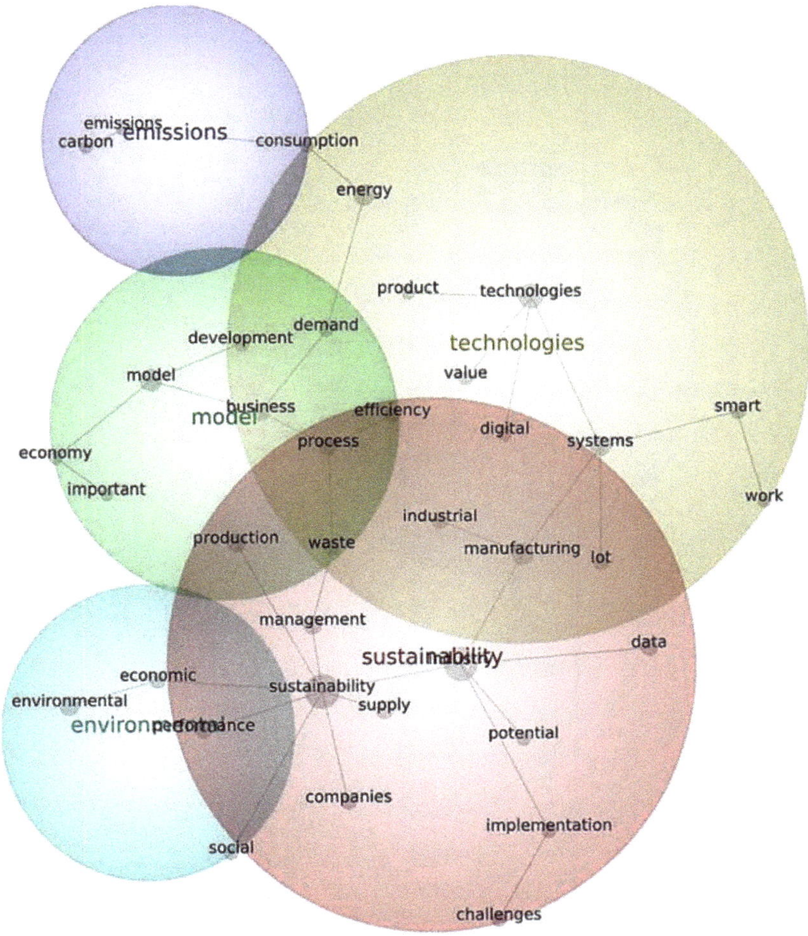

Figure 2. Concept map of the chosen scientific journal papers published between 2010 and June 2020 Source: Authors' work.

4.2. The Thematic Concerns in the Newspapers and Magazines Articles

The authors used the slider % visible concepts to 100% and changed the number of concepts visible on the map from 50% (automatically) to 100%. The theme size was moved from 33% (automatic) to 53%. A theme is defined as a group or cluster of concepts that are correlated based on certain commonalities or connections. The commonality is gleaned from the proximity of concepts (represented by colored circles) on the concept map. The names of the themes are taken from the concept that is the most prominent in the group of interrelated concepts for each theme. The number of hits found in the excerpts was calculated based on the frequency of occurrence and the correlation to the concepts [17]. Table 2 presents the themes, hits, and related concepts.

Figure 3 represents a detailed view of the concept map. The four themes with the most significant number of hits have »sustainability«, »energy«, »businesses«, and »companies«.

Figure 3 shows that the circles of specific themes overlapping with the circles of other themes. For example, there is an overlap among the generated theme »sustainability«, »energy«, and »businesses«. There is also a thematic overlap among »businesses«, »sustainability«, and »energy«. The theme

»energy« overlaps with the themes »sustainability«, »businesses«, »companies«, »oil«, and »world«. The theme »companies« overlaps with the theme »energy«. The theme »world« overlaps with the themes »energy« and »oil« and the theme »oil« overlaps with the themes »world« and »energy«. It can also be seen in Figure 3 that the concept »data« intersects with »sustainability« and »business«, whilst »industry« can be found between the thematic of » business« and »energy«. The concept of »policy« is situated between the intersection of the themes »world« and »energy«. The concept »carbon« lies between the intersection of the themes »oil« and »energy«, and the concept »supply« lies between the intersection of the themes »companies« and »energy«.

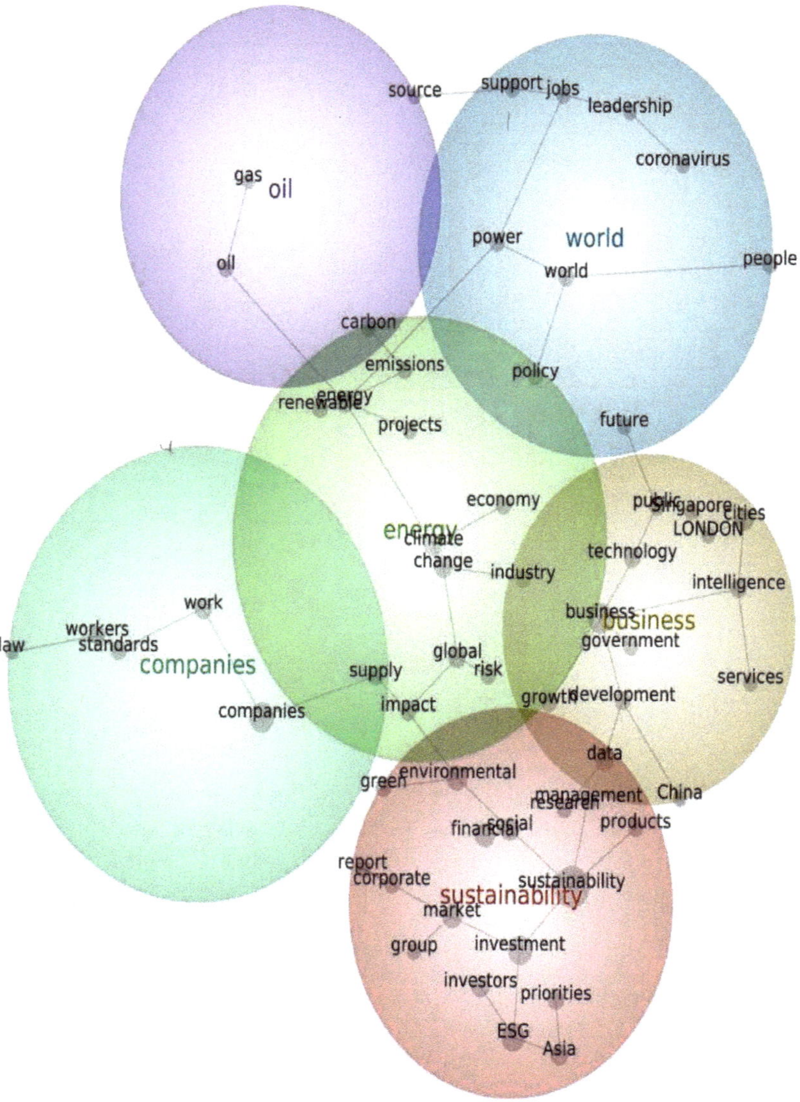

Figure 3. Concept map of the chosen newspapers and journals articles published between 2015 and June 2020.

Table 2. Themes and concepts in the journals in the years from 2015 to June 2020.

Theme	Hits	Concepts
Sustainability	1825	sustainability, investment, ESG, environmental, financial, market, social, investors, Asia, priorities, green, research, management, report, corporate, products
Energy	1098	energy, change, climate, global, impact, supply, industry, renewable, risk, economy, emissions
Business	991	business, data, cities, technology, development, intelligence, government, growth, services
Companies	681	companies, work, standards, workers
World	394	world, people, power
Oil	98	oil

Source: Authors' work.

5. Discussion

5.1. Research Theme through the Research Source Characteristics

In the first part of this chapter, papers from the scientific journals are analyzed, followed by papers from the newspapers and journals are analyzed.

5.2. Research Theme from Scientific Journals Articles

Based on the primary concepts that can be derived from the analysis of scientific journal articles, it can be concluded that the five themes (see Figure 2), regarding the treatment of I4.0 and sustainability with a focus on technology, clean production, and the environment are: Sustainability, technologies, model, environmental, and emissions.

The manufacturing sector (theme sustainability and technology) is undergoing fundamental transformations [59] focusing on reducing greenhouse gas emissions and waste generation by implementing the low-carbon and recycling technologies in order to enhance sustainability [60]. I4.0 affects the sustainability aspect of manufacturing, but not only the ecological aspect (i.e., renewable energy, resource efficiency), but also technical, social, and organizational aspect of sustainability [61]. The central paradigm within the technical aspect of I4.0 is the Internet of Things, which is evolving together with the Internet of Space on the Internet of Everything. Internet of Things plays a crucial role in connecting a variety of everyday objects to the Internet. Internet of Things applications have become indispensable in the areas of communication, service provision, information, and process management. IoT applications play a pivotal role in the development of smart cities, more proper healthcare, efficiency in agriculture, as well as industry and production. For example, the implementation of information and technology systems in agriculture through the intensive use of data can improve agricultural productivity [62]. Informatization in I4.0 creates effective monitoring and control of the material world utilizing information technology. The unbridled growth and development in the devices that are connected in the IoT applications, as well as the heterogeneity of and lack of uniformity of network technologies, raise concerns about the sustainability of Internet of Things. Solutions are sought by introducing a decentralized Internet of Things platform, and controlling the Internet of Things environment during implementation, monitoring the quality of the system, and making efficient use of available resources. Therefore, solutions must be tendered to ensure the Internet of Things, as well as the Industrial Internet of things and all its applications, will have longevity [63].

The basis of the development paradigms of I4.0 is the digitalization, informatization, and connection of industrial and other social processes. The development and establishment of data analysis, machine learning, and artificial intelligence, as well as business operations, is crucial for the implementation of the processes, which will also affect the emergence of the potential of all three dimensions of sustainability (Triple Bottom Line concept of sustainability). I4.0 enables the Triple Bottom Line by improving productivity and product quality [64]. The technological transformation enables the emergence of the cyber-physical principles, which arose as a result of the informatization processes of transformation in manufacturing, logistics, and supply chain [65]. In recent years, manufacturing and research in this field are oriented towards digitalization, informatization, data analysis [66,67], and the exposure of the digital twins [68], which can help to identify real-time events and predict future events.

Manufacturing companies are also faced with the need to reorganize their supply chains in line with United Nations Sustainability Goals and market situation (e.g., in doing so, they are following various practices such as lean, green, circular, and I4.0, which lead in finding value chain partners that meet both sustainability criteria and the required technological know-how [69]. Regarding manufacturing (economic aspect) that is continuously monitoring energy consumption, environmentally sound manufacturing (e.g., implementing technologies that enable cleaner production, use of renewable energy, and reduction of pollution from changing polluting raw materials with alternative environmentally friendly materials (ecological) [70,71] leads to a safer working environment, less intensive workload, job enrichment, and social equality (social) [72]. It can be concluded that due to sustainable production, I4.0 is proven to be beneficial for societal development [59]. For example, I4.0 can improve social welfare with corporate sustainability activities [73]. The organizational aspect of sustainability starts with the organizational culture comprising sustainability vision, mission, and values [74]. It is also important that business strategies are sustainable [75] and that the main drivers of I4.0 can generate high triple bottom line profits (environmental-economic-social) in emerging markets as it is becoming a significant factor in emerging markets because they still use unsustainable approaches: Child labor, exploitation of workers, use of fossil fuels, and inadequate environmental legislation allowing the use of obsolete, ecologically questionable technologies and raw materials [76].

For these reasons, I4.0 is reshaping business models and production processes of the organizations fundamentally [77] (theme model). The traditional business model in the manufacturing sector is changing. Production processes are oriented toward energy optimization, reuse, and recycling of wasted materials [78]. Adopting a novel approach to business modelling is, therefore, mainly focused on business process automation (adaption of I4.0) and principles of circular economy [79]. Digital and connected manufacturing technologies are the main enabling factors for new business models.

Moreover, I4.0 requires that the business models of other organizations within the ecosystem are in equilibrium [80]. Digital platforms are an essential business model within I4.0 platforms, and they are emerging in different sectors, including retail, education, healthcare, transport, aggrotech, fintech, and real estate. These platforms are characterized by the fact that they change the labor market with the emergence of non-standard forms of work, from platform work to freelance work. Platform work is characterized by mass work, which can also be offered from home. The emergence of digital platforms for working from home becomes particularly crucial during the outbreak of COVID-19 when quarantine forced many people to stay at home. This period showed how important digital platforms are for global achievements (e.g., work, study, health, businesses) and how they can be used better and sustainably [81].

For a company, business models represent a form of timely reaction to market requirements. The markets are changing fast, and companies are managed and organized, and their supply chains are also changing. Markets are becoming more dynamic with the new digital technologies, as evidenced by the fact that the average lifetime of a company listed in the S & P 500 has fallen from 60 to 18 years. Companies can thus establish themselves in a short time [82]. Business models are also being adapted concerning the way production processes and supply chain organization are designed. For example, the digitalization of production processes, together with the connection of smart devices and systems, enables real-time control and enables fast, decentralized decision-making [83]. The advantages of virtual and augmented reality technologies are increasingly used to optimize production processes. Simulations aim to prevent the occurrence of bottlenecks in advance and to save time. Shorter production lines are also to be achieved through additive production [84]. Sustainable business models promote transparency and change consumer behavior by involving consumers more closely in the creation of new products and services. I4.0 thus encouraged a shift from mass production to personal (individual) centralized manufacturing. The manufacturing transformation encourages factories to use digital and information technologies (mass data, cyberspace, artificial intelligence) and to adapt their methods of advertising, delivery and customer service, product design, manufacturing, or testing [82].

Economic performance, such as production efficiency and labor productivity, tends to be the immediate outcome of I4.0, which will transform future manufacturing and business models for sustainability [15] (theme environmental). The transformation will change the dominant linear economic model into economic sustainability. One example is the utilization of renewable sources of energy as well as existing underutilized sources of energy (e.g., geothermal energy), which in the long run will invariably be instrumental in development. The possibility is endless, as local economies and the value they generate can also benefit (fiscal returns, cost reduction, increase in purchasing power, job creation, increased tax revenues); other benefits also include social, ecological, and ethical aspects, all as a result of an increase in revenue retention in the region [85]. The mining industry represents the second case where the introduction of sustainable environmental criteria has a notable impact on the economy, environment, water pollution, as well as the varying social issues found within the mining environment. For these reasons, it is essential for the global mining industry in terms of how it will deal with sustainability. Ranängen and Lindman [86] for example, based on research in the Nordic mining industry, recommend that it should follow sustainable guidelines in the areas of "corporate governance, fair business practices, economic aspects, human rights, labor practices, society and the environment."

The manufacturing sector is a high polluter of greenhouse gases, mainly carbon emissions (theme emissions), which affect global environmental sustainability and contributes to extreme weather conditions and pollution [87]. In order to reduce air pollution, some countries implemented carbon tax levies (for this reason, organizations have the motive to implement the green technologies (e.g., energy-saving and low-carbon technologies) [88,89]. The pollution is only one of the reasons why it is vital to introduce new business models in industries like shoes, textiles, or metals. Notably, the labor shortage is rectified, as well as the need to control costs and reduce the deleterious effects that the product has on the environment, especially with carbon emissions. Though the list is more complex, these are essential factors in the shifting perspective of the footwear industry, which has led to the adaptation of better production models. Hence, the output leads to better quality and highly competitive products. At the same time, all manufacturing industries require higher costs for them to introduce carbon tax rates that all companies should aim to invest in low-carbon technologies [89]. In the field of corporate sustainability, raising the importance of sustainable reporting, which includes information on the environment, society, and governance (ESG), is becoming increasingly important in the financial markets. There is a risk that by reporting on sustainability, companies are planning a speculative effect on the value of their shares at the time of publication. The last challenge is a challenge for policymakers in formulating rules for ESG reporting. The aim of the shift in policy seeks to incentivize transparency as value-relevant; the findings have indirectly pointed in the same direction as the ESG. Thus, reporting is likely having the benefit of an improvement in market efficiency and strengthening of the growing confidence of the various stakeholders: Investors, companies, institutions, and practitioners in the materiality of ESG information [90].

5.3. Research Theme from Newspapers and Magazines Articles

After conducting an ACA of (see Figure 3) magazine and newspaper articles with the content on I4.0 and sustainability, focusing on technology, clean production, and the environment, it is possible to identify and explain the most critical and recent global socio-economic and technological trends.

Concern for the environment and people's demands for environmental protection (theme world) came to the fore with the development of socially responsible movements in the 1960s. Nevertheless, the natural and urban environment is increasingly threatened by the effects of weather. In Singapore, for example, they face rising sea levels and rising tropical temperatures. The outbreak of the COVID-19 pandemic in 2020 confronted the city with an increase in waste of up to 40% of normal levels. At the time of the pandemic, people stayed at home, which led to an increase in the amount of waste, which consisted mainly of food packaging waste. However, the COVID-19 pandemic in 2020 became an indicator of the impact of human lifestyle and business activities on the environment. Declining economic activity and increased homeworking have led to falling carbon dioxide emissions. Analysts have made conservative estimates that there could be a 4% fall in carbon dioxide emissions in 2020 compared to 2019 owing mostly in part to the global pandemic. However, even if this reduction had remained constant each throughout the decade, we would still fall short of the 7.6% annual reduction that has been predicted to meet the 1.5-degree Celsius global warming target. Based on these results, we see that even repeated shutdowns of countries cannot slow down climate change. How can this be achieved through individual measures such as reduction, reuse, and recycling? The second problem, which is also evident in Singapore, is that their government is trying to develop sustainability-oriented environmental programs. At the same time, other Asian neighboring countries are not aware of the environmental problems because they have no regulated legislation in the field of environmental protection and even less suitable environmental projects supported by local governments [91].

Based on the theme of sustainability, it can be stated that the policy of investing in sustainable projects in the ESG sector represents an important measure of confidence of companies in sustainable projects, and investors in research and development capacities of companies and countries that promote sustainable investments [92]. Over the last decade, passive assets and responsible investment have stimulated the growth of passive sustainable funds. They have promoted active equity activities with passive fund providers not only in Europe and the United States (US), but this investment area is also gaining ground in Asian countries. In the period after 2015, the US even denies the importance of sustainable environmental protection and rejects United Nations Agenda 21. The US government administration did this because, in their opinion, it was prepared to weaken the economic interests of both companies and the state (e.g., Trump administration withdraws from the US Paris Climate Agreement, stops Obama's Clean Power Plan and the reopening of coal mines, reduces the importance of toxic air pollution regulations) [93].

The European Union (EU) is aware of the importance of developing a green investment. The EU is, therefore, encouraging the issuance of green bonds, which are playing an increasing role in financing the investments needed to make the low-carbon transition a success. However, due to internal disagreements between members on whether nuclear energy is a sustainable energy source, there are currently no standard European rules for green investment in the EU. These rules will define what is considered a sustainable investment. The rules will, therefore, cover all types of energy, including nuclear energy. Coal investments will be excluded from any sustainability definitions following the rules. The rules will make it possible to communicate how investors deal with assets such as green bonds, bank loans, and investment products. The rules will help to eliminate the so-called "greenwashing" that occurs when countries and companies want to show their environmental capabilities in a better light than they are. Once the rules are adopted, it is expected that more capital will be redirected to activities that are consistent with the Paris climate convention. The rules are expected to take effect in 2021 [94].

The upturn in ESG in Asia in recent years has been driven by increased regulatory pressure in the context of environmental protection targets (e.g., China). Analysts see another reason in the increasing investment by Japanese and Thai state pension funds in sustainable projects, as they have identified potential growth in ESG approach in the region. It was the state pension funds that also increased the interest of other investors in the region in ESG. National and multilateral development banks have also taken significant steps in the field of ESG by adopting new recommendations as policy. Regulation of the ESG has also been undertaken by capital market regulators (both government agencies and the stock exchange themselves) [95]. Over the past three years, both investors and companies in the region have become aware of the importance and promotion of principles of responsible investment (PRI), as the significant problem with ESG is the often-subjective investment ratings. As a result, there have been frequent abuses in recent years in cases of nonsufficient corporate reporting on sustainable development. Investors point to the problem of lower quality and inadequate data, which makes it impossible to make well-considered investment decisions [96].

In Asia, despite the extremely fragmented regulatory environment, they perceive the importance of the ESG approach to promote values and ethical principles, but not the financial value or returns themselves. ESG, therefore, has led to the increase of ethical investment practices, which has provided a framework for sustainability that not only focuses on the damage to the planet as a whole, but passes on that responsibility to corporations as well. These practices have bolstered the influence of non-financial factors, such as social and environmental factors, on the stock market and the company's goodwill. It has become essential for ESG to influence risk-adjusted returns [97]. However, time will tell, based on an analysis of stock market indices and the value of company shares, whether the choice of ESG strategy has contributed to better business results. The companies have to ensure that their sustainable corporate reports include quality data about their exposure to climate change risks. In the field of sustainable corporative reporting, it is expected that initiatives such as the EU's taxonomy will provide a better definition of sustainability and thus help companies to ensure the right data for reporting. It is also expected that the Global Task Force on Climate-Related Financial Disclosures will control the corporate sustainability reporting, which will reduce investment risks and stabilize the ESG market in the future [98]. Awareness of the importance of sustainability and corporate social responsibility is also in line with customer expectations and attitudes. Research on consumer behavior shows that consumers do not reject products or services based on price and quality preferences. Modern consumers are increasingly aware of the importance of social and moral values, and the consequences of this awareness are visible in the extraordinary growth of the global market for organic and environmentally friendly products, which have well-known and socially responsible members of the supply chain [99,100]. Socially responsible companies work without plastic packaging and sell products without animal testing, without leather (except in the form of reuse), without genetically modified food, without palm oil and its environmentally deleterious reputation, the aim is to offer completely natural products for the health of their customers. Consumerism is, thus increasingly associated with people's social-ethical behavior and awareness [101].

Regarding themes of energy, oil, and companies, the following can be concluded. In 2019 and 2020, global investment companies, banks, and state pension funds (e.g., Credit Suisse, Deutsche Bank, the Dutch Fund) announced a reduction or even a withdrawal of investments from fossil fuel production (thermal coal mining, oil production from tar sands). Companies have become aware of the importance of sustainable financing and are shifting their investments to sustainable projects [74,75]. The use of renewable energies and clean energy is increasing, but the demand for oil and gas is expected to continue for decades to come. The change in the field of reducing the carbon intensity of energy is taking place as a result of pressure from investors, growing consumer awareness, and governments adopting stricter environmental legislation due to climate change. For example, California will ban the use of diesel trucks from 2045 [102].

Global markets are changing due to the emergence of new clean technologies, sustainable materials, work standards, and the awareness of customers and companies of the sustainability of production and consumption. Companies (from the food industry to the fashion industry) are preparing plans to reduce their carbon dioxide emissions, eliminate disposable plastics in packaging and sustainably sell more products (e.g., digital platforms) [103]. In the field of agriculture, too, increasing attention is being paid to sustainable principles. For example, farmers are moving from conventional to ecological practices, which involves reducing the excessive use of fertilizers and thus preventing nutrient depletion of the soil. They are investing more and more in technology and work processes, with the result that pesticides and fertilizers are being displaced from production processes. Drones with sensitive sensors and software are already being used in the agricultural sector, which can predict potential problems due to the health of the soil or the plants themselves, vegetables, and vines. The results of using cleaner technologies and smarter equipment are healthier soils, healthier agricultural products, higher yields per hectare, and a significantly reduced ecological footprint [104].

It is predicted that the development of sustainable measures will have an impact on future commercial success. The sustainably conscious have set themselves the goal of becoming "carbon neutral" along the entire value chain (production, packaging, logistics). Companies are increasingly turning to the use of renewable energies and are switching their logistics to electric vehicles and drones for delivery. Carbon netting, which results from the integration of sustainability standards into business processes, thus enables the financing of emission reductions in other areas, such as the prevention of deforestation. Due to the growing demand for sustainable and ethically produced products and services, companies have begun to create brands that emphasize product relevance under international standards for social, environmental, and animal welfare practices [103,105].

The technological development in I4.0 requires oil companies to diversify their investments and reinvest in solar power generation in recent years (after a failed investment cycle at the beginning of the 21st century), buy companies to sell electricity to electric vehicles, take over power utilities, and even invest in projects such as floating wind farms [106]. However, it is also necessary to be aware of the consequences of the transformation in the field of abandoning internal combustion engines and replacing them with electric vehicles and probably also with hydrogen-powered vehicles and other options that are still being developed in the laboratories today. Leaders in the automotive industry face the challenge of building the car of the future. They are expected to act responsibly in the context of change, the consequences of which will be visible in the short term as the disappearance of traditional roles in the automotive industry and its supply chains. Supply chains could be reduced by 20 to 30%, which in turn would mean a loss of 20 to 30% of jobs. It is estimated that by the end of the I4.0 (around 2030), one-third of the workforce in the automotive industry will become redundant. In terms of production, this will be the most significant change in history. Companies will also need to redesign their supply chain management, procurement, and human resources management [107].

Chocolate manufacturers are one such example. They are aware that consumers are increasingly demanding products of ethical origin from chocolate products. They are therefore strengthening all their sustainability programs, which have so far had only a negligible impact on reducing child labor in cocoa production in West Africa. The company Nestle decided in 2019 to invest in the promotion of sustainable cocoa extraction in the supply chain in West Africa. The company will begin to reduce the number of underage children working in cocoa supply chains. Together with the International Cocoa Initiative, it aims to eliminate child labor in Côte d'Ivoire by 2025. In doing so, they expect government support to promote the training of farmers and other programs to address this problem [108]. Humankind has been used to these consequences of industrial transformation since the last three industrial revolutions. All three past industrial revolutions have made it possible to create more jobs than technological developments have eliminated. More significant growth was achieved, which led to a reduction in the poverty rate and enabled people to lead a better life. The question is, however, what will the industrial revolution bring to society, which will have or already has a strong influence on production, which today accounts for 70% of GDP on a global scale? Governments, together with companies, will probably have to find solutions for workers in sustainable social policy programs, whose training and experience will become irrelevant due to advanced technological developments [109]. In 2020, for example, Mercedes Benz announced the layoff of only 15,000 employees by 2022, which is due to the digital transformation of production and the attempt to adapt production processes to the production of electric cars. The entire automotive industry announced the layoff of 80,000 employees for a short period of three years (2020–2023). As we can see, the social and economic impact of the digital transformation will be decisive. In the entire value chain (not only in the automotive industry but also in other sectors), the middleman will become superfluous [107].

The introduction of sustainability into the traditional concept of business operations is based on a thematic approach (theme business). For example, sustainability is connected to climate change, diversity, and the awareness of corporate leaders that human rights must be respected (these are the rights of employees in global supply chains). In Europe and Asia (especially in China), both business schools and companies are increasingly aware of the importance of taking sustainability into account in business. Business schools so try to the introduce the new technologies and products that do not pollute the environment and use renewable energy and through sustainable business models (e.g., circular economy, green transformation, social entrepreneurship) that include business ethics. On the other hand, the US does not pursue sustainable concepts and does not invest in the training of professionals in this field [110]. The People's Republic of China, however, has addressed the problem of how to solve environmental problems resulting from rapid economic growth. With increasing investment in research and development, they want to become world leaders in the development of environmental technologies. China has overtaken the USA as a leader in the production of and exporter of technologies that are more environmentally friendly.

Additionally, China joins Europe and North America for the large share of patent applications in the space of environmentally conscious technology and lead all the above in the development of batteries, automobile production, and solar technologies [111]. The China Corporate Responsibility and Sustainability Council notes, however, that China is still confronted with social and economic constraints resulting from its rapid economic development. For example, a large number of people moved from the countryside to more than 10 million cities, which are now confronted with unregulated sewage, traffic problems, too high levels of pollution of the atmosphere and water resources, etc. China needs to address social and environmental issues sustainably. Otherwise, it will increase costs and damage the reputation and growth of companies in China [112].

The importance of digitizing the human environment and business became even more critical when the outbreak of COVID-19 showed that human life must be further digitized. The need to establish creative online content and online business services are increasingly coming to the fore. The most digitized cities are London, Paris, San Francisco, Los Angeles, and Singapore. Part of the answer of what is a smart city probably lies in the Internet of Things, which is part of the idea of smart technologies or even a link between them. Remote-controlled operations, technological innovations, and self-propelled cars are only a part of the whole story. Internet of Things provides more data that can help improve many aspects of our daily lives. They can even help create jobs. For example, according to Forbes, Barcelona has saved €75 million a year as well as created employment for 47,000 in technology development by taking strategic steps in the provision of high-speed Internet connections, smart lighting, smart irrigation, and parking management. In Amsterdam, the city uses the Internet of Things-based infrastructure to monitor traffic flow, energy consumption, and public safety using real-time data. In Boston and Baltimore, the development of smart garbage bins create efficiency by utilizing smart technology to indicate how full they are, which is relayed to the sanitary worker for more efficient collection route mapping [113].

The appearance of COVID-19 will have a lasting effect and have a notable impact on ensuring that sustainable cities and smart cities are not mutually exclusive. The concept of a smart city is a model of developing a sustainable city. When we talk about the emergence of the sustainable cities, their development may take longer if it is based on the use of traditional planning tools, without the necessary high-tech solutions that are part of the smart city. A smart city (e.g., green Vienna or Lisbon projects) has become more sustainable (e.g., sensors controls the energy and water consumption and provide better living conditions (more green areas), which it achieves by implementing high-tech solutions in the urban fabric. The emergence of COVID-19 is changing public transport in the cities, and in spring 2020, cities have launched projects to promote walking and cycling. With the increase in e-bikes, scooters, and transport, the focus has shifted to reach. Mayors around the world are already taking the opportunity to redesign lanes for pedestrians and cyclists. It is expected that the car-sharing model will expand in cities. The relationship between illness and density is also perceived as complex.

Some high-density cities, such as Seoul, Taiwan, and Singapore, have suppressed mainly the virus. In other lower-density cities, such in the Italian region Lombardy and U.S. country Louisiana, the virus has spread rapidly. London now has a much lower rate of infection than the north-east of Great Britain [114].

A study published in The Economist [115] shows that on a global level, residents not only want to be informed about the course of these processes but also want to participate in the formulation of urban development policies. At the global level, companies involved in these projects also expect cities to draw up long-term development plans. The main differences between the population's expectations of urban development occur at the micro-level. In cities such as Los Angeles, San Francisco, Dubai, and Zurich, residents need to develop a smart city initiative that supports the development of citizen participation processes in the preparation and adoption of the city budget. In London, Riyadh, Stockholm, and Sydney, residents expect more long-term planning. In less developed cities like Johannesburg and Mumbai, people want to create initiatives that promote the development of small businesses. In São Paulo and Singapore, residents want to ensure fairer access to smart services, while in Hong Kong São Paulo and Singapore, residents are most in favor of better protection of personal data.

5.4. The Comparison of the Analysis's Results

The analyses of both types of literature, both scientific and professional, show that there are common topics they write about, which are related to the field of clean production, emissions, renewable energy, climate change, sustainable investments and corporate sustainability. An urgent global issue that extends all over the world is the promotion of energy-saving technologies and reduction of carbon dioxide emissions [60].

We also looked for differences in the topics covered by the literature of different types. We found that the scientific literature focuses more on changes in business models, production processes, and technologies that enable sustainable development. Newspapers and magazines articles write more about sustainable or green investment, sustainable standards, and sustainable reporting. It is going for themes that are directly important for current sustainable business development and encouraging the research and development of cleaner and smarter technologies and processes.

The relevant and current theme nowadays is oriented towards COVID-19 and its impact on the economy and society. The newspapers and some latest research journals include articles of the COVID-19 outbreak and its effect on the economy and the environment. Indeed, the outbreak of the virus brings new thinking about the reorganization of the complex relationships between consumers, businesses, and the state [116]. The question is whether, in rescuing the economy, we will unreasonably seek to return to the old patterns as quickly as possible, or whether we will use the moment to reshape and restructure the national economy. It offers a unique opportunity to solve two crises at once, with prudent behavior and wise action [117]. The response to the health and economic crisis can be enhanced by tackling the environmental crisis, which can also disrupt the food chain [118]. The latter, however, first requires an ambitious restructuring plan that transitions from a linear to a circular economy. The ideas of green transformation are not entirely new at such times [119]. During the previous crisis, state aid for the car giants was made conditional on their moving to stricter emission standards. However, the current situation allows the implementation of much more ambitious plans compared to then, as the social climate is much more favorable today than it was in 2008 [120]. Companies will have to thoroughly rethink their existing business models, organizational structure, and the way they work during and after the crisis. Intertwined global supply chains, marketing approaches that respond to pre-crisis consumer habits, and mandatory physical presence in the workplace are just some of the critical factors already under discussion and the reason for the forthcoming transformation of the business environment. The coming year or two will bring many new insights into labor productivity as one of the critical factors in production. Management discussions will revolve around the possibilities and opportunities of the digital transformation of companies. Internal policies will adapt to the new situation in the areas of business travel, contractual relations, and security. The wait for state aid will not solve several challenges, so companies must use this time to make a radical transformation. The latter requires a thorough reflection on the strategy of the company and its role in society as a whole, the business and profit model, and, last but not least, the role of employees in this process [121,122].

Finally, the role of the states themselves is, and remains, extremely important. They will bear a considerable part of the financial burden of the crisis, so timely planning and smart conditioning of measures are crucial. Rescuing companies cannot and should not be aimed at returning to a pre-crisis state. The latter is not only impossible because the world has changed considerably in a few months, but also pointless because we could miss one of the few opportunities for an extensive, green, circular transformation of the economy. State-sponsored financial instruments should, therefore, be based on the principles of green financing, grants should include a commitment to meet ambitious environmental standards, and measuring the success of a national rescue package should be based not only on traditional macroeconomic indicators, but also on broader social and environmental impacts [123]. Sustainable development and recycling must become more than just theoretical concepts, which means that they must be operationalized through sector-specific and fact-based measures. Countries can build on existing commitments and solutions; within the EU these include, for example, the efforts of European climate and energy policy or reporting and taxonomy standards for sustainable activities in the EU [124].

6. Conclusions

We conducted an ACA of scientific journals papers and newspaper and magazines papers with the content on I4.0 and sustainability, focusing on technology, clean production, and the environment. According to the comparison of the analyzed themes between both research, groups of papers were identified and explained the most critical and recent global socio-economic and technological trends.

The results revealed that there is some overlapping between the concepts that emerged from the newspaper and scientific journal papers, such as various aspects of sustainability. However, newspaper papers also investigated some concepts that are substantially different from scientific journal papers, and these are mostly related to business and corporate issues. The analyses research found out that common topics mentioned in the newspaper and scientific journal papers covered clean production, emissions, renewable energy, climate change, sustainable investments, and corporate sustainability. According to the content analyses, it can be concluded that an urgent global issue that extends all over the world is the promotion of energy-saving technologies and reduction of carbon dioxide emissions [60].

The results of the ACA of the scientific journals papers show that their thematic is focused more on changes in business models, production processes, and technologies that enable sustainable development. The themes of newspapers and magazines papers are more focused on sustainable or green investment, sustainable standards, and sustainable reporting. They are going for themes that are directly important for current sustainable business development and encouraging the research and development of cleaner and smarter technologies and processes.

Based on the ACA results, we find that the digital transformation in both manufacturing and the products themselves (e.g., electric cars) and the requirements to adapt the business to and operation of products following the United Nations Sustainability Goals lead to a reorganization of their supply chains, what has positive and negative consequences for society itself. The positive consequences are particularly important for developing countries, where strict environmental protection criteria are not enforced, and companies' sustainability standards are not met, leading to non-compliance with both occupational safety and child labor. The negative consequences are visible or will be visible in the reduction of jobs because the digital transformation will result in the middleman becoming redundant. Here, both scientists and governments are facing a solution to the consequences of the digitalization of companies, for which it will be necessary to find a social consensus, which will have to be supported by concrete research both among companies and the public [125]. This research should focus on finding solutions to mitigate the effects of both digitization and the requirements of the United Nations Sustainability Goals, which include new sustainable policy programs that will offer various new forms of employment in the Universal Basic Income (UBI). Thus, in Germany in August 2020, they started a test phase under which 120 citizens will receive €1200 a month for three years [126]. It can be concluded that future researches should be focused on monitoring of the CSR and sustainable development impact on the economy. It will also be necessary to take into account the issues related to COVID-19, which has a significant impact on changes in the field of work organization, economics, and the operation of the company itself and citizen wellbeing, especially in cities that have to be transformed not only in smarter and more sustainable ways [127], but also safe from crime, natural disasters, pandemic, and other catastrophes [128,129]. The virus pandemic is leading to a faster digital transformation, and various digital platforms are increasingly being developed to enable work and education from home [130]. In the health industry, service digitalization becomes important for the monitoring of the patient from home, and it also enabled control and transparency of medical-epidemiological research and mental health service, which is especially important in pandemic time [131,132]. In the field of public administration, communication with the city administration is established, and their concerns are reported. Online citizens' participation in discussion about cities policies and budgeting is enabled. More and more administrative procedures can be done online, such as business registration, dog registration, information about historical certificates or registration, citizens id cards have been expanded, and e-voting is enabled [133,134]. We think that more research will be needed on adapting people to new conditions, such as working from home, as well as research on trust and security in digital business and the implementation of administrative procedures and e-democracy.

The next important question is how we will gain energy in the future? As we can see, opinions are very divided across countries, and practice shows that countries that have opted for predominantly green energy are having increasing problems with electricity shortages because they are no longer producing enough to shut down nuclear and fossil fuel power plants to meet the needs of both industry and households [135]. Research also shows that Germany's renewable energy project failed because the German CO2 Emissions is 10 times higher than nuclear-powered in France [136].

In September 2020, we witnessed the electricity crisis in Germany, which was forced to import energy from countries such as Poland, Bosnia and Herzegovina, the Czech Republic, and Slovenia [137]. The paradox is that such an "energy green country" cannot produce sufficient quantities of electricity with green technology and must import it from countries where it is produced in nuclear and thermal power plants. The insecure energy supply, which depends on domestic green electricity, is threatened for the national economy [135,138,139]. In the field of energy, research is emerging on the applicability of new modular nuclear power plants and the very meaning of dependence on only the so-called green energy produced by solar power plants, windmills, etc.

Finally, let us mention a relatively new concept: Sustainable business finance. Within companies, sustainable reporting is becoming more critical. Companies must therefore add information on the environment, society, and governance (ESG) to their business reports. ESG information shows a picture of the company and is therefore becoming increasingly important for financial markets. Investors have become aware of the importance of sustainable development, which is increasingly gaining its place in the laws of individual countries, and among end customers, who are increasingly in demand for sustainable products and services [140,141]. In this field, it will be necessary to research sustainable reporting and, based on the analysis of company reports, to determine how this reporting changes over time and what information it contains. On the other hand, it will be necessary to analyze the responses of financial markets to insufficient sustainable reporting and the consequences for these companies.

The research paper limitation is based on the procedures of article selection, analytical methodology, and the purpose of the analyzed results. Thus, we acknowledge that the articles analyzed in the given selected period from a specific database possibly do not contain all scientific research and professional views on the selected research content, especially when only English-language articles were processed.

The main idea, however, is that the present analysis gives us an insight into the critical issues of Industry 4.0 and sustainable development. The limitations above represent the challenge that future researchers would need to overcome.

Author Contributions: The authors V.R., O.T., M.P.B., A.J., and M.M. contributed equally. All authors have read and agreed to the published version of the manuscript.

Funding: This research received no external funding.

Conflicts of Interest: The authors declare no conflict of interest.

References

1. United Nations. Transforming Our World: The 2030 Agenda for Sustainable Development. United Nations 2015. Available online: https://sustainabledevelopment.un.org/post2015/transformingourworld (accessed on 22 September 2020).
2. Ejsmont, K.; Gladysz, B.; Kluczek, A. Impact of Industry 4.0 on Sustainability—Bibliometric Literature Review. *Sustainability* **2020**, *12*, 5650. [CrossRef]
3. Piccarozzi, M.; Aquilani, B.; Gatti, C. Industry 4.0 in Management Studies: A Systematic Literature Review. *Sustainability* **2018**, *10*, 3821. [CrossRef]
4. Jerman, A.; Bach, M.P.; Bertoncelj, A. A Bibliometric and Topic Analysis on Future Competences at Smart Factories. *Machines* **2018**, *6*, 41. [CrossRef]
5. Jerman, A.; Erenda, I.; Bertoncelj, A. The Influence of Critical Factors on Business Model at a Smart Factory: A Case Study. *Bus. Syst. Res. J.* **2019**, *10*, 42–52. [CrossRef]
6. Birkel, H.S.; Veile, J.W.; Müller, J.M.; Hartmann, E.; Voigt, K.-I. Development of a Risk Framework for Industry 4.0 in the Context of Sustainability for Established Manufacturers. *Sustainability* **2019**, *11*, 384. [CrossRef]
7. Müller, J.M.; Däschle, S. Business Model Innovation of Industry 4.0 Solution Providers towards Customer Process Innovation. *Processes* **2018**, *6*, 260. [CrossRef]
8. Oztemel, E.; Gursev, S. Literature review of Industry 4.0 and related technologies. *J. Intell. Manuf.* **2018**, *31*, 127–182. [CrossRef]
9. Van Dijk, T.A. *News Analysis*; Routledge: Hillsdale, NJ, USA, 2013.
10. Reah, D. *The Language of Newspapers*; Routledge: London, UK, 1998.
11. Lööw, J.; Abrahamsson, L.; Johansson, J. Mining 4.0—The Impact of New Technology from a Work Place Perspective. *Min. Met. Explor.* **2019**, *36*, 701–707. [CrossRef]
12. Xu, S.; Stienmetz, J.; Ashton, M. How will service robots redefine leadership in hotel management? A Delphi approach. *Int. J. Contemp. Hosp. Manag.* **2020**, *32*, 2217–2237. [CrossRef]
13. Baldwin, R. *The Globotics Upheaval: Globalization, Robotics, and the Future of Work*; Oxford University Press: Oxford, UK, 2019.
14. Nuvolari, A. Understanding successive industrial revolutions: A "development block" approach. *Environ. Innov. Soc. Transit.* **2019**, *32*, 33–44. [CrossRef]
15. Ghobakhloo, M. Industry 4.0, digitization, and opportunities for sustainability. *J. Clean. Prod.* **2020**, *252*, 119869. [CrossRef]
16. Cheng, M.; Edwards, D. A comparative automated content analysis approach on the review of the sharing economy discourse in tourism and hospitality. *Curr. Issues Tour.* **2017**, *22*, 35–49. [CrossRef]
17. Pucihar, A. The digital transformation journey: Content analysis of Electronic Markets articles and Bled eConference proceedings from 2012 to 2019. *Electron. Mark.* **2020**, *30*, 29–37. [CrossRef]
18. Pool, I.D.S.; Berelson, B. Content Analysis in Communication Research. *Am. Sociol. Rev.* **1952**, *17*, 515. [CrossRef]
19. Berelson, B.R. Content analysis. In *Handbook of Social Psychology*; Lindsey, G., Ed.; Addison-Welsley: Reading, MA, USA, 1954; pp. 488–522.
20. Kracauer, S. The Challenge of Qualitative Content Analysis. *Public Opin. Q.* **1952**, *16*, 631. [CrossRef]
21. Schreier, M. *Qualitative Content Analysis in Practice*; Sage: London, UK, 2012.
22. Parker, E.B.; Holsti, O.R. Content Analysis for the Social Sciences and Humanities. *Am. Soc. Rev.* **1970**, *35*, 356. [CrossRef]

23. George, A.L. Quantitative and qualitative approaches to content analysis. In *Trends in Content Analysis*; Pool, I.D.S., Ed.; University of Illinois Press: Urbana, IL, USA, 1959; pp. 7–32.
24. Mayring, P. Qualitative content analysis. *Qual. Soc. Res.* **2000**, *1*, 12–19.
25. Kohlbachter, F. The use of qualitative content analysis in case study research. *Qual. Soc. Res.* **2005**, *7*, 1–23.
26. Krippendorff, K. Measuring the Reliability of Qualitative Text Analysis Data. *Qual. Quant.* **2004**, *38*, 787–800. [CrossRef]
27. Hsieh, H.-F.; Shannon, S.E. Three Approaches to Qualitative Content Analysis. *Qual. Health Res.* **2005**, *15*, 1277–1288. [CrossRef]
28. Nunez-Mir, G.C.; Iannone, B.V.; Pijanowski, B.C.; Kong, N.; Fei, S. Automated content analysis: Addressing the big literature challenge in ecology and evolution. *Methods Ecol. Evol.* **2016**, *7*, 1262–1272. [CrossRef]
29. Shapiro, G.; Markoff, J. A Matter of Definition. *Text Anal. Soc. Sci.* **2020**, *1*, 9–32. [CrossRef]
30. Mehl, M.R.; Gill, J.A. Computerized content analysis. In *Advanced Methods for Behavioral Research on the Internet*; Gosling, S., Johnson, J., Eds.; American Psychological Association Publications: Washington, DC, USA, 2010; pp. 34–59.
31. Blei, D.M. Topic modeling and digital humanities. *J. Dig. Hum.* **2012**, *2*, 8–11.
32. Usai, A.; Pironti, M.; Mital, M.; Mejri, C.A. Knowledge discovery out of text data: A systematic review via text mining. *J. Knowl. Manag.* **2018**, *22*, 1471–1488. [CrossRef]
33. McNamara, D.S.; Graesser, A.; McCarthy, P.M.; Cai, Z. *Automated Evaluation of Text and Discourse with Coh-Metrix*; Cambridge University Press: New York, NY, USA, 2014.
34. Blaschke, T. Object based image analysis for remote sensing. *ISPRS J. Photogramm. Remote. Sens.* **2010**, *65*, 2–16. [CrossRef]
35. Shelley, M.; Krippendorff, K. Content Analysis: An Introduction to its Methodology. *J. Am. Stat. Assoc.* **1984**, *79*, 240. [CrossRef]
36. Vaismoradi, M.; Turunen, H.; Bondas, T. Content analysis and thematic analysis: Implications for conducting a qualitative descriptive study. *Nurs. Health Sci.* **2013**, *15*, 398–405. [CrossRef]
37. Smith, A.E.; Humphreys, M.S. Evaluation of unsupervised semantic mapping of natural language with Leximancer concept mapping. *Behav. Res. Methods* **2006**, *38*, 262–279. [CrossRef]
38. Amini, M.; Bienstock, C.C.; Narcum, J.A. Status of corporate sustainability: A content analysis of Fortune 500 companies. *Bus. Strat. Environ.* **2018**, *27*, 1450–1461. [CrossRef]
39. Kim, D.; Kim, S. Sustainable Supply Chain Based on News Articles and Sustainability Reports: Text Mining with Leximancer and DICTION. *Sustainability* **2017**, *9*, 1008. [CrossRef]
40. Lock, I.; Araujo, T. Visualizing the triple bottom line: A large-scale automated visual content analysis of European corporations' website and social media images. *Corp. Soc. Responsib. Environ. Manag.* **2020**. [CrossRef]
41. Paolone, F.; Sardi, A.; Sorano, E.; Ferraris, A. Integrated processing of sustainability accounting reports: A multi-utility company case study. *Meditari Account. Res.* **2020**. [CrossRef]
42. Roblek, V.; Meško, M.; Pejic-Bach, M.; Thorpe, O.; Šprajc, P. The Interaction between Internet, Sustainable Development, and Emergence of Society 5.0. *Data* **2020**, *5*, 80. [CrossRef]
43. Sullivan, K.; Thomas, S.; Rosano, M. Using industrial ecology and strategic management concepts to pursue the Sustainable Development Goals. *J. Clean. Prod.* **2018**, *174*, 237–246. [CrossRef]
44. Keller, T.R.; Hase, V.; Thaker, J.; Mahl, D.; Schäfer, M.S. News Media Coverage of Climate Change in India 1997–2016: Using automated content analysis to assess themes and topics. *Environ. Commun.* **2020**, *14*, 219–235. [CrossRef]
45. Bednarek, M. *Evaluation in Media Discourse: Analysis of a Newspaper Corpus*; Continuum: London, UK, 2006.
46. McCombs, B.L. The Learner-Centered Model: Implications for Research Approaches. In *Interdisciplinary Handbook of the Person-Centered Approach*; Springer Science and Business Media LLC: New York, NY, USA, 2013; pp. 335–352.
47. Schmidt, A.; Ivanova, A.; Schäfer, M.S. Media attention for climate change around the world: A comparative analysis of newspaper coverage in 27 countries. *Glob. Environ. Chang.* **2013**, *23*, 1233–1248. [CrossRef]
48. Schweinsberg, S.; Darcy, S.; Cheng, M. The agenda setting power of news media in framing the future role of tourism in protected areas. *Tour. Manag.* **2017**, *62*, 241–252. [CrossRef]
49. Thetela, P. Evaluated entities and parameters of value in academic research articles. *Engl. Specif. Purp.* **1997**, *16*, 101–118. [CrossRef]

50. Evangelista, P.; Santoro, L.; Thomas, A. A Systematic literature review from 2000–2016. *Sustainability* **2018**, *10*, 1627. [CrossRef]
51. Moher, D.; Liberati, A.; Tetzlaff, J.; Altman, D.G.; Prisma Group. Preferred reporting items for systematic reviews and meta-analyses: The PRISMA statement. *PLoS Med.* **2009**, *6*, e1000097. [CrossRef]
52. Proquest. News & Newspapers. Available online: https://about.proquest.com/libraries/academic/news-newspapers/?page=2 (accessed on 18 August 2020).
53. Bach, M.P.; Krstić, Ž.; Seljan, S.; Turulja, L. Text Mining for Big Data Analysis in Financial Sector: A Literature Review. *Sustainability* **2019**, *11*, 1277. [CrossRef]
54. Leximancer. Leximancer User Guide. 2020. Available online: https://info.leximancer.com/ (accessed on 2 May 2020).
55. Watson, M.; Smith, A.; Watter, S. Leximancer Concept Mapping of Patient Case Studies. In *Lecture Notes in Computer Science*; Springer Science and Business Media LLC: Berlin, Germany, 2005; pp. 1232–1238.
56. Angus, D.; Rintel, S.; Wiles, J. Making sense of big text: A visual-first approach for analysing text data using Leximancer and Discursis. *Int. J. Soc. Res. Methodol.* **2013**, *16*, 261–267. [CrossRef]
57. Campbell, C.; Pitt, L.F.; Parent, M.; Berthon, P.R. Understanding Consumer Conversations around Ads in a Web 2.0 World. *J. Advert.* **2011**, *40*, 87–102. [CrossRef]
58. Esser, F.; Hanitzsch, T. (Eds.) *The Handbook of Comparative Communication Research*; Routledge: Abingdon, UK, 2013.
59. Beier, G.; Niehoff, S.; Ziems, T.; Xue, B. Sustainability aspects of a digitalized industry – A comparative study from China and Germany. *Int. J. Precis. Eng. Manuf. Technol.* **2017**, *4*, 227–234. [CrossRef]
60. Fujii, M.; Fujita, T.; Dong, L.; Lu, C.; Geng, Y.; Behera, S.K.; Park, H.S.; Chiu, A.S. Possibility of developing low-carbon industries through urban symbiosis in Asian cities. *J. Clean. Prod.* **2016**, *114*, 376–386. [CrossRef]
61. Beier, G.; Ullrich, A.; Niehoff, S.; Reißig, M.; Habich, M. Industry 4.0: How it is defined from a sociotechnical perspective and how much sustainability it includes—A literature review. *J. Clean. Prod.* **2020**, *259*, 120856. [CrossRef]
62. Lytos, A.; Lagkas, T.; Sarigiannidis, P.G.; Zervakis, M.; Livanos, G. Towards smart farming: Systems, frameworks and exploitation of multiple sources. *Comput. Netw.* **2020**, *172*, 107147. [CrossRef]
63. Mocnej, J.; Pekar, A.; Seah, W.K.; Papcun, P.; Kajati, E.; Cupkova, D.; Koziorek, J.; Zolotova, I. Quality-enabled decentralized IoT architecture with efficient resources utilization. *Robot. Comput. Manuf.* **2021**, *67*, 102001. [CrossRef]
64. Braccini, A.M.; Margherita, E.G. Exploring Organizational Sustainability of Industry 4.0 under the Triple Bottom Line: The Case of a Manufacturing Company. *Sustainability* **2018**, *11*, 36. [CrossRef]
65. Da Xu, L.; Xu, E.L.; Li, L. Industry 4.0: State of the art and future trends. *Int. J. Prod. Res.* **2018**, *56*, 2941–2962. [CrossRef]
66. Altay, N.; Gunasekaran, A.; Dubey, R.; Childe, S.J. Agility and resilience as antecedents of supply chain performance under moderating effects of organizational culture within the humanitarian setting: A dynamic capability view. *Prod. Plan. Control.* **2018**, *29*, 1158–1174. [CrossRef]
67. Papadopoulos, T.; Gunasekaran, A.; Dubey, R.; Altay, N.; Childe, S.J.; Fosso-Wamba, S. The role of Big Data in explaining disaster resilience in supply chains for sustainability. *J. Clean. Prod.* **2017**, *142*, 1108–1118. [CrossRef]
68. Frank, A.G.; Dalenogare, L.S.; Ayala, N.F. Industry 4.0 technologies: Implementation patterns in manufacturing companies. *Int. J. Prod. Econ.* **2019**, *210*, 15–26. [CrossRef]
69. Yadav, G.; Luthra, S.; Jakhar, S.K.; Mangla, S.K.; Rai, D.P. A framework to overcome sustainable supply chain challenges through solution measures of industry 4.0 and circular economy: An automotive case. *J. Clean. Prod.* **2020**, *254*, 120112. [CrossRef]
70. de Sousa Jabbour, A.B.L.; Jabbour, C.J.C.; Foropon, C.; Godinho Filho, M. When titans meet–Can industry 4.0 revolutionize the environmentally-sustainable manufacturing wave? The role of critical success factors. *Technol. Forecast. Soc. Chang.* **2018**, *132*, 18–25. [CrossRef]
71. Reddy, N.; Chen, L.; Zhang, Y.; Yang, Y. Reducing environmental pollution of the textile industry using keratin as alternative sizing agent to poly(vinyl alcohol). *J. Clean. Prod.* **2014**, *65*, 561–567. [CrossRef]
72. Stock, T.; Obenaus, M.; Kunz, S.; Kohl, H. Industry 4.0 as enabler for a sustainable development: A qualitative assessment of its ecological and social potential. *Process. Saf. Environ. Prot.* **2018**, *118*, 254–267. [CrossRef]

73. Kaymak, T.; Bektas, E. Corporate Social Responsibility and Governance: Information Disclosure in Multinational Corporations. *Corp. Soc. Responsib. Environ. Manag.* **2017**, *24*, 555–569. [CrossRef]
74. Kantabutra, S.; Ketprapakorn, N. Toward a theory of corporate sustainability: A theoretical integration and exploration. *J. Clean. Prod.* **2020**, *270*, 122292. [CrossRef]
75. Santana, M.; Cobo, M. What is the future of work? A science mapping analysis. *Eur. Manag. J.* **2020**. [CrossRef]
76. Luthra, S.; Kumar, A.; Zavadskas, E.K.; Mangla, S.K.; Garza-Reyes, J.A. Industry 4.0 as an enabler of sustainability diffusion in supply chain: An analysis of influential strength of drivers in an emerging economy. *Int. J. Prod. Res.* **2019**, *58*, 1505–1521. [CrossRef]
77. Ramakrishna, S.; Ngowi, A.; De Jager, H.; Awuzie, B.O. Emerging Industrial Revolution: Symbiosis of Industry 4.0 and Circular Economy: The Role of Universities. *Sci. Technol. Soc.* **2020**. [CrossRef]
78. Nascimento, D.L.M.; Alencastro, V.; Quelhas, O.L.G.; Caiado, R.G.G.; Garza-Reyes, J.A.; Rocha-Lona, L.; Tortorella, G. Exploring Industry 4.0 technologies to enable circular economy practices in a manufacturing context. *J. Manuf. Technol. Manag.* **2019**, *30*, 607–627. [CrossRef]
79. Munsamy, M.; Telukdarie, A.; Fresner, J. Business process centric energy modelling. *Bus. Process. Manag. J.* **2019**, *25*, 1867–1890. [CrossRef]
80. Kohtamäki, M.; Parida, V.; Oghazi, P.; Gebauer, H.; Baines, T. Digital servitization business models in ecosystems: A theory of the firm. *J. Bus. Res.* **2019**, *104*, 380–392. [CrossRef]
81. Kramer, A.; Kramer, K.Z. The potential impact of the Covid-19 pandemic on occupational status, work from home, and occupational mobility. *J. Vocat. Behav.* **2020**, *119*, 103442. [CrossRef]
82. Li, G.; Hou, Y.; Wu, A. Fourth Industrial Revolution: Technological drivers, impacts and coping methods. *Chin. Geogr. Sci.* **2017**, *27*, 626–637. [CrossRef]
83. Ritter, T.; Pedersen, C.L. Digitisation capability and the digitalization of business models in business-to-business firms: Past, present, and future. *Ind. Mark. Manag.* **2020**, *86*, 180–190. [CrossRef]
84. Obiso, J.J.A.; Himang, C.M.; Ocampo, L.A.; Bongo, M.F.; Caballes, S.A.A.; Abellana, D.P.M.; Deocaris, C.C.; Padua, R.R.A., Jr. Management of Industry 4.0–reviewing intrinsic and extrinsic adoption drivers and barriers. *Int. J. Tech. Manag.* **2019**, *81*, 210–257. [CrossRef]
85. Peura, P.; Haapanen, A.; Reini, K.; Törmä, H. Regional impacts of sustainable energy in western Finland. *J. Clean. Prod.* **2018**, *187*, 85–97. [CrossRef]
86. Ranängen, H.; Lindman, Å. A path towards sustainability for the Nordic mining industry. *J. Clean. Prod.* **2017**, *151*, 43–52. [CrossRef]
87. Lu, C.-J.; Yang, C.-T.; Yen, H.-F. Stackelberg game approach for sustainable production-inventory model with collaborative investment in technology for reducing carbon emissions. *J. Clean. Prod.* **2020**, *270*, 121963. [CrossRef]
88. Tsai, W.-H.; Lu, Y.-H. A Framework of Production Planning and Control with Carbon Tax under Industry 4.0. *Sustainability* **2018**, *10*, 3221. [CrossRef]
89. Tsai, W.-H.; Jhong, S.-Y. Production decision model with carbon tax for the knitted footwear industry under activity-based costing. *J. Clean. Prod.* **2019**, *207*, 1150–1162. [CrossRef]
90. Aureli, S.; Gigli, S.; Medei, R.; Supino, E. The value relevance of environmental, social, and governance disclosure: Evidence from Dow Jones Sustainability World Index listed companies. *Corp. Soc. Responsib. Environ. Manag.* **2019**, *27*, 43–52. [CrossRef]
91. Lali, L. Beyond the 3 Rs: How Can Singapore Move Forward on Sustainability? *The Business Times*, 30 May 2020. Available online: https://www.businesstimes.com.sg/brunch/beyond-the-3-rs-how-can-singapore-move-forward-on-sustainability (accessed on 6 July 2020).
92. Stevenson, D. Are ESG and Sustainability the New Alpha Mantra? *Financial Time*, 1 April 2020. Available online: https://www.ft.com/content/6cee0b48-7760-46a5-9759-243aaaff7f8a (accessed on 10 July 2020).
93. Gibbens, S. 15 Ways the Trump Administration Has Changed Environmental Policies. *National Geographics*, 1 February 2019. Available online: https://www.nationalgeographic.com/environment/2019/02/15-ways-trump-administration-impacted-environment/ (accessed on 9 July 2020).
94. Khan, M. Brussels Ramps up Emissions Goals as MEPs Declare 'Climate Emergency'. *Financial Time*, 28 November 2019. Available online: https://www.ft.com/content/cadca90e-11da-11ea-a7e6-62bf4f9e548a (accessed on 12 September 2020).

95. Wincuinas, J. Sustainable and Actionable: A Study of Asset-Owner Priorities for ESG investing in Asia. Available online: https://eiuperspectives.economist.com/sustainability/sustainable-and-actionable-study-asset-owner-priorities-esg-investing-asia (accessed on 9 July 2020).
96. Hay, G. Breakingviews—Sustainable Investing Will Wind up in the Dock. *Reuters*, 2019. Available online: https://www.reuters.com/article/us-global-sustainability-breakingviews-idUSKBN1YZ0PC (accessed on 7 July 2020).
97. Landrum, S. Millennials Driving Brands to Practice Socially Responsible Marketing. *Forbes Magazin*, 17 March 2017. Available online: https://www.forbes.com/sites/sarahlandrum/2017/03/17/millennials-driving-brands-to-practice-socially-responsible-marketing/ (accessed on 5 July 2020).
98. Wong, K. How to Be a More Conscious Consumer, Even If You're on a budget. *The New York Times*, 1 October 2019. Available online: https://www.nytimes.com/2019/10/01/smarter-living/sustainabile-shopping-conscious-consumer.html (accessed on 7 July 2020).
99. The Economist. Big Business is Beginning to Accept Broader Social Responsibilities. Available online: https://deliverchange.economist.com/article/big-business-beginning-accept-broader-social-responsibilities/ (accessed on 9 July 2020).
100. Jessop, S. Deutsche Bank Targets 200 Billion Euros of Sustainable Investment by 2025. *Reuters*, 2020. Available online: https://www.reuters.com/article/us-deutsche-bank-sustainability-idUSKBN22P132 (accessed on 7 July 2020).
101. Neghaiwi Hughes, B. Credit Suisse Sets Up Investment Banking Sustainability Advisory. *Reuters*, 2019. Available online: https://uk.reuters.com/article/uk-credit-suisse-gp-sustainability/credit-suisse-sets-up-investment-banking-sustainability-advisory-idUKKBN21L2Z6 (accessed on 9 July 2020).
102. Gitlin, M.J. California Set to Ban All Heavy Diesel Trucks and Vans by 2045. *Ars Technica*, 26 June 2020. Available online: https://arstechnica.com/cars/2020/06/california-set-to-ban-all-heavy-diesel-trucks-and-vans-by-2045/ (accessed on 9 July 2020).
103. Thomasson, E. Zalando to Push 'Sustainable' Fashion, Cut Emissions. *Reuters*. 2019. Available online: https://www.reuters.com/article/us-zalando-environment/zalando-to-push-sustainable-fashion-cut-emissions-idUSKBN1X90X8 (accessed on 9 July 2020).
104. Cobburn, C. Why Industry is Going Green on the Quiet. *The Guardian*, 8 September 2019. Available online: https://www.theguardian.com/science/2019/sep/08/producers-keep-sustainable-practices-secret (accessed on 9 July 2020).
105. Washington Post. Transcript: A World in Balance—Solutions for Sustainability. Available online: https://www.washingtonpost.com/blogs/post-live/wp/2017/11/16/transcript-a-world-in-balance-solutions-for-sustainability/ (accessed on 9 July 2020).
106. Vaughan, A. BP Aims to Invest More in Renewables and Clean Energy. *The Guardian*, 6 February 2018. Available online: https://www.theguardian.com/business/2018/feb/06/bp-aims-to-invest-more-in-renewables-and-clean-energy (accessed on 9 July 2020).
107. Sonnemaker, T. Mercedes-Benz Parent Company Daimler is Preparing to Lay Off 15,000 Workers as It Tries to Adapt to Electric Cars. *Business Insider*, 10 February 2020. Available online: https://www.businessinsider.com/mercedes-benz-parent-daimler-reportedly-planning-15000-layoffs-2020-2 (accessed on 9 July 2020).
108. Angel, M. Corrected-Nestle Invests $45 Mln a Year in Cocoa Sustainability. *Reuters*, 2019. Available online: https://in.reuters.com/article/nestle-cocoa-sustainability-idINL8N28J4Z0 (accessed on 8 July 2020).
109. Monaghan, A. Businesses Must Address Impact of Next Industrial Revolution, Says Siemens Boss. *The Guardian*, 16 July 2018. Available online: https://www.theguardian.com/business/2018/jul/15/global-workforce-will-be-decimated-by-fourth-revolution-says-siemens-boss (accessed on 8 July 2020).
110. Thomspon, J. Big Investors' Sustainability Push Drives Demand for Environmental Expertise. *Financial Time*, 2020. Available online: https://www.ft.com/content/362fdc36-3b97-11ea-b84f-a62c46f39bc2 (accessed on 9 July 2020).
111. Davidson, H. China on Track To Lead in Renewables as US Retreats, Report Says. *The Guardian*, 10 January 2018. Available online: https://www.theguardian.com/environment/2018/jan/10/china-on-track-to-lead-in-renewables-as-us-retreats-report-says (accessed on 9 July 2020).
112. Friedman, R. Sustainability: Moving the Conversation Forward. *The Economist*, 3 December 2018; Inteligence Unit. Available online: https://eiuperspectives.economist.com/sustainability/sustainability-moving-conversation-forward (accessed on 10 July 2020).

113. McFarlane, C. Are Smart Cities The Pathway To Blockchain And Cryptocurrency Adoption? *The Forbes*, 18 October 2019. Available online: https://www.forbes.com/sites/chrissamcfarlane/2019/10/18/are-smart-cities-the-pathway-to-blockchain-and-cryptocurrency-adoption/#7a10b7846093 (accessed on 5 July 2020).
114. Rogers, B. Cities are Not Dead—They Will Get Younger. *Financial Times*, 24 May 2020. Available online: https://www.ft.com/content/d4fff7a2-9b63-11ea-871b-edeb99a20c6e (accessed on 9 July 2020).
115. Gold, M. Accelerating Urban Intelligence: People, Business and the Cities of Tomorrow. *The Economist*, 27 April 2019; Intelligence Unit. Available online: https://eiuperspectives.economist.com/technology-innovation/accelerating-urban-intelligence-people-business-and-cities-tomorrow/white-paper/accelerating-urban-intelligence-people-business-and-cities-tomorrow (accessed on 7 July 2020).
116. Rudolph, C.W.; Allan, B.; Clark, M.; Hertel, G.; Hirschi, A.; Kunze, F.; Shockley, K.; Shoss, M.; Sonnentag, S.; Zacher, H. Pandemics: Implications for Research and Practice in Industrial and Organizational Psychology. *Ind. Org. Psy.* **2020**. [CrossRef]
117. Van den Heuvel, M. How COVID Turned a Spotlight on Weak Worker Rights. *The Harvard Gazzete*. 23 June 2020. Available online: https://news.harvard.edu/gazette/story/2020/06/labor-law-experts-discuss-workers-rights-in-covid-19/ (accessed on 15 September 2020).
118. Gustin, G. Think Covid-19 Disrupted the Food Chain? Wait and See What Climate Change Will Do. *Inside Climate News*, 7 July 2020. Available online: https://insideclimatenews.org/news/06072020/coronavirus-agriculture-food-chain-future-climate-change (accessed on 7 July 2020).
119. Ghisellini, P.; Cialani, C.; Ulgiati, S. A review on circular economy: The expected transition to a balanced interplay of environmental and economic systems. *J. Clean. Prod.* **2016**, *114*, 11–32. [CrossRef]
120. Harari, Y.N. The World after Coronavirus. *The Financial Times*, 20 March 2020. Available online: https://www.ft.com/content/19d90308-6858-11ea-a3c9-1fe6fedcca75 (accessed on 17 July 2020).
121. Bapuji, H.; de Bakker, F.G.; Brown, J.A.; Higgins, C.; Rehbein, K.; Spicer, A. Business and Society Research in Times of the Corona Crisis. *Buss. Soc.* **2020**, *59*, 1067–1078. [CrossRef]
122. Hua, J.; Shaw, R. Coronavirus (Covid-19) "infodemic" and emerging issues through a data lens: The case of China. *Int. J. Environ. Res. Public Health* **2020**, *17*, 2309. [CrossRef] [PubMed]
123. Boot, A.W.A.; Carletti, E.; Kotz, H.-H.; Krahnen, J.P.; Pelizzon, L.; Subrahmanyam, M.G. Corona and Financial Stability 3.0: Try Equity—Risk Sharing For Companies, Large and Small. *SAFE Policy Lett.* **2020**, *81*. Available online: https://safe-frankfurt.de/policy-center/policy-publications/policy-publ-detailsview/publicationname/corona-and-financial-stability-30-try-equity-risk-sharing-for-companies-large-and-small.html/ (accessed on 4 September 2020).
124. Trippel, E. How green is green enough? The changing landscape of financing a sustainable European economy. *ERA Forum* **2020**, 1–16. [CrossRef]
125. Goerzen, A.; Iskander, S.P.; Hofstetter, J.S. The effect of institutional pressures on business-led interventions to improve social compliance among emerging market suppliers in global value chains. *J. Int. Bus. Policy* **2020**, 1–21. [CrossRef]
126. Payne, A. Germany is Beginning a Universal-Basic-Income Trial with People Getting $1400 a Month for 3 Years. *The Business Insider*, 2020. Available online: https://www.businessinsider.com/germany-begins-universal-basic-income-trial-three-years-2020-8 (accessed on 9 October 2020).
127. Grah, B.; Dimovski, V.; Peterlin, J. Managing Sustainable Urban Tourism Development: The Case of Ljubljana. *Sustainability* **2020**, *12*, 792. [CrossRef]
128. Bartoli, G.; Fantacci, R.; Gei, F.; Marabissi, D.; Micciullo, L. A novel emergency management platform for smart public safety. *Int. J. Commun. Syst.* **2013**, *28*, 928–943. [CrossRef]
129. Yigitcanlar, T.; Butler, L.; Windle, E.; DeSouza, K.C.; Mehmood, R.; Corchado, J.M. Can Building "Artificially Intelligent Cities" Safeguard Humanity from Natural Disasters, Pandemics, and Other Catastrophes? An Urban Scholar's Perspective. *Sensors* **2020**, *20*, 2988. [CrossRef]
130. Verma, S.; Gustafsson, A. Investigating the emerging COVID-19 research trends in the field of business and management: A bibliometric analysis approach. *J. Bus. Res.* **2020**, *118*, 253–261. [CrossRef]
131. Keesara, S.; Jonas, A.; Schulman, K. Covid-19 and Health Care's Digital Revolution. *N. Engl. J. Med.* **2020**, *382*, e82. [CrossRef]
132. Taylor, C.B.; E. Fitzsimmons-Craft, E.; Graham, A.K. Digital technology can revolutionize mental health services delivery: The COVID-19 crisis as a catalyst for change. *Int. J. Eat. Disord.* **2020**, *53*, 1015–1181. [CrossRef]

133. Roblek, V.; Pejić Bach, M.; Meško, M.; Bertoncel, T. Best practices of the social innovations in the framework of the e-government evolution. *Amfiteatru Econ.* **2020**, *22*, 275–302. [CrossRef]
134. Oni, A.; Oni, S.; Mbarika, V.; Ayo, C.K. Empirical study of user acceptance of online political participation: Integrating Civic Voluntarism Model and Theory of Reasoned Action. *Gov. Inf. Q.* **2017**, *34*, 317–328. [CrossRef]
135. Elavarasan, R.M.; Afridhis, S.; Vijayaraghavan, R.R.; Subramaniam, U.; Nurunnabi, M. SWOT analysis: A framework for comprehensive evaluation of drivers and barriers for renewable energy development in significant countries. *Energy Rep.* **2020**, *6*, 1838–1864. [CrossRef]
136. Stop This Thing. Available online: https://stopthesethings.com/2019/01/06/germanys-renewable-energy-fail-german-co2-emissions-10-times-higher-than-nuclear-powered-france/ (accessed on 10 October 2020).
137. Blackmon, D. Renewables Won't Save Us If the Electric Grid Is Not Ready. *Forbes*, 30 September 2020. Available online: https://www.forbes.com/sites/davidblackmon/2020/09/30/renewables-wont-save-us-if-the-electric-grid-is-not-ready/?fbclid=IwAR0u12ZVqn5kFh0Kd51_o65MKHTJvZojI070gc7iqJm62gAVGzAOa3oLTu8#31d8a1f87abf (accessed on 10 October 2020).
138. Bersano, A.; Segantin, S.; Falcone, N.; Panella, B.; Testoni, R. Evaluation of a potential reintroduction of nuclear energy in Italy to accelerate the energy transition. *Electr. J.* **2020**, *33*, 106813. [CrossRef]
139. Shellenberger, M. Renewables Threaten German Economy & Energy Supply, McKinsey Warns in New Report. *Forbes*, 5 September 2019. Available online: https://www.forbes.com/sites/michaelshellenberger/2019/09/05/renewables-threaten-german-economy-energy-supply-mckinsey-warns-in-new-report/#619733688e48 (accessed on 10 October 2020).
140. Masum, M.H.; Latiff, A.R.A.; Osman, M.N.H. Voluntary Reporting, Sustainable Reporting and Transition Economy. *Int. Bus. Account. Res. J.* **2020**, *4*, 81–88. [CrossRef]
141. Man, M.; Bogeanu-Popa, M.-M. Impact of Non-Financial Information on Sustainable Reporting of Organisations' Performance: Case Study on the Companies Listed on the Bucharest Stock Exchange. *Sustainability* **2020**, *12*, 2179. [CrossRef]

sustainability

MDPI

Article

Environmental Impact of Subsidy Concepts for Stimulating Car Sales in Germany

Malte Scharf , Ludger Heide *, Alexander Grahle , Anne Magdalene Syré and Dietmar Göhlich

Department of Methods for Product Development and Mechatronics, Technical University of Berlin, 10623 Berlin, Germany; malte.scharf@campus.tu-berlin.de (M.S.); alexander.grahle@tu-berlin.de (A.G.); a.syre@tu-berlin.de (A.M.S.); dietmar.goehlich@tu-berlin.de (D.G.)

* Correspondence: ludger.heide@tu-berlin.de; Tel.: +49-(0)30-314-73-858

Received: 6 November 2020; Accepted: 28 November 2020; Published: 1 December 2020

Abstract: In 2020, vehicle sales decreased dramatically due to the COVID-19 pandemic. Therefore, several voices have demanded a vehicle subsidy similar to the "environmental subsidy" in Germany in 2009. The ecological efficiency of vehicle subsidies is controversially discussed. This paper establishes a prognosis of the long-term environmental impacts of various car subsidy concepts. The CO_2 emissions of the German car fleet impacted by the purchase subsidies are determined. A balance model of the CO_2 emissions of the whole car life cycle is developed. The implementation of different subsidy scenarios directly affects the forecasted composition of the vehicle population and, therefore, the resulting life-cycle assessment. All scenarios compensate the additional emissions required by the production pull-in within the considered period and, hence, reduce the accumulated CO_2 emissions until 2030. In the time period 2019–2030 and for a total number of 0.72 million subsidized vehicles—compensating the decrease due to the COVID-19 pandemic—savings of between 1.31 and 7.56 million t CO_2 eq. are generated compared to the scenario without a subsidy. The exclusive funding of battery electric vehicles (BEVs) is most effective, with an ecological break-even in 2025.

Keywords: subsidy; automotive industry; prognosis; COVID-19; environmental impact; life-cycle analysis

1. Introduction

As a result of the containment measures against the COVID-19 pandemic, international vehicle sales collapsed dramatically in the first half of 2020. The German Association of the Automotive Industry (Verband der Automobilindustrie, VDA) predicts a decline of −23% of new passenger car registrations in Germany compared to the prior year [1]. With more than 800,000 employees and an annual turnover of 435 billion EUR in 2019, the German automotive industry is essential for prosperity and employment in Germany [2]. According to the VDA president Hildegard Müller, the massively reduced production will lead to a decrease in employment [1].

From an economic point of view, the COVID-19 pandemic shows some similarities to the 2008 financial crisis: At that time, the passenger car registrations in Germany decreased to the lowest level since the German reunification. For the following year, the forecast without any car sales stimulation predicted 2.8 million new registrations, almost 0.3 million less than the already historically low number of 3.09 million new registrations in 2008.

In reaction, the German government decided to introduce a subsidy program in 2009, in which a purchase bonus for new cars could be earned if the old ones were scrapped. One target of the "environmental bonus" was to replace old cars with high specific emissions with new and more efficient ones. The government's goal was to reduce pollution and stimulate car sales at the same time.

Thus, the number of new registrations in 2009 rose by 23.3% to 3.81 million vehicles and provided, according to Höpfner et al. [3], reduced pollution due to the rejuvenation of the German passenger car fleet.

Due to the renewed economic challenges as a result of the COVID-19 pandemic, a considerable number of voices from the automotive industry [4], automotive lobby [5], and parts of German politics [6] demand a subsidy concept similar to the "environmental bonus" introduced in 2009. The scope of this potential subsidy has not yet been worked out. Unlike in 2009, electromobility has found its way into the automotive market, and therefore, specific funding of battery electric vehicles (BEVs) and plug-in hybrid electric vehicles (PHEVs) is conceivable to further reduce German car fleet emissions. Contrary to the concept proposed in this paper, existing subsidy concepts for electric and hybrid cars [7] do not require a replacement of conventional cars. An additional promising concept includes rewarding purchases of smaller vehicles.

The goal of this paper is a quantified evaluation of various subsidy concepts with regards to the ecological aspects, focusing on the CO_2 emissions. To provide a holistic view on the environmental impact of the various subsidy concepts, the cradle-to-grave life cycle, including production, operation, and end-of-life (EoL) emissions, is considered.

2. Literature Review

As presented in Section 1, the financial crisis of 2008 and the introduced car subsidy show some similarities to the current situation due to the COVID-19 pandemic.

The "environmental bonus" introduced in Germany has been analyzed extensively. Shortly after the "environmental bonus" was introduced, Höpfner et al. [3] determined a positive ecological effect of the subsidy. In addition, they demonstrated that the CO_2 emissions of the pulled-forward manufacturing were compensated after 6000 km driving distance due to the reduced use-phase emissions of the new cars.

Klößner et al. [8] examined the impact of European car scrappage programs on new vehicle registrations and respective CO_2 emissions. Using a multivariate synthetic control method with time series of economic predictors, they found that the German subsidy had a positive effect on stabilizing the car market. However, the economic benefit caused 2.4 million tons of additional CO_2 emissions according to their work.

Various economic examinations of the "environmental bonus" have been made. The micro- and macroeconomic effects of the "environmental bonus" were investigated by Läufer et al. [9]. They concluded that the "environmental bonus" was not very effective, but created macroeconomic stability.

Müller et al. [10] analyzed the impact of the subsidy on the overall car sales by using a dataset provided by the Organisation for Economic Co-operation and Development (OECD) for 23 countries and found a positive effect of car scrappage programs on overall car sales as long as the subsidy was in place.

The pull-forward effect of the German car scrappage scheme was examined by Böckers et al. [11], who created a monthly dataset of new car registrations owned by private consumers. According to them, the small and upper small car segments benefited specifically from the scrappage program, as they made up 84% of the newly registered cars during the program.

In response to the financial crisis, other countries besides Germany introduced car subsidy programs. The "Summary of the Consumer Assistance to Recycle and Save Act of 2009" (CARS) was launched in 2009 by the US government [12] and has been broadly reviewed. Lenski et al. [13] analyzed the net effect of CARS on greenhouse gas emissions from a full vehicle-life-cycle perspective. They found that CARS had a one-time effect of preventing 4.4 million metric tons of CO_2 eq. emissions, about 0.4% of US annual light-duty vehicle emissions.

By comparing the predicted fuel economy without the existence of the program and the actual data, Sivak et al. [14] determined an improved average fuel economy of the US passenger car fleet in July and August 2009.

Li et al. [15] investigated the effects of the CARS program on new vehicle sales and the environment. By using Canada as the control group in a difference-in-differences framework, they determined that CARS increased new vehicle sales only by about 0.37 million during July and August of 2009, implying that approximately 45% of the spending went to consumers who would have purchased a new vehicle anyway. They calculated a reduction of the CO_2 emissions by 9–28.2 million tons.

In summary, the literature shows disagreement about the environmental impact of the car subsidy in 2009, and does not answer whether a new subsidy would have a positive impact on the environment.

To the best of our knowledge, there is no study that predicts the environmental impact of a car subsidy that addresses the decrease of vehicles sales due to the COVID-19 pandemic. We perform a life-cycle analysis of the German passenger fleet with a level of detail that no previous study has shown.

3. Methodology

The model developed in this paper calculates the annual CO_2 emissions of the German passenger car fleet iteratively starting in 2019. New registrations and decommissions of the present year lead to the fleet data of the subsequent year. For the baseline of the CO_2 emissions, the impact of the COVID-19 pandemic on the car sales profile is considered. Three different subsidy concepts are applied to the baseline scenario, and variations of the CO_2 emissions are determined.

3.1. Scope

To analyze the environmental impacts of various subsidy concepts, the full life cycles of passenger cars are taken into account. Therefore, temporal and geographical system boundaries must be defined. To predict the long-term impacts, the time period covers the years from 2019 to 2030. Due to the rapid technological progress, the reliability of the predicted data after 2030 decreases significantly.

As effects that extend beyond the national borders are subject to large uncertainties, the system boundary is drawn around Germany. After deregistration from the German transport system, only the EoL (decommissioning) emissions and no further driving performances will be accounted for in the life-cycle analysis, as shown in Figure 1.

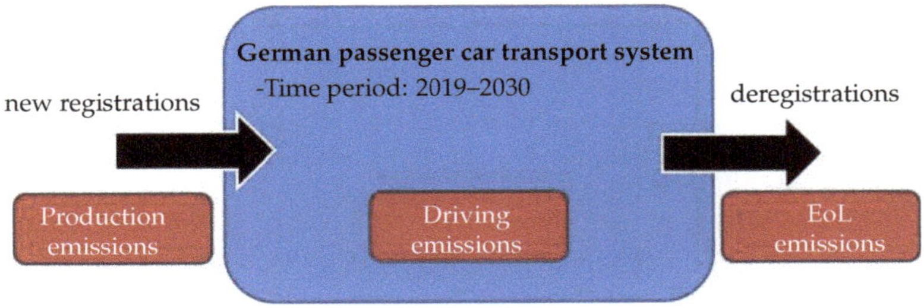

Figure 1. System boundaries.

The car population is categorized according to two main attributes: segment and drive train technology. The segment categorization is based, amongst other parameters, on the weight and size of the vehicle, according to the Federal Motor Transport Authority (Kraftfahrtbundesamt, KBA) clustering method [16]. In this work, we use an English translation of the segment names as used by the KBA (Table 1).

Table 1. Segment translations.

Segment Name According to the KBA [16]	English Segment Name
Minis	mini class
Kleinwagen	small class
Kompaktklasse	compact class
Mittelklasse	middle class
Obere Mittelklasse	upper middle class
Oberklasse	upper class
Sport Utility Vehicles (SUVs)	sport utility vehicles (SUVs)
Mini-Vans	mini vans
Großraum-Vans	large vans
Sportwagen	sports cars
Geländewagen	off-roaders
Wohnmobile	caravans

In this analysis, there is no consideration of the segments of sports cars and caravans. This is because of their exclusive use for recreational or leisure purposes, which makes the expenditure of public funds for these vehicles unjustifiable. Furthermore, commercial vehicles like trucks are excluded, since the subsidy presented here is focused solely on private customers. Finally, off-roaders are excluded, as they are used either commercially (e.g., forestry) or for leisure purposes. This limits the scope to 84% of the whole German passenger car population.

Furthermore, we focus only on the following drive train technologies: gasoline, diesel, electric, and plug-in hybrid. In this context, a plug-in hybrid vehicle contains a gasoline engine and an electric drive train.

3.2. Current Vehicle Distribution

At first, a model was defined that describes the German passenger car population for the next ten years. Therefore, 2019 is considered as a baseline with real data. From 2020–2030, we take recourse to prognosis data, as shown in Section 3.3.

To obtain a sufficient dataset for the 2019 baseline, the vehicle data from the 2017 "Mobility in Germany" (Mobilität in Deutschland, MID) study by the German Federal Ministry of Transport and Digital Infrastructure [17] were categorized into segments and drive train technologies according to Section 3.1. Given the negligible change of 0.2 years in the average vehicle age between 2017 and 2019 [18], we assumed that the overall age distribution of the German vehicle stock did not change significantly. Therefore, a linear transformation of the 2017 data was made to obtain the age distribution of the segments in 2019. Figure 2 presents the resulting age distribution of the German passenger car fleet.

The same data are also available for the vehicle drive train technology and are linked with the year of manufacture and the segment category.

Combined with the number of the total vehicle stock [19], the baseline for 2019 was built.

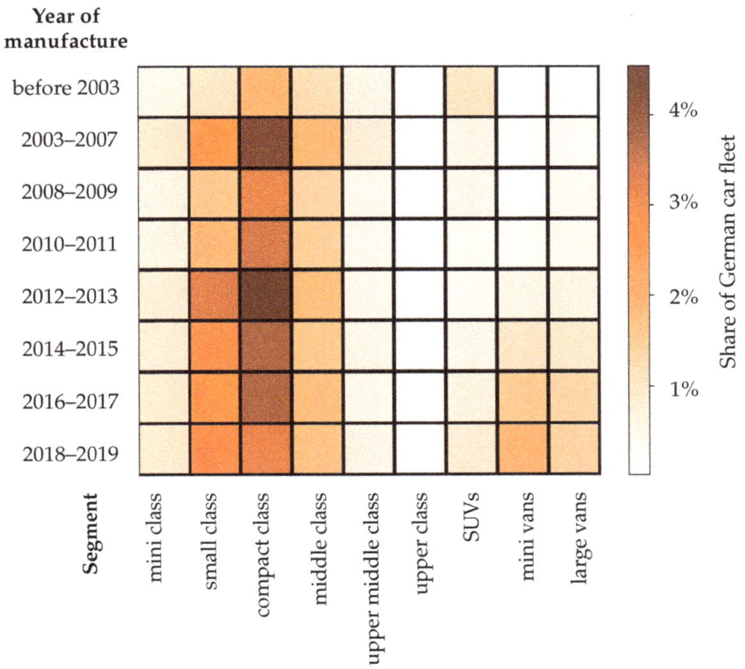

Figure 2. Age distribution of the German vehicle fleet in 2019 (vehicles in scope: 82% of overall vehicle stock).

3.3. German Car Population Development until 2030

In order to predict the car population until 2030, estimations for new registrations and deregistrations must be made.

3.3.1. New Passenger Car Registrations

To determine new passenger car registrations, the total number and their distribution in the categories defined in Section 3.1 are needed. The annual number of new registrations and their classification regarding the drive train technology were obtained from the "proKlima" scenario of Agora 2018 [20] (Figure 3).

Additionally, an estimation concerning the segments was created based on the trend of the new registration numbers from previous years. For the large van, mini van, upper class, and upper middle class segments, we assumed that the share of new registrations remains constant. In addition, we assumed that the new registration share of segments not considered in this analysis will stay constant, which results in a non-varying share of the overall scope of 82%. A noticeable growth of the sport utility vehicle (SUV) segment has been observed over the last years [21]. The vehicle segment development is visualized in Figure 4.

This trend is expected to continue and is anticipated to reach an annual new registration share of 40% in 2030. Since this value is based on an assumption, a sensitivity analysis is performed in Section 4.4.2.

The growth in the SUV segment was subtracted proportionally from the other segments depending on their share in the reference year 2019. The shares of the large van and mini class segments were not modified, as customers are unlikely to change to an SUV due to the completely different usage profiles.

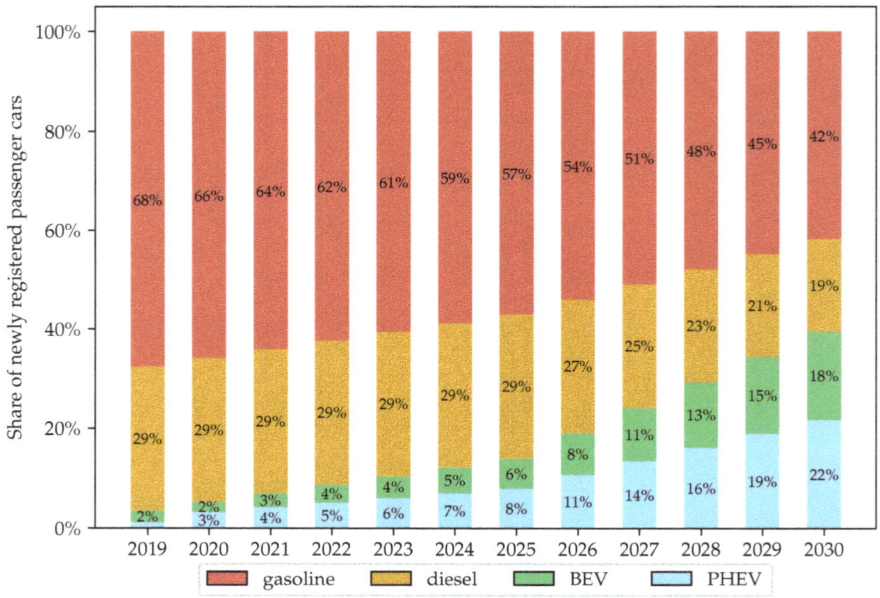

Figure 3. Linear interpolation of the "proKlima" drive drain technology scenario.

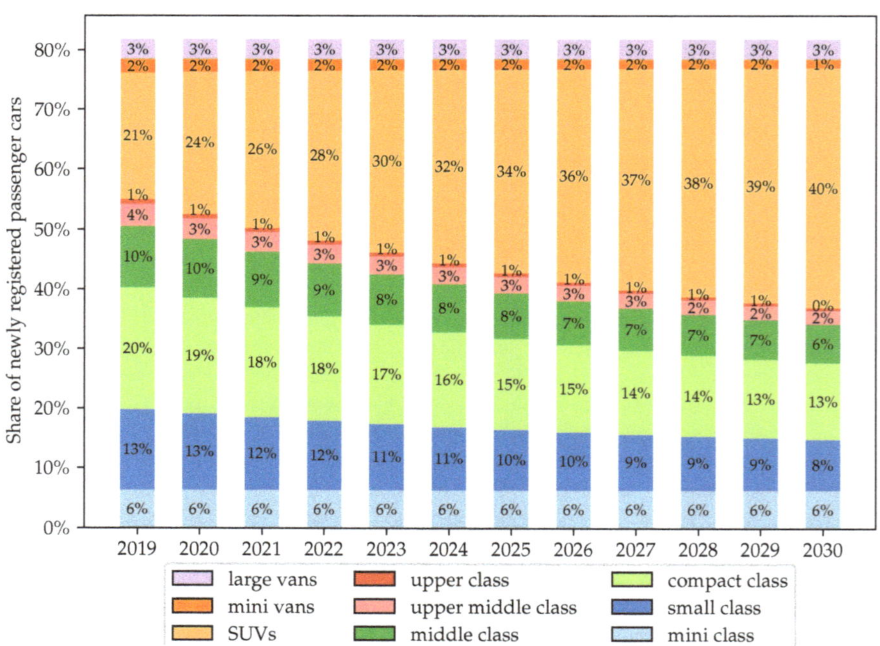

Figure 4. Segment scenario.

3.3.2. Vehicle Disposal

In this analysis, we assumed that every vehicle that is deregistered from the German transport system, as defined in Section 3.1, will immediately be disposed of, meaning scrapped or recycled. As usage profiles of exported vehicles are unknown, the vehicles are disposed of simultaneously with their deregistration in this model.

For the last four years, a consistent age-dependent deregistration trend can be seen in Figure 5. Therefore, the future deregistration rates are assumed to be identical to those in 2019. The peak after three years can be explained by the end of many leasing contracts, which can contribute to exportation of vehicles. Additionally, the first legally required technical inspection ("Hauptuntersuchung", HU) of the vehicles and the associated phasing out of early defective vehicles take place in this period. This might add to the observed peak. The peaks that occur every two years, especially for older vehicles, are also due to the interval of the mandatory technical inspections in Germany. As technical defects are detected and repairs may become necessary, it is often cheaper to scrap or export the vehicles than to continue their operation.

As explained in Section 3.4, vehicle production is modeled separately from deregistrations, and only for new vehicles. The negligible number of new registrations (negative deregistration) for very old vehicles (probably as vintage vehicles) is ignored.

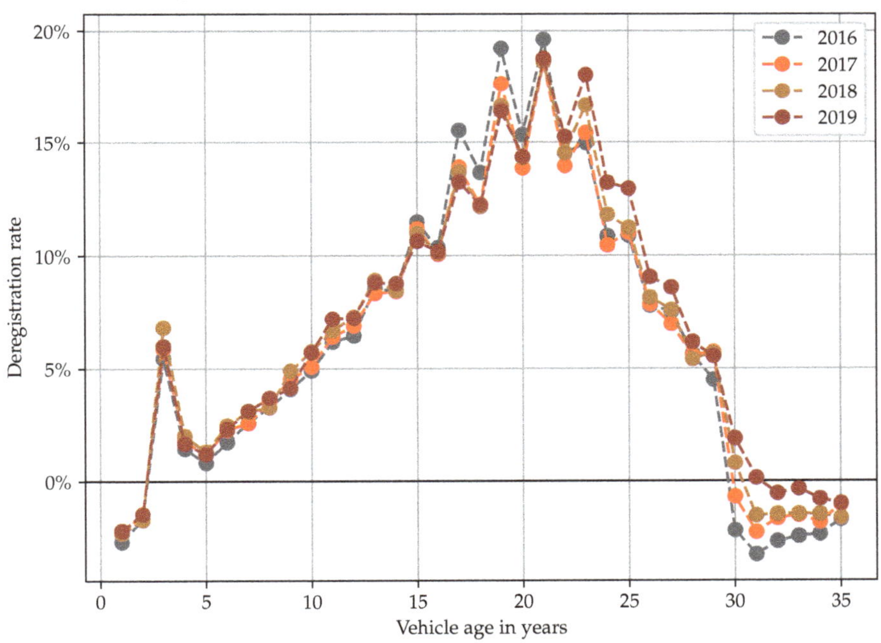

Figure 5. Age-dependent deregistration data for 2016–2019 (based on [22]).

3.4. Modeling Car Population Development

The development of the car population for different scenarios is calculated using an iterative deregistration and production process. In the first step, the vehicle population (starting with the 2019 population from [22]) for each year of manufacture is reduced according to the deregistration rate for the respective age. Subsequently, the age groups are shifted back one year, and the vehicle production of the considered year is set to maintain a constant total vehicle population. The vehicle deregistration rate is scaled from the real data shown in Figure 5 in order to match the prognosis of newly registered cars between 2020 and 2030 and of the "proKlima" [20] scenario.

In the next step, the modified scenarios (shown in Figure 6) are derived from the baseline by adapting the deregistrations only. First, the deregistration rate is decreased to reflect the reduced registrations due to owners' financial uncertainty during the COVID-19 pandemic. The total vehicle registrations in 2020 are decreased by 23% compared to 2019 [1]. To simulate a subsidy in 2021, the deregistration of vehicles older than 10 years is increased to match the total number of subsidized vehicles. For the baseline scenario, this number is set to the predicted new registration decrease due to the COVID-19 pandemic (0.72 million). In Section 4.4, we discuss the impact of the number of subsidized cars; therefore, additional analyses of twice and four times the baseline amount (1.43 and 2.87 million) were made.

This approach considers the reduced vehicle production in subsequent years after the subsidy is introduced (pull-out effect) and the increased total vehicle production due to early deregisterations compared to a scenario without a subsidy.

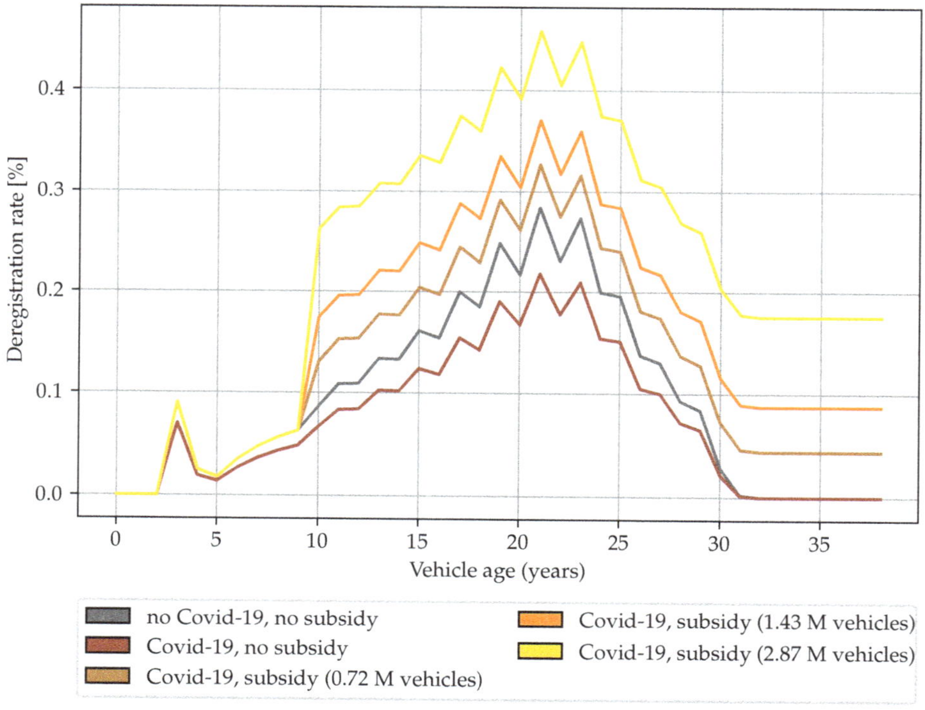

Figure 6. Deregistration rate distributions of different scenarios.

Figure 7 shows the annual and the accumulated vehicle registrations for different scenarios. The economic uncertainty due to the COVID-19 pandemic reduces the total number of vehicle registrations by forcing owners to keep their vehicles longer. On the contrary, the subsidy leads to earlier replacements and, therefore, increases the new registrations. As already explained for Figure 5, there is a first peak of vehicle decommissioning after three years.

This can also be seen in the new registrations, as the additional vehicles produced due to the subsidy also lead to an increased number of vehicles being taken out of service three years after the subsidy. In turn, this leads to an increase in new purchases that can be seen four years after the subsidy.

This time delay is due to the fact that we model the transition from one year to the next, and thus, the ages of the vehicles in the previous year are taken into account for the scrapping and new purchases in the year under consideration.

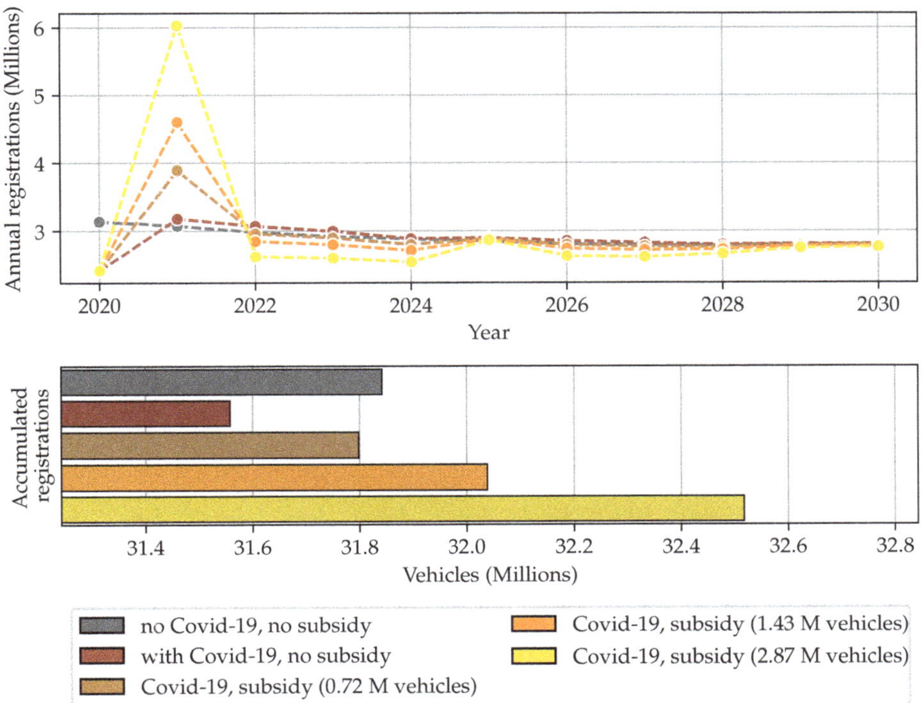

Figure 7. Annual and accumulated vehicle registrations.

3.5. Life-Cycle Emissions

To perform a cradle-to-grave life-cycle analysis, the production, use, and EoL phases are considered. The focus is primarily on the calculation of greenhouse gas emissions (CO_2 equivalent emissions, 20-year global warming potential (GWP 20)). In this analysis, the direct CO_2 emissions of the combustion engines in the use phase are set identically to their CO_2 equivalent emissions.

The amounts of production and EoL emissions are assumed to be constant for the upcoming years. They are only dependent on the segment and power train type of the vehicle and do not change with the production and disposal date.

In order to analyze the life-cycle emissions of the drive train technologies per segment, specific input data (weight, direct CO_2 emissions, fuel, and electric energy consumption) are needed. Complete data for gasoline and diesel vehicles are available. For PHEVs and BEVs, the required input is derived from defined reference vehicles. Currently, in some segments, there are no vehicles available. In these cases, input data are interpolated from adjacent segments.

3.5.1. Production

All production analyses were performed with the Ecoinvent 3.5 database [23]. The cutoff allocation was used and all location settings were set to global.

Petrol and Diesel

To perform a life-cycle analysis for gasoline and diesel vehicles, the curb weight was used as the input value (Table 2).

Table 2. Weight distribution of vehicles with combustion engines [24].

Segment	Curb Weight [kg]
mini class	1038
small class	1191
compact class	1389
middle class	1617
upper middle class	1831
upper class	2035
SUVs	1506
mini class—vans	1514
large vans	1754

It was assumed that there is no curb weight difference between gasoline and diesel vehicles. The emissions were calculated with the processes *passenger car production, gasoline* and *passenger car production, diesel* in Ecoinvent. To consider production only, the Ecoinvent data (flow) *manual dismantling of a used passenger car with internal combustion engine* were excluded.

Battery Electric Vehicles (BEVs)

In order to perform the production analysis for the BEVs, the curb weight is divided into the battery weight and the remaining vehicle weight. The battery weight is calculated based on the battery capacity. Therefore, a constant energy density for all BEVs is defined.

A representative energy density at a battery packaging level is defined by averaging the values of the VW E-golf (113 Wh/kg) [25] and the 2017 Tesla Model S P100D (160 Wh/kg) [26]. This results in an energy density of 136 Wh/kg, which is used to calculate the battery weight of all BEVs based on the battery capacity. For each segment, a reference BEV is defined based on the number of newly registered vehicles in 2019 [27] (Table 3). In the SUV segment, the Hyundai Kona Electro and the Audi E-Tron both had nearly the same registration numbers in 2019. Since the share of compact SUVs is expected to continue growing [27], the Hyundai Kona Electro is chosen as the reference model.

Table 3. Weight distribution of battery electric vehicles (BEVs).

Segment	Curb Weight (kg)	Battery Capacity (kWh)	Battery Weight (kg)	Remaining Curb Weight (kg)	Reference
mini class	1095	17.6	129	966	smart EQ fortwo. [28]
small class	1345	42	308	1037	BMW i3 [29]
compact class	1545	40	293	1252	Nissan Leaf [30]
middle class	1611	55	404	1207	Tesla Model 3 [31]
upper middle class	-	-	-	-	-
upper class	2290	100	734	1556	Tesla Model S [31]
SUVs	1593	39	288	1305	Hyundai Kona Elektro [32]
mini vans	1610	39	288	1322	Kia e-soul [33]
large vans	-	-	-	-	-

The emissions emitted from the BEV battery pack production were determined with the process *battery production, Li-ion, rechargeable, prismatic*. This process also considers the transportation of the single cells from Peking to Amsterdam by ship and a 1000 km transportation route within Europe by truck.

For the remaining electric vehicles, the process *passenger car production, electric, without battery* [34] was used with the remaining curb weight from Table 3.

Plug-in Hybrid Electric Vehicles (PHEVs)

In this analysis, the PHEV is a combination of a conventional and an electric vehicle. Therefore, the PHEV consists of a battery, electric power train components, and the remaining vehicle, which contains all components of the combustion power train. This analysis is based on manufacturers' data from reference vehicles (Table 4).

Table 4. Manufacturers' data for plug-in hybrid electric vehicles (PHEVs).

Segment	Curb Weight (kg)	Battery Capacity (kWh)	Maximum Power of Electric Drive Train (kW)	Reference
mini class	-	-	-	-
small class	1660	7.6	65	Mini Cooper SE Countryman [35]
compact class	1750	8.8	65	BMW 225xe Active Tourer [36]
middle class	1780	9.8	50	Kia Optima Plug-In Hybrid [37]
upper middle class	1910	11.2	62	BMW 530 e [38]
upper class	2170	14.1	100	Porsche Panamera 4 E-Hybrid [39]
SUVs	1971	13.8	70	Mitsubishi Outlander [40]
mini vans	1725	15.6	75	Mercedes B 250 e [41]
large vans	-	-	-	-

Similarly to the BEV batteries, the PHEV battery weight is calculated based on the battery capacity with a constant energy density for all types of plug-in vehicles. We consider all batteries at the packaging level. Since the PHEV batteries are significantly smaller than the BEV ones, an assumption of the same energy density is not suitable.

Therefore, the mean value of the energy densities of the Kia Optima (75 Wh/kg) [37] and the Mercedes B 250 e (104 Wh/kg) [42] of 89.5 Wh/kg are assumed for all PHEVs.

In addition, the electric power train components, such as an electric motor, a charger, and cables, are considered. Due to lacking data on the actual weight of these components, the weight is scaled with the maximum power of the electric power train. In this analysis, the baseline is made according to the default value in Ecoinvent of a 100 kW power train with a weight of 70 kg. The resulting parameters (Table 5) are used in Ecoinvent for the production analysis.

Table 5. Calculated weight distribution of PHEVs.

Segment	Remaining Curb Weight (kg)	Battery Weight (kg)	Electric Power Train Weight (kg)	Reference
mini class	-	-	-	-
small class	1525	85	50	Mini Cooper SE Countryman [35]
compact class	1601	99	50	BMW 225xe Active Tourer [36]
middle class	1632	110	39	Kia Optima Plug-In Hybrid [37]
upper middle class	1724	125	50	BMW 530 e [38]
upper class	1935	158	77	Porsche Panamera 4 E-Hybrid [39]
SUVs	1763	154	54	Mitsubishi Outlander [40]
mini vans	1393	174	58	Mercedes B 250 e [41]
large vans	-	-	-	-

For the battery pack production of the PHEVs, the same method as that used for the BEVs was applied. The process *electric motor production, vehicle* was used to determine the emitted emissions to produce the components of the electric power train.

The remaining part of the vehicle is assumed to be a gasoline engine car. Therefore, the process *passenger car production, petrol* was applied to analyze the production.

3.5.2. Use Phase

The emissions emitted during the use phase are composed of the emissions based on fuel and electric energy consumption, the well-to-tank emissions, and the emissions due to maintenance of the vehicle and the road infrastructure.

In 2019, the average annual mileage of passenger cars was 13,602 km [43]. In this model, it is assumed that, in the future, every car will have the 2019 average annual mileage.

Combustion Engine Vehicles

The use phase emissions of the gasoline- and diesel-powered vehicles are dependent on the vehicle's segment and age. Based on these criteria, the KBA [44] provides a database of the average direct CO_2 emissions and fuel consumption, which are displayed in the supplementary materials in Table S1. In this work, they are assumed to be identical to the CO_2 equivalent emissions.

Battery Electric Vehicles (BEVs)

In this analysis, the BEVs cause indirect CO_2 emissions because of their consumed electric energy during the use phase.

For each segment, a reference BEV is defined based on the number of newly registered vehicles in 2019 [27]. If a model in this segment would have a dominant share of the newly registered cars, it is considered as the reference BEV. Otherwise, the two models with the highest number of newly registered vehicles in the segment are selected, and a weighted average of the energy consumption (weight factor) according to their new registration numbers is built (Table 6).

Table 6. Energy consumption of BEVs.

Segment	Reference Model	Energy Consumption (kWh/100 km)	Weight Factor	Weighted Average Energy Consumption (kWh/100 km)
mini class	smart EQ fortwo. [28]	14.0	1	14.0
small class	BMW i3 [29]	13.1	0.66	14.5
	Renault Zoe [45]	17.2	0.34	
compact class	VW E Golf [46]	12.9	0.64	14.4
	Nissan Leaf [30]	17.1	0.36	
middle class	Tesla Model 3 [31]	16.0	1	16.0
upper middle class	-	-	-	-
upper class	Tesla Model S [31]	19.0	1	19.0
SUVs	Hyundai Kona Electro [32]	15.0	0.51	18.6
	Audi E-Tron [47]	22.4	0.49	
mini vans	Kia e-soul [33]	15.6	1	15.6
large vans	-	-	-	-

In 2019, there were not any BEVs newly registered in the upper middle class or large van segments. Since the market for electric vehicles will grow [20], we assume that future electric vehicles will also be available in these segments. Based on the curb weight of the combustion engine vehicles, we estimate that upper middle class and large vans have the same energy consumption. The electric energy consumption of upper middle class is assumed as an average of those of the middle class and upper class.

Plug-In Hybrid Electric Vehicles (PHEVs)

To determine comparable CO_2 emissions from plug-in hybrids, a combination of the direct exhaust emissions and the energy consumption of the electric drive train is made (Table 7).

The manufacturers' data for the reference vehicles either rely completely on the New European Driving Cycle (NEDC) or were measured with the Worldwide Harmonized Light Vehicles Test Procedure (WLTP) and were transferred back to the NEDC. The NEDC considers half of the driving cycle using the combustion engine and the other half using the electric power train. In these data, the electric energy consumption is considered emission-free. In order to correct this effect, we add the appropriate indirect emissions for electric power consumption.

Table 7. CO_2 emissions and energy consumption of PHEVs.

Segment	CO_2 Emissions (Combined) (g/km)	Fuel Consumption (L/100 km)	Energy Consumption (kWh/100 km)	Reference Model
mini class	-	-	-	-
small class	45	2,0	14.0	Mini Cooper SE Countryman
compact class	42	1.9	13.5	BMW 225xe Active Tourer
middle class	37	1.6	12.2	Kia Optima Plug-In Hybrid
upper middle class	42	1.8	14.8	BMW 530 e
upper class	62	2.7	16.1	Porsche Panamera 4 E-Hybrid
SUVs	40	1.8	14.8	Mitsubishi Outlander
mini vans	32	1.4	14.7	Mercedes B 250 e
large vans	-	-	-	-

Indirect Emissions

In this model, the consumed electric energy causes indirect emissions based on the specific carbon dioxide emissions of the German electricity mix. In 2020, the generation of one kWh electric energy is predicted to cause 432 g CO_2 eq. emissions [48].

Pehnt et al. [48] calculated 318 g CO_2 eq. emissions for the electricity mix in 2030; the value here decreases linearly for the years 2020 to 2030. The well-to-tank factors are used to determine the upstream emissions of consuming fuel. Here, Schallaboeck et al. [49] stated that 685 g or 408 g CO_2 equivalent emissions produce one liter of gasoline or diesel, respectively. The data for the fuel consumption are displayed in the supplementary materials in Table S2. All PHEVs are assumed to have a gasoline engine.

Maintenance

During the use phase, the replacement of spare vehicle parts and the maintenance of the road network are considered. To calculate the vehicle maintenance, the flow *passenger car maintenance* was used in Ecoinvent and was adapted according to the vehicle's weight and power train type [50]. We followed the same approach as in the Agora (2019b) [51] "basic scenario", assumed for each vehicle a lifespan of 150,000 km, and did not consider any battery exchange for BEVs and PHEVs.

In order to determine the maintenance of the road network, the flow *road maintenance* is used. This introduces a weight-dependent contribution during the use phase.

3.5.3. End-of-Life

In this model, it is assumed that every deregistered vehicle is eventually disposed of. The equivalent CO_2 emissions due to this process were determined in Ecoinvent using the flows *manual dismantling of used passenger car with internal combustion engine, treatment of used glider, passenger car, shredding, treatment of used internal combustion engine, shredding, treatment of used powertrain for electric passenger car, dismantling*, and *market for used Li-ion battery* according to [34].

3.6. Subsidy Concepts

Three subsidy concepts are introduced in this chapter. They reflect aspects of the current discussions about the possible subsidy scope. Höpfner et al. [3] showed that 84% of the subsidy from the "environmental bonus" in 2009 was used to purchase small cars (mini class, small class, and compact class). Therefore, 84% of all new registrations due to the subsidy will happen in the mini class, small class, and compact class categories. For this distribution, the predicted new car registration shares for 2021 are used. All segment shares are evenly scaled in order to reach an 84% share in the small cars category. The impact of this effect is discussed in Section 4.4.1.

The distribution shown in Figure 8 is used as a baseline for all subsidy concepts. In Section 4.4.1, the influence of this distribution is discussed.

Similarly to the "environmental bonus" in 2009, a mandatory requirement to earn the subsidy is a minimum vehicle age (in this case, 10 years).

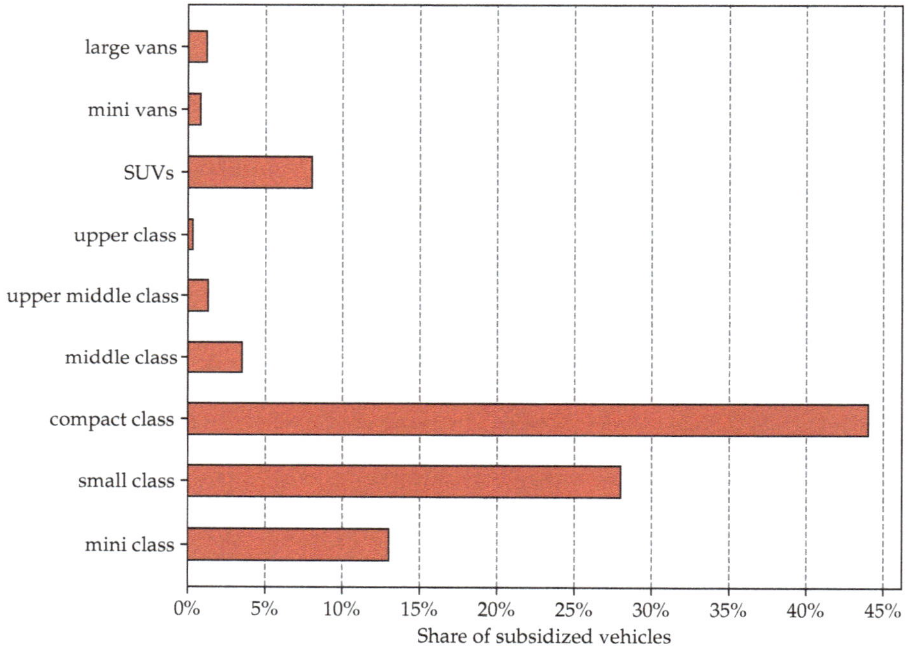

Figure 8. Subsidized vehicle distribution of the "broad funding" concept.

3.6.1. "Broad Funding" Concept

In this concept, the subsidy is not restricted to a certain vehicle type. Except for the higher demand for the mini, small, and compact class segments, the same composition of newly registered cars as predicted without the subsidy will occur. The subsidy only increases the total number of newly registered and deregistered cars. For every newly registered car due to the subsidy, a car in the same segment, manufactured in 2010 or earlier, is disposed of. Vehicle purchases in higher segments are not subsidized.

3.6.2. "Innovation" Concept

With this concept, only BEVs are subsidized. For the baseline model, the same distribution of subsidized cars as in Section 3.6 is considered, except only BEVs will be newly registered due to the subsidy.

3.6.3. "Downsizing" Concept

Lastly, a concept is shown that only allows a subsidy if the new car is at least one segment smaller (defined by the segment's CO_2 emissions) than the one traded in. The respective segment shifts in the "downsizing" concept are displayed in Table 8.

Table 8. Segment shifts in "downsizing" concept.

Segment of Vehicle Traded in	Segment of New Vehicle
mini class	-
small class	mini class
compact class	small class
middle class	compact class
upper middle class	middle class
upper class	upper middle class
SUVs	middle class
mini vans	middle class
large vans	mini vans

As per the CO_2 emission-based downsizing definition, the SUVs and mini vans are shifted to the middle class segment.

4. Results

The goal of this work is to analyze the impact of a subsidy on the CO_2 eq. emissions of the German passenger car fleet. This is achieved by considering categorized vehicle emissions and fleet composition data.

4.1. *Life-Cycle Analysis*

The life cycle of a vehicle consists of production, maintenance, use phase, and EoL processes. For each element, the respective emissions are calculated per segment and per drive train technology.

4.1.1. Production

Using the process described in Section 3.5.1, an analysis of the production emissions for each vehicle is made. The results for each vehicle class and drive train technology are shown in Figure 9.

The production of BEVs and PHEVs results in higher CO_2 equivalent emissions than the production of conventional vehicles. This phenomenon emerges in every segment and can be attributed, in addition to the higher vehicle weight of BEVs and PHEVs, to the greater effort of producing batteries and electrical parts. The BEV production effort increases substantially for larger segments as more battery capacity is implemented. This effect is less pronounced for PHEVs.

4.1.2. Maintenance

The emissions from the vehicle and road maintenance per driven kilometer (Figure 10) scale mainly with the vehicle weight.

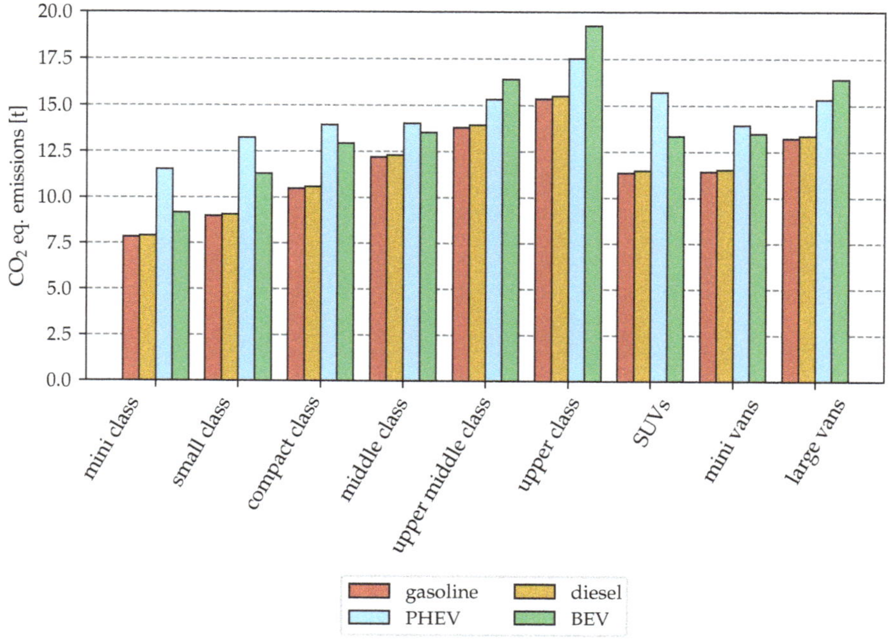

Figure 9. Production emissions per vehicle, segmented by class and drive train technology.

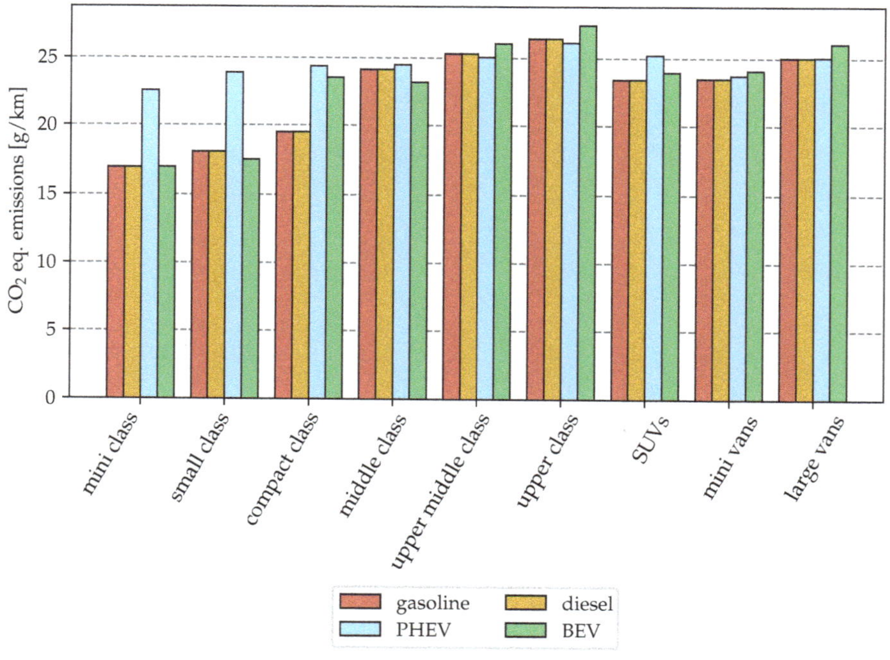

Figure 10. Maintenance emissions.

4.1.3. Use Phase

The CO_2 equivalent emissions depend on the vehicle specifications. In addition to the direct exhaust emissions, vehicles with combustion engines cause indirect emissions via the procurement of the fuel. Figure 11 shows the combined specific CO_2 eq. emissions per kilometer in 2019.

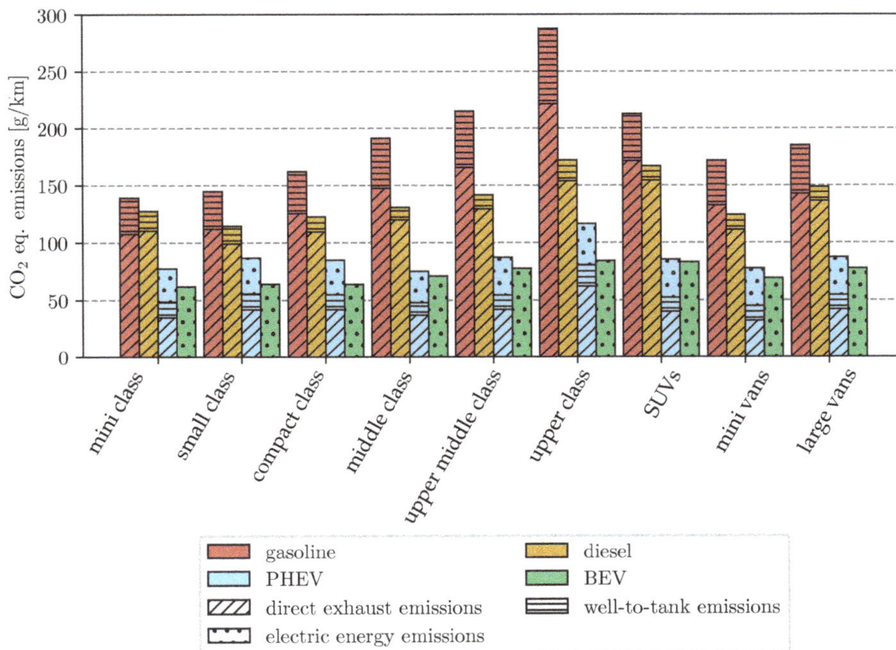

Figure 11. Composition of CO_2 eq. emissions per driven kilometer.

Vehicles with a gasoline engine have the highest specific emissions across every segment, followed by diesel engine vehicles. The BEVs have the lowest specific emissions of the considered drive train technologies in all segments.

4.1.4. End of Life

Due to the high recycling effort for the batteries, we see increased EoL emissions for the BEVs and PHEVs compared to the conventionally powered vehicles (Figure 12). The higher battery capacity in larger segments leads to higher EoL emissions for the BEVs compared to the PHEVs. Again, the vehicle weight is an important driver of EoL emissions.

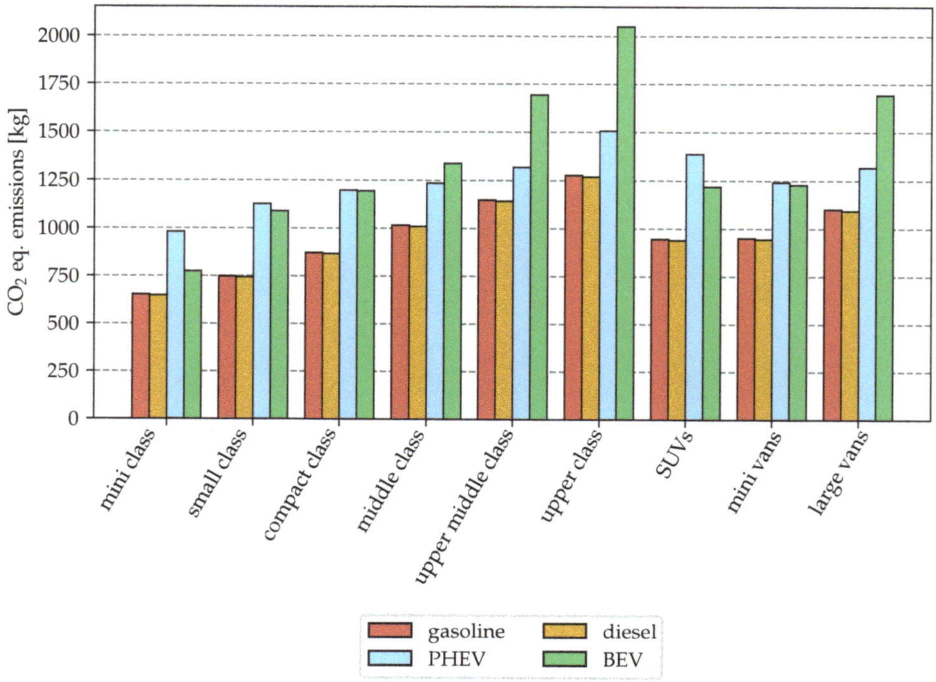

Figure 12. End-of-life emissions per vehicle, segmented by class and drive train technology.

4.2. Subsidy's Impact on Fleet Emissions for 2019–2030

Using a combination of the results from Section 4.1 and the German car population scenario from Section 3.3, a prediction of the CO_2 equivalent emissions emitted from the German passenger car system until 2030 is made. The emissions decrease in 2020 due to reduced vehicle production (Figure 13). As shown in Figure 7, the vehicle production increases in the subsequent years due to the subsidy. Therefore, higher emissions are expected before the vehicle fleet benefits from the increased efficiency that leads to reduced annual emissions.

The cumulative CO_2 eq. emissions of the baseline scenario without a subsidy (Figure 13) are compared to the cumulative emissions of the various subsidy concepts. Each concept is calculated with 0.72 million subsidized vehicles. The developed model enables the analysis of any number of subsidized vehicles.

Figure 14 shows the comparison of the different subsidy concepts relative to the baseline scenario. As a result, the accumulated differences of the CO_2 equivalent emissions are shown for the respective scenarios.

The cumulative difference shows significant additional emissions in 2021 for all subsidy concepts. The effect displays the higher number of new vehicles and the related production emissions for this year. The emission savings are dependent on the subsidy concept.

The "innovation" subsidy clearly lowers the CO_2 equivalent emissions of the German passenger car transport system until 2030 the most out of all considered concepts. Even with the highest CO_2 equivalent emissions in 2021, the "innovation" concept reaches the break-even earliest in the year 2025. The additional CO_2 equivalent emissions of the "innovation" concept in 2021 are 8.6% higher than the "broad funding" and 20.8% higher than the "downsizing" concepts. This effect can be attributed to the higher production effort for electric vehicles, as shown in Figure 9.

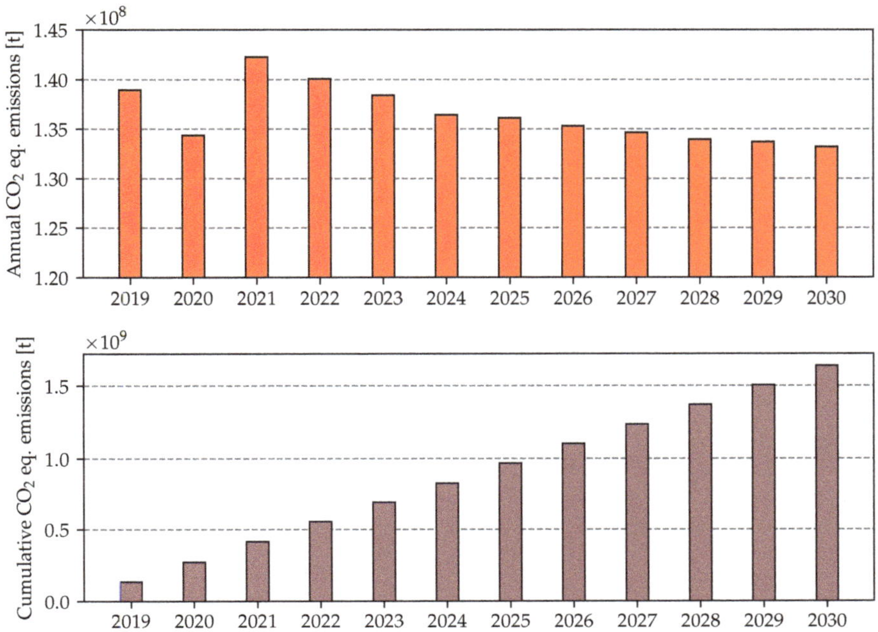

Figure 13. Life-cycle CO_2 eq. emissions of the baseline scenario without a subsidy.

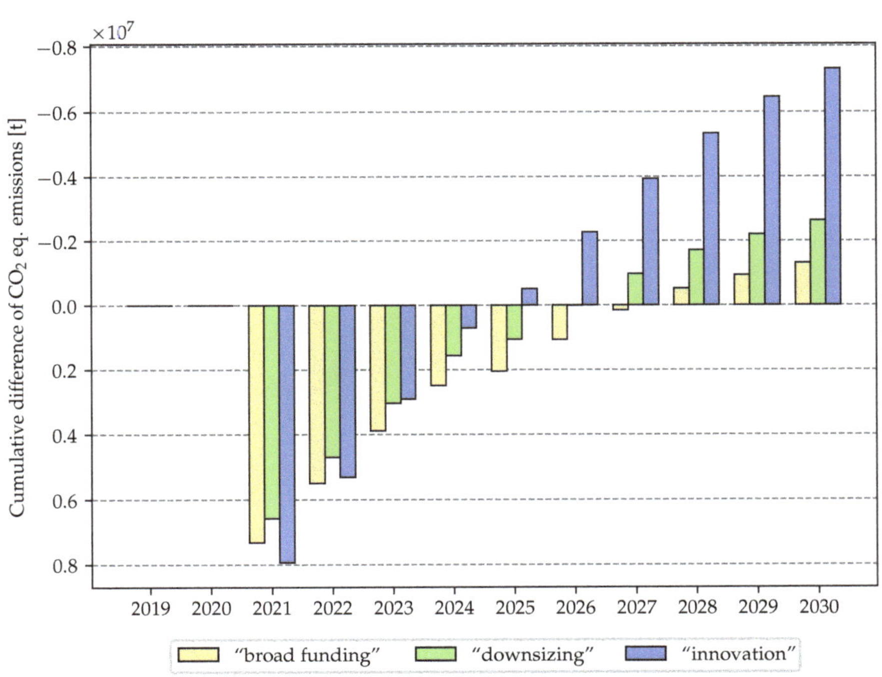

Figure 14. Cumulative difference of CO_2 eq. emissions compared to the scenario without a subsidy.

Despite having the highest CO_2 equivalent emissions in 2021, the accumulated CO_2 equivalent emission savings until 2030 of the "innovation" concept are 285.7% higher than the "downsizing" and 576.7% higher than the "broad funding" concepts. Compared with the scenario without a subsidy, the "innovation" concept reaches a cumulative difference of -7.56×10^6 t CO_2 equivalent emissions until 2030.

In the "downsizing" concept, compared to the "broad funding" concept, smaller vehicles are produced. Therefore, in 2021, the CO_2 equivalent emissions of the "downsizing" concept are lower than the emissions in the "broad funding" concept. In addition, the "downsizing" concept reaches the break-even point in 2027, one year earlier than the "broad funding" subsidy concept.

4.3. Sensitivities

Due to the complexity of the model, dependencies on key parameters are reviewed and their respective impacts on the subsidy concepts are determined.

4.4. Number of Subsidized Cars

As shown in Section 3.4, we assume that the number of subsidized cars matches the predicted registration decrease in 2020 due to the COVID-19 pandemic. To determine the impact of the number of subsidized cars (~0.7 million), this value is doubled (~1.4 million) and quadrupled (~2.8 million).

As expected, the cumulative difference of CO_2 eq. emissions in 2030 is proportional to the number of subsidized cars, which can be seen in Figure 15.

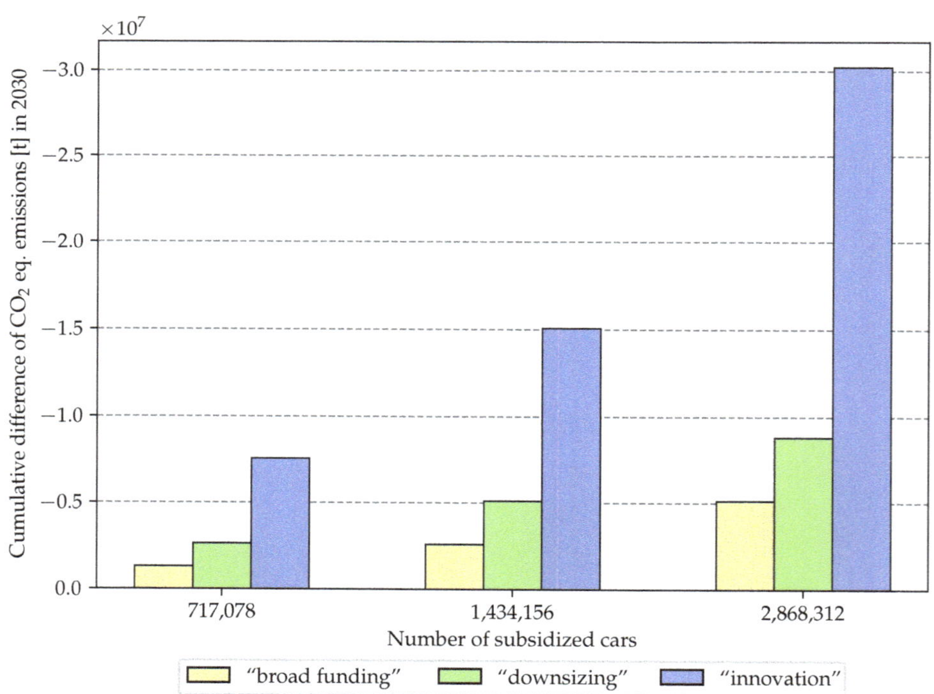

Figure 15. Number of subsidized cars.

4.4.1. Vehicle Distribution

As discussed in Section 3.6, the "environmental bonus" in 2009 was primarily used to purchase smaller vehicles. This can be attributed to the fairly high share of the subsidy compared to the purchase price.

It would be conceivable to design the subsidy in a way that the amount is determined by a fixed percentage of the purchase price. It is assumed that this leads to a vehicle distribution of the subsidized cars proportional to the predicted new passenger car registrations in 2021.

Figure 16 shows the difference between a scenario proportional to 2021 and a scenario shifted to smaller cars. The modified distribution of subsidized vehicles has an impact on the results.

As shown in Table 9, all subsidy concepts are less efficient when using the distribution proportional to 2021. In this context, the "broad funding" concept even increases the CO_2 eq. emissions compared to the no-subsidy baseline. As a result, subsidizing smaller cars significantly increases the efficiency of the subsidy.

Table 9. Difference of cumulative CO_2 eq. emissions compared to the scenario without a subsidy in 2030.

Subsidy Concept	Difference to "Without Subsidy" Baseline	
	Shifted Distribution to Smaller Cars	Proportional to 2021 Distribution
"broad funding"	−0.080%	0.037%
"downsizing"	−0.161%	−0.017%
"innovation"	−0.462%	−0.401%

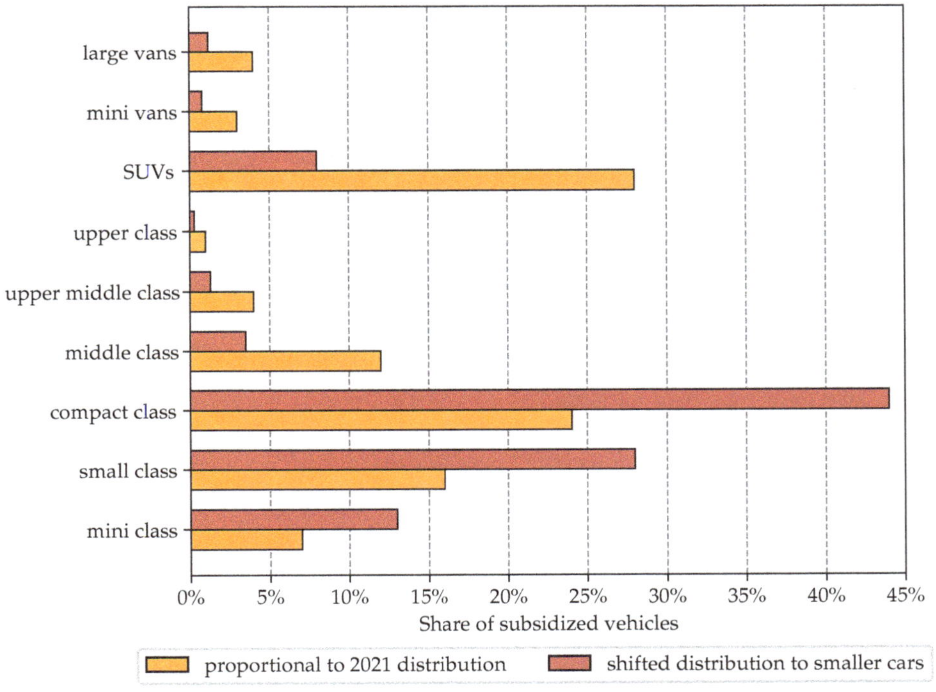

Figure 16. Subsidized car distribution for two scenarios under investigation.

4.4.2. New Vehicle Registrations

Drive Train Technologies

To predict the German car population until 2030, we rely on Section 3.3 and on the "ProKlima" scenario stated in Agora (2018) [20] to describe the trend of the drive train technologies. To estimate the impact of the subsidy for a baseline scenario with a higher share of newly registered PHEVs and BEVs, an analysis with the "ProKlima Plus" scenario is made.

In Table 10, the values for the drive train technology development of the different scenarios can be seen. To determine the share of newly registered cars for each year, a linear interpolation between the displayed values is made.

Table 10. Annual share of newly registered cars.

Drive Train Technology	"ProKlima" 2025	"ProKlima" 2030	"ProKlima Plus" 2025	"ProKlima Plus" 2030
gasoline	57%	42%	27%	28%
diesel	29%	19%	28%	10%
PHEV	8%	22%	20%	29%
BEV	6%	18%	25%	47%

Figure 17 shows the cumulative CO_2 eq. emission difference of the "innovation" concept with the two different assumptions for the annual newly registered cars until 2030. It is shown that the subsidy loses efficiency if the baseline share of PHEVs and BEVs rises. The cumulative CO_2 eq. emission difference will be smaller due to the overall more efficient German passenger car fleet with the "proKlima Plus" scenario.

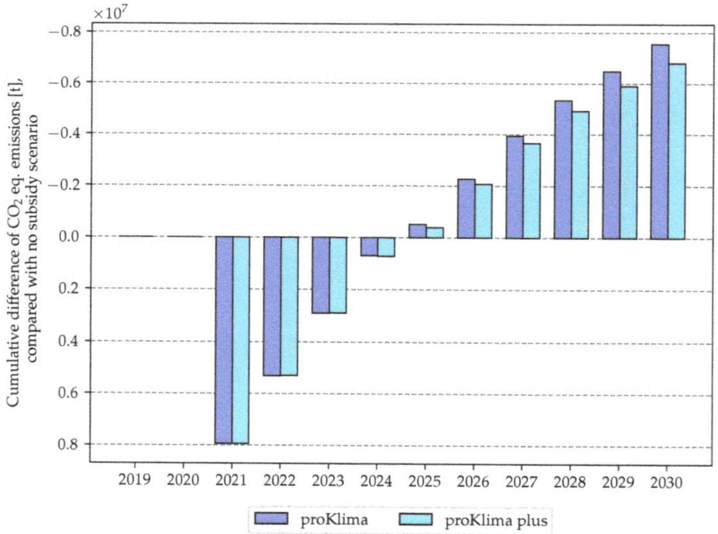

Figure 17. Cumulative CO_2 eq. emission difference between the "no subsidy" baseline and the "innovation" concept using the "proKlima" and "proKlima Plus" registration predictions.

Segments

In Section 3.3, an assumption of the development of the segment composition of newly registered cars is introduced. Here, we assume that the SUV segment will reach a 40% share of the annual newly registered cars in 2030. Since this parameter is uncertain, analysis with either a 30% or 50% proportion of SUVs of the newly registered cars in 2030 is made. For the "innovation" concept, the 30% SUV share scenario generates 2.40% less CO_2 eq. emissions savings compared to the 40% SUV share baseline due to the overall more efficient car fleet. On the contrary, the cumulative CO_2 eq. emissions savings of the "innovation" concept in the 50% SUV share scenario are 0.84% higher compared to the baseline.

This shows that the results of the study are robust against this parameter because the trend and the statement of the results are not changed by a variation of the SUV share.

5. Discussion

The absolute CO_2 eq. emissions savings are proportional to the number of subsidized cars. For the results shown, we set the subsidized cars in the model to about 0.72 million cars. This is more conservative than the actual replacements of 1.95 million [3] during the 2009 "environmental bonus".

Our model is able to deal with a variable number of cars, as shown in the sensitivity analysis in Section 4.4. The actual number of subsidized cars would depend on micro- and macroeconomic factors, as well as technology-specific concerns, with different scenarios needing different funding to reach the same total number of replaced vehicles. Additionally, an investigation of the basic conditions, e.g., charging facilities, is needed to match the implementation of the number of subsidized vehicles. However, since we are not conducting an economic analysis, quantifying these factors reliably is out of the scope of this work. Our model can be easily updated to calculate the effects of arbitrary numbers of subsidized vehicles. In case the 2020 program would have the same impact on car sales, the absolute savings grow accordingly.

In the absence of detailed prognosis data for the technological improvements, we assume no efficiency gains regarding the CO_2 eq. emissions. The trend of the combustion engines indicates a saturation for the CO_2 efficiency [44]. Assuming that the electric drive train technologies will still gain efficiency [52] and more electricity will be generated from renewable sources, the total savings for the "innovation" concept would be higher.

Since only the direct CO_2 emissions for the driving during the use phase of combustion engines are available, the actual CO_2 eq. emissions are slightly higher. As this effect is not relevant to BEVs, since CO_2 eq. emissions are considered for the electricity production, the "innovation" concept would generate marginally higher savings when considering all emissions from the combustion engines.

This paper shows that all considered subsidy concepts eventually have a positive effect on reducing the CO_2 equivalent emissions of the German passenger car fleet. This corresponds with both Höpfner et al. [3] and Lenski et al. [13], who stated that the 2009 car subsidy had a positive effect on the environmental performance of the various car fleets.

Contrary to Klößner et al. [8], the long-term effect on the CO_2 emissions is positive, as windfall gains (subsidized purchases that would happen without the subsidy) are not included and BEVs and PHEVs are considered. We only expect additional CO_2 emissions to occur immediately after the subsidy is introduced due to the increased production, which is overcompensated by savings during the use phase.

To calculate the LCA emissions, we clustered the vehicle fleet into segments to reflect the actual distribution. Agora (2019) [53] used only one reference vehicle for each drive train technology. For comparable segments, the production emissions of the BEVs show similar results. In this work, the gasoline and diesel vehicles result in higher production emissions compared to those in Agora (2019) [53]. This can be attributed to a deviation in parameter settings.

6. Conclusions

In order to calculate the cumulative CO_2 eq. emissions of the German passenger car fleet from 2019–2030, a granular model was developed. Three different subsidy concepts were introduced and the impact on the CO_2 eq. emissions was determined.

Subsidizing the German passenger car system due to the COVID-19 pandemic shows long-term CO_2 emissions savings for nearly all investigated scenarios. Only in the sensitivity analysis, the "broad funding" concept leads to slightly increased emissions when using a vehicle distribution proportional to that of 2021. Taking the sensitivity analyses into account, the "innovation" concept shows the most significant emissions savings in the German passenger car system.

Considering the time period 2019–2030 and a total number of 0.72 million subsidized vehicles, the "innovation" concept generates about 7.56 million t less CO_2 eq. emissions compared to the scenario without a subsidy. This equates to 0.46% of the total CO_2 eq. emissions of the addressed segments in this period. The "downsizing" and "broad funding" concepts create savings of 0.16% and 0.08%, respectively.

In 2021, the increased vehicle production leads to higher CO_2 emissions for all subsidy scenarios compared to the scenario without a subsidy. The ecological break-even is reached in 2025 for the "innovation" concept, in 2027 for the "downsizing" concept, and in 2028 for the "broad funding" concept.

If, from an economic and political point of view, a subsidy program for passenger cars in Germany is considered to be desirable, we clearly recommend the exclusive funding of BEVs because the "innovation" concept achieves the highest positive climate impact at the earliest time.

7. Outlook

To further investigate the environmental impact of passenger car subsides in Germany, additional greenhouse gases in the use phase and air pollutants must be considered. It seems reasonable to add greenhouse gases with a high GWP. Due to the current discussion on driving bans in German city centers, the subsidy's impact on nitrogen oxides would enhance the model's result.

It is conceivable to examine further subsidy concepts, such as focusing on PHEVs or particular segments.

In the current model, the technological status of the vehicles is assumed to remain at the level of 2019. The indirect emissions of only BEVs and PHEVs will decrease until 2030 due to the slight reduction of the specific carbon dioxide emissions of the German electricity mix. This means that the vehicle weight and direct exhaust emissions of each newly registered vehicle will remain constant in the current model.

In our model, we define a reference vehicle to describe the BEVs and PHEVs due to a lack of data provided by the KBA. To determine the detailed PHEV and BEV model parameters more precisely, the data of all vehicles available need to be consolidated.

As Höpfner et al. [3] notes, the vehicle subsidy in 2009 led to reduced used car sales. This was not implemented in the current model and might affect the disposal rate and the number of newly registered cars.

This paper did not investigate the economic impacts of subsidy scenarios. A detailed market analysis is needed to estimate consumers' buying behavior. To improve the model, it is necessary to further examine the subsidy's impact on the newly registered cars and to address economic effects, like on-top sales and windfall gains.

Supplementary Materials: The following are available online at http://www.mdpi.com/2071-1050/12/23/10037/s1: Table S1: Direct CO_2 emissions per kilometer, Table S2: Fuel consumption per 100 km.

Author Contributions: Conceptualization, L.H. and A.G.; methodology, M.S. and L.H.; software, A.M.S., L.H., and M.S.; validation, A.M.S., A.G., M.S., L.H., and D.G.; investigation, M.S., A.G., A.M.S., and L.H.; resources, A.M.S. and D.G.; data curation, M.S.; writing—original draft preparation, M.S.; writing—review and editing, A.M.S., A.G., L.H., and M.S.; visualization, L.H. and M.S.; supervision, D.G.; funding acquisition, D.G. All authors have read and agreed to the published version of the manuscript.

Funding: This research was funded by the Deutsche Forschungsgemeinschaft (DFG, German Research Foundation), grant number 398051144, project title: "Analysis of strategies to fully de-carbonize urban transport".

Acknowledgments: We acknowledge the support from the German Research Foundation and the Open Access Publication Fund of TU Berlin.

Conflicts of Interest: The authors declare no conflict of interest. The funders had no role in the design of the study; in the collection, analyses, or interpretation of data; in the writing of the manuscript, or in the decision to publish the results.

References

1. VDA. VDA Expects Car Sales in Germany and Europe to Lose around One Quarter in 2020. 2020. Available online: https://www.vda.de/en/press/press-releases/200703-VDA-expects-car-sales-in-Germany-and-Europe-to-lose-around-one-quarter-in-2020 (accessed on 20 July 2020).
2. German Federal Ministry for Economic Affairs and Energy. Automobilindustrie. 2020. Available online: https://www.bmwi.de/Redaktion/DE/Textsammlungen/Branchenfokus/Industrie/branchenfokus-automobilindustrie.html (accessed on 1 July 2020).
3. Höpfner, U. Abwrackprämie und Umwelt—Eine Erste Bilanz. 2009. Available online: http://docplayer.org/40629988-Abwrackpraemie-und-umwelt-eine-erste-bi-lanz.html (accessed on 30 November 2020).
4. Frankfurter Allgemeine Zeitung. Weil Will Autoindustrie mit Abwrackprämie Stärken. 2020. Available online: https://www.faz.net/aktuell/wirtschaft/auto-verkehr/corona-krise-weil-will-autoindustrie-mit-abwrackpraemie-staerken-16724366.html (accessed on 11 June 2020).
5. Verband der Automobilindustrie. Corona-Krise: Starthilfe für die Automobilindustrie. 2020. Available online: https://www.vda.de/de/themen/automobilindustrie-und-maerkte/Coronavirus-Update/Corona-Krise--Starthilfe-f-r-die-Autoindustrie-.html (accessed on 11 June 2020).
6. Lamparter, D.H. Eine Abwrackprämie für die Wirtschaft oder Hilfe für die Umwelt? 2020. Available online: https://www.zeit.de/mobilitaet/2020-04/autobranche-abwrackpraemie-coronavirus-rezession-konsum (accessed on 11 June 2020).
7. Bundesamt für Wirtschaft und Ausfuhrkontrolle. Merkblatt für Anträge von Elektrisch betriebenen Fahrzeugen: Richtlinie zur Förderung des Absatzes von Elektrisch Betriebenen Fahrzeugen (Umweltbonus) vom 25.06.2020. 2020 Available online: https://www.bafa.de/SharedDocs/Downloads/DE/Energie/emob_merkblatt_2020-0708.html (accessed on 27 May 2020).
8. Klößner, S.; Pfeifer, G. Synthesizing Cash for Clunkers: Stabilizing the Car Market, Hurting the Environment: Beiträge zur Jahrestagung des Vereins für Socialpolitik 2015: Ökonomische Entwicklung—Theorie und Politik—Session: Automobiles and the Environment, No. F13-V1, ZBW. 2015. Available online: http://hdl.handle.net/10419/113207 (accessed on 30 November 2020).
9. Läufer, Nikolaus K. A.. Mikro- und makroökonomische Effekte der Abwrackprämie. *Wirtschaftsdienst* **2009**, *89*, 410–418. [CrossRef]
10. Müller, A.; Heimeshoff, U. *Evaluating the Causal Effects of Cash-for-Clunkers Programs in Selected Countries: Success or Failure?* Beiträge zur Jahrestagung des Vereins für Socialpolitik 2013: Wettbewerbspolitik und Regulierung in einer Globalen Wirtschaftsordnung—Session: Empirics: Markets and Media F13-V3; 2013. Available online: http://hdl.handle.net/10419/79802 (accessed on 30 November 2020).
11. Böckers, V.; Heimeshoff, U.; Müller, A. *Pull-Forward Effects in the German Car Scrappage Scheme: A Time Series Approach*; DICE Discussion Paper; DICE: Düsseldorf, Germany, 2012; Volume 56, ISBN 9783863040550.
12. National Highway Traffic Safety Administration. *Consumer Assistance to Recycle and Save Act of 2009 (CARS): Summary of the Consumer Assistance to Recycle and Save Act of 2009 and Notice of Upcoming Rulemaking Proceeding*; National Highway Traffic Safety Administration: Washington, DC, USA, 2009.
13. Lenski, S.M.; Keoleian, G.A.; Bolon, K.M. The impact of 'Cash for Clunkers' on greenhouse gas emissions: A life cycle perspective. *Environ. Res. Lett.* **2010**, *5*, 044003. [CrossRef]
14. Sivak, M.S.; Schoettle, B. *The Effect of the "Cash for Clunkers" Program on the Overall Fuel Economy of Purchased New Vehicles*; Report No. UMTRI-2009-34; University of Michigan, Ann Arbor, Transportation Research Institute: Ann Arbor, MI, USA, 2009.
15. Li, S.; Linn, J.; Spiller, E. Evaluating "Cash-for-Clunkers": Program effects on auto sales and the environment. *J. Environ. Econ. Manag.* **2013**, *65*, 175–193. [CrossRef]
16. Kraftfahrtbundesamt. Fachartikel: Marken und Modelle. 2011. Available online: https://www.kba.de/SharedDocs/Publikationen/DE/Statistik/Fahrzeuge/FZ/Fachartikel/marken_modelle_20110515.pdf?__blob=publicationFile&v=6 (accessed on 20 July 2020).

17. German Federal Ministry of Transport and Digital Infrastructure. Mobilität in Deutschland—Mobilität in Tabellen. 2017. Available online: https://mobilitaet-in-tabellen.dlr.de/mit/ (accessed on 27 May 2020).
18. Kraftfahrtbundesamt. Fahrzeugzulassungen—Bestand an Kraftfahrzeugen und Kraftfahrzeuganhängern nach Fahrzeugalter 1. Januar 2019: FZ 15. 2019. Available online: https://www.kba.de/DE/Statistik/Produktkatalog/produkte/Fahrzeuge/fz15_b_uebersicht.html (accessed on 21 June 2020).
19. Kraftfahrtbundesamt. Personenkraftwagen am 1. Januar 2019 nach Ausgewählten Merkmalen. 2019. Available online: https://www.kba.de/DE/Statistik/Fahrzeuge/Bestand/Jahresbilanz/fz_b_jahresbilanz_archiv/2019/2019_b_barometer.html?nn=2601598 (accessed on 7 May 2020).
20. Agora Verkehrswende. Die Fortschreibung der Pkw-CO_2-Regulierung und ihre Bedeutung für das Erreichen der Klimaschutzziele im Verkehr. 2018. Available online: https://www.agora-verkehrswende.de/veroeffentlichungen/die-fortschreibung-der-pkw-co2-regulierung-und-ihre-bedeutung-fuer-das-erreichen-der-klimaschutzzie/ (accessed on 30 November 2020).
21. Brandt, M. Neuzulassungsrekord bei SUVs. 2020. Available online: https://de.statista.com/infografik/19572/anzahl-der-neuzulassungen-von-suv-in-deutschland// (accessed on 10 May 2020).
22. Kraftfahrtbundesamt. Fahrzeugzulassungen—Bestand an Kraftfahrzeugen und Kraftfahrzeuganhängern nach Fahrzeugalter 1. Januar 2020: FZ 15. 2020. Available online: https://www.kba.de/SharedDocs/Publikationen/DE/Statistik/Fahrzeuge/FZ/2020/fz15_2020_pdf.pdf?__blob=publicationFile&v=6 (accessed on 14 October 2020).
23. Wernet, G.; Bauer, C.; Steubing, B.; Reinhard, J.; Moreno-Ruiz, E.; Weidema, B. The ecoinvent database version 3 (part I): Overview and methodology. *Int. J. Life Cycle Assess.* **2016**, *21*, 1218–1230. [CrossRef]
24. Ahlswede, A. Leergewicht von Pkw-Neuzulassungen in Deutschland nach Segmenten 2018. 2019. Available online: https://de.statista.com/statistik/daten/studie/239565/umfrage/leergewicht-von-pkw-neuzulassungen-in-deutschland-nach-segmenten/ (accessed on 13 June 2020).
25. ADAC. Technische Daten VW e-Golf (04/17–05/20). 2018. Available online: https://www.adac.de/rund-ums-fahrzeug/autokatalog/marken-modelle/vw/golf/vii-facelift/266575/#technische-daten (accessed on 21 May 2020).
26. Tesla Inc. Request for Issuance of a New Certificate of Conformity—Running Change to Add Model S 100D and New 75 kWh FW Limited Battery Pack Configuration. 2017. Available online: https://iaspub.epa.gov/otaqpub/display_file.jsp?docid=39836&flag=1 (accessed on 29 October 2020).
27. Kraftfahrtbundesamt. Neuzulassungen von Kraftfahrzeugen und Kraftfahrzeuganhängern—FZ 8. Available online: https://www.kba.de/DE/Statistik/Produktkatalog/produkte/Fahrzeuge/fz8/fz8_gentab.html (accessed on 25 July 2020).
28. Smart. smart fortwo.—Technische Daten. 2020. Available online: https://www.smart.com/de/de/modelle/technische-daten-eq (accessed on 25 August 2020) .
29. BMW. Technical Specifications BMW i3 (120 Ah). 2018. Available online: https://www.press.bmwgroup.com/global/article/detail/T0285608EN/technical-specifications-of-the-bmw-i3-120-ah-and-the-bmw-i3s-120-ah-valid-from-11/2018?language=en (accessed on 25 August 2020).
30. Nissan. NISSAN LEAF—Abmessungen und Technische Daten. 2018. Available online: https://www.nissan.de/fahrzeuge/neuwagen/leaf/abmessungen-technische-daten.html (accessed on 25 August 2020).
31. TESLA. Verbrauchsangaben und Energieeffizienz. 2018. Available online: https://www.tesla.com/de_CH/support/european-union-energy-label (accessed on 25 August 2020) .
32. Hyundai. Technische Daten—Hyundai Kona Elektro. 2020. Available online: https://www.hyundai.news/de/pressemappen-modelle/der-hyundai-kona-elektro-technische-daten/ (accessed on 25 August 2020).
33. Kia. Technische Daten—Kia e-soul. 2020. Available online: https://www.kia.com/de/modelle/e-soul/technische-daten/ (accessed on 25 August 2020).
34. Del Duce, A.; Gauch, M.; Althaus, H.J. Electric passenger car transport and passenger car life cycle inventories in ecoinvent version 3. *Int. J. Life Cycle Assess.* **2016**, *21*, 1314–1326. [CrossRef]
35. BMW. Technische Daten MINI Countryman. 2018. Available online: https://www.press.bmwgroup.com/deutschland/article/detail/T0284410DE/technische-daten-mini-countryman-gueltig-ab-09/2018?language=de (accessed on 25 August 2020).
36. BMW. BMW 2er Active Tourer: Motoren & Technische Daten. 2020. Available online: https://www.bmw.de/de/neufahrzeuge/2er/activetourer/2020/bmw-2er-active-tourer-technische-daten.html#tab-0 (accessed on 25 August 2020).

37. Kia. Technische Daten Kia Optima Sportswagon PHEV. 2020. Available online: https://www.kia.com/de/modelle/optima-sportwagon-phev/technische-daten/ (accessed on 25 August 2020).
38. BMW. 5er Limousine—Technische Daten. 2020. Available online: https://www.bmw.de/de/neufahrzeuge/5er/limousine/2020/bmw-5er-limousine-technische-daten.html#tab-0 (accessed on 25 August 2020).
39. Dr. Ing. h.c. F. Porsche AG. Daten & Ausstattung Panamera E-Hybrid Modelle. 2020. Available online: https://www.porsche.com/germany/models/panamera/panamera-e-hybrid-models/panamera-4-e-hybrid/featuresandspecs/ (accessed on 25 August 2020).
40. Mitsubishi Motors. Der Outlander Plug-in Hybrid. 2020. Available online: https://www.mitsubishi-motors.de/outlander (accessed on 25 August 2020).
41. Daimler AG. Mercedes-Benz B 250 e: Der Alleskönner. 2020. Available online: https://media.daimler.com/marsMediaSite/de/instance/ko/Mercedes-Benz-B-250-e-Der-Alleskoenner.xhtml?oid=44351229 (accessed on 25 August 2020).
42. Mercedes-Benz Niederlassung Berlin. A 250 e und B 250 e. Available online: https://www.mercedes-benz-berlin.de/de/desktop/passenger-cars/vehicle-type/new-cars/alternativer-antrieb/elektromobilitaet/eq-power-modelle/a-250-e-und-b-250-e.html (accessed on 25 August 2020).
43. Kraftfahrtbundesamt. Verkehr in Kilometern—Inländerfahrleistung (VK), Entwicklung der Fahrleistungen nach Fahrzeugarten Seit 2015. 2020. Available online: https://www.kba.de/DE/Statistik/Kraftverkehr/VerkehrKilometer/verkehr_in_kilometern_node.html (accessed on 14 July 2020).
44. Kraftfahrtbundesamt. Neuzulassungen nach Umwelt-Merkmalen (FZ 14). 2020. Available online: https://www.kba.de/DE/Statistik/Produktkatalog/produkte/Fahrzeuge/fz14_n_uebersicht.html (accessed on 11 July 2020).
45. Renault. Renault ZOE—Technisches Datenblatt. 2020. Available online: https://www.renault.de/elektrofahrzeuge/zoe/technische-daten.html (accessed on 25 August 2020).
46. Volkswagen AG. Der e-Golf. Kataloge und Preislisten. 2020. Available online: https://www.volkswagen.de/de/modelle-und-konfigurator/e-golf.html/__layer/carfeatures/features/models/golf/e-golf/technical_data/technical-data-layer/master.layer (accessed on 25 August 2020).
47. Audi AG. Audi e-tron Technische Daten. 2020. Available online: https://www.audi.de/de/brand/de/neuwagen/tron/audi-e-tron.html (accessed on 25 August 2020).
48. Pehnt, M.; Mellwig, P.; Blömer, S.; Hertle, H.; Nast, M.; von Oehsen, A.; Lempik, J.; Langreder, N.; Thamling, N.; Hermelink, A.; et al. 7–03–17 Untersuchung zu Primärenergiefaktoren Endbericht. 2018. Available online: https://www.gih.de/wp-content/uploads/2019/05/Untersuchung-zu-Prim%C3%A4renergiefaktoren.pdf (accessed on 19 October 2020).
49. Schallaböck, K.O.; Carpantier, R. *Umweltbegleitforschung fuer PWK und Leichte Nutzfahrzeuge*; FKZ 03KP5003; FKZ: Wuppertal, Germany, 2012.
50. Syré, A.M.; Heining, F.; Göhlich, D. Method for a Multi-Vehicle, Simulation-Based Life Cycle Assessment and Application to Berlin's Motorized Individual Transport. *Sustainability* **2020**, *12*, 7302. [CrossRef]
51. Agora Verkehrswende. Klimabilanz von Elektroautos. 2019. Available online: https://www.agora-verkehrswende.de/fileadmin/Projekte/2018/Klimabilanz_von_Elektroautos/Agora-Verkehrswende_22_Klimabilanz-von-Elektroautos_WEB.pdf (accessed on 10 June 2020).
52. Woodward, M. Battery Electric Vehicles: New Markets. New Entrants. New Challenges. Available online: https://www2.deloitte.com/uk/en/insights/industry/automotive/battery-electric-vehicles.html (accessed on 27 August 2020).
53. Agora Verkehrswende. Klimabilanz von Strombasierten Antrieben und Kraftstoffen. 2019. Available online: https://www.agora-verkehrswende.de/fileadmin/Projekte/2019/Klimabilanz_Batteriefahrzeugen/32_Klimabilanz_strombasierten_Antrieben_Kraftstoffen_WEB.pdf (accessed on 21 June 2020).

sustainability

MDPI

Article

Removal of Basic Brown 16 from Aqueous Solution Using Durian Shell Adsorbent, Optimisation and Techno-Economic Analysis

Yashni Gopalakrishnan [1], Adel Al-Gheethi [1,*], Marlinda Abdul Malek [2,*], Mawar Marisa Azlan [1], Mohammed Al-Sahari [1], Radin Maya Saphira Radin Mohamed [1,*], Sadeq Alkhadher [1] and Efaq Noman [3,4]

1 Micropollutant Research Centre (MPRC), Faculty of Civil Engineering & Built Environment, Universiti Tun Hussein Onn Malaysia, Parit Raja 86400, Batu Pahat, Johor, Malaysia; yashni_g@yahoo.com (Y.G.); mawarmarisa@yahoo.com (M.M.A.); mohammedalsahari@gmail.com (M.A.-S.); sadeqalkhather@yahoo.com (S.A.)

2 Institute of Sustainable Energy, Universiti Tenaga Nasional, Selangor 43000, Malaysia

3 Department of Applied Microbiology, Faculty of Applied Science, Taiz University, Taiz City 00967, Yemen; eanm1984@gmail.com

4 Faculty of Applied Sciences and Technology, University Tun Hussein Onn Malaysia (UTHM), Pagoh Higher Education Hub, KM 1, Jalan Panchor, Panchor 84000, Johor, Malaysia

* Correspondence: adel@uthm.edu.my (A.A.-G.); Marlinda@uniten.edu.my (M.A.M.); maya@uthm.edu.my (R.M.S.R.M.); Tel.: +60-01-111-098-362 (A.A.-G.); +6019-33-22-775 (M.A.M.); Fax: +603-8921-2116 (M.A.M.)

Received: 2 September 2020; Accepted: 27 September 2020; Published: 27 October 2020

Abstract: Azo dyes including C. I. Basic Brown 16 (BB16) are one of the coloured organic compounds that have adverse effects on human health and the environment. The current work aims to optimise the adsorption of C.I BB16 in aqueous solution using durian (Durio zibethinus murray) shell as a low-cost green adsorbent. Durian shell was characterised by Fourier transform infrared spectroscopy (FTIR) and scanning electron microscopy (SEM). The adsorption process was optimised with response surface methodology (RSM) based on pH (4–8), time (30–240 min), durian shell dosage (0.1–1.0 g/L) and initial concentration of C.I BB16 (10–20 ppm). The removal efficiency was determined based on the reduction of chemical oxygen demand (COD) and the decolourisation of C.I BB16. The techno-economic analysis was described in the current work to know the economic feasibility of durian shells as an adsorbent. The SEM images showed that durian shell adsorbent has a smooth surface with no pores. FTIR spectra confirmed the presence of -C-O, =C–H, C=C, -C-O-C and O-H bonds in durian shell. Maximum decolourisation (77.6%) and COD removal (80.6%) for C.I BB16 was achieved with the interaction between pH, time and adsorbent dose and initial concentration of C.I BB16. The optimal operating factors for adsorption of C.I BB16 recorded at pH 8, time (30 min), durian shell dosage (1 g/L) and 15 mg /L of C.I BB16 concentrations were 77.61 vs. 74.26 (%) of C.I BB16 removal and 80.60 vs. 78.72 (%) of COD removal with an R^2 coefficient of 0.94 at $p < 0.05$. The specific cost of durian shell coagulant production is USD 172.71 per ton which is lower than the market price of honeydew peels-activated carbon (HDP-AC) (USD 261.81) and the commercial market price of activated carbon which is USD 1000.00/tons. These findings indicated that the durian adsorbent provides alternative methods for treating hair dye wastewater. These findings indicated that durian shells have a high potential for the adsorption of C.I BB16 in aqueous solution.

Keywords: Basic Brown 16; adsorption; durian shell; mechanism; techno-economic analysis

1. Introduction

Azo dyes are coloured organic compounds bearing the functional group R−N=N−R′, in which R and R′ are aryl or heteroaryl. More than 0.7 million tons of these compounds are produced every year from which 10–15% enters into the environment through industrial wastewater, especially textile industries [1]. The improper disposal of these wastewaters is the main problem of environmental pollution because of its high toxicity and stability to withstand degradation even under extreme circumstances [2]. The main concerns related to the presence of azo dyes in water ecosystems are its effect on water transparency, photosynthetic activity reduction and the harmful consequences on aquatic flora and fauna. Moreover, most of the azo dyes have been documented to be mutagens or carcinogens to human beings [3]. C. I. BB16 is one of the most applicable dyes in the textile and biological dyeing industries. It contains (-N=N-) as chromophore connected to aromatic systems with lateral groups, including CH_3, which can be very harmful and resistant to conventional treatment processes. Although C.I BB16 has caused health hazards because of which it has been prohibited in many areas, it is still widely practised in several countries [4].

Several treatment technologies for removing azo dyes from industrial wastewater have been used such as coagulation-flocculation [5], chemical and electrochemical oxidation [6], ozonation [7], ion exchange [2] and anaerobic-aerobic system [8]. However, most of the azo dyes are still persistent in the treated wastewater because of their high stability. Therefore, the removal of azo dyes required an advanced removal technology such as adsorption which is one of the most favourable processes in comparison to other practices because of its simplicity, practicality, productivity and inexpensiveness [9]. Adsorption is also known as physical attachment or binding of ions and molecules onto the surface of other molecular substances [10]. The exploration for new adsorbents has been concentrated on biomaterials because of their high availability, eco-friendliness, economic feasibility and source i.e., from renewable materials. These resources are locally and easily accessible in considerable quantities in the agroindustry. Thus, they are typically inexpensive and have little or no economic significance. For instance, these resources are wastes with no industrial or agricultural use, with their accumulation being problematic in many cases [11]. The adsorbents comprise proteins, polysaccharides, cellulose and lignin, which contain various functional groups such as phenolic and amino groups that are accountable for pollutant adsorption. Among several plants, pineapple plant stem [12], leaves of Prunus Dulcis [13], tomato waste and apple juice residue [14], grape peel [15], *Thapsia transtagana* stems [16], mushroom [17] have been reported to produce biocompatible adsorbents.

In this study, Durio Zibethinus Murray (durian) shell was utilised for the adsorption of C.I BB16 because of its richness in proteins, polysaccharides, cellulose and lignin that contain various functional groups such as phosphate, carboxyl, sulphate, phenolic and amino groups that are accountable for azo dyes adsorption [18]. Moreover, the durian shell contains high fibre content that contributes to the adsorption process [19]. Durian peels are broadly obtainable as a waste from the durian processing business at a cheaper rate in comparison to other natural supplies. This is because only one-third of durian is digestible (20–30% flesh), whereas the shell and the seed (55–66%) are generally disposed of [20]. The productivity of durian in Malaysia alone was 3800 metric tons in the year 2016 and 22,000 metric tons in the year 2020. With every 10 tons, more than 7 tons of durian shells are produced [21]. Although durian shell as adsorbent was investigated by previous researchers [21,22], no comprehensive study has been conducted on the optimisation of adsorption of C.I BB16 by using a response surface methodology (RSM) and this emphasises the novelty of the current work.

Figure 1 presents the bibliometric analysis of 43 papers published on the durian shell according to the Scopus database using specific keywords such as "durian" AND "adsorption". It was noted that the durian shell was used as an adsorbent for removing methylene blue, basic dyes, brilliant green, basic green four dye, crystal violet, gentian violate, acid green 25 as well as heavy metals such as chromium, lead, zinc from different wastewater sources. The main factors affecting the adsorption process included pH, time, the concentration of pollutants, moisture and particle size of the durian shell powder. The durian shells have been subjected to acid treatment to prepare charcoal, biochar and carbon active which are the most common adsorbents.

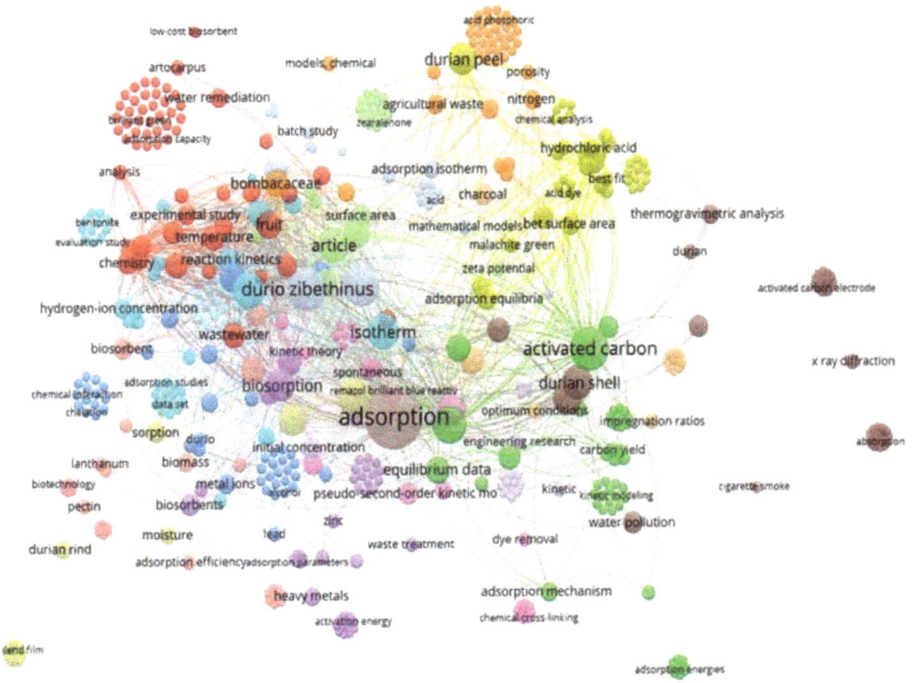

Figure 1. Bibliometric analysis for 43 paper published on the durian shell according to the Scopus database using specific keywords included "durian" AND "adsorption".

The factors affecting adsorption need to be optimised to reinforce the process of adsorption. Response surface methodology (RSM) is a tool that uses the mathematical and statistical approach to develop, optimise and enhance products and industrial processes. The main aim of RSM is to lessen the number of experimental runs and attain maximum data output. It allows changing numerous variables at the same time to plan fewer number of experiments needed to optimise the desired process [23]. RSM can be used to study the effects of independent variables, multiple factors and their interaction effects [24]. The effects of two or more variables are determined by factorial design. Meanwhile, central composite design (CCD) with RSM is used for the optimisation process. Besides, the application of CCD prevents deceiving conclusions and grant conclusions that are appropriate for several experimental conditions. Only one study has been optimised for the durian seed prepared as activated carbon for removing C.I methylene blue [25]. However, RSM is not used as a modelling and optimisation tool for the adsorption of C. I BB16. Moreover, techno-economic analysis is used for evaluating the potential of agriculture wastes as an alternative adsorbent for the commercial activated carbon (AC). No study has investigated the final price of the durian shell compared with that of commercial products in the market. The current price of 1 ton AC from coal ranges from USD 1190 to 16343/tons AC [26]. Therefore, the current study aims to optimise the factors affecting the adsorption of C.I BB16 by durian shell as an adsorbent by RSM. Moreover, techno-economic analysis for the preparation and application of durian shell adsorbent was conducted and compared with that of commercial AC products.

2. Materials and Methods

2.1. Preparation of Durian Shell as an Adsorbent

The durian shells were gathered locally from the neighbourhood shops. The durian shell was rinsed two to three times by water for the removal of dirt and impurities followed by open-air drying for 48 h. Thereafter, it was oven-dried at 150 °C for 24 h. The dried durian shell was then ground and sieved to the particle size <150 μm. The ground durian shell was kept in the airtight plastic container for the adsorption studies [22]. It was noted that 1 kg of durian shell generates 500 g of the powder adsorbent with particle size <150 μm.

2.2. Characterisation of Adsorbent

2.2.1. Field Emission Scanning Electron Microscope (FESEM)

The surface texture and morphology of the durian shell was analysed by field emission scanning electron microscopy (FESEM; Ion Company, New York, NY, USA). The sample was prepared by mixing two drops of 100% ethanol with 0.1 g durian shell on clean silicon wafers and then dried at 50 °C for 2 min to eliminate the solvent, coated with gold using the sputtering technique, fractured and observed at a beam energy of 20 kV.

2.2.2. Fourier Transform Infrared (FTIR)

The functional groups in the durian shell were identified by Fourier transform infrared (FTIR) spectroscopy using Nicolet iS50 (Thermo Scientific, Waltham, MA, USA) with a scanning range of 400–4000 cm^{-1} using attenuated total reflection (ATR) technique. First, the crystal plate (sample holder) was cleaned with acetone. The durian shell sample (5 mg) was placed directly on the crystal plate and then pressed using the metallic pressing device and scanned to get the IR spectrum [27].

2.3. Preparation of C.I BB16 in Aqueous Solution

C.I BB16 was supplied by Sigma-Aldrich (M) Sdn Bhd, Malaysia, and used as received without further purification. A stock solution of 1000 mg/L was prepared by dissolving the appropriate amount (1000 mg) of C.I BB16 in a litre of deionised water. The working solution was prepared by diluting the stock solution with deionised water to give the appropriate concentration of the working solutions [28].

2.4. Experimental Set-up

The experimental set-up in this study includes the preparation of durian shell and C.I BB16 solution, independent factors and optimisation process. Factorial complete randomised design (CRD) (3 × 3 × 3) in triplicate was executed to investigate the independent factors that affect the adsorption of C.I BB16. Four independent factors that are pH (4–8), time (30–240 min), durian shell dosage (0.1–1.0 g) and initial C.I BB16 concentration (10–20 ppm) were investigated. The range for each independent factor was chosen based on previous investigations [29,30].

2.4.1. Adsorption of C.I BB16

The adsorption of C.I BB16 experiments were performed using a 250 mL Erlenmeyer conical flask that holds 100 mL of an aqueous solution of C.I BB16. To ascertain the pH condition, the pH was modified deliberately for the solution, ranging between 4 and 8. These were conducted by using 0.1 M sodium hydroxide (NaOH) and 0.1 hydrochloric acid (HCl) to acquire the desired pH value [31]. The mixture was stirred using the orbital shaker at 150 rpm and the contact time was specified between 30 min and 4 h. Treated C.I BB16 samples were filtered using the membrane filter (0.2 μm) [16]. The efficiency of the durian shell in the adsorption of C.I BB16 was investigated by determining the absorbance of C.I BB16 sample before and after each experimental run using UV-Vis spectrophotometer

(DR6000) at 661 and 620 nm. The adsorption of C.I BB16 efficiency was determined by calculating the decolourisation according to Equation (1) [32]

$$Decolorisation\ (\%) = \frac{A_0 - A}{A_0} \quad (1)$$

where A_0 is the initial absorbance of C.I BB16 in samples before the adsorption and A is the absorbance of C.I BB16 after the adsorption—moreover, the mineralisation of C.I BB16 was confirmed by COD analysis, according to Gonçalves et al., [11]. *COD* reduction was calculated using Equation (2)

$$COD\ (\%) = \frac{C_0 - C_f}{C_0} \quad (2)$$

where C_0 is the initial *COD* concentration (mg/L) and C_f is the final *COD* concentration (mg/L) after the adsorption.

2.4.2. Optimisation of Adsorption of C.I BB16

The experimental set-up of the current work consists of C.I BB16 solutions and durian shell adsorbent as well as pseudo-first and second-order kinetic model for adsorption process using response surface methodology (RSM). The factorial complete randomised design (CRD) (4 × 2) in triplicates was used to study the optimal factors affecting the adsorption process with four (4) independent factors ((pH 4–8), x_2 (time, 30–240 min), x_3 (adsorbent dosage (0.1–1 g/L), x_4 (initial concentration of C.I BB16) (10–20 mg/L) and two dependent variables y_1 (decolourisation of C.I BB16) (%), y_2 (COD removal) (%). The independent factors were selected and used for designing the experimental runs using Design Expert 6.0.10, central composite design (CCD) (Stat-Ease Inc., Minneapolis, MN, USA) software that suggested 25 experimental runs. The experimental runs (one run recommended) were carried out to authenticate the efficiency of the results of the evaluated elements. Four independent variables including pH, time, durian shell dosage and initial concentration of C.I BB16 were optimised for adsorption of C.I BB16. The optimisation experiments were also operated in three replicates to enhance the estimated data accuracy by inferring that no bias and systematic error exist. Moreover, the variables' ranges (high and low) in this study are depicted as (low, medium and high, being coded as −1, 0 and +1). The efficiency of adsorption of C.I BB16 was verified by the percentage of the removal (y) which was denoted as the responses of the parameters of the experimental design. The parameters examined for adsorption of C.I BB16 were decolourisation (%) (y_1) and COD removal (%) (y_2) and the working variables were x_1, x_2, x_3 and x_4. The quadratic model used to get the optimisation of the responses as a function of the independent process variables is as follows:

$$y = \beta 0 + \sum_{i=1}^{k} \beta_i x_i + \sum_{i=1}^{k} \beta_{ii} {x_i}^2 + \sum_{i=1}^{n} \sum_{i<j}^{n} \beta_{ij} x_i x_j \quad (3)$$

where y is the expected response for the adsorption, while $\beta 0$, β_i, β_{ii} and β_{ij} are the linear model coefficients, x_i is the coded independent factors, and x_i is the independent factors in a coded form, while k represents independent variable numbers.

The relationship between the experimental data and the coded variable is represented as:

$$x_i = \varepsilon_i \frac{[\mathrm{HL} + \mathrm{LL}]/2}{[\mathrm{HL} - \mathrm{LL}]/2} \quad (4)$$

where x_i represents the coded variable, ε_i is the experimental laboratory data, HL and LL represent the maximum and minimum values of the independent factors respectively.

2.5. Techno-Economic Analysis for Hair Dye Greywater Treatment Using Durian Shell Adsorbent

The annual time for the fill operation including preparation and application is 5760 h/year (which is equivalent to 240 operating days). The TCI for a proposed preparation and application processes includes WCC and FCE (Equation (5)) [33,34].

$$TCI = WCC + FCE \tag{5}$$

where *TCI* is the total capital investment, *WCC* is the working capital cost and *FCE* is the fixed capital estimation.

The annual operation cost (*AOC*) for the preparation of durian shell as adsorbent and then the application of the adsorbent in the hair dye greywater and the costs of raw materials (C_{RM}), waste generated from the preparation process (C_{WG}), utilities (C_U) and extra cost (C_E) were calculated on an annual basis according to Equation (6).

$$AOC = C_{RM} + C_{WG} + C_U + C_E) \tag{6}$$

According to the mathematical model designed by the specific cost of durian shell adsorbent production C_Z (kg/USD) is calculated based on the annual costs of capital C_C, utilities cost C_U, raw material cost C_{RM} and extra cost C_E divided by the annual durian shell adsorbent production (Equation (7)) [35].

$$C_Z = \frac{(C_C + C_U + C_{RM} + C_E)}{E_P} \tag{7}$$

2.6. Statistical Analysis

The significant role of the independent factors as well as the interactions between these factors was analysed using ANOVA ($p < 0.05$). The contribution of independent factors in the adsorption as a function for the quadratic model was identified by the adjusted coefficient of determination (R^2 adj). The relationship between the independent factors and their effects on the adsorption was represented in a three-dimensional graphical representation, which is RSM.

3. Results and Discussion

3.1. Characteristics of Durian Adsorbent

The surface morphology of the durian shell before and after the adsorption process is presented in Figure 2. It was noted that the adsorbent surface having a smooth surface with no pores might be due to the existence of pectin and lignin. These findings are inconsistent with those reported by Lazim et al. [22]. In contrast, the durian shell after the adsorption process have rough surface morphology with dye residues.

The FT-IR spectrum of the durian shell is shown in Figure 3. The FTIR spectra showed a broad peak at 3413.86 cm^{-1} which specifies the existence of phenol group of lignin and cellulose. The peaks at 1623.92 are assigned as the C=C aromatic bond or amide areas that are characteristics of proteins and enzymes. The spectrum exhibited at 1437.20 is classified as the C-O stretching vibrations of carboxylic acids (-COOH) and the spectrum at 1054.42 cm^{-1} corresponds to the asymmetric stretch of C-N bonds from substituted amines [36]. The small band at 616.98 cm^{-1} owing to the C-Cl stretching of alkyl halides [37]. Similarly, Tham et al. [38] observed 3334 cm^{-1} which is related to the hydroxyl (O-H) group such as alcohol and phenols. They also observed peaks at 1024 cm^{-1} and 1420 cm^{-1} respectively representing coupled C–O stretching and amides functional groups of primary amides and aliphatic amides respectively.

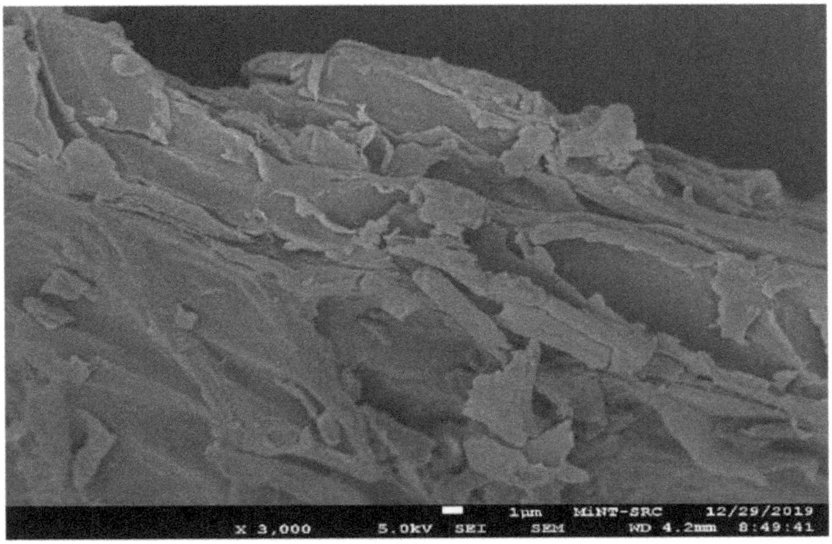

(A)

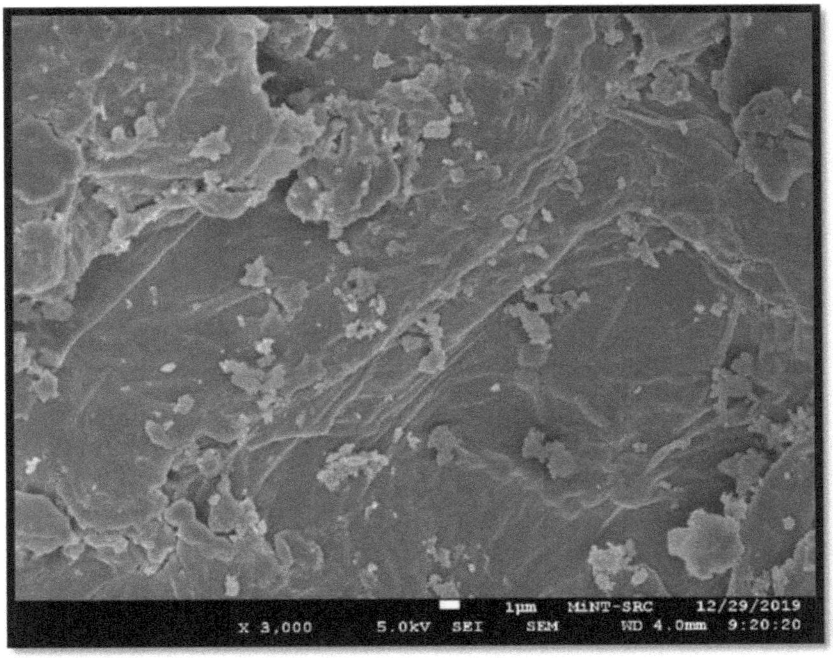

(B)

Figure 2. Surface morphology of durian shell adsorbent using FESEM; (**A**) before (**B**) after.

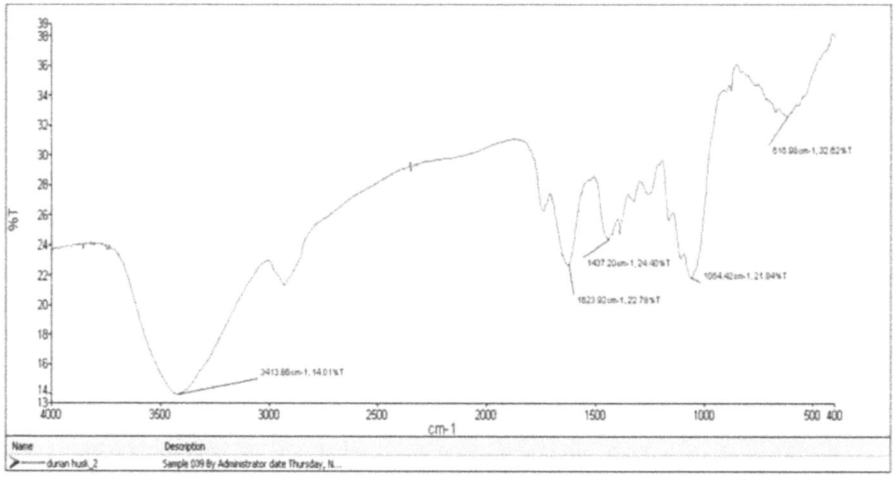

Figure 3. FTIR analysis of durian shell.

3.2. *Optimisation of Adsorption of C.I BB16*

The parameter for optimisation of the adsorption of C.I BB16 using durian shell as adsorbent was performed in the current study by Design-Expert software—the maximum decolourisation of C.I BB16 80.7% recorded at pH 4, 30 min, 0.1 g/L of the adsorbent dosage and 10 mg/L of the initial concentration of C.I BB16. In contrast, the maximum reduction of COD was 86.2% at pH 6, after 30 min, with 1 g/L of adsorbent dosage and 10 mg/L of C.I BB16 initial concentrations (Table 1).

Table 1. Central composite design arrangement and responses for adsorption of C.I BB16 in aqueous solution.

Run	x_1	x_2	x_3	x_4	y_1	y_2
1	4.00	30.00	0.10	10.00	80.65	57.52
2	8.00	30.00	0.10	10.00	77.36	63.82
3	8.00	30.00	1.00	15.00	72.81	78.09
4	8.00	30.00	0.10	20.00	72.01	55.14
5	8.00	30.00	0.10	20.00	77.24	59.79
6	4.00	30.00	1.00	20.00	82.07	62.06
7	4.00	30.00	0.10	10.00	76	78.22
8	6.00	30.00	1.00	10.00	71.04	86.18
9	6.00	30.00	0.55	15.00	72.82	79.54
10	8.00	30.00	1.00	15.00	77.61	80.6
11	4.00	135.00	0.55	15.00	72.78	30.11
12	4.00	135.00	1.00	10.00	70.26	68.33
13	8.00	135.00	1.00	20.00	72.07	53.08
14	8.00	135.00	1.00	20.00	72.54	52.02
15	6.00	135.00	0.10	15.00	40.03	18.08
16	4.00	135.00	0.10	20.00	50	10.64
17	8.00	135.00	0.55	15.00	56.28	29.53
18	6.00	135.00	1.00	15.00	57.64	58.85
19	8.00	240.00	0.10	10.00	46	28.54
20	8.00	240.00	0.10	20.00	38.82	26.23
21	8.00	240.00	1.00	10.00	51.22	56.74
22	4.00	240.00	0.10	10.00	40.61	24.88
23	4.00	240.00	1.00	20.00	69.34	48.65
24	8.00	240.00	1.00	10.00	62.86	67.94
25	6.00	240.00	0.55	15.00	52.05	32.46

x_1 (pH), x_2 (time), x_3 (adsorbent dosage (g/L), x_4 (initial concentration of C.I BB16) (mg/L), y_1 (decolourisation of C.I BB16) (%), y_2 (COD removal) (%).

In comparison Lazim et al. [22] revealed that a durian peel had removed 69.60% of bisphenol with an adsorption capacity of 4.20 mg/g for 24 h. These differences would be related to the types of pollutants treated in their investigations. Lazim et al. [22] studied the bisphenol, while in this study, C.I BB16 dye was treated. Moreover, they treated the durian peel with sulfuric acid before the adsorption bisphenol. Besides, Abd Latib et al. [39] revealed that 99% of methylene blue had been removed using durian shell activated carbon at the conditions 0.60 M H_2O_2 at 700 °C for 30 min. The findings in the present study indicated that the investigated factors play an important role in the adsorption and mineralisation rate of C.I BB16. Decolourisation and mineralisation in the adsorption process are affected by pH, time, durian shell dosage and initial C.I BB16 concentration. In this study, the effect of pH on the adsorption and mineralisation of C.I BB16 was studied at different pH values ranging from 4.0 to 8.0. It is well accepted that the pH of the solution plays a vital part in the adsorption process, which influences the surface charge of the adsorbent and the degree of ionisation and specification of the adsorbate. Generally, it can be deduced that adsorption of C.I BB16 dye by durian shell adsorbent was low at low pH. Therefore, the findings recommended that the adsorption of C.I BB16 on durian shell is favourable in alkaline solutions compared to in neutral and acidic conditions. This is because at the low pH environment, the competition between excess hydroxyl ions, $H^{+,}$ and the cationic groups on the C.I BB16 dye for adsorption sites will make the surface of the adsorbent to exist as a positively charged molecule [40,41].

The adsorption of C.I BB16 investigated the contact time of 30 min to 240 min. It can be deduced that the adsorption and mineralisation rate of C.I BB16 increases as contact time increases. However, a further increase in contact time did not enhance the C.I BB16 decolorisation process and mineralisation rate. The quick absorption rate at the beginning is attributed to the availability of plentiful active binding sites on the adsorbent surface. Nevertheless, the adsorption process develops an attachment-controlled route owing to less accessibility of adsorption sites as contact time increases. Besides, different durian shell dosages were investigated, ranging from 0.1 to 1 g. The results indicated the decolourisation mineralisation rate of C.I BB16 increased when the durian shell dosage increased. This is assigned to a rise in the adsorptive surface area and the accessibility of more active binding sites on the surface of the durian shell. However, the presence of excess durian shell decreases the adsorption capacity of C.I BB16—the decrease in total surface area available for C.I BB16 from aggregation or overlying of adsorption sites [42]. Hence, as the dosage of durian shell adsorbent increases, the amount of C.I BB16 adsorbed onto durian shell adsorbent gets reduced, thus causing a decrease in adsorption capacity. In some cases, a further increase in adsorbent dose did not remarkably alter the adsorption capacity. This is owing to the non-accessibility of active sites on the adsorbent and establishment of equilibrium between the C.I BB16 and the durian shell adsorbent [43]. The rate of adsorption and mineralisation rate is a function of the initial concentration of the adsorbate, which makes it an important factor to be considered for effective adsorption—the percentage removal of C.I BB16 decreased with increasing the initial C.I BB16 concentration (Figures 4 and 5). It seems that all adsorbents have a limited number of active sites and at a certain concentration, the active sites become saturated. The initial concentration may provide the necessary driving force to overcome the mass transfer resistance of C.I BB16 between the aqueous and the solid phase [44]—the increase in the initial C.I BB16 concentration also enhances the interaction between the C.I BB16 anion in the aqueous phase and the biomass surface; resulting in higher uptake of C.I BB16 for the given amount of biomass [29,42].

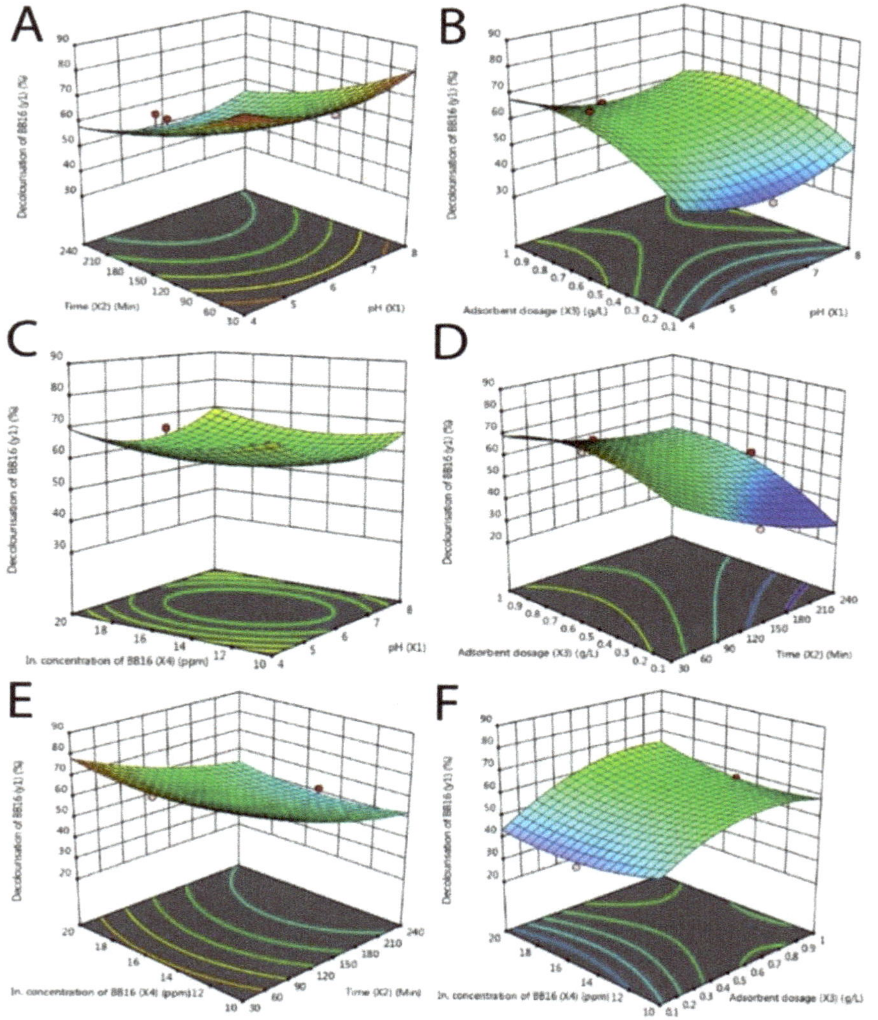

Figure 4. Three-dimensional response surface plot for adsorption of C.I BB16 in aqueous solution;. y_1: decolourisation; as a response of interaction between independent factors X_1 (pH), X_2 $(Time)$, X_3 (adsorbent dosage) and X_4 (BB16 conc.). (**A**) interaction between X1,X2; (**B**) interaction between X1,X3; (**C**) interaction between X1,X4; (**D**) interaction between X2,X3; (**E**) interaction between X2,X4; (**F**) interaction between X3,X4.

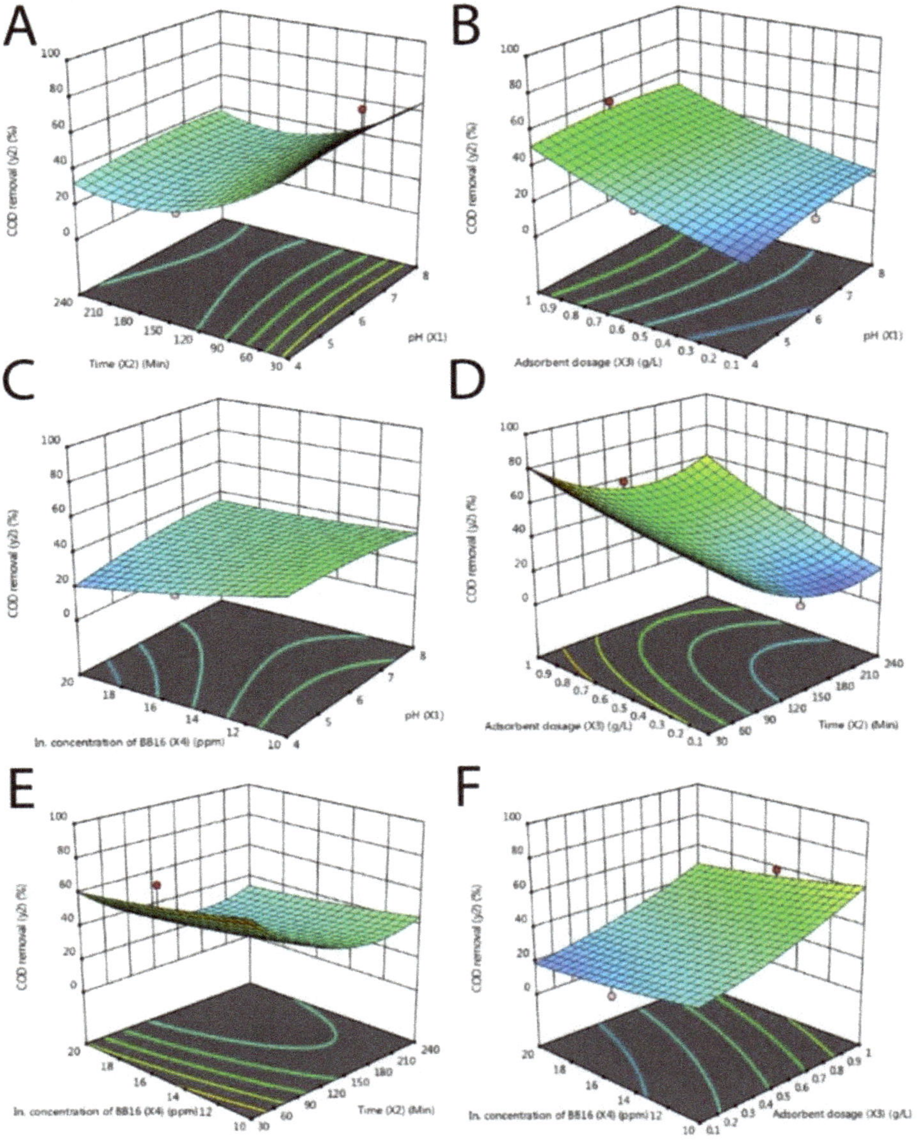

Figure 5. Three-dimensional response surface plot for adsorption of C.I BB16 in aqueous solution;. y_2: COD removal; as a response of interaction between independent factors X_1, X_2, X_3 and $X_4 X_1$ (pH), X_2 $(Time)$, X_3 (adsorbent dosage) and X_4 (BB16 conc.). (**A**) interaction between X1,X2; (**B**) interaction between X1,X3; (**C**) interaction between X1,X4; (**D**) interaction between X2,X3; (**E**) interaction between X2,X4; (**F**) interaction between X3,X4.

The regression coefficients for the independent factors are illustrated in Table 2. The interactions between these factors and the linear and quadratic regression coefficients were identified by the least square method. p-value with 95% of the confidence level was used as a sign for the importance of each independent factor. The results noted that factors x_1, x_2, x_3 and x_4 have a significant quadratic effect on the decolourisation of C.I BB16 ($p < 0.0002$), as well as for COD removal (%). Moreover, these factors have a synergistic effect on the decolourisation of C.I BB16, final pH and COD removal (%) ($p < 0.05$). The second-order model described the relationship between decolourisation and COD reduction and selected factors as given by final equations in terms of actual factors as following:

$$\begin{aligned} y_1 = {} & 57.76 - 1.09x_1 - 12.41x_2 + 7.73x_3 + 0.20x_4 + 7.55x_1^2A^2 + 3.90x_2^2 - 7.97x_3^2 \\ & +4.20x_4^2 - 0.26x_1x_2 - 1.63x_1x_3 + 1.32x_1x_4 + 4.65x_2x_3 + 0.36x_2x_4 \\ & +2.83x_3x_4 \end{aligned} \tag{8}$$

$$\begin{aligned} y_2 = {} & 35.79 + 2.130x_1 - 14.64x_2 + 15.55x_3 - 6.73x_4 - 2.66x_1^2 + 16.90x_2^2 + 3.31x_3^2 \\ & +1.41x_4^2 + 0.87x_1x_2 - 1.42x_1x_3 + 5.00x_1x_4 + 4.65x_2x_3 + 1.37x_2x_4 \\ & -0.48x_3x_4 \end{aligned} \tag{9}$$

Table 2. Regression coefficient and their significance of the quadratic model for the adsorption of C.I BB16 in aqueous solution.

Term	Coefficient		Standard Error		*F* Value		*p* Value	
	y_1	y_2	y_1	y_2	y_1	y_2	y_1	y_2
Model	57.76	35.79	3.37	5.26	11.35	11.55	0.0002	0.0002
X_1	−1.09	2.13	1.31	2.05	0.69	1.08	0.4246	0.3234
X_2	−12.41	−14.64	1.43	2.23	75.67	43.05	<0.0001	<0.0001
X_3	7.73	15.55	1.25	1.96	38.04	62.86	<0.0001	<0.0001
X_4	0.20	−6.73	1.43	2.24	0.020	9.04	0.8895	0.0132
X_1^2	7.55	−2.66	3.27	5.11	5.33	0.27	0.0436	0.6146
X_2^2	3.90	16.90	2.58	4.04	2.28	17.54	0.1619	0.0019
X_3^2	−7.97	3.31	4.08	6.38	3.83	0.27	0.0790	0.6154
X_4^2	4.20	1.41	4.04	6.32	1.08	0.050	0.3230	0.8275
X_1X_2	−0.26	0.87	1.60	2.51	0.025	0.12	0.8766	0.7353
X_1X_3	−1.63	−1.42	1.42	2.22	1.32	0.41	0.2772	0.5354
X_1X_4	1.32	5.00	1.49	2.32	0.79	4.64	0.3937	0.0566
X_2X_3	4.65	4.65	1.48	2.31	9.91	4.04	0.0104	0.0721
X_2X_4	0.36	1.37	1.58	2.48	0.053	0.31	0.8234	0.5920
X_3X_4	2.83	−0.48	1.36	2.13	4.33	0.052	0.0641	0.8250

x_1 (pH), x_2 (time), x_3 (adsorbent dosage (g/L), x_4 (initial concentration of C.I BB16), y_1 (decolourisation of C.I BB16) (%), y_2, (COD removal) (%).

The summary of the analysis of variance (ANOVA) for the quadratic model is shown in Table 3. It can be deduced that the linear model for decolourisation (%) and COD removal (%) was significant at a confidence level of ($p < 0.05$) with determination coefficients (R^2 adj) equal to R^2 (adj) are 0.8879 and 0.8603 respectively.

Table 3. Analysis of the variance (ANOVA) of the response surface quadratic model for the adsorption of C.I BB16 in aqueous solution.

Source	Degree of Freedom	Sum of Squares		Mean Square		*F* Value		*p*-Value	
		y_1	y_2	y_1	y_2	y_1	y_2	y_1	y_2
Model	14	4248.69	10578.36	303.48	755.60	11.35	11.55	0.0002 significant	0.0002 significant
Residual error	10	267.35	654.04	26.74	65.40				
Lack-of-fit	5	163.49	362.55	32.70	72.51	1.57	1.24	0.3154 non-significant	0.4083 non-significant
Pure error	5	103.86	291.49	20.77	58.30				
Total	24	4516.04	11232.40						

y_1 (decolourisation of C.I BB16) (%), y_2 (COD removal) (%); y_1 $R^2 = 0.9408$; R^2 (adj) $= 0.8579$; y $R^2 = 0.9418$; R^2 (adj) $= 0.8603$.

These results ascertain the suitability of the model. The lack of fit for the model was non-significant value ($p > 0.05$), which specifies that the model was predicted for the decolourisation (%) and COD removal (%), while was suitable for prediction within the range of the independent factors explored in the current investigation. Moreover, the small residual values are an emblem for the worthy agreement of the experimental data with the mathematical model. RSM with CCD analysis was executed to examine the interactions between the variable factors within the tested range when one factor is an actual factor at the centre point. For the decolourisation (%) and COD removal (%), a significant ($p < 0.05$) interaction between x_1 (4–8), x_2 (30–240 min), x_3 (0.1–1 g) and x_4(10–20 ppm) as actual factors is shown in Figures 4 and 5. The correlation plot between model-predicted value vs. actual value and residuals vs. run number; A; decolourisation (y_1), B; COD removal (y_2) is depicted in Figure 6. The results revealed a significant correlation between the independent factors and the removal percentage of the dependent variables.

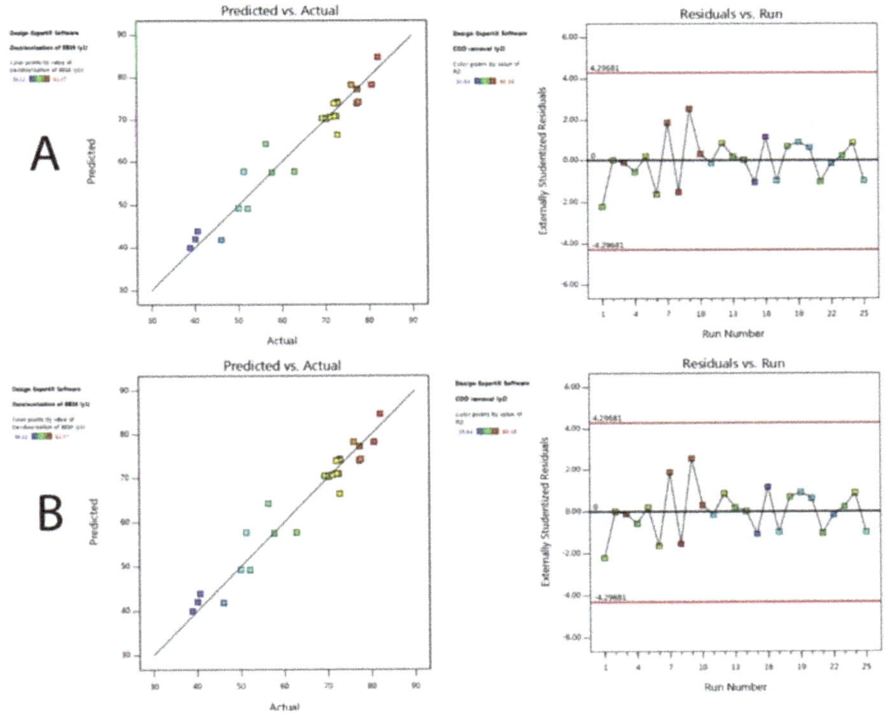

Figure 6. Correlation plot between model-predicted value vs. actual value and residuals vs. run number; (**A**) decolourisation (y_1), (**B**) COD removal (y_2).

Validation of the Optimal Parameters

The possible direction for maximising the adsorption of C.I BB16 is by using point optimisation technique with expert design software. The optimal operation parameters of the adsorption of C.I BB16 were recorded at pH 8 (x_1), 30 min (x_2), 1.0 g of adsorbent (x_3) and 15 ppm of C.I BB16 (x_4). At the optimal conditions of the adsorption of C.I BB16, the independent factors interacted significantly ($p < 0.05$) (Table 4).

Table 4. The best operating parameters for the adsorption of C.I BB16 in aqueous solution.

Responses	x_1	x_2	x_3	x_4	Predicted Results	Experimental Results
y_1	8.00	30.00	1.00	15.00	77.61	74.26
y_2					80.60	78.72

x_1 (pH), x_2 (time), x_3 (adsorbent dosage (g/L), x_4 (initial concentration of C.I BB16), y_1 (Decolourisation of C.I BB16) (%), y_2 (COD removal) (%).

This specifies that interaction effects occur as a consequence of one factor influenced by another factor. The pH, time, durian shell adsorbent dosage and initial concentration of C.I BB16 act together very well to attain the maximum decolourisation and COD removal rate. The percentage error among the actual and predicted values was recorded less than 5% (confidence level 95%) which indicates that no substantial difference was observed and therefore, affirms that the RSM method was suitable to optimise the operational settings of adsorption of C.I BB16. In comparison with the previous studies by Mohammed et al. [19], it was noted that durian shell sorbents achieved 95.91% and 97.81% decolourisation of C. I Methylene Blue and C. I Brilliant Green, respectively. The differences in the results occurred as these factors were investigated individually, whereas, in their study, the interactions between these independent factors were not considered. In the current investigations, the optimisation of the adsorption of C.I BB16 achieved more than 77.61%, which designate the role of RSM design in attaining the maximum and optimum conditions for the adsorption of C.I BB16. The high effectiveness of adsorption of C.I BB16 in reaction to the examined independent variables would be described based on the occurrence of a significant synergic relationship between the parameters. These outcomes are constant with the authors in the literature, who directed that the investigated factors play a significant part in the adsorption of C.I BB16, Mondal et al. [29]. The optimisation process using RSM regulates the best operation parameters to reach high adsorption efficiency.

3.3. Mechanism of Adsorption

To describe the nature of adsorption, the mechanism of adsorption was studied by interpretation of the intraparticle diffusion (Figure 7).

The key factors that control the rate of adsorption can be affected by the adsorption at the internal and external surface which depends on the binding process, transfer of mass through the external boundary and the adsorbate diffusion to the adsorption site across liquid pores or solid surface. Therefore, the decolourisation (77.61%) and COD removal (80.60%) at pH 8, 30 min, 1 g/L of durian shell dosage and 15 ppm of C.I BB16 were achieved based on independent factors explored in the current investigation. The competition between hydroxyl ions, H+ and the cationic groups on the C.I BB16 dye for adsorption sites will change the surface of adsorbent charged molecule positively or negatively depends on pH value, which will improve the positively charged dye cations through electrostatic forces of attraction. Durian is the member of the Bombacaceae family which is largely found in Malaysia and Indonesia of Southeast Asia. Durian peels are the waste materials of durian and consist of a lot of polysaccharides, cellulose and lignin. Those ingredients have different active functional groups that relatively affect the adsorption of analyses (dye, metal), via interactions occurred between solid surfaces and adsorbate (Ahmad and Danish, 2018); structure of the adsorbate and the functional groups of the adsorbent surfaces. Durian adsorption of C.I BB16 molecules results from hydrogen bonding between OH and NH_2 groups of dye molecules and OH groups of durian peels and electrostatic interaction between the negatively charged durian surface and cationic dye. Referred to FTIR spectra, the results showed that carboxylic acids (-COOH) and the hydroxyl groups were available on the sites because of interactions of the surface of the peel. These functional groups could control the uptake of positively charged C.I BB16 molecule. Whereas, electrostatic attraction, h-bonding could occur between the negatively charged surface and positively charged +NH-CH_3-C.I BB16 molecules [45,46].

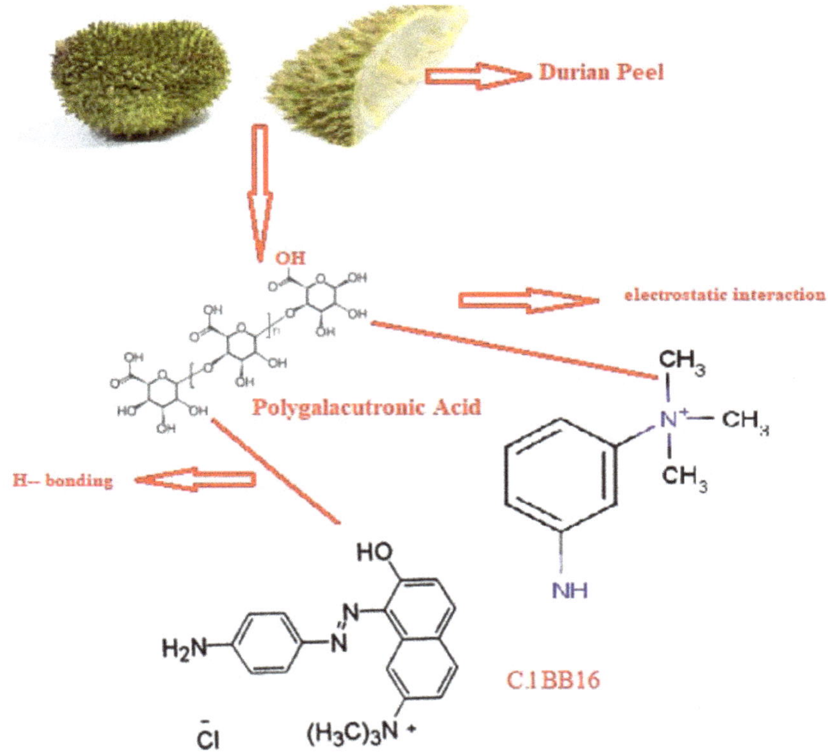

Figure 7. Removal mechanism of C.I BB16 using durian shell adsorbent.

3.4. *Techno-Economic Analysis for Preparation and Application of Durian Shell for Hair Dye Greywater Treatment*

The durian shell adsorbent preparation and application diagram was designed using SuperPro Designer, as presented in Figure 8b. The suggested adsorbent preparation and application unit is proposed to have consisted of four stages. In stage A, the raw materials of durian shell adsorbent are collected, washed and granulated as well as adsorbent dose preparation is done. In stage B, the hair dye greywater is subjected to primary treatment to remove the large suspended solids, which might affect the adsorption process negatively. The main treatment process of the hair dye greywater by the adsorption using durian shell adsorbent is supposed to be performed in Stage C. Finally, the stage proposes the post-treatment process for the treated hair dye wastewater which is safe for disposal while the durian shell might be used as the substrate for fuel or biohydrogen production because of the high contents of the carbohydrates.

The FCE represents all equipment purchase cost, process piping, equipment installation, electrical systems, instrumentation and controls, buildings, construction and yard improvements as well as the *WCC* which might be up to 6.5% of FCE as described by Han et al. [46]. Therefore, the FCE for designing a preparation and coagulation unit for wastewater treatment with 1000 m^3/day capacity could be up to USD 32,000.00 (Table 5).

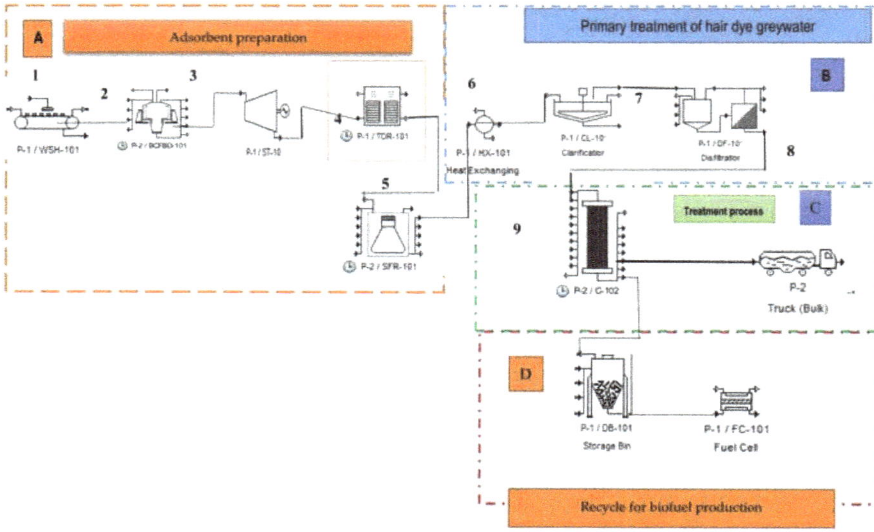

(a)

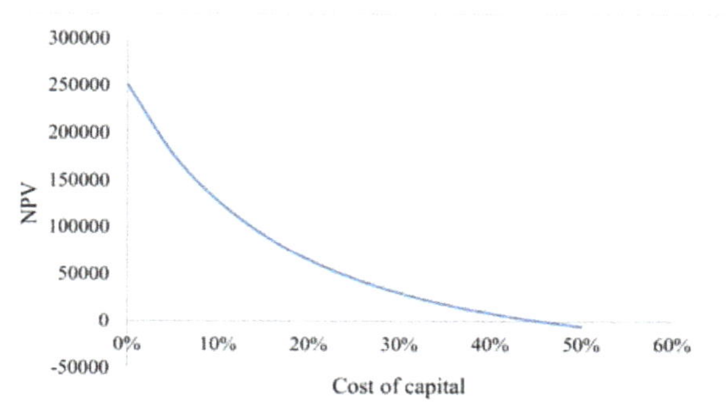

(b)

Figure 8. (**a**) The flowchart for preparation and application of durian shell adsorbent for C.I BB16 wastewater treatment. (**b**) Internal rate of return (IRR) of durian shell adsorbent production.

The additional costs for the treatment process might reach 50% of the equipment cost, which was equivalent to USD 16,000.00 [47]. The TCI of the plant might go up to USD 48,000.00 WCC, which was estimated to be 6.5% of the FCE to the tune of about USD 3120.00 Therefore, according to Equation (1), the TCI of USD 51,120.00 could be calculated. The techno-economic analysis is one of the best methods for determining the effectiveness of any alternative method for treating environmental pollution because this process provides more details for the total cost of the treatment process for specific pollution in comparison to the current technologies [47]. It has also been used to estimate the H_2 production by fungi from food wastes and provided very accurate information on the total cost [44,48].

Table 5. Fixed capital estimate (FCE) for preparation and application of durian shell adsorbent for hair dye wastewater treatment.

<table>
<tr><th>No</th><th>Item Code *</th><th>Item</th><th>Percentage of FCE</th><th>Cost</th></tr>
<tr><td>1</td><td>P-1/WSH-101</td><td>Washing tank</td><td rowspan="9">62.5%</td><td rowspan="5">10,000.00</td></tr>
<tr><td>2</td><td>P-2/BCFBD-101</td><td>Basket centrifuge</td></tr>
<tr><td>3</td><td>P-1/ST-10</td><td>Straight flow steam Turbine Generator</td></tr>
<tr><td>4</td><td>P-1/TDR-101</td><td>Tray Drying</td></tr>
<tr><td>5</td><td>P-2/SFR-101</td><td>Shake flasks</td></tr>
<tr><td>6</td><td>P-1/HX-101</td><td>Heat Exchange</td><td>1000.00</td></tr>
<tr><td>7</td><td>P-1/CL-101</td><td>Clarification</td><td>2000.00</td></tr>
<tr><td>8</td><td>P-1/DF-101</td><td>Dia-filtration</td><td>3000.00</td></tr>
<tr><td>9</td><td>P-2/C-102 PBA</td><td>Adsorption unit</td><td>4000.00</td></tr>
<tr><td></td><td colspan="3">Total Equipment purchase cost</td><td>20,000.00</td></tr>
<tr><td>10</td><td colspan="2">Equipment installation</td><td>3.13</td><td>1000.00</td></tr>
<tr><td>11</td><td colspan="2">Process piping</td><td>3.13</td><td>1000.00</td></tr>
<tr><td>12</td><td colspan="2">Instrumentation and controls</td><td>6.25</td><td>2000.00</td></tr>
<tr><td>13</td><td colspan="2">Electrical systems</td><td>3.13</td><td>1000.00</td></tr>
<tr><td>14</td><td colspan="2">Buildings</td><td>15.63</td><td>5000.00</td></tr>
<tr><td>15</td><td colspan="2">Yard improvements</td><td>3.13</td><td>1000.00</td></tr>
<tr><td>16</td><td colspan="2">Construction</td><td>3.13</td><td>1000.00</td></tr>
<tr><td></td><td colspan="3">TOTAL</td><td>32,000.00</td></tr>
</table>

* Refer the item code to Figure 8a.

3.4.1. Annual Operation Cost

The C_{RM} is the cost of durian shell. The durian shell as well as the chemicals required for preparation of the durian shell to be used as a coagulant was supplied by a local supplier. The C_U include the cost of electricity and water which are required for the preparation of durian shell adsorbent and the wastewater treatment and was estimated based on the price for each unit in the local currency. The C_{WG} represents the final solid waste generated from the coagulation treatment unit of the wastewater which is considered as a biomass yield with high concentrations of carbohydrates and could be used as bio-sorbents for wastewater treatment as well as for fuel production. Moreover, the generated wastewater from the adsorption unit could be safe for disposal.

Table 6 shows the annual operation cost for the preparation of durian shell adsorbent. The raw material cost (C_{RM}) was USD zero/year (since the durian shells are agriculture wastes and no chemical are required for the preparation). In the current study, the best dose for achieving high COS, turbidity and TSS removal are 1 g/L. The suggested design here manages with 1000 m^3 per day for 240 days per year. The experiments confirmed that 1 kg of raw durian shell generates 500 g pure coagulant. In order to deal with 240,000 m^3 per year, the process should deal with 480 tons of raw durian shell to generate 240 tons (1000 kg/day) of pure coagulant to treat the quantities required per year. The total cost of utilities is estimated to be USD 5000.00/year as estimated by Han et al. [48] in China, which has a similar cost for water and electricity as Malaysia. The costs of maintenance, insurance and labour as well as the operating labour cost, were based on the estimation that each process operation using 100 kg of raw durian shell to produce 50 kg of pure adsorbent daily required only one operator with an average salary of 6000.00 USD/year per operator (max 3 h/day). The maintenance and insurance were estimated at 2% and 1% of the FCC, respectively, according to Vlysidis et al. [47] and Ljunggren and Zacchi [49]. It can be concluded that the AOC is USD 11,960.00/year.

Table 6. Annual operation cost of the preparation and application of durian shell adsorbent for hair dye wastewater treatment.

	Component	Price	Unit	Quantity	Cost (USD)
Raw material (chemicals)	Cassava peels	Free	USD/kg	48 ton/year	0
Utilities	Electricity	0.04	USD/kWh	100,000	4000.00
	Water	0.01	USD/m^3	100,000	1000.00
Other costs	Labour	6000.00	USD/employee	1	6000.00
	Maintenance	2	% of FCE		640.00
	Insurance	1	% of FCE		320.00
		Total			**11,960.00**

3.4.2. Annual Profitability and Revenue

The revenue for the preparation and application of durian shell adsorbent is generated from selling the adsorbent for the treatment of different wastewater. The current price of chemical adsorbent might reach USD 261.81.00 per tons of Honeydew peels-activated carbon (HDP-AC) and the commercial market of the activated carbon price is USD 1000.00/tons [50]. The machines in Table 5, have 100 kg of the capacity for each operation which takes place within a day. Therefore, the total capacity per year is to produce 240 tons of pure durian shell adsorbent. The total net of the raw durian shell in the current work was 500 g per 1000 g of the raw materials. Therefore, the revenue for each production process is 1000 kg of the pure durian shell adsorbent which gives USD 62834.00 of the revenue. The annual revenue of USD 62,834.00 could be estimated. Considering the 15% local tax in Malaysia, this is estimated to be USD 9425.00 and the annual operation cost is USD 11,960.00, the annual profitability (after the tax) of the durian shell adsorbent is USD 41,499.24 (Table 7a). Based on Equation (9), the specific cost of pure durian shell adsorbent production is estimated to be USD 172.71 per ton which is lower than the market price of honeydew peels-activated carbon (HDP-AC) (USD 261.81) and the commercial market of activated carbon price which is USD 1000.00/tons.

Table 7. (**a**) Annual profitability of the preparation and application of durian shell adsorbent for hair dye wastewater treatment. (**b**) Cost factors associated with the adsorption of C.I BB16 from the hair dye wastewater by durian shell adsorbent for hair dye wastewater in comparison to the physical and chemical treatment process.

(a)

	Price	Quantity	Value
Durian shell	USD 261.81/Ton	240 Ton	62,834.00
Annual revenue			62,834.00
Annual operation cost			11,960.00
Local tax			9425.00
Annual profitability			41,499.24

(b)

Cost Factor	Physical	Chemical	Durian Shell Adsorbent
Heat and energy requirement	2	1	0
Chemicals (acid/alkali/ammonia)	0	2	0
pH neutralisation	0	2	0
Detoxification/conditioning	1	2	0
Special reactor construction	2	2	0
Total	5	9	0

In this study, the pure durian shell adsorbent production unit was economically evaluated according to the internal rate of return (IRR), net present value (NPV) as well as the payback period (PBP) of 10 years of the lifetime. The IRR (the efficiency of the investment) was 45.62% in the current study (Figure 8b). Based on the analysis given above, the durian shell exhibited high efficiency for adsorption of hair dye wastewater. Therefore, the techno-economic analysis as described in the current work is economically feasible by 45.62% of the IRR. It might provide alternative methods for chemical and physical coagulation for wastewater treatment (Table 7b).

4. Conclusions

It can be concluded that an adsorbent was successfully produced via the durian shell and successfully showed high efficiency for the adsorption of C.I BB16—the optimal operating factors for adsorption C.I BB16 recorded at pH 8, time (30 min), durian shell dosage (1 g/L), and 15 mg /L of C.I BB16 concentrations where 77.61 vs. 74.26 (%) of C.I BB16 removal and 80.60 vs. 78.72 (%) of COD removal was the observed and predicted results recorded with an R^2 coefficient of 0.94, respectively. The process is very economically feasible with 45.62% of the internal rate of return (IRR). These findings indicate that the durian shell adsorbent has great potential to be applied as a low cost and eco-friendly adsorbent in the adsorption process of dyes in aqueous solution. The utilisation of durian shells as an adsorbent also helped in reducing solid waste in agriculture.

Author Contributions: Conceptualization, A.A.-G.; methodology, Y.G. and M.M.A.; software, M.A.-S.; validation, A.A.-G., M.A.M. and R.M.S.R.M.; formal analysis, A.A.-G.; investigation, S.A.; resources, A.A.-G.; data curation, Y.G.; writing—E.N., A.A.-G.; writing—review and editing, M.A.M.; visualization, A.A.-G.; supervision, R.M.S.R.M.; project administration, M.A.M. All authors have read and agreed to the published version of the manuscript.

Funding: This research received no external funding.

Acknowledgments: This research was funded by the technical & financial support from UNITEN RMC Internal Research Grant: RJO 10517919/iRMC/Publication and Ministry of Education Malaysia (KPM) through the Fundamental Research Grant Scheme (FRGS) with reference code: FRGS/1/2019/WAB05/UTHM/02/5 (Modification of Bead Adsorbents with Ceramic Sanitary Ware Waste (CSWW) and Chitosan for Laundry Greywater (LGW) Safe Disposal) as financial support for this research project.

Conflicts of Interest: The authors declare no conflict of interest.

References

1. Fernandes, N.C.; Brito, L.B.; Costa, G.; Taveira, S.F.; Cunha-Filho, M.; Oliveira, G.A.R.; Marreto, R. Removal of azo dye using Fenton and Fenton-like processes: Evaluation of process factors by Box–Behnken design and ecotoxicity tests. *Chem. Interact.* **2018**, *291*, 47–54. [CrossRef] [PubMed]
2. Joseph, J.; Radhakrishnan, R.C.; Johnson, J.K.; Joy, S.P.; Thomas, J. Ion-exchange mediated removal of cationic dye-stuffs from water using ammonium phosphomolybdate. *Mater. Chem. Phys.* **2020**, *242*, 122488. [CrossRef]
3. Rosu, C.M.; Vochita, G.; Mihăşan, M.; Avădanei, M.; Mihai, C.; Gherghel, D. Performances of Pichia kudriavzevii in decolorization, biodegradation, and detoxification of C.I. Basic Blue 41 under optimized cultural conditions. *Environ. Sci. Pollut. Res.* **2018**, *26*, 431–445. [CrossRef]
4. Aguilar, Z.G.; Brillas, E.; Salazar, M.; Nava, J.L.; Sirés, I.; Sadornil, I.S. Evidence of Fenton-like reaction with active chlorine during the electrocatalytic oxidation of Acid Yellow 36 azo dye with Ir-Sn-Sb oxide anode in the presence of iron ion. *Appl. Catal. B Environ.* **2017**, *206*, 44–52. [CrossRef]
5. Obiora-Okafo, I.A.; Onukwuli, O.D. Optimization of Coagulation-Flocculation for Colour Removal from Azo Dye Using Natural Polyemers: Response Surface Methodological Approach. *Niger. J. Technol.* **2017**, *36*, 482–495. [CrossRef]
6. Isarain-Chávez, E.; Baro, M.D.; Rossinyol, E.; Morales-Ortiz, U.; Sort, J.; Brillas, E.; Pellicer, E. Comparative electrochemical oxidation of methyl orange azo dye using Ti/Ir-Pb, Ti/Ir-Sn, Ti/Ru-Pb, Ti/Pt-Pd and Ti/RuO2 anodes. *Electrochim. Acta* **2017**, *244*, 199–208. [CrossRef]

7. Castro, F.D.; Bassin, J.P.; Dezotti, M. Treatment of a simulated textile wastewater containing the Reactive Orange 16 azo dye by a combination of ozonation and moving-bed biofilm reactor: Evaluating the performance, toxicity, and oxidation by-products. *Environ. Sci. Pollut. Res.* **2016**, *24*, 6307–6316. [CrossRef]
8. Bahia, M.; Passos, F.; Adarme, O.F.H.; Aquino, S.F.; Silva, S.Q. Anaerobic-Aerobic Combined System for the Biological Treatment of Azo Dye Solution Using Residual Yeast. *Water Environ. Res.* **2018**, *90*, 729–737. [CrossRef]
9. Zong, E.; Liu, X.; Jiang, J.; Fu, S.; Chu, F. Preparation and characterization of zirconia-loaded lignocellulosic butanol residue as a biosorbent for phosphate removal from aqueous solution. *Appl. Surf. Sci.* **2016**, *387*, 419–430. [CrossRef]
10. Pathak, P.; Gupta, D.K. Strontium contamination in the environment. In *The Handbook of Environmental Chemistry*; Springer International Publishing: Cham, Switzerland, 2020. [CrossRef]
11. Gonçalves, J.A.C.; Schwantes, D.; Campagnolo, M.A.; Dragunski, D.C.; Tarley, C.R.T.; Silva, A.K.D.S. Removal of toxic metals using endocarp of açaí berry as biosorbent. *Water Sci. Technol.* **2018**, *77*, 1547–1557. [CrossRef]
12. Chan, S.-L.; Tan, Y.P.; Abdullah, A.H.; Ong, S.-T. Equilibrium, kinetic and thermodynamic studies of a new potential biosorbent for the removal of Basic Blue 3 and Congo Red dyes: Pineapple (Ananas comosus) plant stem. *J. Taiwan Inst. Chem. Eng.* **2016**, *61*, 306–315. [CrossRef]
13. Jain, S.N.; Gogate, P.R. Acid Blue 113 removal from aqueous solution using novel biosorbent based on NaOH treated and surfactant modified fallen leaves of Prunus Dulcis. *J. Environ. Chem. Eng.* **2017**, *5*, 3384–3394. [CrossRef]
14. Heraldy, E.; Lestari, W.W.; Permatasari, D.; Arimurti, D.D. Biosorbent from tomato waste and apple juice residue for lead removal. *J. Environ. Chem. Eng.* **2018**, *6*, 1201–1208. [CrossRef]
15. Ma, L.; Jiang, C.; Lin, Z.-Y.; Zou, Z. Microwave-Hydrothermal Treated Grape Peel as an Efficient Biosorbent for Methylene Blue Removal. *Int. J. Environ. Res. Public Heal.* **2018**, *15*, 239. [CrossRef]
16. Machrouhi, A.; Farnane, M.; Elhalil, A.; Abdennouri, M.; Tounsadi, H.; Barka, N. Heavy metals adsorption by Thapsia transtagana stems powder: Kinetics, equilibrium and thermodynamics. *Moroc. J. Chem.* **2019**, *7*, 98–110.
17. Yildirim, A.; Acay, H.; Baran, F. Synthesis and characterisation of mushroom-based nanocomposite and its efficiency on dye biosorption via antimicrobial activity. *Int. J. Environ. Anal. Chem.* **2020**, 1–18. [CrossRef]
18. Saueprasearsit, P. Adsorption of chromium (Cr+6) using durian peel. *Int. Conf. Biotechnol. Environ. Manag.* **2011**, *18*, 33–38.
19. Mohammed, S.A.; Najib, N.W.A.Z.; Muniandi, V. Durian rind as a low cost adsorbent. *Int. J. Civ. Environ. Eng.* **2012**, *12*, 51–56.
20. Foo, K.Y.; Hameed, B. Transformation of durian biomass into a highly valuable end commodity: Trends and opportunities. *Biomass Bioenergy* **2011**, *35*, 2470–2478. [CrossRef]
21. Koay, S.C.; Subramanian, V.; Chan, M.Y.; Pang, M.M.; Tshai, K.Y.; Cheah, K.H. Preparation and Characterization of Wood Plastic Composite Made Up of Durian Husk Fiber and Recycled Polystyrene Foam. *MATEC Web Conf.* **2018**, *152*, 02019. [CrossRef]
22. Lazim, Z.M.; Hadibarata, T.; Othman, M.H.D.; Yusop, Z.; Wirasnita, R.; Nor, N.M. Utilization of durian peel as potential adsorbent for bisphenol a removal in aquoeus solution. *J. Teknol.* **2015**, *74*, 109–115. [CrossRef]
23. Méndez, A.A.; Pena, L.B.; Curto, L.M.; Sciorra, M.D.; Ulloa, R.M.; Aguilar, S.M.G.; Ramos, J.M.V.; Gallego, S.M. Optimization of recombinant maize CDKA;1 and CycD6;1 production in Escherichia coli by response surface methodology. *Protein Expr. Purif.* **2019**, *165*, 105483. [CrossRef] [PubMed]
24. Ciric, A.R.; Krajnc, B.; Heath, D.; Ogrinc, N. Response surface methodology and artificial neural network approach for the optimization of ultrasound-assisted extraction of polyphenols from garlic. *Food Chem. Toxicol.* **2020**, *135*, 110976. [CrossRef]
25. Ahmad, M.A.; Hamid, S.R.A.; Yusop, M.F.M.; Aziz, H.B.A. Optimization of microwave-assisted durian seed based activated carbon preparation conditions for methylene blue dye removal. In Proceedings of the International Conference of Global Network for Innovative Technology and Awam International Conference in Civil Engineering (Ignite-Aicce&Rsquo;17): Malaysia Sustainable Technology and Practice for Infrastructure and Community Resilience, Penang, Malaysia, 8–10 August 2017; Volume 1892, p. 040019.
26. Reshadi, M.A.M.; Bazargan, A.; McKay, G. A review of the application of adsorbents for landfill leachate treatment: Focus on magnetic adsorption. *Sci. Total. Environ.* **2020**, *731*, 138863. [CrossRef] [PubMed]

27. Fahimmunisha, B.A.; Ishwarya, R.; AlSalhi, M.; Devanesan, S.; Govindarajan, M.; Vaseeharan, B. Green fabrication, characterization and antibacterial potential of zinc oxide nanoparticles using Aloe socotrina leaf extract: A novel drug delivery approach. *J. Drug Deliv. Sci. Technol.* **2020**, *55*, 101465. [CrossRef]
28. Barrera, H.; Cruz-Olivares, J.; Frontana-Uribe, B.A.; Gómez-Díaz, A.; Reyes-Romero, P.G.; Barrera-Diaz, C.E. Electro-Oxidation–Plasma Treatment for Azo Dye Carmoisine (Acid Red 14) in an Aqueous Solution. *Materials* **2020**, *13*, 1463. [CrossRef]
29. Mondal, N.K.; Samanta, A.; Roy, P.; Das, B. Optimization study of adsorption parameters for removal of Cr(VI) using Magnolia leaf biomass by response surface methodology. *Sustain. Water Resour. Manag.* **2019**, *5*, 1627–1639. [CrossRef]
30. Sanati, A.M.; Kamari, S.; Ghorbani, F. Application of response surface methodology for optimization of cadmium adsorption from aqueous solutions by Fe3O4@SiO2@APTMS core–shell magnetic nanohybrid. *Surf. Interfaces* **2019**, *17*, 100374. [CrossRef]
31. Petrović, M.S.; Šoštarić, T.; Stojanović, M.D.; Milojković, J.; Mihajlović, M.; Stanojević, M.; Stanković, S. Removal of Pb^{2+} ions by raw corn silk (*Zea mays* L.) as a novel biosorbent. *J. Taiwan Inst. Chem. Eng.* **2016**, *58*, 407–416. [CrossRef]
32. Ben Amar, M.; Walha, K.; Salvadó, V. Evaluation of Olive Stones for Cd(II), Cu(II), Pb(II) and Cr(VI) Biosorption from Aqueous Solution: Equilibrium and Kinetics. *Int. J. Environ. Res.* **2020**, *14*, 193–204. [CrossRef]
33. Kwan, T.H.; Pleissner, D.; Lau, K.Y.; Venus, J.; Pommeret, A.; Lin, C.S.K. Techno-economic analysis of a food waste valorization process via microalgae cultivation and co-production of plasticizer, lactic acid and animal feed from algal biomass and food waste. *Bioresour. Technol.* **2015**, *198*, 292–299. [CrossRef] [PubMed]
34. Han, W.; Fang, J.; Liu, Z.; Tang, J. Techno-economic evaluation of a combined bioprocess for fermentative hydrogen production from food waste. *Bioresour. Technol.* **2016**, *202*, 107–112. [CrossRef] [PubMed]
35. Han, W.; Hu, Y.; Li, S.; Huang, J.; Nie, Q.; Zhao, H.; Tang, J. Simultaneous dark fermentative hydrogen and ethanol production from waste bread in a mixed packed tank reactor. *J. Clean. Prod.* **2017**, *141*, 608–611. [CrossRef]
36. Bulgariu, L.; Bulgariu, D. Functionalized soy waste biomass—A novel environmental-friendly biosorbent for the removal of heavy metals from aqueous solution. *J. Clean. Prod.* **2018**, *197*, 875–885. [CrossRef]
37. Sharma, D.; Sabela, M.; Kanchi, S.; Bisetty, K.; Skelton, A.A.; Honarparvar, B. Green synthesis, characterization and electrochemical sensing of silymarin by ZnO nanoparticles: Experimental and DFT studies. *J. Electroanal. Chem.* **2018**, *808*, 160–172. [CrossRef]
38. Tham, Y.; Latif, P.A.; Abdullah, A.M.; Shamala-Devi, A.; Taufiq-Yap, Y. Performances of toluene removal by activated carbon derived from durian shell. *Bioresour. Technol.* **2011**, *102*, 724–728. [CrossRef]
39. Latib, E.H.A.; Mustfha, M.S.; Sufian, S.; Shaari, K.Z.K. Methylene Blue Dye Adsorption to Durian Shell Activated Carbon. *Key Eng. Mater.* **2013**, *594–595*, 350–355. [CrossRef]
40. Naushad, M.; Alqadami, A.A.; Alothman, Z.A.; Alsohaimi, I.H.; Algamdi, M.S.; Aldawsari, A.M. Adsorption kinetics, isotherm and reusability studies for the removal of cationic dye from aqueous medium using arginine modified activated carbon. *J. Mol. Liq.* **2019**, *293*, 111442. [CrossRef]
41. Albadarin, A.B.; Collins, M.N.; Naushad, M.; Shirazian, S.; Walker, G.; Mangwandi, C. Activated lignin-chitosan extruded blends for efficient adsorption of methylene blue. *Chem. Eng. J.* **2017**, *307*, 264–272. [CrossRef]
42. Bahrami, M.; Amiri, M.; Bagheri, F. Optimization of the lead removal from aqueous solution using two starch based adsorbents: Design of experiments using response surface methodology (RSM). *J. Environ. Chem. Eng.* **2019**, *7*, 102793. [CrossRef]
43. Sun, Y.; Yang, Y.; Yang, M.; Yu, F.; Ma, J. Response surface methodological evaluation and optimization for adsorption removal of ciprofloxacin onto graphene hydrogel. *J. Mol. Liq.* **2019**, *284*, 124–130. [CrossRef]
44. Abukhadra, M.R.; Adlii, A.; Bakry, B.M. Green fabrication of bentonite/chitosan@cobalt oxide composite (BE/CH@Co) of enhanced adsorption and advanced oxidation removal of Congo red dye and Cr (VI) from water. *Int. J. Boil. Macromol.* **2019**, *126*, 402–413. [CrossRef] [PubMed]
45. Lou, T.; Yan, X.; Wang, X. Chitosan coated polyacrylonitrile nanofibrous mat for dye adsorption. *Int. J. Boil. Macromol.* **2019**, *135*, 919–925. [CrossRef] [PubMed]
46. Ahmad, R.; Kumar, R. Adsorption studies of hazardous malachite green onto treated ginger waste. *J. Environ. Manag.* **2010**, *91*, 1032–1038. [CrossRef] [PubMed]

47. Vlysidis, A.; Binns, M.; Webb, C.; Theodoropoulos, C. A techno-economic analysis of biodiesel biorefineries: Assessment of integrated designs for the co-production of fuels and chemicals. *Energy* **2011**, *36*, 4671–4683. [CrossRef]
48. Han, W.; Yan, Y.; Gu, J.; Shi, Y.; Tang, J.; Li, Y. Techno-economic analysis of a novel bioprocess combining solid state fermentation and dark fermentation for H2 production from food waste. *Int. J. Hydrogen Energy* **2016**, *41*, 22619–22625. [CrossRef]
49. Ljunggren, M.; Zacchi, G. Techno-economic evaluation of a two-step biological process for hydrogen production. *Biotechnol. Prog.* **2009**, *26*, 496–504. [CrossRef]
50. Yunus, Z.M.; Al-Gheethi, A.; Othman, N.; Hamdan, R.; Ruslan, N.N. Removal of heavy metals from mining effluents in tile and electroplating industries using honeydew peel activated carbon: A microstructure and techno-economic analysis. *J. Clean. Prod.* **2020**, *251*, 119738. [CrossRef]

Publisher's Note: MDPI stays neutral with regard to jurisdictional claims in published maps and institutional affiliations.

sustainability

MDPI

Article

Diversity of Carbon Storage Economics in Fertile Boreal Spruce (*Picea Abies*) Estates

Petri P. Kärenlampi

Lehtoi Research, 81235 Lehtoi, Finland; petri.karenlampi@professori.fi

Abstract: A "normal forest", an idealized estate with a uniform distribution of stand ages, can be used in the study of sustainable management practices. As the normal forest contains a variety of stand ages, the characteristics of the stands can be represented in terms of a "normal stand", with properties known as a function of age. This paper takes seven never-thinned stands as seven "normal stands", which describe seven estates of normal forest. The intention is to study the robustness of carbon storage microeconomics to varying estate characteristics. It was found that the economically optimal rotation ages vary. The state sums of volume and capitalization, corresponding to any optimal rotation, also vary significantly. Growth rates vary more than the optimal expected stand volumes. Consequently, any excess volume related to carbon storage adds on to an almost unified basic volume. For all seven normal estates, the most economical way of increasing carbon storage is to increase the size of trees retained in thinning from above.

Keywords: capitalization; capital return rate deficiency; expected value; carbon storage; estate fertility; carbon rent

check for updates

Citation: Kärenlampi, P.P. Diversity of Carbon Storage Economics in Fertile Boreal Spruce (*Picea Abies*) Estates. *Sustainability* **2021**, *13*, 560. https://doi.org/10.3390/su13020560

Received: 15 December 2020
Accepted: 6 January 2021
Published: 8 January 2021

Publisher's Note: MDPI stays neutral with regard to jurisdictional claims in published maps and institutional affiliations.

1. Introduction

Boreal forests constitute a significant carbon sink. A particular benefit of boreal regions is carbon storage in the soil. The amount of soil carbon may exceed the carbon storage in living biomass [1–4]. However, living biomass produces the litter that results in soil carbon accumulation, and consequently the rate of carbon storage depends on the rate of biomass production on the site. The biomass production rate is related to the amount of living biomass [2,5–7].

In the occurrence of clearcutting and soil preparation, the net release of carbon from the soil to the atmosphere begins [8–11]. However, recent unpublished datasets indicate that maintenance of canopy cover may or may not be essential in carbon sequestration.

Human activity related to the sequestration and release of carbon into the atmosphere forms a complex system of industries and disciplines [12–16]. Instead of trying to propose any interdisciplinary optimum, we here discuss microeconomics of carbon sequestration in forestry, focusing on the estate level.

In principle, it would be straightforward to apply a carbon trade policy in forestry, possibly administered by public agents [17]. An unbiased carbon trade, however, would incorporate a high initial expense [18]. However, it has been recently shown that a carbon rental system is equivalent to the carbon trade system, without much of an initial expense [18].

From the viewpoint of generic instructions or policy actions, it might be beneficial to reduce the variety of initial estate states by adopting some kind of unifying boundary conditions. A tempting candidate is the normal forest principle [19]. This principle simply refers to the postulation that stand ages are evenly distributed, and stand characteristics are uniquely determined by stand age. Such a postulation, even if often departing from reality, simplifies many treatments, producing idealized systems that are stationary in time. Application of the normal forest principle allows for the determination of an optimal rotation age, as well as expected values of estate characteristics corresponding to the financial optimum.

Microeconomically optimal rotation age and the expected value of stand volume are not necessarily optimal from the viewpoint of national economics or the viewpoint of carbon sequestration. It has been recently shown that microeconomics often favors solutions with relatively low capitalization [20–23]. Low capitalization reduces volumetric growth rate, as well as litter accumulation rate [21,23,24].

A rather recent paper clarified the microeconomics of carbon sequestration within a boreal spruce forest estate [23]. It found that the best capital return rate is achieved by multiple repeated thinning from above to the transition diameter where sawlogs are gained instead of pulpwood only [21–23]. Increased carbon storage was found to be most economically achieved by increasing the harvesting limit diameter [23]. However, carbon rent derivable from present carbon market prices appeared not high enough to compensate for the induced capital return deficiency [23].

A significant revision to the recent approach [23] has been presented still more recently [25]. Literature values of the yield of sawlogs and pulpwood [26,27] were replaced by empirical observations [25]. The resulting value increment of a typical spruce tree as a function of breast-height diameter is a much smoother function than the literature values would indicate [25]. Some small trees have yielded sawlogs. On the other hand, the expected sawlog content of larger trees is smaller than the literature values would indicate, as operators are trained not to produce sawlogs that may be rejected at the mill [25].

The consequences of smoother value increments are significant [25]. Due to the price premium of sawlogs from clearcutting, all feasible silvicultural treatments on fertile spruce estates terminate in clearcutting instead of multiple repeated thinnings from above [21–23,25]. Financially optimal expected values of timber stocks per hectare are greater when the empirical log yield function is used, in comparison to the literature values of sawlog yield [21,23,25].

The recent treatments have been based on field observations, a growth model, and financial theory. The growth model, as well as the financial treatment, apply probability theory. However, the outcome of the growth model is deterministic, since it is composed in terms of expected values only [21,22,28]. The financial treatment, even if probabilistic as such, applies the normal forest principle, thus applying the postulate of constant stand age distribution within an estate [21–23,25,29,30]. Consequently, any density function of stand age becomes determined by one parameter (rotation age), and the financial treatment becomes deterministic.

The field observations naturally contain statistical scatter. However, recent papers extended the normal forest principle in a somewhat novel manner [23,25]. Instead of discussing the variety of stand properties in detail, a representative "normal stand" was composed as a combination of two stands of average fertility and younger-than-average age. Application of this normal stand procedure has effectively neglected statistical scatter in field observations and rendered the entire treatment deterministic [23,25].

This paper intends to clarify the effects of statistical scatter in field observations. It is of interest whether the results gained when such a scatter is neglected are robust in the presence of the scatter. However, we do not abandon the normal stand - principle. Instead, we applied that principle separately for any of the seven experimental stands where field data has been collected. Each of the seven unthinned stands is used as a normal stand characterizing an estate. The seven normal forest estates is then discussed in order to clarify the robustness of recent results to statistical scatter in field observations. It is worth noting that discussion of any of the seven constituted estates is deterministic—it is of interest whether some of them significantly differ from results gained by neglecting the scatter in field observations.

First, experimental materials are presented, as well as related computational methods. Then, microeconomic methods, as well as carbon rent formulae are introduced. The methods largely unify with a very recent paper [25], apart from the consideration of the scatter in the characteristics of the normal stands. Third, results are reported for four different kinds of treatment schedules. Finally, the outcome is discussed, and conclusions regarding the robustness of carbon sequestration systems to estate characteristics are given.

2. Materials and Methods

2.1. Empirical Observations

Eleven circular plots of 314 square meters in area were taken in November 2018 from typical spots of 11 spruce-dominated forest stands in Vihtari, Eastern Finland. The area of the Vihtari estate is 30 ha, with an elevation from 115 to 125 m above sea level. Seven of the stands had experienced only young stand cleaning, whereas four of the stands were previously thinned commercially. Breast-height diameters were recorded as well as tree species, and a quality class was visually determined for any measured tree, reflecting its suitability for adding value in the future.

On measured plots on sites without any previous commercial thinning, the basal area of tree trunks at breast-height varied from 32 to 48 m^2/ha, and the stem count from 1655 to 2451 per hectare. On measured plots on stands previously thinned commercially, the basal area varied from 29 to 49 m^2/ha, and the stem count from 891 to 955 per hectare. Further details of the experimental stands are reported in an earlier paper [21].

The Vihtari estate was characterized by measurements from the 11 sample plots. However, such a sample did not necessarily represent the entire estate accurately. More importantly, the sampling did not conform to the normal forest principle, with assumptions of constant age distribution and stand characteristics uniquely determined by stand age. Here, we utilize the normal forest principle in terms of establishing a "normal stand" based on the observations and then approximating the development of this "normal stand" as a function of stand age. Introducing an even stand age distribution produces a "normal estate" on the basis of the normal stand, following the normal forest principle [20].

In recent studies [23,25], the normal stand has been established based on two never-thinned sample plots of medium site fertility and a younger range of stand age. The intention was to establish one normal stand representing the typical characteristics of the 11 sample plots measured at the Vihtari estate. The selection of such a representative stand simplified the treatments and rendered the entire analysis deterministic. In other words, it neglected the variation between the 11 sample plots.

The intention of this paper being to investigate the variability among the sample plots, any of the seven never-thinned sample plots is taken as a normal stand. Commercially-thinned stands are neglected since the thinning revenue i not known. Any of the normal stands iss taken to represent a normal estate, representing a "normal forest" [20]. The robustness of recent results to the scatter among sample plots is then investigated in terms of the seven normal estates.

Instead of using literature values [21–23,26] as an estimate of the yield of sawlogs and pulpwood from any tree of particular breast-height diameter, a recently collected dataset of 6123 spruce trees is used [25]. The data was collected by four different single-grip harvesters operated by six individuals at four harvesting sites. Log cutting instructions provided by three different sawmilling companies were applied. One thinning site was located at Vihtari, and two thinning sites and one clearcutting site at Ilomantsi, all in eastern Finland. The area of the sampling territory was 100 km (west to east) by 30 km (south to north). Elevation varied from 115 to 205 m above sea level.

The proportion of sawlogs of the total harvester-measured volume, as a function of breast-height diameter, is shown in Figure 1 [25]. Figure 1 also shows the literature values used in [21–23,26]. It is found that the empirical sawlog yield function is smoother than the function taken from the literature. Some sawlogs were gained from trees of a diameter of 175 mm, unlike in the literature data. On the other hand, the sawlog proportion of trees thicker than 200 mm is smaller in the empirical function. The latter probably is due to harvester operators being trained not to produce sawlogs that would be rejected at the mill.

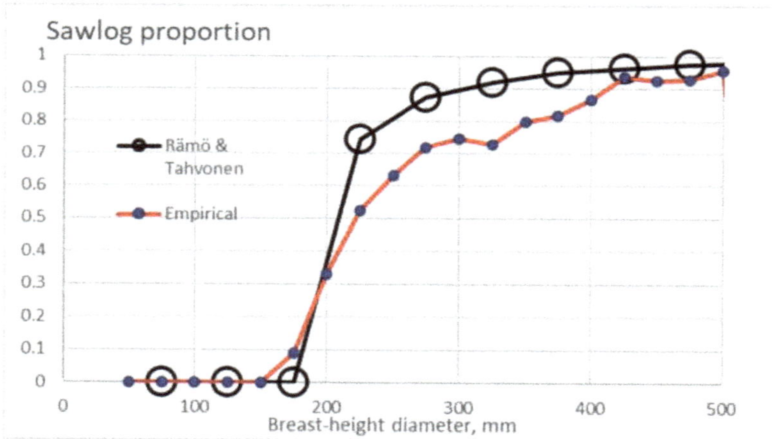

Figure 1. Literature values as well as empirical values of the sawlog proportion within the commercial section of spruce stems of different sizes.

2.2. Growth Model

For prognostication of further development of the normal stand, some kind of growth model is needed. The growth model of Bollandsås et al. [21,22,28] was adopted, discussing not only growth but also mortality and recruitment. The original growth model [21,22,28] discussed 50 mm breast-height diameter classes within a temporal resolution of five years. Any diameter class was represented by its central tree, and the process of growth was described in terms of the probability of any tree to transfer to the next diameter class [21,22,28]. The underlying idea of the description of growth is that any tree either remains in the same diameter class or transfers to the next diameter class within the five-year time interval. This underlying idea naturally greatly simplifies computation.

It's found from Figure 1 that according to the literature values of sawlog yield, there is a sharp transition between diameter classes centered at 175 and 225 mm. Such a sharp transition would induce a huge value increment. With the empirical yield function being smoother, there is a need for a greater resolution in the tree size description. It is not too complicated to modify the growth model from the size resolution of 50 mm to 25 mm [25]. However, to retain the underlying principle of the growth model, this requires a corresponding change in the temporal resolution, from 5-year time steps to 30 months.

The simultaneous change in the time and size resolutions retains the probability of any tree to transfer to the next diameter class. The same is not true in the case of recruitment and mortality. The latter two quantities are affected by time resolution only. Correspondingly, the recruitment and mortality values become scaled along with the time step by a factor of 1/2.

For any 25-mm diameter class of trees, the volumetric amount of two assortments, pulpwood and sawlogs are taken from the empirical observations as the expected values for any diameter class.

Description of stand development until the time of observation in 2018 required another kind of approach. An exponential growth function was fitted between an approximated worthless initial volume of 15 m^3/ha and the 2018 commercial volume estimate for any of the seven sample plots.

Possibly, the simplest way to approximate financial history is to determine an internal rate of operative return for the period from stand establishment to 2018. In other words, we require

$$V(\tau_3)e^{-s\tau_3} - R(\tau_1)e^{-s\tau_1} - C(\tau_2)e^{-s\tau_2} = 0 \quad (1)$$

where R is regeneration expense at regeneration time τ_1, C is young stand cleaning expense at cleaning time τ_2, and V is stumpage value of trees at observation time τ_3. In this study, the observation time τ_3 corresponds to November 2018, regeneration time is clarified according to known stand age, and young stand cleaning is assumed to have occurred ten years after regeneration. It is assumed that prices and expenses do not evolve in real terms, and thus presently valid expenses can be used in Equation (1). The regeneration expense is taken as 1250 Euro/ha, and young stand cleaning as 625 Euro/ha.

It is worth noting that the operative internal rate of return s in Equation (1) does not correspond to the capital return rate in the entire activity, as the latter depends on nonoperative capitalization such as bare land value.

2.3. Financial Considerations

To determine a momentary capital return rate, we need to discuss the amount of financial resources occupied [21,23,24,31]. This is done in terms of a financial potential function, defined in terms of capitalization per unit area K. The momentary capital return rate becomes

$$r(t) = \frac{d\kappa}{K(t)dt} \tag{2}$$

where κ in the numerator considers value growth, operative expenses, interests, and amortizations but neglects investments and withdrawals. In other words, it is the change of capitalization on an economic profit/loss basis. K in the denominator gives capitalization on a balance sheet basis, being directly affected by any investment or withdrawal. It is worth noting that timber sales do not enter the numerator of Equation (2); selling trees at market price levels does not change the amount of wealth, it only converts wealth from trees into the form of cash. However, harvesting naturally changes capitalization appearing in the denominator of Equation (2). Harvesting also likely changes the change rate of capitalization occurring after the harvest.

Equation (2) gives a momentary capital return rate, not necessarily sufficient for management considerations. By definition, the expected value of capitalization per unit area is

$$\langle K \rangle = \int_{-\infty}^{\infty} p(K)KdK \tag{3}$$

where $p(K)$ is the probability density function of capitalization K. By change of variables we get

$$\langle K \rangle = \int_{0}^{\tau} p(K)K\frac{dK}{da}da = \int_{0}^{\tau} p(a)K(a)da \tag{4}$$

where a is stand age (or time elapsed since latest regeneration harvesting), and τ is rotation age. The expected value of the change rate of capitalization is

$$\left\langle \frac{d\kappa}{dt} \right\rangle = \int_{0}^{\tau} p(a)\frac{d\kappa(a)}{dt}da \tag{5}$$

Correspondingly, the expected momentary rate of relative capital return is

$$\langle r(t) \rangle = \frac{\left\langle \frac{d\kappa}{dt} \right\rangle}{\langle K \rangle} = \frac{\int_{0}^{\tau} p(a)\frac{d\kappa(a,t)}{dt}da}{\int_{0}^{\tau} p(a)K(a,t)da} = \frac{\int_{0}^{\tau} p(a)K(a,t)r(a,t)da}{\int_{0}^{\tau} p(a)K(a,t)da} \tag{6}$$

We find from Equation (6) that the expected value of capital return rate within an estate generally evolves in time as the probability density of stand ages evolves. However, Equation (6) can be simplified to be independent of time by adopting the normal forest principle, where stand age probability density is constant [20]. Besides, the constancy of the expected value of capital return rate in time requires that prices and expenses do not evolve in real terms. Then, the expected value of the capital return rate becomes

$$\langle r \rangle = \frac{\int_0^{\tau} \frac{d\kappa(a)}{dt} da}{\int_0^{\tau} K(a) da} = \frac{\int_0^{\tau} K(a) r(a) da}{\int_0^{\tau} K(a) da} \quad (7)$$

It has been recently shown that Equation (7) corresponds to the ratio of the partition functions of the change rate of capitalization and capitalization itself [32]. It also has been recently shown that the maximization of the net present value of future revenues may result in financially devastating consequences [33]. Momentary capital return rate as given in Equation (2) was introduced in 1860 [31]; an expected value was mentioned in 1967 [29,30]; however, applications have been introduced only recently [21–23,25,32,33].

In the context of the improvement harvesting, 20% of the stem count of good-quality trees is removed in all diameter classes due to the establishment of striproads.

The stumpage value is determined in terms of roadside price, deduced by harvesting expense. We here use the roadside prices recently applied by Parkatti et al. [34], 34.04 Euro/m^3 for spruce pulpwood, and 58.44 Euro/m^3 for sawlogs. We further use the same harvest-expense function as Parkatti et al. [34], stated to be based on a productivity study of Nurminen et al. [35]. In addition to the expense function, we include a fixed harvesting entry expense per hectare. The justification of the latter is that some sites require at least partial pre-harvest cleaning. The entry expense is approximated as 200 Euro/ha. Again, it is assumed that prices and expenses do not evolve in time. In other words, the capital return rate is discussed in real terms.

The original version of the growth model operates in five-year time steps [21–23,28], as discussed above. Consequently, an eventual harvesting entry may take place every five years. According to recent investigations, it often is favorable to harvest every five years [21,23]. However, the fixed harvesting entry cost of 200 Euro/ha restricts low-yield harvesting entries. Numerical investigations indicated that it is not reasonable to harvest if the yield would be less than 20–30 m^2/ha. Such an entry limit is in concert with practices applied in the area.

The feasibility of harvesting is investigated in five-year intervals, always considering two options: thinning or clearcut. Clearcutting expenses are lower than thinning harvesting costs, according to Parkatti et al. [34], stated to be based on a productivity study of Nurminen et al. [35]. Moreover, a 15% clearcutting premium for the roadside price of sawlogs is applied, following local tradition. The premium, as well as the harvesting cost reduction, is applied within the last 30-month period before eventual clearcutting.

Thinning procedures are iteratively designed to maximize the expected value of capital return rate, up to the rotation age providing the maximum expected value. After the maximum expected value, any thinning procedure is designed to maximize the capital return rate within the next five years.

Four alternative thinning strategies are investigated for any of the seven normal estates. Two of the strategies are based on thinning from above. The first does not apply any restriction to cutting limit diameters, but the thinning procedure is designed in order to maximize the expected value of capital return rate according to Equation (7). The second does apply a minimal 238 mm diameter limit for thinning from above as a restriction Ref. [25]. The remaining two thinning strategies refer to proportional thinning. In other words, the probability of tree removal is the same in any diameter class. The first of the two proportional thinning strategies is taken as heavy proportional thinning, where 50% of acceptable-quality trees are removed. Gentle proportional thinning removes 30% of acceptable-quality trees in any diameter class.

Application of Equation (7) results in an expected value of capital return rate for any treatment schedule investigated. Treatment schedules not corresponding to the most economical one induce some amount of capital return rate deficiency, in terms of percentage per annum. Such deficiency is related either to the amount of timber stock deviating from the optimal timber stock, or the rotation age deviating from the optimal rotation age, or both. In the former case, the deficiency often is due to an excess standing volume, which allows for expressing the deficiency per excess volume unit.

2.4. Carbon Rent Considerations

The last issue in this section of methods regards carbon trade and carbon renting. It is worth noting that carbon prices or rents do not enter the analysis described above in any way. Carbon prices and rents are discussed to enable a comparison of the capital return deficiency per excess volume with any hypothetical carbon rent.

It has been recently shown that policies based on carbon rent are equivalent to policies based on carbon sequestration subsidies and taxies [18]. Unbiased carbon sequestration trade would require a huge initial investment; correspondingly, mostly carbon rent procedures are practically feasible [18]. We here present a brief derivation of the equivalency of the two principles of subsidies, although adopting boundary conditions possibly less restrictive than those of Lintunen et al. [18].

Let us establish a carbon sequestration subsidy system at a particular time τ_1. Within a time range up to time τ_2, the total carbon trade compensation is

$$p_{\tau_1}C_{\tau_1} + \int_{\tau_1}^{\tau_2} \frac{d(pC)}{dt}dt = p_{\tau_1}C_{\tau_1} + \int_{\tau_1}^{\tau_2} \left[p\frac{dC}{dt} + C\frac{dp}{dt}\right]dt \quad (8)$$

where p_t is carbon price at time t, and C_t is carbon inventory at time t. On the other hand, the revenue from carbon rentals is

$$\int_{\tau_1}^{\tau_2} uCdt \quad (9)$$

where u is the rent rate per carbon unit. Now, to establish equivalency between the carbon storage trade and rent, Equations (8) and (9) must become equal. Equality naturally should apply in any possible circumstance. One of the circumstances is that the time change rate of prices, as well as inventories, is zero. In such a case, the latter term of Equation (8) vanishes. Consequently, the long-term flow of carbon rents should equal a one-time initial storage purchase payment. If the duration of the rent payments extends towards infinity, the only possibility is that the present value of rent payments forms a contracting series. One possibility of such a contracting series is

$$u_{\tau_1}C_{\tau_1}\int_{0}^{\infty} e^{-qt}dt = p_{\tau_1}C_{\tau_1} \quad (10)$$

where q is a discount rate. The corresponding solution for the carbon rent rate is

$$u_t = qp_t \quad (11)$$

One can readily show that Equation (11) applies not only to a steady-state of Equations (8) and (9) but also to any incremental carbon price and inventory.

3. Results

The expected value of capital return rate as a function of rotation age, integrated over the stand lifespan according to Equation (7) is shown in Figure 2. In Figure 2, no restriction is applied in the maximization or the capital return rate. Any feasible treatment cycle terminates in clearcutting, and thus utilizes the premium in clearcutting price, as well as the harvesting expense reduction. In most cases, thinnings from the above are made to 188 mm cutting diameter limit, even if also 213 mm limits appear. The number of thinnings up to the financially optimal rotation age varies from one to three. After the optimal rotation age, one thinning is triggered in five of the seven cases, in the absence of prior clearcutting.

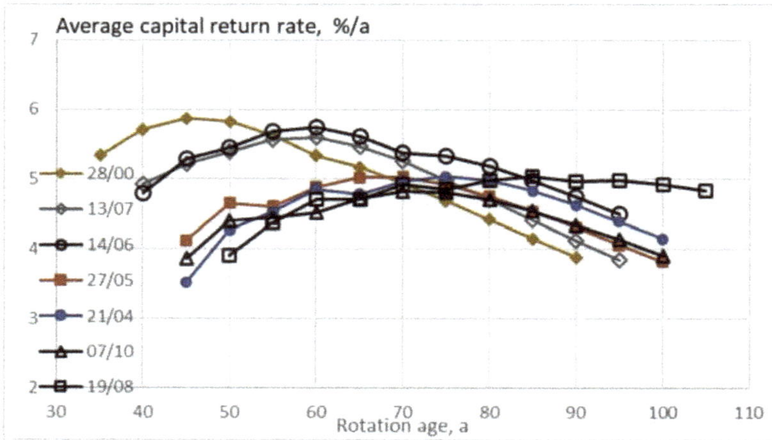

Figure 2. Capital return rate according to Equation (7) for seven different normal estates, without any restriction in the design of the applied thinning procedure. The legend identifies the experimental plot identifying any normal estate.

It is found from Figure 2 that the normal estates, with properties determined from the corresponding normal stands, differ in terms of the expected value of capital return rate but they differed more in the optimal rotation age. The amplitudes of variation are 9.8% and 31%, respectively. The optimal rotation ages span from 45 to 85 years. This is due to fertility, as well as the number and size of trees observed. A third significant contributor is the age of any normal stand in the occurrence of data sampling. The oldest stand was of age 45 years. The capital return rate probably would be higher and the corresponding rotation age lower if the stand had been thinned earlier.

The expected value of capital return rate as a function of rotation age is shown in Figure 3, with the restriction of a minimum cutting limit diameter of 238 mm in thinnings from above. Again, any feasible treatment cycle terminates in clearcutting, and thus utilizes the premium in clearcutting price, as well as the harvesting expense reduction. The restriction induces gentler thinnings and greater capitalizations. The restriction naturally reduces achievable capital return rates. The gentler thinnings reduce the optimal rotation age in five of the seven cases, even if it increases the number of thinnings. The number of thinnings up to the financially optimal rotation age varies from one to six. The optimal rotation ages now span from 40 to 85 years, the latter corresponding to the case of delayed first thinning. The amplitude of variation in capital return rate is 10.2%, and 36% in optimal rotation age.

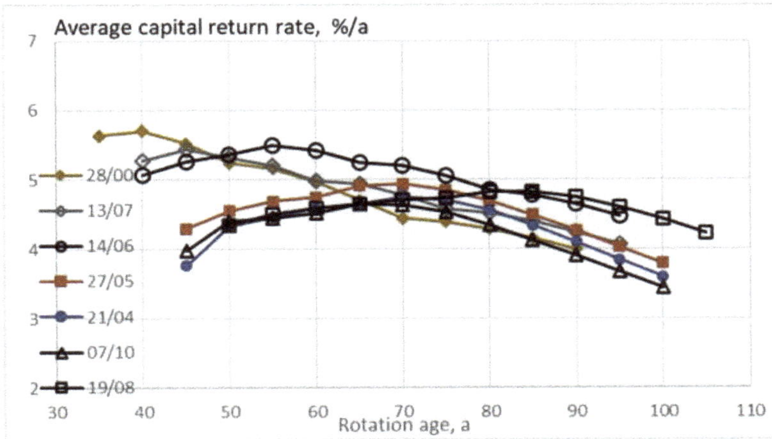

Figure 3. Capital return rate according to Equation (7) for seven different normal estates, with a minimum 238 mm cutting diameter limit in thinnings from above. The legend identifies the experimental plot identifying any normal estate.

The expected value of capital return rate as a function of rotation age is shown in Figure 4, with heavy proportional thinning. Again, any feasible treatment cycle terminates in clearcutting and thus utilizes the premium in clearcutting price, as well as the harvesting expense reduction. The heavy proportional thinning renders lower capital return rates in comparison to Figures 2 and 3, as well as shorter rotation times. The optimal rotation ages now vary from 35 to 60 years. The amplitude of variation in capital return rate is 12.9%, and 26.3% in optimal rotation age. In all cases, there is only one thinning. However, in six of the seven cases the financial result is inferior in comparison to no thinning at all.

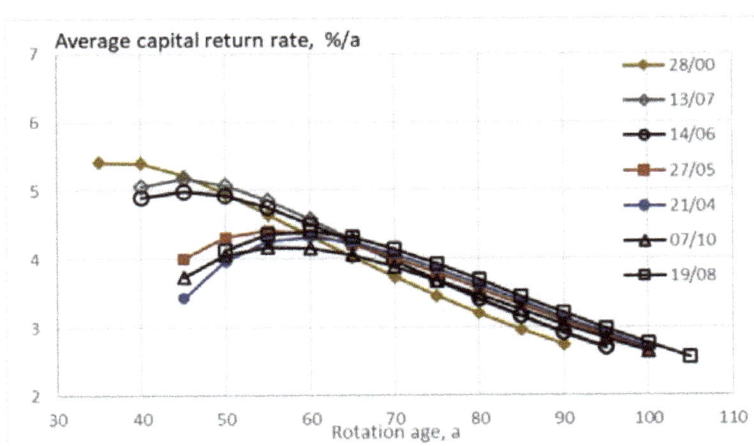

Figure 4. Capital return rate according to Equation (7) for seven different normal estates, with heavy proportional thinning. The legend identifies the experimental plot identifying any normal estate.

The expected value of capital return rate as a function of rotation age is shown in Figure 5, with gentle proportional thinning. Any feasible treatment cycle terminates in clearcutting, and thus utilizes the premium in clearcutting price, as well as the harvesting expense reduction. The gentle proportional thinning renders lower capital return rates in comparison to Figures 2 and 3, as well as shorter rotation times. In comparison to the heavy proportional thinning (Figure 4), capital return rates are higher but optimal rotations shorter. In all cases, there is only one thinning before the rotation age corresponding to the maximum expected capital return rate. Somewhat surprisingly, in five of the seven cases, the financial result is inferior in comparison to no thinning at all. It is, however, worth noting that gaining the clearcutting price premium may be uncertain in the absence of any thinning.

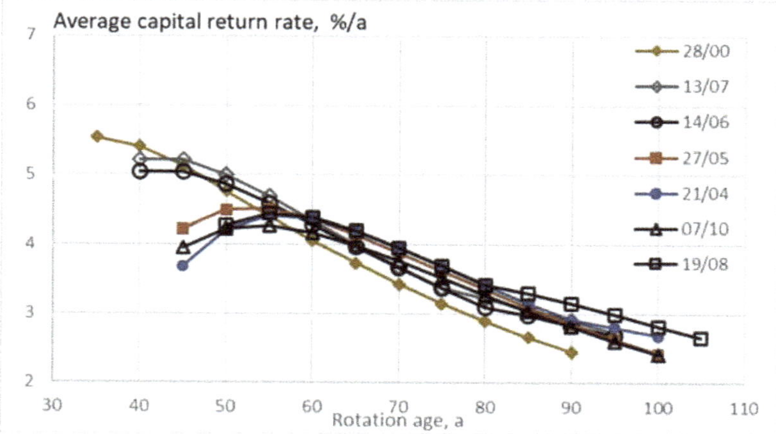

Figure 5. Capital return rate according to Equation (7) for seven different normal estates, with gentle proportional thinning. The legend identifies the experimental plot identifying any normal estate.

The capital return rate deficiency in terms of Euros per excess volume is shown in Figure 6, as a function of excess volume, linking economic and environmental sustainability. No restriction is applied in the maximization or the capital return rate. Only positive values are shown in Figure 6. The rotation age providing the maximum capital return rate corresponds to the origin in any series of observations. All the excess volumes in Figure 6 are due to extended rotation age. It is found that even excess volumes of 10 to 25 cubic meters per hectare induced a deficiency of a Euro per excess cubic meter. In the case of the normal stand with retarded thinning, all excess volumes are small.

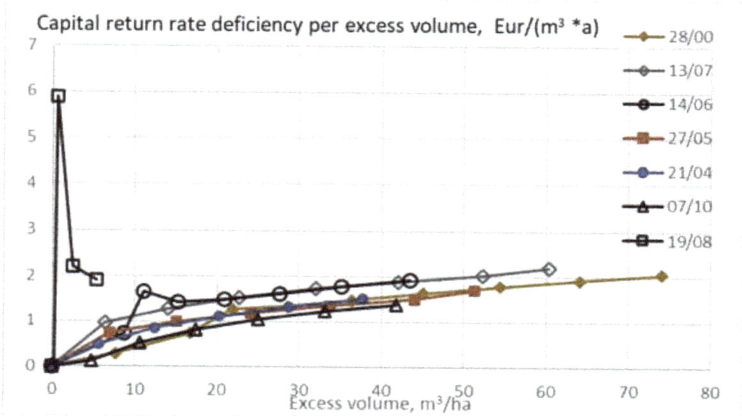

Figure 6. Capital return rate deficiency per excess volume unit for seven different normal estates, without any restriction in the design of the applied thinning procedure. Only positive values are plotted.

The capital return rate deficiency in terms of Euros per excess volume is shown in Figure 7, as a function of excess volume. There is a restriction of the minimum cutting limit diameter of 238 mm in thinnings from above. Small excess volumes tend toward large values of the observable, since the excess volume is in the denominator. Moderate excess volumes tend to induce a much smaller capital return deficiency than in Figure 6. It is found that increasing the size of trees remaining after thinning is an economical way of increasing expected stand volume, in comparison to extending the rotations (Figures 6 and 7).

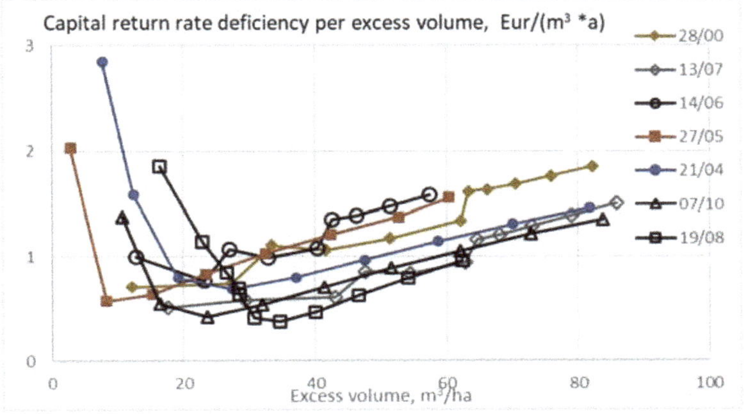

Figure 7. Capital return rate deficiency per excess volume unit for seven different normal estates, with minimum 238 mm cutting diameter limit in thinnings from above.

The capital return rate deficiency in terms of Euros per excess volume is shown in Figure 8, as a function of excess volume, for the case of heavy proportional thinning. Again, small excess volumes tend toward large values of the observable, since the excess volume is in the denominator. The capital return rate deficiency appears to be greater than in thinnings from above (Figures 6 and 7), for moderate excess volumes somewhat below two Euros per excess volume unit.

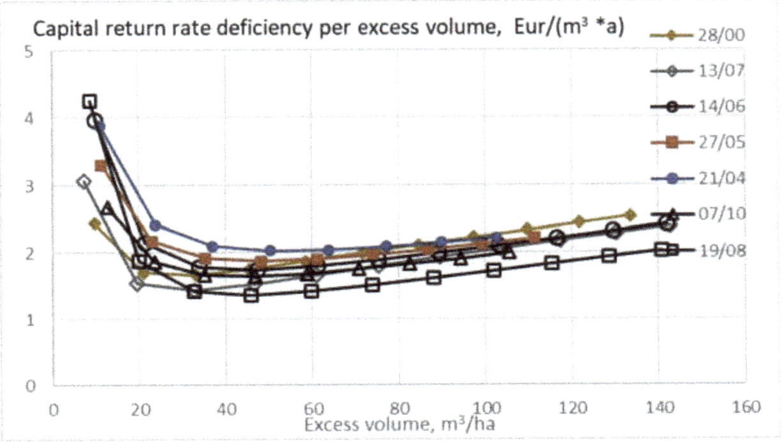

Figure 8. Capital return rate deficiency per excess volume unit for seven different normal estates, with heavy proportional thinning.

The capital return rate deficiency in terms of Euros per excess volume is shown in Figure 9, as a function of excess volume, for the case of gentle proportional thinning. The capital return rate deficiency appears smaller than in heavy proportional thinning (Figure 8), but greater than in restricted minimum cutting limit diameter (Figure 7). For moderate excess volumes, the gentle proportional thinning in Figure 9 is competitive to merely extending the rotation age in Figure 6.

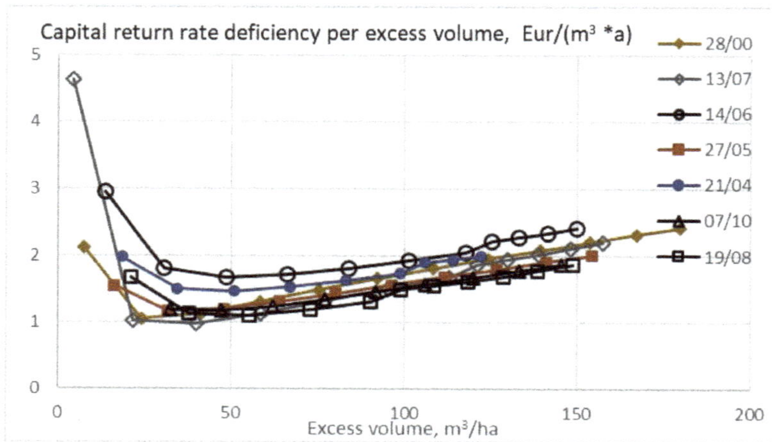

Figure 9. Capital return rate deficiency per excess volume unit for seven different normal estates, with gentle proportional thinning.

4. Discussion

It appeared from Figure 2 that the optimal rotation age varied significantly. More specifically, the amplitude of variation was 31%. It is of interest to consider whether there are other essential features of the normal stands that would unify. Figure 10 shows the expected value of capital return rate as a function of the state sum of annual volumes per hectare. It is found that at the instant of maximal capital return rate, the state sum of volumes varies significantly, as does the state sum of annual capitalizations (not in the figure). The case with retarded thinning shows more than two times the state sums corresponding to the maximum capital return rate, in comparison to the most fertile stand. The amplitudes of variation of optimal capital return rate and the state sum of annual volumes were 9.8% and 36%, respectively. The amplitude of variation of the optimal state sum of annual capitalizations was 26%.

An obvious reason for the variation in Figure 10 is that sites of different fertility have different time scales in their development. Summing either growing time or some state variables does not unify features unless some time-scaling is applied.

It appears, however, that there is a unifying feature in the seven normal estates. The optimal expected stand volume does not vary much (Figure 11). A consequence is that the excess volumes appearing in Figures 6–9 are additive to an almost unified basic volume, with even the case of retarded thinning differing only slightly. In other words, the amplitude of variation of the optimal expected value of stand volume was 9.1%. This greatly simplifies carbon storage applications.

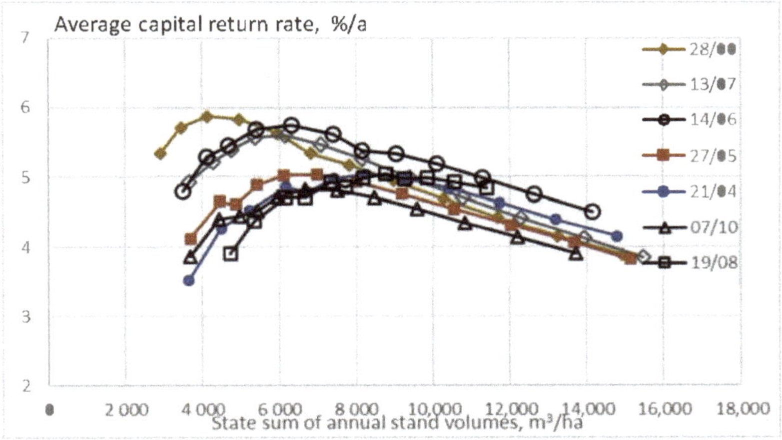

Figure 10. Capital return rate according to Equation (7) for seven different normal estates, without any restriction in the design of the applied thinning procedure, as a function of the state sum of annual volumes.

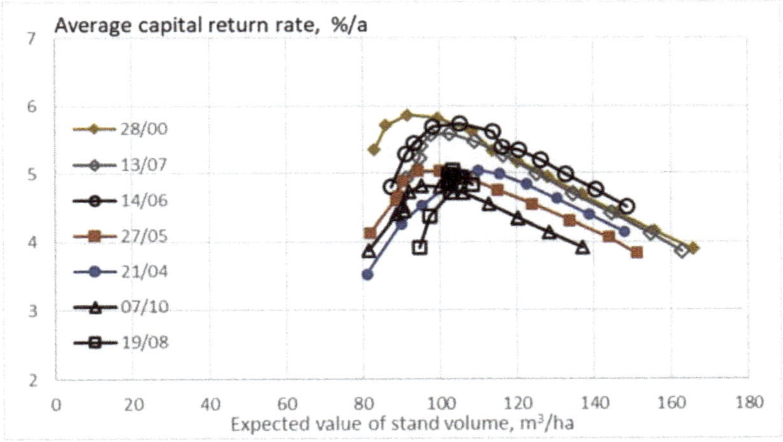

Figure 11. Capital return rate according to Equation (7) for seven different normal estates, without any restriction in the design of the applied thinning procedure, as a function of the expected value of commercial volume per hectare.

We have found from Figures 6–9 that for any of the seven normal estates, the most economical way of increasing carbon storage is to increase the size of trees retained in thinning from above. We even find from Figures 7 and 11 that the economics of carbon storage do not vary widely, even if there are different fertilities and growth rates. Correspondingly, the results of the recent investigation [25] appear to be robust to variations among the sample plots collected in the field.

An iterative search for suitable thinning schedules was done up to the rotation age, giving the greatest capital return rate. After the maximum expected value, any thinning procedure was designed to maximize the expected value of the capital return rate observed within the next five years. Any additional five years was thus considered as a marginal extension of the optimal rotation age.

Within the geographic area, professional practice is not to thin aged spruce stands. This is due to a variety of factors, among them avoidance of wind damage and suspected loss of vigor along with aging. The latter might reduce the recovery of trees remaining after thinning. It is of interest to introduce a boundary condition according to which no thinnings are implemented after the optimal rotation age and then compare it with the results appearing in Figures 2, 3, 6 and 7. It is found from Figures 12 and 13 that excess volumes become greater and the increment of capital return rate deficiency smoother in the absence of post-optimum thinnings.

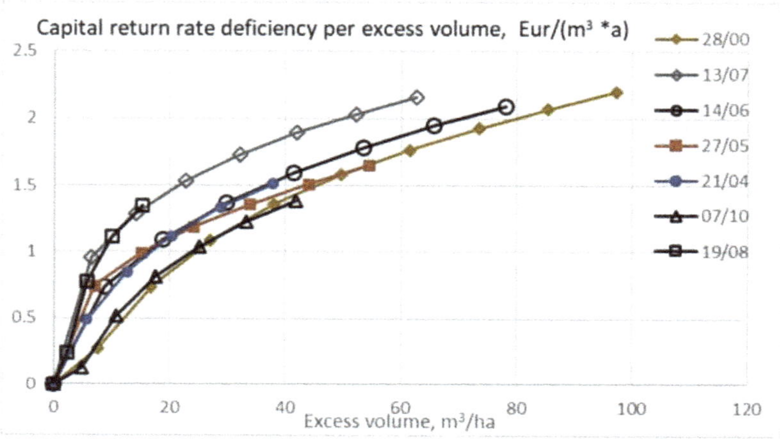

Figure 12. Capital return rate deficiency per excess volume unit for seven different normal estates, without any thinning after the financially optimal rotation age.

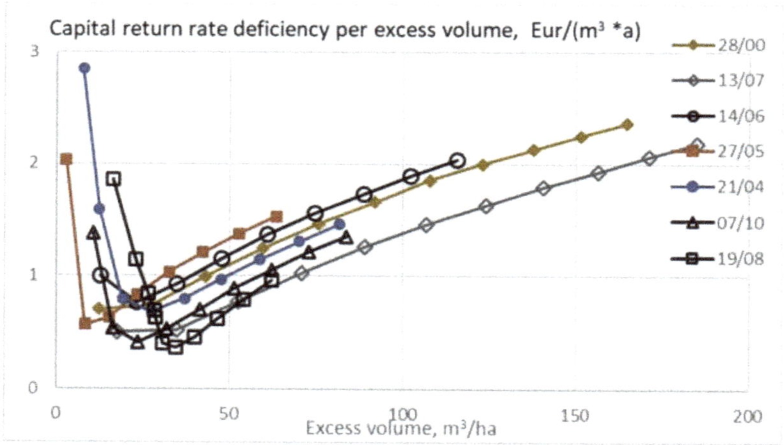

Figure 13. Capital return rate deficiency per excess volume unit for seven different normal estates, with minimum 238 mm cutting diameter limit in thinnings from above, and no thinning after the financially optimal rotation age.

The capital return rate being strongly affected by the clearcutting price premium, the five-year marginal approach either does or does not trigger further thinnings. If one would adopt another boundary condition, for example considering a 15-year margin, a greater number of thinnings would be triggered. After the optimal rotation age, observables would differ. The expected value of the capital return rate would decline more slowly than in Figures 2–5.

If one would maximize the expected value of the capital return rate observed within any 15-year period after the rotation age corresponding to maximal capital return rates, a chart corresponding to Figure 6 would become meaningless. This is because the average stand volume would decrease with time, and the excess volumes would become negative. Instead, setting a restriction in the cutting diameter limit (as in Figure 7) would yield meaningful results. However, with the cutting diameter limit restriction of 238 mm, the excess volumes are still small and partly negative. With 263 mm cutting diameter limit restriction, the excess volumes become positive. The deficiencies are only slightly greater than in Figure 7 and clearly smaller than in Figure 9. A similar result is not achievable by proportional thinning. If one maximizes the expected value of the capital return rate observed within the next 15 years after the rotation age corresponding to maximal capital return rate, the results become inferior in comparison to Figures 8 and 9.

It is further worth noting that multiple thinnings from above would, along with time, transfer the system from the realm of even-aged forestry towards uneven-aged forestry. Such a transition might reduce vigor [36,37], and such a reduced vigor is not considered in the growth model used in this study.

An applicable carbon rent can be derived from carbon storage market price according to Equation (11). At the time of writing, the market price of carbon dioxide emissions is in the order of 25 Euros per ton. This inserted to Equation (11), together with a 3% discount rate [38–40], results as an annual carbon rent of 0.75 Euros per ton of carbon dioxide.

A cubic meter of commercial trunk volume in the boreal forest stores a ton of carbon dioxide (in living biomass, litter, and soil) [1,2,4,7,41,42]. Correspondingly, one might expect a carbon rent of 0.75 Euros per standing cubic meter. In accordance with a recent study [25], that is enough only up to an excess volume of 30 m^3/ha. In other words, the recent result is now confirmed considering variation among sample plots measured in the field. For greater levels of carbon storage, the deficiency of capital return rate per standing cubic meter is greater than the appropriate rent.

Any carbon sequestration trade compensation must be proportional to the carbon storage on the estate level [18]. If it would not be, agents with large carbon inventories would initially suffer heavy and unjustified release taxes. Unlike the carbon trade, there is no particular reason why the carbon rent should be proportional. If the carbon rent is made nonproportional, the marginal rent may reasonably reach or exceed the capital return rate deficiency of Figure 7, also on the right of the figure.

It was recently claimed that changing soil fertility as well as changing temperature sum would change the values of all time derivatives appearing in any growth model, as well as in any carbon storage model [23,25]. Correspondingly, capital return rates, as well as annual capital return deficiencies, would change along with all observables resulting from the integration of time derivatives. This statement is confirmed by Figures 2–5 and 10, where both of the axes contain time-dependent quantities. There is a significant scatter among normal estates corresponding to normal stands of different fertility. On the other hand, quantities not involving time derivatives should be robust to fertility. This is found from the horizontal axes of Figures 6–9 and Figures 11–13, which do not contain time derivatives or their integrals. The vertical axis of the latter figures do contain time-dependent quantities, and correspondingly display significant scatter among the normal estates in relative terms. It is, however, worth noting that the variation of optimal capital return rate in Figures 2, 10 and 11 is surprisingly small. The amplitude was 9.8%, in contrast to 9.1% optimal stand volume. The much bigger variations in other time-dependent quantities may be not only fertility-related but possibly also stand age-related.

In addition to soil fertility, there are other aspects of how the seven normal stands differ. Stem counts and stem size distributions differ, as well as stand age at the time of observation in 2018. Such factors do contribute to the characteristics of the normal estates that the normal stands represent. The results above indicate that for the stand of the oldest age, the first thinning was clearly overdue. Apart from that, the present material does not indicate what would be an optimal stage for the first thinning.

The seven normal stands, representing seven normal estates, were collected from one physical estate in eastern Finland. Correspondingly, the variation among the normal estates reflects variation among stands within the physical estate. The physical estate was completely on mineral soil and relatively productive throughout.

The yield of logs from trees was modeled based on an empirical dataset of 6123 harvested spruce trees. There is no guarantee that such an empirical dataset would apply to all circumstances. The quality of tree trunks varies by stand. Quality requirements of sawlogs vary by sawmill and are subject to change in time. Geographic areas differ, as well as tree species. The yield of logs from trees does differ depending on the type of harvesting (clearcutting, thinning from above/below, continuous cover operations, etc.), stand age and so on.

Many of the quantitative results in this paper depend on the market prices of roundwood assortments, bare land, silvicultural and harvesting expenses, and carbon emission as well as the discount rate applied in Equations (10) and (11). Considering eventual changes in the latter two is straightforward in terms of Equations (10) and (11). Changes in the former quantities contribute trivially if they are proportional. In the case of significant nonproportional changes in prices, most of the figures of this paper will have to be redrawn.

5. Conclusions

The robustness of recent results regarding microeconomics of carbon storage in fertile boreal forest estates is investigated. For all investigated seven normal estates, carbon stocking can be increased at the expense of a capital return rate deficiency. The most economical way of increasing carbon storage is to increase the size of trees retained in thinning from above, in all seven cases.

Normal stands and normal estates show different characteristics in time-dependent features. However, they unify in time-independent characteristics. In particular, the financially optimal expected value of stand volume per hectare does not vary much, which greatly simplifies carbon storage applications.

Funding: This research was partially funded by the Niemi Foundation.

Institutional Review Board Statement: Not applicable.

Informed Consent Statement: Not applicable.

Data Availability Statement: None.

Acknowledgments: The Author is obliged to many individuals who contributed to technical arrangements. In particular, the contribution of Jori Uusitalo was invaluable in the analysis of harvester data. Risto-Matti Räsänen, Ari Haapalainen, Keijo Jormanainen, Tero Luukkainen, and Tuure Korhonen kindly assisted in the collection of the yield data.

Conflicts of Interest: The author declares no conflict of interest.

References

1. Adams, A.; Harrison, R.; Sletten, R.; Strahm, B.; Turnblom, E.; Jensen, C. Nitrogen-fertilization impacts on carbon sequestration and flux in managed coastal Douglas-fir stands of the Pacific Northwest. *For. Ecol. Manag.* **2005**, *220*, 313–325. [CrossRef]
2. Lal, R. Forest soils and carbon sequestration. *For. Ecol. Manag.* **2005**, *220*, 242–258. [CrossRef]
3. Liski, J.; Lehtonen, A.; Palosuo, T.; Peltoniemi, M.; Eggersa, T.; Muukkonen, P.; Mäkipää, R. Carbon accumulation in Finland's forests 1922–2004-an estimate obtained by combination of forest inventory data with modelling of biomass, litter and soil. *Ann For. Sci.* **2006**, *63*, 687–697. [CrossRef]
4. Peltoniemi, M.; Mäkipää, R.; Liski, J.; Tamminen, P. Changes in soil carbon with stand age—An evaluation of a modelling method with empirical data. *Glob. Chang. Biol.* **2004**, *10*, 2078–2091. [CrossRef]
5. Campioli, M.; Vicca, S.; Luyssaert, S.; Bilcke, J.; Ceschia, E.; Iii, F.S.C.; Ciais, P.; Fernández-Martínez, M.; Malhi, Y. Obersteiner, M.; et al. Biomass production efficiency controlled by management in temperate and boreal ecosystems. *Nat. Geosci* **2015**, *8*, 843–846. [CrossRef]

6. Powers, M.; Kolka, R.; Palik, B.J.; McDonald, R.; Jurgensen, M. Long-term management impacts on carbon storage in Lake States forests. *For. Ecol. Manag.* **2011**, *262*, 424–431. [CrossRef]
7. Thornley, J.H.M.; Cannell, M.G.R. Managing forests for wood yield and carbon storage: A theoretical study. *Tree Physiol.* **2000**, *20*, 477–484. [CrossRef]
8. Karhu, K.; Wall, A.; Vanhala, P.; Liski, J.; Esala, M.; Karhu, K. Effects of afforestation and deforestation on boreal soil carbon stocks—Comparison of measured C stocks with Yasso07 model results. *Geoderma* **2011**, *164*, 33–45. [CrossRef]
9. Liski, J.; Palosuo, T.; Peltoniemi, M.; Sievänen, R. Carbon and decomposition model Yasso for forest soils. *Ecol. Model.* **2005**, *189*, 168–182. [CrossRef]
10. Rantakari, M.; Lehtonen, A.; Linkosalo, T.; Tuomi, M.; Tamminen, P.; Heikkinen, J.; Liski, J.; Mäkipää, R.; Ilvesniemi, H.; Sievänen, R. The Yasso07 soil carbon model–Testing against repeated soil carbon inventory. *For. Ecol. Manag.* **2012**, *286*, 137–147. [CrossRef]
11. Tupek, B.; Launiainen, S.; Peltoniemi, M.; Sievänen, R.; Perttunen, J.; Kulmala, L.; Penttilä, T.; Lindroos, A.; Hashimoto, S.; Lehtonen, A. Evaluating CENTURY and Yasso soil carbon models for CO_2 emissions and organic carbon stocks of boreal forest soil with Bayesian multi-model inference. *Eur. J. Soil Sci.* **2019**, *70*, 847–858. [CrossRef]
12. Gustavsson, L.; Haus, S.; Lundblad, M.; Lundström, A.; Ortiz, C.A.; Sathre, R.; Truong, N.; Wikberg, P.-E. Climate change effects of forestry and substitution of carbon-intensive materials and fossil fuels. *Renew. Sustain. Energy Rev.* **2017**, *67*, 612–624. [CrossRef]
13. Van Kooten, G.C.; Bogle, T.N.; De Vries, F.P. Forest Carbon Offsets Revisited: Shedding Light on Darkwoods. *For. Sci.* **2015**, *61*, 370–380. [CrossRef]
14. Van Kooten, G.C.; Johnston, C.M.T. The Economics of Forest Carbon Offsets. *Annu. Rev. Resour. Econ.* **2016**, *8*, 227–246. [CrossRef]
15. Pukkala, T. Carbon forestry is surprising. *For. Ecosyst.* **2018**, *5*, 11. [CrossRef]
16. Seppälä, J.; Heinonen, T.; Pukkala, T.; Kilpeläinen, A.; Mattila, T.; Myllyviita, T.; Asikainen, A.; Peltola, H. Effect of increased wood harvesting and utilization on required greenhouse gas displacement factors of wood-based products and fuels. *J. Environ. Manag.* **2019**, *247*, 580–587. [CrossRef]
17. Pukkala, T. At what carbon price forest cutting should stop. *J. For. Res.* **2020**, *31*, 713–727. [CrossRef]
18. Lintunen, J.; Laturi, J.; Uusivuori, J. How should a forest carbon rent policy be implemented? *For. Policy Econ.* **2016**, *69*, 31–39. [CrossRef]
19. Kilkki, P.; Väisänen, U. Determination of the optimum cutting policy for the forest stand by means of dynamic programming. *Acta For. Fenn.* **1969**, *102*, 1–23. [CrossRef]
20. Leslie, A.J. A review of the concept of the normal forest. *Aust. For.* **1966**, *30*, 139–147. [CrossRef]
21. Kärenlampi, P.P. Harvesting Design by Capital Return. *Forests* **2019**, *10*, 283. [CrossRef]
22. Kärenlampi, P.P. The effect of capitalization on financial return in periodic growth. *Heliyon* **2019**, *5*, e02728. [CrossRef] [PubMed]
23. Kärenlampi, P.P. Estate-Level Economics of Carbon Storage and Sequestration. *Forests* **2020**, *11*, 643. [CrossRef]
24. Kärenlampi, P.P. Stationary Forestry with Human Interference. *Sustainability* **2018**, *10*, 3662. [CrossRef]
25. Kärenlampi, P.P. The Effect of Empirical Log Yield Observations on Carbon Storage Economics. *Forests* **2020**, *11*, 1312. [CrossRef]
26. Rämö, J.; Tahvonen, O. Economics of harvesting boreal uneven-aged mixed-species forests. *Can. J. For. Res.* **2015**, *45*, 1102–1112. [CrossRef]
27. Heinonen, J. Koealojen puu-ja puustotunnusten laskentaohjelma KPL. In *Käyttöohje (Software for Computing Tree and Stand Characteristics for Sample Plots. User's Manual)*; Research Reports; Finnish Forest Research Institute: Vantaa, Finland, 1994. (In Finnish)
28. Bollandsås, O.M.; Buongiorno, J.; Gobakken, T. Predicting the growth of stands of trees of mixed species and size: A matrix model for Norway. *Scand. J. For. Res.* **2008**, *23*, 167–178. [CrossRef]
29. Speidel, G. *Forstliche Betreibswirtschaftslehre*, 2nd ed.; Verlag Paul Parey: Hamburg, Germany, 1967; 226p. (In German)
30. Speidel, G. *Planung in Forstbetrieb*, 2nd ed.; Verlag Paul Parey: Hamburg, Germany, 1972; 270p. (In German)
31. Pressler, M.R. Aus der Holzzuwachlehre (zweiter Artikel). *Allg. Forst-und Jagdztg.* **1860**, *36*, 173–191, Translated by Löwenstein, W.; Wirkner, J.R. as "For the comprehension of net revenue silviculture and the management objectives derived thereof". *J. For. Econ.* **1995**, *1*, 45–87.
32. Kärenlampi, P.P. State-space approach to capital return in nonlinear growth processes. *Agric. Financ. Rev.* **2019**, *79*, 508–518. [CrossRef]
33. Kärenlampi, P.P. Wealth accumulation in rotation forestry–Failure of the net present value optimization? *PLoS ONE* **2019**, *14*, e0222918. [CrossRef]
34. Parkatti, V.-P.; Assmuth, A.; Rämö, J.; Tahvonen, O. Economics of boreal conifer species in continuous cover and rotation forestry. *For. Policy Econ.* **2019**, *100*, 55–67. [CrossRef]
35. Nurminen, T.; Korpunen, H.; Uusitalo, J. Time consuming analysis of the mechanized cut-to-length harvesting system. *Silva Fenn.* **2006**, *40*, 335–363. [CrossRef]
36. Bianchi, S.; Huuskonen, S.; Siipilehto, J.; Hynynen, J. Differences in tree growth of Norway spruce under rotation forestry and continuous cover forestry. *For. Ecol. Manag.* **2020**, *458*, 117689. [CrossRef]
37. Hynynen, J.; Eerikäinen, K.; Mäkinen, H.; Valkonen, S. Growth response to cuttings in Norway spruce stands under even-aged and uneven-aged management. *For. Ecol. Manag.* **2019**, *437*, 314–323. [CrossRef]
38. Pukkala, T. Plenterwald, Dauerwald, or clearcut? *For. Policy Econ.* **2016**, *62*, 125–134. [CrossRef]

39. Pyy, J.; Ahtikoski, A.; Laitinen, E.; Siipilehto, J. Introducing a Non-Stationary Matrix Model for Stand-Level Optimization, an Even-Aged Pine (Pinus sylvestris L.) Stand in Finland. *Forests* **2017**, *8*, 163. [CrossRef]
40. Sinha, A.; Rämö, J.; Malo, P.; Kallio, M.; Tahvonen, O. Optimal management of naturally regenerating uneven-aged forests. *Eur. J. Oper. Res.* **2017**, *256*, 886–900. [CrossRef]
41. Petersson, H.; Holm, S.; Ståhl, G.; Alger, D.; Fridman, J.; Lehtonen, A.; Lundström, A.; Mäkipää, R. Individual tree biomass equations or biomass expansion factors for assessment of carbon stock changes in living biomass–A comparative study. *For. Ecol. Manag.* **2012**, *270*, 78–84. [CrossRef]
42. Schepashenko, D.; Shvidenko, A.; Usoltsev, V.; Lakyda, P.; Luo, Y.; Vasylyshyn, R.; Lakyda, I.; Myklush, Y.; See, L.; McCallum, I.; et al. A dataset of forest biomass structure for Eurasia. *Sci. Data* **2017**, *4*, 170070. [CrossRef]

sustainability

MDPI

rticle

)elineating an Integrated Ecological and Cultural Corridor Network: A Case Study in Beijing, China

'anyan Li [1], Xinhao Wang [2] and Xiaofeng Dong [3,*]

1 School of Civil Engineering, Beijing Jiaotong University, Beijing 100044, China; yylhndx@163.com
2 School of Planning, University of Cincinnati, Cincinnati, OH 45221, USA; wangxo@ucmail.uc.edu
3 College of Architecture & Design, Beijing Jiaotong University, Beijing 100044, China
* Correspondence: dongcity@126.com; Tel.: +86-1881-1791-075

Abstract: This study shows that an integrated ecological and cultural corridor network can help guide city development strategies to better preserve ecological and cultural assets. Traditionally, protection zones and suitable development areas are often identified by separately considering natural elements of the ecosystem and elements of cultural significance. To achieve the purpose of cohesively protecting areas of ecological and/or cultural significance, we have developed a corridor-based spatial framework by integrating ecological and cultural assets. Ecological sources are identified by combining protection prioritization, nature reserves, and water bodies. Ecological corridors are delineated by using the minimum cumulative resistance (MCR) model on a resistance surface constructed from land-use data to connect ecological sources. Ecologically important areas are then delineated by creating a 5-km buffer zone from ecological sources and ecological corridors. Cultural corridors are historical routes and rivers surrounded by abundant cultural nodes. Like ecologically important areas, culturally important areas are delineated by creating a 5-km buffer zone from cultural corridors. Comprehensive regions are the overlap of ecologically and culturally important areas. Finally, the integrated network connects all comprehensive regions following ecological corridors and cultural corridors in such a way that the largest number of ecological sources and cultural nodes are reached. We applied this framework in Beijing, China, and the results show that there are 2011 km^2 of ecological sources, 30 ecological corridors, 423 cultural nodes, seven cultural corridors, and 10 comprehensive regions covering 2916 km^2 in the integrated network. The framework adds new insights to the methodology of considering ecological and cultural assets together in developing protection and development strategies.

Keywords: ecological corridors; cultural corridors; protection prioritization; MCR model; integrated network

:**itation:** Li, Y.; Wang, X.; Dong, X.)elineating an Integrated Ecological ınd Cultural Corridor Network: A :ase Study in Beijing, China. *ustainability* **2021**, *13*, 412. ıttps://doi.org/10.3390/su13010412

ℝeceived: 21 November 2020
\ccepted: 31 December 2020
ʾublished: 5 January 2021

Publisher's Note: MDPI stays neu-ral with regard to jurisdictional clai-ns in published maps and institutio-ıal affiliations.

1. Introduction

Human settlements have evolved from a cluster of cities to metropolitan areas, metropolitan area belts, large metropolitan belts, and megalopolises [1]. The world population has increased exponentially under this massive urbanization. In 2019, the United Nations predicted that the world population will have increased to 9.7 billion by 2050 and 66% of the population will reside in urban areas [2]. The conflicts between economic development and ecological protection have been seriously exacerbated by the rapid and intense changes in the structure and function of the landscape [3,4]. The rapid influx of popular culture from developed countries into developing countries and the widespread homogeneity of cultural practices have marginalized or even led to the disappearance of many indigenous cultures. These factors are the cause of many serious problems in the world today, such as air and water pollution, cultural invasion, biodiversity loss, food insecurity, and rising crime rates. Some scholars have pointed out that the world and its constituent landscapes are on an unsustainable trajectory. The question of how to reduce the effects of urbanization on the ecosystem in order to achieve urban sustainability has become an important focal point in the field of landscape ecology [5–8]. Sustainable development is necessity, not a choice [9].

To achieve sustainable development, human beings must be in harmony with the environment in which they live. Ecosystem services are one of the important factors determining landscape sustainability. Among various definitions of ecosystem services [10–13], the most widely accepted is from the Millennium Ecosystem Assessment: ecosystem services are the benefits people obtain from ecosystems. Ecosystem services have been increasingly considered a crucial bridge between the environment and society, which also safeguard the natural capital for future generations and highlight the contribution of ecosystems to human wellbeing [14–16]. Although ecosystem service evaluation has received increasing attention over the years [17], most studies still use biodiversity conservation, ecological importance, ecological risk evaluation, and resilience assessment to recognize ecological sources when delineating ecological corridors [18,19].

Similar to how modern material civilization comes at the expense of biological diversity, modern spiritual civilization comes at the expense of cultural diversity. With industrialization and human expansion, the earth is losing its biological and cultural diversity. The influence of foreign cultures, the transformation of traditional lifestyles, the mobility of the population, and the over-exploitation of tourism have destroyed various intangible cultural heritage resources. Many cultural heritage resources are the products of the long-term interaction between human beings and the natural environment in indigenous societies, which can enable people to recover the memory of the past and to build new perspectives [20]. However, many ancient buildings, bridges, and other facilities are gradually being replaced by works of modern engineering, and numerous excellent cultural heritage sites are scattered and lack contact with each other, which causes them to disappear over time. It is extremely urgent to establish an integrated conservation strategy to breathe new life into these precious assets.

As for the protection of ecological and cultural resources, inclined research on ecological/cultural conservation strategies focused on the establishment and application of corridors system to solve the relationship between protection and development comprehensively and efficiently [20–25]. In fact, Lewis (1964) put forward the linear concept of "environmental corridor" earlier, which contains surfaces, slopes, rims, and adjacent lands paralleling the corridor. The author also pointed out that the protection of environmental corridor qualities needs the joint efforts of better guidelines, legislation, and volunteered participation, and these corridors can serve as a landscape foil to an ever-advancing urban landscape [26]. Nevertheless, ecological or cultural corridors were always concerned separately rather than simultaneously. The aim of most ecological corridors is to maximize the value of regulating ecosystem services and biodiversity conservation by enhancing the connectivity among landscape elements [27,28], while recreation and tourism, sense of place, and heritage protection are the main functions of cultural corridors [23,29]. The aims, problems addressed, managing approaches, and spatial scale of these two aspects all have significant differences, and the scientific research, management practice and political discourses of ecological conservation and cultural heritage protection have largely been isolated from each other [30]. The participation of local residents is far lower than that of government decision-makers in the planning and management of both ecological corridors and cultural corridors, but the stakeholders and participation manners involved are quite different [30,31]. The combination of ecological corridors and cultural corridors is quite necessary to enrich the study framework and improve spatial integrity.

In this study, we address a few important gaps in the knowledge: (1) in the context of rapid urbanization, which leads to the disappearance of indigenous cultures and ecologically sensitive areas, comprehensive research on the combination of the ecological corridor and cultural corridor is not enough, and (2) current protection studies and practices of ecologically sensitive sites and cultural sites are not linked. We have designed this study to: (1) construct an integrated ecological and cultural corridor-based framework

(2) establish protection priorities by integrating protection values and protection costs, and (3) provide a scientific reference for optimizing the ecological spatial structure and promoting regional sustainable development. The rest of the paper is organized as follows. In Section 2, we describe the study area and our integrated framework. The application of the framework is presented in Section 3. In Section 4, we discuss our findings, and in Section 5, we summarize our conclusions.

2. Materials and Methods

2.1. Study Area

The study region, i.e., the city of Beijing (16,410 km^2), is located in Northern China and is characterized by a variety of landforms and a rich cultural history. The terrain of Beijing slopes downwards from the northwest to the southeast (Figure 1). The city's average population density was 1313 person/km^2, and the population density of the central urban area was 8929 person/km^2 in 2018. Along with rapid urbanization, the area of developed land dramatically increased from 485 km^2 in 1990 to 1525 km^2 in 2018. As a famous historical and cultural city, Beijing has hundreds of key cultural relic protection units, including 6 world heritage sites, such as the Great Wall and the Forbidden City.

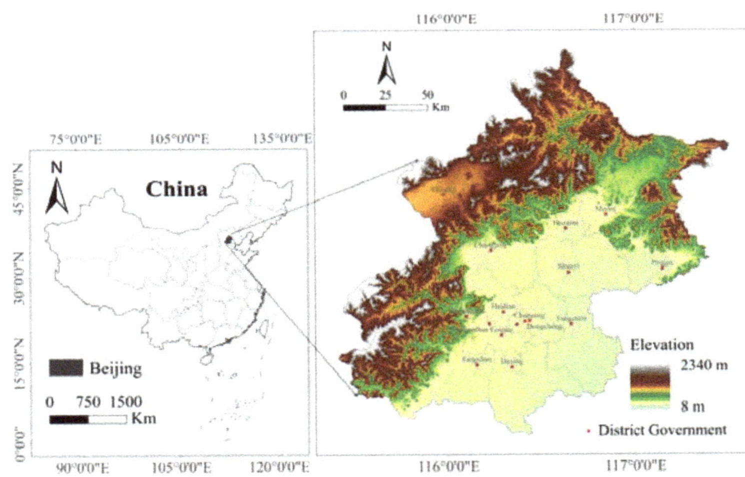

Figure 1. Location of Beijing and its Digital Elevation Model (DEM).

2.2. Data and Data Sources

The data used in this research include: (1) digital elevation model (DEM) data with a spatial resolution of 30 m, (2) land cover raster data with a spatial resolution of 30 m, which is reclassified into 6 types: developed land, forest land, farmland, grassland, water bodies, and unused land, (3) railroads, highways, and national roads in vector data format, (4) water bodies including rivers, lakes, and reservoirs in vector data format, (5) normalized difference vegetation index (NDVI) raster data with a spatial resolution of 1 km, which is resampled to 30 m, (6) the monthly and annual average precipitation of 18 weather stations from 1980 to 2012 in tabular data format, and among these weather stations, there are 6 in Beijing and 12 around Beijing, which can increase the accuracy of created raster in edge area, (7) soil composition and soil texture raster data with a spatial resolution of 1 km, which is resampled to 30 m, (8) nature reserves in vector data format, and (9) cultural heritage in vector data format. The spatial resolution of 30 m is used in all the raster calculations and outcomes. Appendix A Table A1 lists the data and data sources.

2.3. Methodological Framework

Figure 2 illustrates the conceptual framework of delineating the integrated network from ecological corridors and cultural corridors. Identifying ecological sources is the first step in delineating ecological corridors. Ecological sources consist of nature reserves, water bodies, and areas of high protection priority. Factors affecting the protection priority can be grouped into protection values and protection costs. Protection values reflect health and vulnerability, ecological importance, and the existing protection status of the targets [32,33]. Land use types, soil properties, and surface environment affect the ability to maintain biological resources and protection value. In this study, these factors were used in the process of deriving protection values from the amount of soil conservation. All protection interventions have associated costs, which include acquisition costs, management costs, transaction costs, damage costs, and opportunity costs, and protection costs are affected by many factors [34–37]. In this study, we included elevation (ELE), proximity to settlements (PSE), proximity to roads (PRO), and proximity to water bodies (PWA) in calculating the protection costs. Then, taking the protection costs as the resistance surface, paths connecting ecological sources were delineated as ecological corridors using the minimum cumulative resistance (MCR) model. When delineating cultural corridors, we considered designated cultural heritage sites to be cultural nodes. Since many cultural nodes are located along rivers and historical routes, we derived cultural corridors from the spatial distribution characteristics of cultural nodes and historical routes and rivers that have cultural and historical significance. Through a buffer analysis of the ecological sources and ecological corridors, ecologically important areas were obtained. Culturally important areas were acquired through a buffer analysis of cultural corridors. Then, ecologically important areas and culturally important areas were superimposed to obtain the comprehensive regions. Finally, the integrated network was delineated to connect the ecological sources, cultural nodes, ecological corridors, cultural corridors, and comprehensive regions.

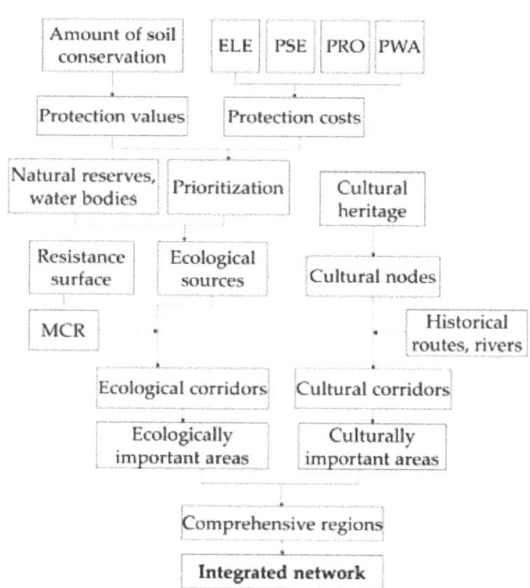

Figure 2. The framework for developing the integrated network.

2.4. Delineation of Ecological Corridors

2.4.1. Identification of Ecological Sources

We derived protection values from the amount of soil conservation. The function of soil conservation is mainly related to climate, soil characteristics, topography, and vegetation. We used the revised universal soil loss equation (RUSLE) to calculate the amount of soil conservation in this study. RUSLE, proposed by Renard, is the most frequently used model [34,38]. It provides an ideal framework for assessing soil erosion and a clear perspective for understanding the interaction between erosion and its contributing factors. The model and its parameters are illustrated below.

$$A = A_0 - A_1, \tag{1}$$

$$A_0 = R \cdot K \cdot L \cdot S, \tag{2}$$

$$A_1 = R \cdot K \cdot L \cdot S \cdot C \cdot P, \tag{3}$$

where A is the amount of soil conservation (t·ha^{-1}·year^{-1}), A_0 is the amount of potential soil loss, and A_1 is the amount of actual soil erosion loss. Other factors are explained as below:

(1) R represents the erosivity factor (MJ·mm·ha^{-1}·h^{-1}·year^{-1}), which is calculated by monthly precipitation and annual precipitation.

$$R = \sum_{i=1}^{12} \left(1.735 \times 10^{\left(1.5 \times lg \frac{P_i^2}{P} - 0.08188\right)} \right), \tag{4}$$

where P_i (mm) is the average precipitation of month i and P (mm) represents the multi-year average precipitation. We converted punctuated weather station data into raster data using the inverse distance weighted (IDW) tool in ArcGIS 10.5.

(2) K represents the soil erodibility factor calculated from the soil content (t·h·MJ^{-1}·mm^{-1}).

$$\begin{gathered} K = \{0.2 + 0.3exp[-0.0256SAN(1 - SIL/100)]\} \times \left[\frac{SIL}{(CLA+SIL)}\right]^{0.3} \times \\ \left\{1.0 - \frac{0.25orgC}{[orgC + \exp(3.72 - 2.95orgC)]}\right\} \times \left[1.0 - \frac{0.7SN1}{SN1 + \exp(-5.51 + 22.9SN1)}\right], \end{gathered} \tag{5}$$

where SAN, SIL, CLA, and $orgC$ represent the proportion of sand, silt, clay, and organic carbon in the soil respectively, $SN1 = 1 - SAN/100$.

(3) L represents the slope length factor.

$$L = (\lambda/22.13)^m, \tag{6}$$

$$m = \beta/(1+\beta), \tag{7}$$

$$\beta = (\sin\theta/0.0896)/[3.0(\sin\theta)^{0.8} + 0.56], \tag{8}$$

where λ is slope length, m is slope length index, and θ is slope measured in percentage.

(4) S represents the slope steepness factor.

$$S = \begin{cases} 10.8\sin\theta + 0.03 \ (\theta < 9\%) \\ 16.8\sin\theta - 0.50 \ (\theta \geq 9\%) \end{cases}, \tag{9}$$

(5) C is the vegetation cover management factor.

$$C = 0.6508 - 0.3436\log_{10} c, \tag{10}$$

$$c = \frac{NDVI - NDVI_{soil}}{NDVI_{veg} - NDVI_{soil}}, \tag{11}$$

where c is vegetation coverage, and $NDVI_{soil}$ and $NDVI_{veg}$ are the values of $NDVI$ when the confidence level is 5% and 95%, respectively.

(6) P is the support practices factor, which is a ratio of the soil loss with a conservation practice to soil loss from straight-row farming up and down the slope [39]. We used a P factor value of 1 in the study.

We normalized A into five classes using the Jenks Natural Breaks Classification method. The five class values represent the protection value (P_v). A P_v of "5" indicates the highest protection value and "1" the lowest. We derived the composite protection costs from elevation (ELE), proximity to settlements (PSE), proximity to roads (PRO), and proximity to water bodies (PWA). In China, land is state-owned or collective owned, and there are 2 types of costs due to conservation: the management cost of direct investment in the establishment and management of protection facilities, and the opportunity cost of abandoning the potential value of land economic use [40]. Management cost is mainly related to the type, area and local economic factors of the reserves [41]. Therefore, the 4 indicators related to elevation and distance in our study mainly affect the opportunity cost. Land in higher elevation areas is less desirable for development and less likely to be used for other economic purposes. Therefore, the opportunity cost is lower. The edge of settlements has a high probability of being used to build housing or other developments, so the opportunity cost lost due to conservation measures decreases as the distance to settlements increases. Roads can cause habitat fragmentation and ecosystem degradation, and land proximity to roads means it is potentially valuable for other economic use, so the opportunity cost is high near roads and decreases as the distance increases from roads. Considering the isolation effect of roads on ecological space, we selected the highways, national roads and railroads in this study. Surface water bodies have a function of conserving water sources, but waterfronts are usually places of human activity. Therefore, the larger the distance from water bodies, the lower the opportunity cost. We used the "Euclidean Distance" tool in ArcGIS 10.5 to obtain the PSE, PRO, and PWR, then we reclassified these indicators to values from "1" to "5" and calculated the protection cost (P_c) as the weighted sum of them (the natural breakpoint method was used to perform the reclassification in this paper). The weights of ELE, PSE, PRO, and PWA are 0.3564, 0.3257, 0.1986, 0.1243 respectively, according to the research of Tao [42]. The protection cost, P_c, was reclassified to five discrete values, with "5" indicating the lowest protection cost and "1" the highest protection cost.

The protection priority was obtained by averaging the protection value and protection cost. In other words, high protection priority areas should have both high protection values and low protection costs. Protection priorities were calculated with the following formula:

$$P_r = \frac{P_v + P_c}{2} \quad (12)$$

where P_v is the protection value, P_c is the protection cost, and P_r is the protection priority with a reclassified value from "1" to "5". Areas with a P_r value of "5" have the highest protection priority level and areas with "1" the lowest.

Finally, ecological sources were determined as areas with priority value of "5", and areas with the priority value of "4" that are located in nature reserves (see details of nature reserves from Appendix A Table A2) and water bodies. Nature reserves consist of wetlands and forests, which can reduce soil erosion, conserve water, adjust the local ecosystem and provide habitats for rare animals, birds, plants, and aquatic wildlife. Water bodies (such as lakes and reservoirs) play an important role in flood prevention and the propagation of aquatic organisms.

2.4.2. Delineation of Ecological Corridors

The MCR model is excellent in terms of expressing the interaction between landscape patterns and ecological processes [43,44]. The first step in the MCR model is to create a resistance surface according to land-use type [45]. Different land-use types have diverse impacts on the ecological resistance coefficient. Distance from developed land partially reflects the impact of human activities on the ecosystem [18]. Opinion on the resistance coefficient is consistent among researchers: the highest resistance coefficient is associated with developed land and decreases as distance to developed land increases [46,47]. We compiled resistance coefficients from the literature, as shown in Tables 1 and 2.

Table 1. Resistance coefficients of different land-use types (R_l, a higher value means higher resistance).

Land-Use Type	Resistance Coefficient (R_l)
Forest land	1
Grassland	10
Rivers and wetland	20
Reservoir	30
Farmland	50
Other land types	80
Developed land	100

Table 2. Resistance coefficients of different distances from developed land (R_d, a higher value means higher resistance).

Distance From Developed Land (m)	Resistance Coefficient (R_d)
0~100	100
100~200	70
200~300	60
300~500	40
500~1000	30
1000~3000	10
3000~5000	5
>5000	1

We calculated the resistance surface using the following formula:

$$R_w = \frac{R_l + R_d}{2}, \tag{13}$$

where R_w is the weighted resistance, R_l is the resistance coefficient of different land-use types, and R_d is the resistance coefficient of different distances from developed land. For example, the weighted resistance value of a grassland grid (R_i = 10) that is 600 m from developed land (R_d = 30) is 20 (= (10 + 30)/2). To improve the accuracy of ecological corridor identification at the boundary, the scope of analysis of the resistance surface is larger than Beijing's administrative boundary.

The second step is to calculate the cost distance to ecological sources through the weighted resistance surface. This process was implemented with the "Cost Distance" tool in ArcGIS 10.5. The input source data raster layer is ecological sources and the input cost raster layer is the weighted resistance surface. The third step is to identify the least-cost paths from any ecological source to other ecological sources using the "Cost Path" tool in ArcGIS 10.5, and this step gives several different paths near ecological sources. Lastly, the paths with the fewest intersections with roads between two sources were selected as ecological corridors.

2.5. Delineation of Cultural Corridors

2.5.1. Identification of Cultural Nodes

People tend to choose cultural spots and scenic spots for social interaction, relaxation, education, and inspiration. There are often rich cultural heritage sites in urban and surrounding areas that have a long history, including material cultural heritage: that of historical, artistic, and scientific value, and intangible cultural heritage: traditional cultures that are related to life in an immaterial form. Material cultural heritage and intangible cultural heritage are always accompanied in space [48]. However, these cultural heritage sites are under pressure from urbanization and tourism development, which threaten their original authenticity and integrity. In this study, we selected the national and municipal cultural heritage sites designated by the State Council of China as cultural nodes. These cultural heritage sites include ancient ruins, historic buildings, ancient tombs, lithoglyphs, historical and cultural towns/villages, etc., of significant historical and cultural value.

2.5.2. Delineation of Cultural Corridors

Cultural corridors are important bonds that link cultural nodes with human activities. They play an important role in tourism development, cultural heritage conservation, and the promotion of a sense of place [22,49,50]. The definition of a cultural corridor in this paper emphasizes its function of connectivity and inheritance. First of all, we identified the areas where cultural nodes are concentrated according to their distribution characteristics. Then, we delineated the routes and rivers that have important historical and cultural significance. We combined these two aspects to obtain the cultural corridors; in other words, the cultural corridors were determined according to whether a route or river has historical and cultural significance and whether there are a large number of cultural nodes around it. That is to say, cultural corridors are historical routes and rivers surrounded by abundant cultural nodes.

2.6. Delineation of the Integrated Network

The width of an ecological corridor has an important impact on the ecological function of the corridor. Some researchers believe that although different edge effects correspond to different corridor widths, generally speaking, the wider the ecological corridor, the better its ecological function [51,52]. On the basis of these research findings, we defined a 5 km buffer zone around ecological sources and ecological corridors as ecologically important areas. Herein, 5 km is a distance for future conservation planning and is not used as the actual corridor width. Similarly, we defined a buffer zone around cultural corridors as culturally important areas. Some researchers believe that the cultural corridor is an integrative concept with the purpose of integrating the conservation of cultural and natural resources [22,53]. Whether it is a corridor along a river valley, the Grand Canal corridor, or China's ancient Silk Road corridor, the conceptual scope is determined according to the objects to be protected [22,54]. Therefore, the buffer distance for cultural corridors in our study was determined by the distribution characteristics of cultural nodes. Comprehensive regions are the overlapping areas of ecologically and culturally important areas, which represent the areas with both ecological and cultural importance. The integrated network was then established by connecting ecological sources, cultural nodes, ecological corridors, cultural corridors, and comprehensive regions.

3. Results

3.1. Ecological Sources and Ecological Corridors

3.1.1. Ecological Sources

As illustrated in Figure 3, we drew out protection prioritization (Figure 3c) based on the protection value (Figure 3a) and protection cost (Figure 3b). The "priority 5" patches, which cover about 4.9% of the study area, are mainly located in the north and the southwest. There are 22 ecological sources after combining "priority 5" patches and "priority 4" patches in nature reserves and water bodies (Figure 3d). The total area of these ecological sources is 2011 km^2: about 12.3% of the study area. The vast majority of ecological sources are located in the northwest and the southwest. The largest ecological source (1150 km^2) is located in the northwest mountainous area and accounts for 57.2% of the ecological sources.

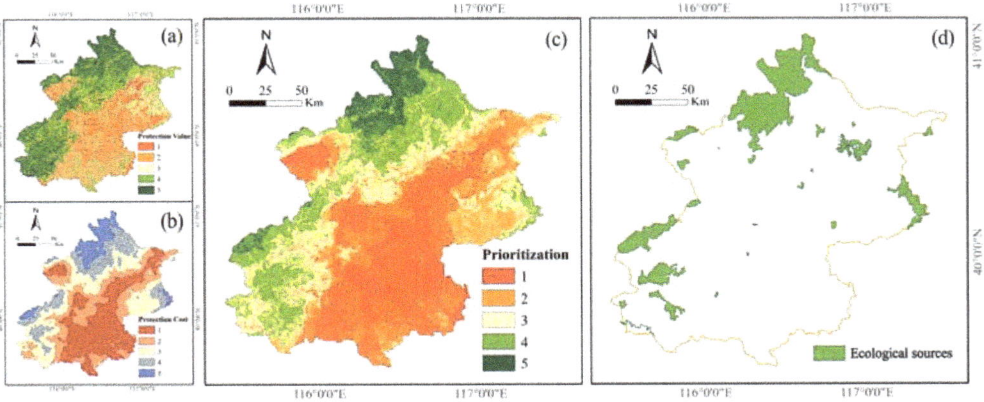

Figure 3. (**a**) Protection values, (**b**) protection costs, (**c**) protection prioritization, and (**d**) ecological sources.

3.1.2. Ecological Corridors

We can see from Figure 4 that ecological corridors connect all ecological sources together and interweave into an ecological network in space. There are 30 ecological corridors with a total length of 228 km, 6 of which extend to the central urban area, and another 7 to Hebei province. The connection to the central urban area encourages greenway construction in the central area of Beijing, and the ecological corridors extending to Hebei province can play an active role in regional collaboration. Hebei province may take these ecological corridors as a reference when they delineate their ecological corridors.

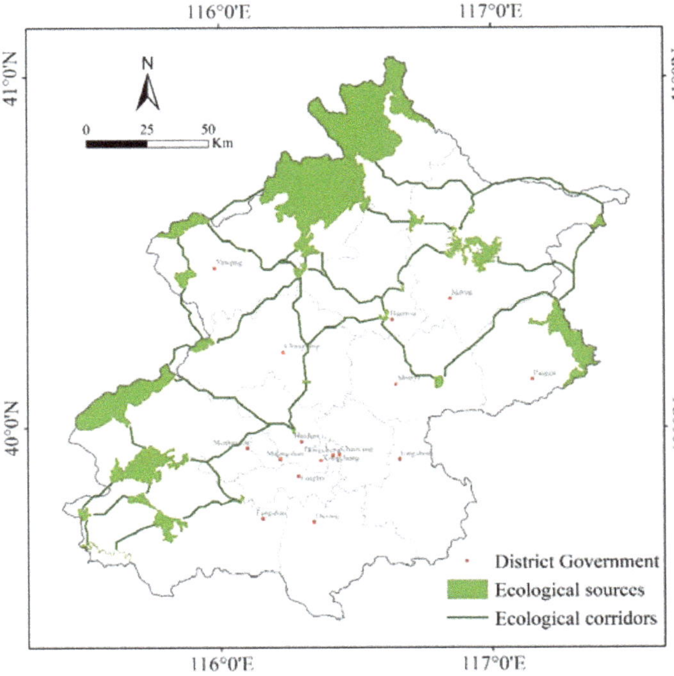

Figure 4. Ecological corridors.

3.2. Cultural Nodes and Cultural Corridors

3.2.1. Cultural Nodes

Figure 5 illustrates the 423 cultural nodes. More than half are in the central urban area and the rest are scattered in the peripheral area (see Appendix A Table A3 for details). More cultural nodes are in the southwest and fewer in the northeast. Six historical and cultural towns/villages are all in peripheral areas with lower levels of urbanization where many high-value historical buildings and ancient features are completely preserved. For example, Lingshui village has extremely rich cultural deposits that originated from the Ming Dynasty. In the north of Beijing, the Great Wall extends from the west to the northeast, along which there are many famous historical relics, such as Badaling, which is famous for its magnificent scenery, facilities, and profound history.

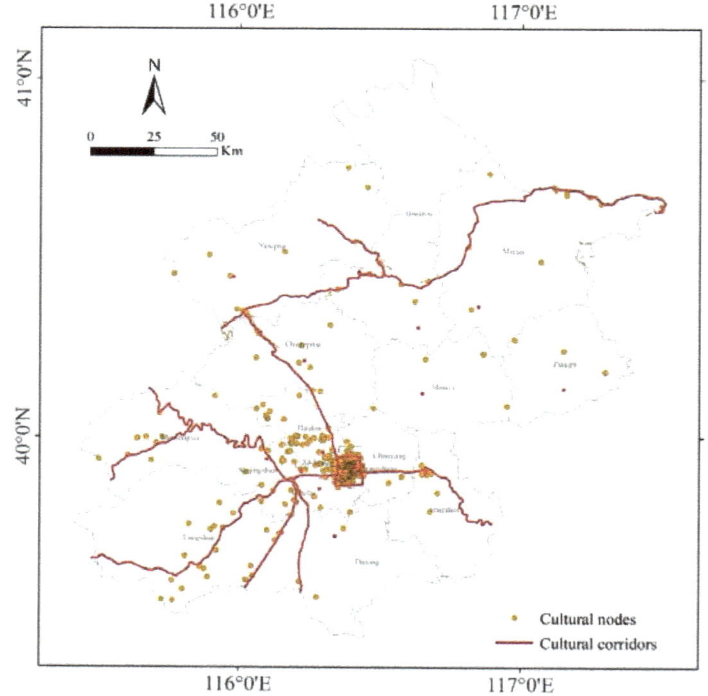

Figure 5. Cultural nodes and cultural corridors.

After observations and trials, we found that a total of 83% of the cultural nodes are within 5 km of the cultural corridors, and these cultural nodes show significant spatial agglomeration characteristics. Therefore, 5 km is an appropriate width to conduct a buffer analysis of cultural nodes in order to delineate the culturally important areas.

3.2.2. Cultural Corridors

Figure 5 presents the cultural corridors in Beijing. The total length of these cultural corridors is 851 km. They are listed below:

(1) The central district cultural corridor is bounded by the second ring road, which encloses 32.93% of Beijing's cultural heritage. There are many cultural nodes in the area along the corridor, which has a large population. Visits to the heritage sites are more frequent than in other cultural corridor areas. At the same time, the impact from outside is greater and protection is more difficult.

(2) The Great Wall cultural corridor extends along the ridgeline of the northern exurbs from west to northeast and covers famous heritage sites such as Badaling, Juyong Pass, Mutianyu, and Simatai. The development of the related cultural industry has promoted economic development and environmental protection in the area.
(3) The Yongding-Qingshui River cultural corridor is mainly in the Mentougou District, extending from the south to the source of the Yongding River and its tributary, Qingshui River. The rich history in this corridor has led to numerous ancient villages and buildings being located there. The cultural corridor plays an important role in promoting the ancient capital, improving the ecological environment, and providing recreational space.
(4) The Grand Canal cultural corridor extends from the city center to Hebei province, along which there are Huitong River, Tonghui River, Wanning bridge, and Dongbuya bridge. As the longest canal in the world, the Grand Canal has played an important role in the cultural and economic development and exchange between the north and the south of China. It was granted World Heritage site status in 2014.
(5) The Beijing–Guangzhou line cultural corridor along the Beijing–Guangzhou railroad to Hebei province begins in the city center. It is an ancient recreational route. There are numerous cultural nodes in the vicinity of the corridor, such as Lugou bridge, Liangxiang tower, and Liuli River bridge.
(6) The Beijing–Taiyuan line cultural corridor along the Beijing–Taiyuan railroad to Hebei province also originates in the city center. It is an ancient recreational route. Famous heritage sites along the cultural corridor include the Peking Man Site at Zhoukoudian, Tantuo temple, and Yao Guangxiao grave.
(7) The Beijing–Baotou line cultural corridor is one of the imperial roads from the Ming Dynasty and Qing Dynasty. It starts from the city center and extends along the Beijing–Baotou railroad to the Badaling Great Wall and Ming Tombs. Famous heritage sites along the cultural corridor include the Beijing–Zhangjiakou Railway and Juyong Pass.

3.3. The Integrated Network

As illustrated in Figure 6, the integrated network visually reveals the relationships between ecological sources, cultural nodes, ecological corridors, and cultural corridors. Ecological sources are mainly distributed in exurbs, while cultural nodes are mainly located in central urban areas. However, the suburbs where ecological sources and cultural nodes coexist, such as Badaling Great Wall, Lingshui village, and Gubei town, are often the most popular tourist destinations, because they have multiple functions including leisure, experiencing traditional culture, and acquiring knowledge. Ecological corridors connecting all ecological sources interweave into a network in the exurbs, while cultural corridors radiate outward from the central urban area. At the same time, ecological corridors and cultural corridors share some intersection areas in the suburbs, which provide convenient positions for us to identify comprehensive regions.

The comprehensive regions that have both ecological and cultural importance cover an area of 2916 km^2, and are mainly distributed in forestland. We numbered these regions from 1 to 10, as illustrated in Figure 6. The Great Wall landscape and the Great Wall culture are the outstanding features of Numbers 1–4. Number 5, a famous tourist destination, is a region with a concentrated distribution of historical and cultural heritage, characterized by royal gardens from the Qing Dynasty. There is a tremendous legacy and beautiful scenery along the Yongding river in Numbers 6–7. Historical and cultural villages and ancient buildings are concentrated in Number 8. There are a lot of celebrated cultural heritage sites in Number 9, such as the Zhoukoudian Peking Man Site, the Jin mausoleum site, etc. Number 10 is located in the famous Juma River scenic spot. In these regions, rich ecological assets and cultural nodes blend well in space, and ecological corridors coexist with cultural corridors. The integration of mountains and cultural heritage is the prominent feature of Chinese mountain culture.

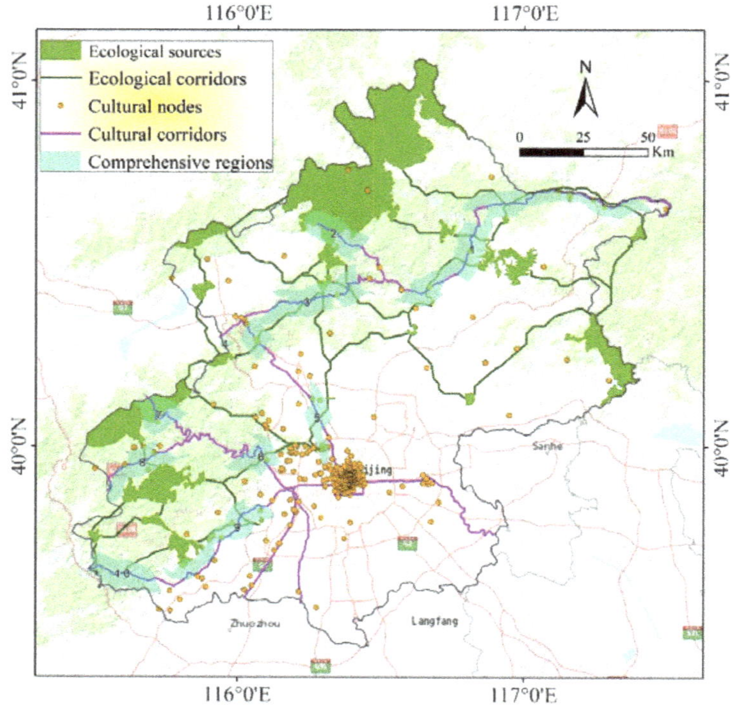

Figure 6. Integrated network.

4. Discussion

The coordination between protection and development is a vital basis for regional planning. Extreme protection or development can lead to unsustainable regional development [3]. Identifying ecological sources based on their protection priority is an effective way to build a smart protection plan. Special guidelines for protection and development can be formulated according to the different characteristics of priority levels, which is also beneficial in terms of the efficient use of funds. Areas in the most urgent state should receive the earliest attention, and urban construction and other development activities should be strictly restricted in these areas. In the case study of Beijing, ecological sources and cultural nodes have shown distinct spatial distribution characteristics. Ecological sources are mainly located in a mountainous region with a large area of forestland and grassland. Cultural nodes are concentrated in the built-up areas, and they are greatly and easily affected by human development activities. However, human beings have been seeking ways to organically integrate with the natural environment. For example, many temples, relics, and traditional villages are integrated into the natural environment, which attracts a large number of tourists every year and provides people with places for leisure activities.

The protection and construction of ecological corridors is still problematic because both ecological benefits and economic development are important for urbanized areas [55]. In the ecosystem, the barrier effect of road networks on biological pathways cannot be ignored. Unlike ecological corridors, cultural corridors in this paper are delineated according to the distribution features of cultural heritage, rivers, and historical routes, which is a subjective process. Although the size and influence radius of cultural heritage have not been taken into account, cultural corridor buffers reveal priority protection areas and systematize the protection of cultural heritage. These culturally important areas deserve priority heritage protection and strict construction control in order for them to retain their local cultural atmosphere.

In the comprehensive regions, in which ecological assets and cultural nodes coexist harmoniously in space, heritage protection, ecological protection, and landscape renovation are the main tasks. These comprehensive regions hold great importance in landscape planning and cultural heritage protection for decision-makers and planners. It is extremely necessary to apply strict and effective measures to reduce the destruction of natural ecology and cultural heritage as a result of human development activities. Some regions have done better in this regard, such as comprehensive region Number 8, and the ecological environment and historical and cultural villages all have been well protected. However, in some other regions, like Number 10, ecological problems such as habitat degradation and water quality deterioration have emerged due to the over-exploitation of tourism activities. In a nutshell, what these regions need is more systematic policies and management strategies. Some engineering and cultural heritage protection measures should be taken into consideration, such as building underground passages, overpasses, cultural squares, relic parks, and cultural attractions. Construction activities should be strictly controlled, and landscape renovation should be carried out in existing construction areas.

The heritage datasets used in our study are the list from official heritage discourse, which is reliant on the knowledge claims of technical and aesthetic experts, and institutionalized in state cultural agencies and amenity societies [56]. Previous studies have shown that public participation can be an effective tool in landscape planning and management [31], but the expression of subaltern discourses of community participation in heritage management and conservation processes in our study is insufficient. In order to reveal the potential elements, participatory strategies and field investigation should be included in future research, and experts involved in research, management, and marketing of cultural heritage as well as local residents should be involved in this process.

Although there are distinctive differences between ecological corridors and cultural ecological corridors, some potential similarities remain. For example, some researchers focused on the topics of cultural heritage, tourism, ecosystem services in landscape corridor study at the same time [57–59]. However, so far, the study of cultural corridors remains relatively weak compared to that of ecological corridors [18,30], and there has not been a systematic methodology framework to combine ecological corridors and cultural corridors in the same context. Therefore, it is quite necessary to enrich the research of corridor framework by combining ecological corridor with cultural corridor. Ecological protection is not a local or personal matter, but one that needs the collaboration of experts from different disciplines and policymakers from different regions. The protection of cultural relics is not only a matter for cultural relic workers but also planners and decision-makers. The integrated analysis of ecological corridors and cultural corridors is significant for the implementation of an integrated protection strategy.

5. Conclusions

In this paper, we presented a method to delineate an integrated network that combines ecological corridors and cultural corridors. Along the integrated framework, we located many cultural heritage sites in areas that would be considered suitable for development using the traditional method, which only considers ecological elements when defining protection zones. We identified these areas as not being suitable for large-scale development, as it will lead to the destruction or even disappearance of local traditional features. Ecological and cultural corridors perform different functions, but they serve the same purpose of improving the quality of life. By considering them together, we can demonstrate the integral connection between them and cohesively protect areas of ecological and/or cultural significance. The integrated network can provide more reasonable suggestions for the optimization of the urban spatial structure. Different well-directed protection strategies can be adopted for three types of regions: ecologically important areas, culturally important areas, and ecologically and culturally important areas (comprehensive regions), which were obtained by buffer analysis and superposition analysis of ecological sources, ecological corridors, and cultural corridors.

This research method has no strict limit in terms of study scale, so it can be applied on larger (urban agglomeration) and smaller (county) scales as long as the data accuracy is acceptable, making it easy for decision-makers to implement at different levels. Our framework on the integrated network combines ecological corridors and cultural corridors in the same context, which makes the research on the two no longer isolated. Based on the result, we promote coordination of diversified aims, such as ecosystem services evaluation, biodiversity conservation, recreation and ecotourism, and heritage protection.

There is still room for improvement in our integrated network. Because the ecological corridors identified in our study are lines, we plan to explore practical quantitative methods to determine the width of corridors to guide the construction of an ecological network in the future. We also plan to include more views of local communities in the next phase of the study. The current setting of cultural corridor buffer distance is uniformly applied to all corridors. We realize that the influence radii of different cultural heritage sites can be quite different and plan to further explore more suitable methods for delineating variable cultural corridor buffer distances, which could be a combination of quantitative methods and qualitative methods to incorporate site specific information. In addition, we plan to collect local input of cultural sites to complement the heritage datasets. Finally, comprehensive regions are currently delineated from overlay analysis. We plan to enhance the method with stakeholder input and field investigation. We anticipate that accurate and practical delineation of the integrated network will support sustainable development goals.

Author Contributions: Conceptualization, X.W., Y.L., and X.D.; methodology, Y.L. and X.W.; formal analysis, X.W. and Y.L.; investigation, Y.L.; data curation, Y.L.; writing—original draft preparation, Y.L.; writing—review and editing, X.W.; visualization, Y.L.; supervision, X.D.; project administration, X.D. All authors have read and agreed to the published version of the manuscript.

Funding: This research was financially supported by the Fundamental Funds for Humanities and Social Sciences of Beijing Jiaotong University (2016jdzd02).

Institutional Review Board Statement: Not applicable.

Informed Consent Statement: Not applicable.

Data Availability Statement: Data can be provided upon request from the corresponding author.

Acknowledgments: The authors also acknowledge with gratitude Liam Foley and Yanyan Zhu for their advice in editing this paper.

Conflicts of Interest: The authors declare no conflict of interest.

Appendix A

Table A1. Data sources in this study.

Data Name	Data Source
Digital Elevation Model (DEM)	Geospatial Data Cloud site, Computer Network Information Center, Chinese Academy of Sciences. (http://www.gscloud.cn)
Land cover data (2018) Roads (2018) Water bodies (2018) NDVI (2018)	International Scientific & Technical Data Mirror Site, Computer Network Information Center, Chinese Academy of Sciences. (http://www.resdc.cn)
Monthly average meteorological data (1980–2017)	China Meteorological Administration (http://cdc.nmic.cn/home.do)
Soil data	Cold and Arid Regions Sciences Data Center at Lanzhou (http://westdc.westgis.ac.cn)
Nature reserves (to 2017)	Beijing Municipal Environmental Protection Bureau (http://www.bjepb.gov.cn/bjhrb/index/index.html)
Cultural heritages (to 2019)	State Administration of Cultural Heritage (http://www.sach.gov.cn/)

Table A2. Nature reserves in Beijing (to 2017).

Name	Main Protection Objects	Type
Song mountain	Wild animals, such as golden eagle, natural oil pine forest	Forest ecosystem
Baihua Mountain	Temperate secondary forest, such as brown eared pheasant, arethusa and Dahurian larch	Forest ecosystem
Labagoumen	Natural secondary forest	Forest ecosystem
Wild duck lake	Wetland and migratory bird	Wetland
Yunmeng Mountain	Secondary forest	Forest ecosystem
Yunfeng Mountain	Secondary Pinus Tabulaeformis	Forest ecosystem
Wuling mountain	Valuable and rare animals and plants, natural secondary forests and typical forest ecosystems.	Forest ecosystem
Sizuolou	Natural secondary forest and national protected plants (wild soybean, Amur corktree, tilia amurensis and Acanthopanax)	Forest ecosystem
Yudu mountain	Forest and wild animals and plants	Forest ecosystem
Lianhua mountain	Wild animals and plants	Forest ecosystem
Datan	Natural secondary forest and wild animals and plants	Forest ecosystem
Jinniu lake	Wetland	Wetland
Baihebao	Water conservation forest	Forest ecosystem
Taian mountain	Forest and wild animals and plants	Forest ecosystem
Shuitou	Forest and wild animals and plants	Forest ecosystem
Puwa	Forest ecosystem	Forest ecosystem
Hanshiqiao	Wetland and migratory bird	Wetland
Juma river	Aquatic wildlife, such as Giant salamander	Wetland
Huaisha and Huaijiu river	Aquatic wildlife, such as Giant salamander, needle-mackerel and mandarin duck	Wetland
Shihuadong	Karst caverns	Geological heritage
Chaoyang temple	Fossil wood	Geological heritage

Table A3. List of Cultural Heritage Sites in Beijing.

Number	Name	Level
1	Ancient Cliff Dwelling Site	Beijing Municipal
2	Anti-Japanese War Sites of Yuzi Mountain	Beijing Municipal
3	Architectural Heritage of Beijing Normal University	Beijing Municipal
4	Architectural Heritage of Leshan Park	Beijing Municipal
5	Architecture Remains of Daci Yanfu Palace	Beijing Municipal
6	Architecture Remains of Nations' Affairs Office	Beijing Municipal
7	Back Hall of Capital City Temple	Beijing Municipal
8	Bai Yihua Martyr Cemetery	Beijing Municipal
9	Baipu Temple	Beijing Municipal
10	Baoguo Temple and Gu Tinglin Temple	Beijing Municipal
11	Beiguan Dragon Temple	Beijing Municipal
12	Beihai, Tuancheng	Beijing Municipal
13	Beijia Park	Beijing Municipal
14	Beijing Babaoshan Revolutionary Cemetery	Beijing Municipal
15	Beijng Newspaper Museum	Beijing Municipal
16	Bi Xia Yuanjun Temple Site of Yahuan Mountain	Beijing Municipal
17	Cai Yuanpei's Former Residence	Beijing Municipal
18	Changxindian "Twenty-seven" Revolutionary Sites	Beijing Municipal
19	Chaozhong Bridge	Beijing Municipal
20	Charity Temple	Beijing Municipal
21	Chen Duxiu's Former Residence	Beijing Municipal
22	Cheng Yanqiu's Former Residence	Beijing Municipal
23	Chengze Park	Beijing Municipal
24	Chinese Episcopal Church	Beijing Municipal
25	Church of St. Michael	Beijing Municipal
26	Clay City Site at Caizhuang	Beijing Municipal

Table A3. *Cont.*

Number	Name	Level
27	Coloured Glaze Factory of the Ministry of Works of Qing Dynasty	Beijing Municipal
28	Confucius Temple	Beijing Municipal
29	Congress of the Republic of China	Beijing Municipal
30	Cross-Street Building of Sanguan Pavilion	Beijing Municipal
31	Cuandixia Ancient Residential village	Beijing Municipal
32	Da Park	Beijing Municipal
33	De Shoutang Pharmacy	Beijing Municipal
34	Diaoyutai and Yangyuan Temple	Beijing Municipal
35	Dinghui Temple	Beijing Municipal
36	Dongsi Mosque	Beijing Municipal
37	DongYue Temple	Beijing Municipal
38	Dongyue Temple, Shangzhuang	Beijing Municipal
39	Doudian Clay City	Beijing Municipal
40	Drama Stage of Anhui Guide Hall	Beijing Municipal
41	Drama Stage of Yangping Guide Hall	Beijing Municipal
42	Early Buildings of Beijing Hotel	Beijing Municipal
43	Eight Sites of Xishan Mountain	Beijing Municipal
44	Fanzi Stone Inscription	Beijing Municipal
45	Female Normal College of Former National Beiping University	Beijing Municipal
46	Fire God Temple of Huashi	Beijing Municipal
47	Fomer Site of Continential Bank (Beijing)	Beijing Municipal
48	Forked Road City	Beijing Municipal
49	Former Furen Univeisity	Beijing Municipal
50	Former Peking Union Medical College	Beijing Municipal
51	Former Sino-France University	Beijing Municipal
52	Former Site of American Embassy	Beijing Municipal
53	Former Site of Austrian Embassy	Beijing Municipal
54	Former Site of Bank of Communications	Beijing Municipal
55	Former Site of Bank of Gold	Beijing Municipal
56	Former Site of Banque Indosuez (Beijing)	Beijing Municipal
57	Former Site of Baoshang Bank	Beijing Municipal
58	Former Site of Bazaar	Beijing Municipal
59	Former Site of Beijing Origin	Beijing Municipal
60	Former Site of Belgian Embassy	Beijing Municipal
61	Former Site of Branch College of Beijing Normal University	Beijing Municipal
62	Former Site of British Embassy	Beijing Municipal
63	Former Site of Central Bank	Beijing Municipal
64	Former site of China Geology Investigation Institute	Beijing Municipal
65	Former Site of Chinese Bible Society	Beijing Municipal
66	Former Site of Citibank	Beijing Municipal
67	Former Site of Duan Qirui Government	Beijing Municipal
68	Former Site of Dutch Embassy	Beijing Municipal
69	Former Site of Exhibition Hall of Geology, Beijing University	Beijing Municipal
70	Former Site of French Embassy	Beijing Municipal
71	Former Site of French Post Office	Beijing Municipal
72	Former Site of General Post Office of the Qing Dynasty	Beijing Municipal
73	Former Site of Japanese Embassy	Beijing Municipal
74	Former Site of Jiaoshi Building and Baiyou Building of Fuyu Female School	Beijing Municipal
75	Former Site of Nanyuan Army Headquarters	Beijing Municipal
76	Former Site of North Telephone Subexchang of Beiping	Beijing Municipal
77	Former Site of Notre Dame French School	Beijing Municipal
78	Former Site of Salt Industry Bank	Beijing Municipal
79	Former Site of Tongzhou Army	Beijing Municipal
80	Former site of work study program in France in Chang Xindian	Beijing Municipal
81	Former Site of Zhengyangmen East Station of Beijing Fengtian Railway	Beijing Municipal
82	Former Teaching Building of Luhe Middle School	Beijing Municipal
83	Fuguo Street Quadrangle, Xicheng District	Beijing Municipal

Table A3. *Cont.*

Number	Name	Level
84	Fusheng Temple	Beijing Municipal
85	Fuyou Temple	Beijing Municipal
86	Girderless Pavilion	Beijing Municipal
87	Glacial Striae	Beijing Municipal
88	Gold Hall of Huguo Temple	Beijing Municipal
89	Gonghua City	Beijing Municipal
90	Gonghua Palace	Beijing Municipal
91	Gongjian Ice Cellar	Beijing Municipal
92	Great Hall of Lingyan Temple	Beijing Municipal
93	Guangfu Temple	Beijing Municipal
94	Guanghua Temple	Beijing Municipal
95	Guangji Bridge	Beijing Municipal
96	Heilong Pool and Longwang Temple	Beijing Municipal
97	Heping Temple	Beijing Municipal
98	Hongluo Temple	Beijing Municipal
99	House at East Imperial Root South Street, Dongcheng District	Beijing Municipal
100	House at North Buzong Hutong, Dongcheng District	Beijing Municipal
101	House at Weijia Hutong, Dongcheng District	Beijing Municipal
102	Huguang Guide Hall	Beijing Municipal
103	Huixian Hall	Beijing Municipal
104	Hunan Guide Hall	Beijing Municipal
105	Huoshen Temple	Beijing Municipal
106	Imperial Ancestral Temple	Beijing Municipal
107	Imperial City Wall Site	Beijing Municipal
108	Imperial College Street	Beijing Municipal
109	Jade Emperor Tower	Beijing Municipal
110	Ji Xiaolan's Former Residence	Beijing Municipal
111	Jiaolao Tai	Beijing Municipal
112	Jiemin Hall	Beijing Municipal
113	Jinghua Publishing House	Beijing Municipal
114	Jingming Park	Beijing Municipal
115	Jingyi Park (Xiangshan Moutain)	Beijing Municipal
116	Jintai Academy	Beijing Municipal
117	Jiufeng Seismic Station	Beijing Municipal
118	Juesheng Temple	Beijing Municipal
119	Kang Youwei's Former Residence	Beijing Municipal
120	Lao She's Former Residence	Beijing Municipal
121	Lejia Garden	Beijing Municipal
122	Li Dazhao's Former Residence	Beijing Municipal
123	Liangxiang Tower	Beijing Municipal
124	Lingzhao Temple	Beijing Municipal
125	Liuyang Guide Hall	Beijing Municipal
126	Long'an Temple	Beijing Municipal
127	Longquan Temple of Bailong Pool	Beijing Municipal
128	Lotus Pool	Beijing Municipal
129	Lu Xun's Former Residence	Beijing Municipal
130	Lumi Warehouse	Beijing Municipal
131	Lvzu Pavilion	Beijing Municipal
132	Main Building of Beijng Library	Beijing Municipal
133	Main Building of Italian Embassy Site	Beijing Municipal
134	Mansion of Beile Tao	Beijing Municipal
135	Mansion of Crown Prince of Ning County	Beijing Municipal
136	Mansion of Crown Prince of Shuncheng County	Beijing Municipal
137	Mansion of King Chun	Beijing Municipal
138	Mansion of King Fu	Beijing Municipal
139	Mansion of King Heng	Beijing Municipal
140	Mansion of King Li	Beijing Municipal

Table A3. *Cont.*

Number	Name	Level
141	Mansion of King Qing	Beijing Municipal
142	Mansion of King Seng	Beijing Municipal
143	Mansion of King Zheng	Beijing Municipal
144	Mansion of Princess Hejing	Beijing Municipal
145	Mao Dun's Former Residence	Beijing Municipal
146	Martyr Li Dazhao cemetery	Beijing Municipal
147	Mei Lanfang's Former Residence	Beijing Municipal
148	Memorial of Sun Yat Sen's Death	Beijing Municipal
149	Monument to the Luanzhou Uprising	Beijing Municipal
150	Monument to the Martyrs who Died in the Anti-Japanese War in Wanping County	Beijing Municipal
151	Nangangwa Bridge	Beijing Municipal
152	Niangniang Temple of North Peak	Beijing Municipal
153	Nianhua Temple	Beijing Municipal
154	Ninghe Temple	Beijing Municipal
155	Niujie Street Mosque	Beijing Municipal
156	No.36 Fuxue Hutong, Dongcheng District	Beijing Municipal
157	No.63–65 Quadrangle, Dongsiliutiao, Dongcheng District	Beijing Municipal
158	North Guide Hall of Tingzhou, Fujian	Beijing Municipal
159	North New Warehouse	Beijing Municipal
160	Old Messuage	Beijing Municipal
161	Old Style Shops	Beijing Municipal
162	Public Elder Longevity Tower	Beijing Municipal
163	Publishing Factory Site of Ministry of Finance of the Republic of China	Beijing Municipal
164	Puzhao Temple	Beijing Municipal
165	Qi Baishi's Former Residence	Beijing Municipal
166	Qinghe Hancheng Site	Beijing Municipal
167	Qingyin Pavilion of Yunhui Building	Beijing Municipal
168	Randeng Tower	Beijing Municipal
169	Remains of Mansion of Zhaohui	Beijing Municipal
170	Remains of the School Department in Qing Dynasty	Beijing Municipal
171	Residence Group of Union Hospital	Beijing Municipal
172	Riverside City and Enemy Platform	Beijing Municipal
173	Rong Tomb Site at Yuhuangmiao Mountain	Beijing Municipal
174	Ruins of Shang, Zhou Dynasty at Liu Lihe	Beijing Municipal
175	Sansheng Temple	Beijing Municipal
176	Shangzhai Cultural Site	Beijing Municipal
177	Shaoxing Guide Hall	Beijing Municipal
178	Sheng Pingshu Drama Stage	Beijing Municipal
179	Shifang Buddists Tower	Beijing Municipal
180	Shijia Hutong, Dongcheng District	Beijing Municipal
181	Shuanglin Temple	Beijing Municipal
182	Shuiguan Great Wall	Beijing Municipal
183	Shuntian State-run School	Beijing Municipal
184	Shuqu Square Stele of Zhengyang Bridge	Beijing Municipal
185	Sibei Temple, Tao ranting	Beijing Municipal
186	No.11 Quadrangle, Neiwubu Street, Dongcheng District	Beijing Municipal
187	No.11 Quadrangle, Xisibei 3, Xicheng District	Beijing Municipal
188	No.129 Quadrangle, Lishi Hutong, Dongcheng District	Beijing Municipal
189	No.13,15 Quadrangle, Fangjia Hutong, Dongcheng District	Beijing Municipal
190	No.15 Quadrangle, Qian Gongyong Hutong, Xicheng District	Beijing Municipal
191	No.19 Quadrangle, Xisibei 3, Xicheng District	Beijing Municipal
192	No.2 Quadrangle, Guoxiang Hutong, Dongcheng District	Beijing Municipal
193	No.20 Quadrangle, Xinkai Road, Chongwen District	Beijing Municipal
194	No.23 Quadrangle, Xisibei Avenue 6, Xicheng District	Beijing Municipal
195	No.7,9 Quadrangle, Back Yuan'en Temple Street, Dongcheng District	Beijing Municipal
196	No.15 Quadrangle and its Tile Carving, Dongmianhua Hutong, Dongcheng District	Beijing Municipal
197	No.25 Quadrangle, Art Museum East Street, Dongcheng District	Beijing Municipal

Table A3. *Cont.*

Number	Name	Level
198	No.153 Quadrangle, Di'anmenxi Avenue, Xicheng District	Beijing Municipal
199	No.93 Quadrangle, Fuchengmennei Street, Xicheng District	Beijing Municipal
200	No.255 Quadrangle, Gulou East Avenue, Dongcheng District	Beijing Municipal
201	No.13 Quadrangle, Heizhima Hutong, Dongcheng District	Beijing Municipal
202	No.5 Quadrangle, Maoer Hutong, Dongcheng District	Beijing Municipal
203	No.7,9 Quadrangle, Qiangulouyuan Hutong, Dongcheng District	Beijing Municipal
204	No.7,9 Quadrangle, Qianyongkang Hutong, Dongcheng District	Beijing Municipal
205	No.15 Quadrangle, Shajing Hutong, Dongcheng District	Beijing Municipal
206	Quadrangle, Xijiaominxiang Street 87, Beixinhua Street 112, Xicheng District	Beijing Municipal
207	No.25–37 Quadrangle, Xitangzi Hutong, Dongcheng District	Beijing Municipal
208	Site of Chinese Communist Delegation of the Ministry of Military Transfer in 1946	Beijing Municipal
209	Site of European and American students' Association	Beijing Municipal
210	Site of Hebei-Rehe-Chahaer Advance Army Commander's Headquarter of Eight Route Army	Beijing Municipal
211	Site of Taiye Pool at the Mid-capital of the Kin Dynasty	Beijing Municipal
212	Site of the Farmer and Worker Bank of China	Beijing Municipal
213	Site of Tuanhe Palace	Beijing Municipal
214	Site of Xizhi Men Station of Pingsui	Beijing Municipal
215	Sites of Bai Fuquan	Beijing Municipal
216	Sites of Da Baotai Tombs of Western Han Dynasty	Beijing Municipal
217	Songzhu Temple and Zhizhu Temple	Beijing Municipal
218	South Mansion of King Chun	Beijing Municipal
219	South New Warehouse	Beijing Municipal
220	Southeastern Corner Tower of the City	Beijing Municipal
221	Stage and Guandi Temple at Huapen Village	Beijing Municipal
222	Stone Buddha in Baishui Temple	Beijing Municipal
223	Stone Sculptures of Yi Sanga	Beijing Municipal
224	Temple of Empress of Fengtai	Beijing Municipal
225	Temples and Yunshui Cave at Shangfang Moutain	Beijing Municipal
226	The Bell Tower	Beijing Municipal
227	The Chairman Mao Memorial Hall	Beijing Municipal
228	The Drum Tower	Beijing Municipal
229	The Former Chartered Bank	Beijing Municipal
230	The Front of Former Site of Qian Xiangyi	Beijing Municipal
231	The Front of Former Site of Rui Fuxiang	Beijing Municipal
232	The Front of Quanjude Roast Duck	Beijing Municipal
233	The Great Hall of Pudu Temple	Beijing Municipal
234	The Great Hall of Shuntian Mansion	Beijing Municipal
235	The Interior Office of Inspecting the Imperial Government in Qing Dynasty	Beijing Municipal
236	The Land Altar (Zhongshan Park)	Beijing Municipal
237	The Remains of Beijing City Wall in Ming Dynasty	Beijing Municipal
238	The statue of Wei Taihe	Beijing Municipal
239	The Tenth Hotel at Grain Shop Street	Beijing Municipal
240	Three Eighteen Martyr Monument	Beijing Municipal
241	Tianli Coal Factory Site	Beijing Municipal
242	Tiewa Temple	Beijing Municipal
243	Tomb of Crown Prince of Fu County	Beijing Municipal
244	Tomb of Laoshan of Han Dynasty	Beijing Municipal
245	Tomb of Li Zhuowu	Beijing Municipal
246	Tomb of Liang Qichao	Beijing Municipal
247	Tomb of Sunyue	Beijing Municipal
248	Tomb of Tianyi	Beijing Municipal
249	Tomb, Stele and Status of Zhan Tianyou	Beijing Municipal
250	Tombs of Lu Huixiang's Family	Beijing Municipal
251	Tombs of Soldiers and Men Killed in the Battle of Gubeikou	Beijing Municipal
252	Tongyun Bridge and Remains of Zhang Jiawan City Wall	Beijing Municipal
253	Tongzhou Mosque	Beijing Municipal
254	Tower of Zen Master Wuai	Beijing Municipal
255	Tuancheng Fortress	Beijing Municipal
256	Tucheng	Beijing Municipal

Table A3. *Cont.*

Number	Name	Level
257	Wanning Bridge	Beijing Municipal
258	Waterworks of Qing Dynasty	Beijing Municipal
259	Weiming Lake District, Former Yanjing University	Beijing Municipal
260	Wofo Temple	Beijing Municipal
261	Wuta Tower	Beijing Municipal
262	Xianliang Temple	Beijing Municipal
263	Xianying Temple	Beijing Municipal
264	Xihuang Temple	Beijing Municipal
265	Xishiku Church	Beijing Municipal
266	Xiuyun Temple	Beijing Municipal
267	Xuanren Temple	Beijing Municipal
268	Xuechi Ice Cellar	Beijing Municipal
269	Xuhua Pavilion and Song Hall	Beijing Municipal
270	Yandun	Beijing Municipal
271	Yang Jiaoshan Temple	Beijing Municipal
272	Yaowang Temple of Fengtai	Beijing Municipal
273	Yinshan Tower	Beijing Municipal
274	Yiyuan Park	Beijing Municipal
275	Yongning Catholic Church	Beijing Municipal
276	Yuansheng Palace	Beijing Municipal
277	Yun Tai	Beijing Municipal
278	Yuqian Temple	Beijing Municipal
279	Zhao Tower	Beijing Municipal
280	Zhaoxian Temple	Beijing Municipal
281	Zhengyang Gate and Arrow tower	Beijing Municipal
282	Zhengyi Temple	Beijing Municipal
283	Zhongshan Guide Hall	Beijing Municipal
284	Zhou Jixiang Tower	Beijing Municipal
285	Zhu Yizhun's Former Residence	Beijing Municipal
286	Cuandixia Village, Zhaitang Town	National
287	Gubeikou Town	National
288	Jiaozhuanghu Village, Longwantun Town	National
289	Lingshui Village, Zhaitang Town	National
290	Liuliqu Village, Longquan Town	National
291	Shuiyu Village, Nanjiao Town	National
292	Ancient Building Group in Cuandixia Village	National
293	Ancient Cliff Dwelling in Yanqing	National
294	Ancient Weather Station	National
295	Anhui Guide Hall	National
296	Badaling Great Wall	National
297	Baiyun Temple	National
298	Baoguo Temple	National
299	Beihai and Tuancheng	National
300	Beijing Bell Tower, Drum Tower	National
301	Beijing City Walls of the Ming Dynasty	National
302	Beijing Huguang Guide Hall	National
303	Biyun Temple	National
304	Bolin Temple	National
305	Bridge of Liuli River	National
306	Buddists Hall, Stone Inscription and Tower of Kongshui Cave	National
307	Changchun Temple	National
308	Cheng'en Temple	National
309	Chinese Episcopal Church	National
310	Chongli Residence	National
311	Church of Xi Shiku	National
312	Cishou Temple Tower	National
313	Commerce Building in Dashila Area	National
314	Confucius Temple	National

Table A3. *Cont.*

Number	Name	Level
315	Da Gaoxuan Hall	National
316	Dahui Temple	National
317	Dajue Temple	National
318	Desheng Gate Archery Tower	National
319	Dongyue Temple, Beijing	National
320	Duobao Buddist Tower of Liangxiang	National
321	Early Architectures inTshinghua Univeisity	National
322	East Hall	National
323	Embassy Architecture Group of Dongjiaominxiang	National
324	Fahai Temple	National
325	Fayuan Temple	National
326	Former Site of Agricultural Experimental Farm of Qing Dynasty	National
327	Former Site of Beijing Branch of China Bible Church	National
328	Former Site of Beijing Female Normal College	National
329	Former Site of Beijing Parliament	National
330	Former Site of Beiping Library	National
331	Former Site of Branch College of Beijing Normal University	National
332	Former Site of Exhibition Hall of Geology, Beijing University	National
333	Former Site of Main Campus of Furen University	National
334	Former Site of National Mongolian Tibetan School	National
335	Former Site of Peking Union Medical College	National
336	Former Site of Radio 491	National
337	Former Sites of Shengxin Middle School and Youzhen Female Middle School	National
338	Former Sites of the Army and Navy Departments of the Qing Dynasty	National
339	Gate of Heavenly Peace	National
340	Guangji Temple	National
341	Guanyue Temple	National
342	Guide Hall of Sun Yat-sen	National
343	Guo Moruo's Former Residence	National
344	Imperial Ancestral Temple	National
345	Imperial Archives	National
346	Imperial College	National
347	Jianruiying Martial Arts Hall	National
348	Jiaozhuanghu Tunnel Battle Site	National
349	Jietai Temple	National
350	Jingming Park	National
351	Juesheng Temple	National
352	Ke Park	National
353	Li Dazhao's Former Residence	National
354	Lingyue Temple	National
355	Lu Xun's Former Residence in Beijing	National
356	Lugou Bridge	National
357	Mansion of Crown Prince of Keqin County	National
358	Mansion of King Chun	National
359	Mansion of King Fu	National
360	Mei Lanfang's Former Residence	National
361	Memorial Park of Luanzhou Uprising of Xinhai Year	National
362	Modern Bank Building Group in Xijiaominxiang	National
363	Modern School Buildings Group in Tongzhou	National
364	Moke Temple	National
365	Mout Jing	National
366	Nankou-Badaling Section of Jing-Zhang Raiway	National
367	Niujie Street Mosque	National
368	Peking Man Site at Zhoukoudian	National
369	Prince Gong's Mansion and Park	National
370	Publishing House Site of Ministry of Finance of the Republic of China	National
371	Pudu Temple	National
372	Qingjinghuacheng Tower	National
373	Red Building in Peking University	National

Table A3. *Cont.*

Number	Name	Level
374	Remains of Great-Capital City Wall of Yuan Dynasty	National
375	Shangzhai Site	National
376	Shifang Pujue Temple	National
377	Shuangqing Villa	National
378	Shuiguan Site of the Mid-Capital of the Kin Dynasty	National
379	Simatai Section of Great Wall	National
380	Site of Jinling	National
381	Site of Liuli River	National
382	Site of Old Summer Palace	National
383	Site of Shizi Temple	National
384	Song Qingling Children's Science and Technology Museum	National
385	Song Qingling's Former Residence in Beijing	National
386	South Hall	National
387	Southeastern Corner Tower of the Beijing City	National
388	Staiton Building of Beijng Railway Station	National
389	Summer Palace	National
390	Tantuo Temple	National
391	Temple of Heaven	National
392	Temple of the Moon	National
393	Temple of the Past-Ages Emperors	National
394	Temple of Wen Tianxiang	National
395	The Altar to the Sun	National
396	The Forbidden City	National
397	The Lama Temple	National
398	The Land Altar	National
399	The Ming Tombs	National
400	The Site of Atomic Energy	National
401	The Site of the Two-Seven Strike in Changxindian	National
402	The Monument to the People's Heroes	National
403	Tianning Temple Tower	National
404	Tomb and Temple of Yuan Chonghuan	National
405	Tomb of Jingtai	National
406	Tomb of King Chun	National
407	Tombs of Matteo Ricci and Foreign Missionaries of Ming, Qing Dynasties	National
408	Wanshou Temple	National
409	Wansong Elder Tower	National
410	White Tower of Miaoying Temple	National
411	Wuta Temple Tower	National
412	Xiannong Altar	National
413	Yanyuan Buildings of Weiming Lake	National
414	Yao Guangxiao Tomb Tower	National
415	Yasili Hall	National
416	Yinshan Forest of Pagodas	National
417	Yun Terrace in Juyong Pass	National
418	Yunju Temple Tower and Stone Scripture of Fangshan	National
419	Zhengang Tower	National
420	Zhengyang Gate	National
421	Zhihua Temple	National
422	Zhizhu Temple	National
423	Zhong Nan Hai	National

References

1. Fang, C.; Yu, D. Urban agglomeration: An evolving concept of an emerging phenomenon. *Landsc. Urban Plan.* **2017**, *162*, 126–136. [CrossRef]
2. Nations, U. Growing at a Slower Pace, World Population is Expected to Reach 9.7 Billion in 2050 and Could Peak at Nearly 11 Billion around 2100. Available online: https://www.un.org/development/desa/en/news/population/world-population-prospects-2019.html (accessed on 23 March 2020).

3. Li, Y.; Li, Y.; Zhou, Y.; Shi, Y.; Zhu, X. Investigation of a coupling model of coordination between urbanization and the environment. *J. Environ. Manag.* **2012**, *98*, 127–133. [CrossRef] [PubMed]
4. Gunawardhana, L.N.; Kazama, S.; Kawagoe, S. Impact of Urbanization and Climate Change on Aquifer Thermal Regimes. *Water Resour. Manag.* **2011**, *25*, 3247–3276. [CrossRef]
5. Breuste, J.; Qureshi, S.; Li, J. Applied urban ecology for sustainable urban environment. *Urban Ecosyst.* **2013**, *16*, 675–680. [CrossRef]
6. Taylor, L.; Hochuli, D.F. Creating better cities: How biodiversity and ecosystem functioning enhance urban residents' wellbeing. *Urban Ecosyst.* **2015**, *18*, 747–762. [CrossRef]
7. Breuste, J.; Qureshi, S. Urban sustainability, urban ecology and the Society for Urban Ecology (SURE). *Urban Ecosyst.* **2011**, *14*, 313–317. [CrossRef]
8. Wu, J. Urban sustainability: An inevitable goal of landscape research. *Landsc. Ecol.* **2010**, *25*, 1–4. [CrossRef]
9. Wu, J. Landscape sustainability science: Ecosystem services and human well-being in changing landscapes. *Landsc. Ecol.* **2013**, *28*, 999–1023. [CrossRef]
10. Assessment, M.E. *Ecosystems and Human Well-Being: A Framework for Assessment*; Island Press: Washington, DC, USA, 2003.
11. Daily, G.C. *Nature's Services: Societal Dependence on Natural Ecosystems*; Island Press: Washington, DC, USA, 1997.
12. Costanza, R.; d'Arge, R.; de Groot, R.; Farber, S.; Grasso, M.; Hannon, B.; Limburg, K.; Naeem, S.; O'Neill, R.V.; Paruelo, J.; et al. The value of the world's ecosystem services and natural capital. *Nature* **1997**, *387*, 253–260. [CrossRef]
13. Nahlik, A.M.; Kentula, M.E.; Fennessy, M.S.; Landers, D.H. Where is the consensus? A proposed foundation for moving ecosystem service concepts into practice. *Ecol. Econ.* **2012**, *77*, 27–35. [CrossRef]
14. Kubiszewski, I.; Costanza, R.; Anderson, S.; Sutton, P. The future value of ecosystem services: Global scenarios and national implications. *Ecosyst. Serv.* **2017**, *26*, 289–301. [CrossRef]
15. Braat, L.C.; de Groot, R. The ecosystem services agenda:bridging the worlds of natural science and economics, conservation and development, and public and private policy. *Ecosyst. Serv.* **2012**, *1*, 4–15. [CrossRef]
16. Costanza, R.; de Groot, R.; Sutton, P.; van der Ploeg, S.; Anderson, S.J.; Kubiszewski, I.; Farber, S.; Turner, R.K. Changes in the global value of ecosystem services. *Glob. Environ. Chang. Part A Hum. Policy Dimens.* **2014**, *26*, 152–158. [CrossRef]
17. Wolff, S.; Schulp, C.J.E.; Verburg, P.H. Mapping ecosystem services demand: A review of current research and future perspectives. *Ecol. Indic.* **2015**, *55*, 159–171. [CrossRef]
18. Zhang, L.; Peng, J.; Liu, Y.; Wu, J. Coupling ecosystem services supply and human ecological demand to identify landscape ecological security pattern: A case study in Beijing–Tianjin–Hebei region, China. *Urban Ecosyst.* **2017**, *20*, 701–714. [CrossRef]
19. Xie, H.; Yao, G.; Liu, G. Spatial evaluation of the ecological importance based on GIS for environmental management: A case study in Xingguo county of China. *Ecol. Indic.* **2015**, *51*, 3–12. [CrossRef]
20. Campolo, D.; Bombino, G.; Meduri, T. Cultural Landscape and Cultural Routes: Infrastructure Role and Indigenous Knowledge for a Sustainable Development of Inland Areas. *Procedia-Soc. Behav. Sci.* **2016**, *223*, 576–582. [CrossRef]
21. Liang, J.; He, X.; Zeng, G.; Zhong, M.; Gao, X.; Li, X.; Li, X.; Wu, H.; Feng, C.; Xing, W.; et al. Integrating priority areas and ecological corridors into national network for conservation planning in China. *Sci. Total Environ.* **2018**, *626*, 22–29. [CrossRef]
22. Hoppert, M.; Bahn, B.; Bergmeier, E.; Deutsch, M.; Epperlein, K.; Hallmann, C.; Mller, A.; Platz, T.V.; Reeh, T.; Stck, H.; et al. The Saale-Unstrut cultural landscape corridor. *Environ. Earth Sci.* **2018**, *77*, 58. [CrossRef]
23. Oikonomopoulou, E.a.; Delegou, E.T.A.; Sayas, J.B.; Moropoulou, A.A. An innovative approach to the protection of cultural heritage: The case of cultural routes in Chios Island, Greece. *J. Archaeol. Sci. Rep.* **2017**, *14*, 742–757. [CrossRef]
24. Pierik, M.; Dell'Acqua, M.; Confalonieri, R.; Bocchi, S.; Gomarasca, S. Designing ecological corridors in a fragmented landscape: A fuzzy approach to circuit connectivity analysis. *Ecol. Indic.* **2016**, *67*, 807–820. [CrossRef]
25. Ye, H.; Yang, Z.; Xu, X. Ecological Corridors Analysis Based on MSPA and MCR Model—A Case Study of the Tomur World Natural Heritage Region. *Sustainability* **2020**, *12*, 959. [CrossRef]
26. Lewis, P.H. Landscape Resources. In *The Wisconsin Blue Book*; Rupert, T.H., Ed.; Wisconsin Legislative Reference Bureau: Madison, WI, USA, 1964; pp. 130–142.
27. Lee, J.A.; Chon, J.; Ahn, C. Planning Landscape Corridors in Ecological Infrastructure Using Least-Cost Path Methods Based on the Value of Ecosystem Services. *Sustainability* **2014**, *6*, 7564–7585. [CrossRef]
28. Mossman, H.L.; Panter, C.J.; Dolman, P.M. Modelling biodiversity distribution in agricultural landscapes to support ecological network planning. *Landsc. Urban Plan.* **2015**, *141*, 59–67. [CrossRef]
29. Shishmanova, M.V. Cultural Tourism in Cultural Corridors, Itineraries, Areas and Cores Networked. *Procedia-Soc. Behav. Sci.* **2015**, *188*, 246–254. [CrossRef]
30. Xu, H.; Plieninger, T.; Primdahl, J. A Systematic Comparison of Cultural and Ecological Landscape Corridors in Europe. *Land* **2019**, *8*, 41. [CrossRef]
31. Eiter, S.; Vik, M.L. Public participation in landscape planning: Effective methods for implementing the European Landscape Convention in Norway. *Land Use Policy* **2015**, *44*, 44–53. [CrossRef]
32. Ahn, S.; Kim, S. Assessment of watershed health, vulnerability and resilience for determining protection and restoration Priorities. *Environ. Model. Softw.* **2019**, *122*, 103926. [CrossRef]
33. Opdam, P.; Steingröver, E.; Rooij, S.V. Ecological networks: A spatial concept for multi-actor planning of sustainable landscapes. *Landsc. Urban Plan.* **2006**, *75*, 322–332. [CrossRef]

34. Naidoo, R.; Balmford, A.; Ferraro, P.J.; Polasky, S.; Ricketts, T.H.; Rouget, M. Integrating economic costs into conservation planning. *Trends Ecol. Evol.* **2006**, *21*, 681–687. [CrossRef]
35. Prahl, B.F.; Boettle, M.; Costa, L.; Kropp, J.P.; Rybski, D. Damage and protection cost curves for coastal floods within the 600 largest European cities. *Sci. Data* **2018**, *5*, 1–18. [CrossRef] [PubMed]
36. Zhu, M.; Xi, X.; Hoctor, T.S.; Volk, M. Integrating conservation costs into sea level rise adaptive conservation prioritization. *Glob. Ecol. Conserv.* **2015**, *4*, 48–62. [CrossRef]
37. Manhães, A.P.; Loyola, R.; Mazzochini, G.G.; Ganade, G.; Oliveira-Filho, A.T.; Carvalho, A.R. Low-cost strategies for protecting ecosystem services and biodiversity. *Biol. Conserv.* **2018**, *217*, 187–194. [CrossRef]
38. Alexakis, D.D.; Hadjimitsis, D.G.; Agapiou, A. Integrated use of remote sensing, GIS and precipitation data for the assessment of soil erosion rate in the catchment area of "Yialias" in Cyprus. *Atmos. Res.* **2013**, *131*, 108–124. [CrossRef]
39. Renard, K.G.; Foster, F.G.; Weesies, G.A.; McCool, K.K.; Yoder, D.C. *Predicting Soil Loss Erosion by Water: A Guide to Conservation Planning with the Revised Universal Soil Loss Equation (RUSLE)*; US Government Printing Office: Washington, DC, USA, 1997.
40. Yang, Z.; Wu, J. Conservation cost of China's nature reserves and its regional distribution. *J. Nat. Resour.* **2019**, *34*, 839–852.
41. Armsworth, P.R.; Cantú-Salazar, L.; Parnell, M.; Davies, Z.G.; Stoneman, R. Management costs for small protected areas and economies of scale in habitat conservation. *Biol. Conserv.* **2011**, *144*, 423–429. [CrossRef]
42. Tao, G.; Ou, X.; Guo, Y.; Xu, Q.; Yu, Q.; Zhang, Z.; Wang, C. Priority area identification for vegetation in northwest Yunnan, based on protection value and protection cost. *Acta Ecol. Sin.* **2016**, *36*, 5777–5789.
43. Huck, M.; Jędrzejewski, W.; Borowik, T.; Jędrzejewska, B.; Nowak, S.; Mysłajek, R.W. Analyses of least cost paths for determining effects of habitat types on landscape permeability: Wolves in Poland. *Acta Theriol.* **2011**, *56*, 91–101. [CrossRef]
44. Knaapen, J.P.; Scheffer, M.; Harms, B. Estimating habitat isolation in landscape planning. *Landsc. Urban Plan.* **1992**, *23*, 1–16. [CrossRef]
45. Gurrutxaga, M.; Rubio, L.; Saura, S. Key connectors in protected forest area networks and the impact of highways: A transnational case study from the Cantabrian Range to the Western Alps (SW Europe). *Landsc. Urban Plan.* **2011**, *101*, 310–320. [CrossRef]
46. Hepcan, C.C.; Ozkan, M.B. Establishing ecological networks for habitat conservation in the case of Çeşme-Urla Peninsula, Turkey. *Environ. Monit. Assess.* **2011**, *174*, 157–170. [CrossRef] [PubMed]
47. Yu, K.; Wang, S.; Li, D. *Regional Ecological Security Patterns: The Beijing Case. Beijing: China*; Architecture & Building Press: Beijing, China, 2012.
48. UNESCO. What is Intangible Cultural Heritage? Available online: https://ich.unesco.org/en/what-is-intangible-heritage-00003/ (accessed on 15 November 2020).
49. Bozic, S.; Tomic, N. Developing the Cultural Route Evaluation Model (CREM) and its application on the Trail of Roman Emperors. *Tour. Manag. Perspect.* **2016**, *17*, 26–35. [CrossRef]
50. Xu, H.; Zhao, G.; Fagerholm, N.; Primdahl, J.; Plieninger, T. Participatory mapping of cultural ecosystem services for landscape corridor planning: A case study of the Silk Roads corridor in Zhangye, China. *J. Environ. Manag.* **2020**, *264*, 110458. [CrossRef] [PubMed]
51. Zhu, Q.; Yu, K.; Li, D. The width of ecological corridor in landscape planning. *Acta Ecol. Sin.* **2005**, *25*, 2406–2412.
52. Wu, J.; Li, X.; Long, N.; Xiao, J. Ecological Corridor Network Planning for Dense City Areas: A Case Study of Guangzhou Metropolitan Area. *Mod. Urban Res.* **2017**, *2017*, 61–67. [CrossRef]
53. Fabos, J.G. Introduction and overview: The greenway movement, uses and potentials of greenways. *Landsc. Urban Plan.* **1995**, *33*, 1–13. [CrossRef]
54. Li, L.; Shao, M.; Wang, S.; Li, Z. Preservation of earthen heritage sites on the Silk Road, northwest China from the impact of the environment. *Environ. Earth Sci.* **2011**, *64*, 1625–1639. [CrossRef]
55. Peng, J.; Zhao, H.; Liu, Y. Urban ecological corridors construction: A review. *Acta Ecol. Sin.* **2017**, *37*, 23–30. [CrossRef]
56. Smith, L. *Uses of Heritage*; Routeledge: London, UK; New York, NY, USA, 2006.
57. Von Haaren, C.; Reich, M. The German way to greenways and habitat networks. *Landsc. Urban Plan.* **2006**, *76*, 7–22. [CrossRef]
58. Manton, R.; Clifford, E. Identification and classification of factors affecting route selection of cycling routes in Ireland. *Cycl. Res. Int.* **2012**, *3*, 136–153.
59. Ottomano Palmisano, G.; Loisi, R.V.; Ruggiero, G.; Rocchi, L.; Boggia, A.; Roma, R.; Dal Sasso, P. Using Analytic Network Process and Dominance-based Rough Set Approach for sustainable requalification of traditional farm buildings in Southern Italy. *Land Use Policy* **2016**, *59*, 95–110. [CrossRef]

sustainability

MDPI

Article

Biostimulation of Microbial Communities from Malaysian Agricultural Soil for Detoxification of Metanil Yellow Dye; a Response Surface Methodological Approach

Fatin Natasha Amira Muliadi [1], Mohd Izuan Effendi Halmi [1,*], Samsuri Bin Abdul Wahid [1], Siti Salwa Abd Gani [2], Uswatun Hasanah Zaidan [3], Khairil Mahmud [4] and Mohd Yunus Abd Shukor [3]

1 Department of Land Management, Faculty of Agriculture, Universiti Putra Malaysia, Serdang 43400, Selangor, Malaysia; fatin.muliadi@gmail.com (F.N.A.M.); samsuriaw@upm.edu.my (S.B.A.W.)
2 Department of Agricultural Technology, Faculty of Agriculture, Universiti Putra Malaysia, Serdang 43400, Selangor, Malaysia; ssalwaag@upm.edu.my
3 Department of Biochemistry, Faculty of Biotechnology and Biomolecular Sciences, Universiti Putra Malaysia, Serdang 43400, Selangor, Malaysia; uswatun@upm.edu.my (U.H.Z.); mohdyunus@upm.edu.my (M.Y.A.S.)
4 Department of Crop Science, Faculty of Agriculture, Universiti Putra Malaysia, Serdang 43400, Selangor, Malaysia; khairilmahmud@upm.edu.my
* Correspondence: m_izuaneffendi@upm.edu.my; Tel.: +60-3-97694958

Abstract: In the present study, a mixed culture from a local agricultural soil sample was isolated for Metanil Yellow (MY) dye decolorization. The metagenomic analysis confirmed that 42.6% has been dominated by genus *Bacillus*, while *Acinetobacter* (14.0%) is present in the microbial communities of the mixed culture. For fungi diversity analysis, around 97.0% was "unclassified" fungi and 3% was Candida. The preliminary investigation in minimal salt media (MSM) showed that 100% decolorization was achieved after 24 h of incubation. Response surface methodology (RSM) was successfully applied using Box-Behnken design (BBD) to study the effect of four independent parameters—MY dye concentration, glucose concentration, ammonium sulfate concentration, and pH—on MY dye decolorization by the mixed bacterial culture. The optimal conditions predicted by the desirability function were 73 mg/L of MY, 1.934% glucose, 0.433 g/L of ammonium sulfate, and a pH of 7.097, with 97.551% decolorization The correlation coefficients (R^2 and R^2 adj) of 0.913 and 0.825 indicate that the established model is suitable to predict the effectiveness of dye decolorization under the investigated condition. The MY decolorization of the mixed bacterial culture was not affected by the addition of heavy metals in the growth media. Among the 10 heavy metals tested, only copper gave 56.19% MY decolorization, whereas the others gave almost 100% decolorization. The decolorization potential of the mixed bacterial culture indicates that it could be effective for future bioremediation of soil-contaminated sites and treatment solutions of water bodies polluted with the MY dye.

Keywords: bioremediation; dye decolorization; mixed culture; metanil yellow; response surface methodology

Citation: Muliadi, F.N.A.; Halmi, M.I.E.; Wahid, S.B.A.; Gani, S.S.A.; Zaidan, U.H.; Mahmud, K.; Abd Shukor, M.Y. Biostimulation of Microbial Communities from Malaysian Agricultural Soil for Detoxification of Metanil Yellow Dye; a Response Surface Methodological Approach. *Sustainability* **2021**, *13*, 138. https://dx.doi.org/10.3390/su13010138

Received: 30 September 2020
Accepted: 27 November 2020
Published: 25 December 2020

1. Introduction

Azo dyes are widely used in various industries, such as textile, food, paper-making, and cosmetic industries [1]. Azo dyes constitute about one million tons of production, and about 300,000 tons of different dyestuffs have been used per year for the operations of textile dyeing [2]. The increased demand for azo dye textile products has produced effluents that lead to severe water pollution. Metanil Yellow (MY) (monosodium salt of 4-m-sulphophenylazodyephenylamine) is one type of acidic azo dye. MY is commonly used as soap coloring, spirit lacquer, shoe polish, and bloom sheep dip, as well as for the preparation of food stains, leather dyeing, the manufacturing of pigment lakes, and paper staining. It is also used as food colorants in various foodstuffs, mostly in India [2]. However, it was later discovered that MY is carcinogenic [2]. From previous toxicity data, it was shown that when MY was fed to animals, the animals developed testicular lesions, leading to a decrease in the rate of spermatogenesis [2]. According to Saratale (2009), exposure to the dyes leads to potential health hazards, such as asthma, rhinitis, and dermatitis. Apart from the hazards to human health, azo dyes also affect ecosystems. When wastewater effluents are discharged into open water sources, they affect the photosynthetic activity of aquatic organisms [3], and the dissolved dyes may also affect the aquatic organisms since the breakdown products may also be toxic [4].

Removal of dyes from wastewater has gained scientific interest among researchers. As they are relatively recalcitrant to biodegradation, the elimination of colored effluents in wastewater treatment systems is mainly based on physical or chemical procedures such as adsorption, chemical transformation, and incineration [5]. However, these methods are very costly, have high sludge production, and result in the formation of secondary by-products [6]. Therefore, biological methods for removing dyes must be developed because they are environmentally friendly, economical, and more cost-effective compared to physical or chemical procedures [7]. Biodegradation by different microorganisms, which uses bacteria as a dye decolorizing agent, appears to be an attractive alternative [8].

Microbial decolorization occurs under several conditions, such as anaerobic, anoxic and aerobic conditions. For decolorization under aerobic conditions, the bacteria usually require organic carbon sources since they cannot utilize the dye as a growth substrate [8]. Generally, the degradation and decolorization of azo dyes by bacteria proceeds in two stages. The first stage involves reductive cleavage of the dyes' azo linkages, resulting in the formation of generally colorless but potentially hazardous aromatic amines. The second stage involves the degradation of aromatic amines [9]. Reductive cleavage of the –N=N- bond is the initial step in the bacterial degradation of azo dyes. It can occur under different types of mechanisms, such as through enzymatic reactions involving azoreductase and laccase enzymes, low-molecular-weight redox mediators, chemical reduction by biogenic reductants like sulphide, or a combination of these mechanisms [8]. The use of enzymes is beneficial due to their substrate specificity and may be effectively used in textile water pretreatment [10]. During azoreductase, enzymatic dye degradation, the azo bond (–N=N–) is cleaved by the enzyme and four electrons are transferred as the reducing equivalent. In each stage, two electrons are transferred to the azo dye, which is the electron acceptor and decolorization occurs when the colorless solution is formed. The resulting intermediate is a toxic aromatic amine, which is later degraded by the aerobic process or sometimes microaerophilically [10].

It has been reported that bacterial dye decolorization and degradation can occur by a pure culture of bacteria and by a mixed culture of bacteria. According to a previous study it has been reported that mixed bacterial culture can give a better degradation rate than the individual strain [11]. Individual species have the limited metabolic capability to mineralize the dye completely, and in many cases, it has been clearly observed that due to the lack of a catabolic pathway, aromatic amines are not further degraded. Catabolic and syntrophic interactions between indigenous species lead to complete degradation of azo dyes [12]. Azo dyes are not readily metabolized under aerobic conditions, and as a result of metabolic pathways, they degrade into intermediate compounds but not mineralized. They can be completely degraded under coupled aerobic and anaerobic degradation. Therefore, coupled anaerobic treatment followed by aerobic treatment can be an efficient and effective degradation method of azo dyes.

The One Factor at a Time (OFAT) method involves varying a single independent variable while the other variables are constant. This approach, however, is laborious, time-consuming, and incomplete. Thus, response surface methodology (RSM) using Box–Behnken design (BBD), which involves examining the simultaneous, systematic, and efficient variation of the important parameters used to model the decolorization process, identifies the possible interactions and higher-order effects and determines the optimum operational conditions [13].

The Box–Behnken statistical structure is an RSM plan with an autonomous, rotatable quadratic plan with treatment blends at the inside and midpoints at the edges of the procedure space. This structure requires less exploratory runs and less time compared to other RSM models. Thus, it is a more practical system. RSM is a collection of statistical and mathematical techniques that are useful for developing, improving, and optimizing processes. In this study, decolorization of MY by a newly isolated mixed bacterial culture from agricultural soil was optimized using RSM. The BBD matrix was chosen, and MY dye concentration, glucose concentration, ammonium sulfate concentration, and pH were used in this optimization study. This study was carried out to isolate, characterize, and optimize MY decolorization from various agricultural soil samples. It was discovered that most of the bacterial degrading dye was isolated from non-agricultural soil. Thus, the isolation of mixed bacterial cultures from agricultural soil samples improves the bioremediation process. To the best of our knowledge, no study has reported the application of RSM for optimizing the media used in the decolorization of the MY dye. Thus, this finding will provide significant results for MY decolorization using mixed bacterial culture.

2. Materials and Methods

2.1. Isolation and Screening of Mixed Cultures of Bacteria from Agricultural Soil

Forty samples of agricultural soils from different locations were collected for the isolation of bacteria. All soil samples were screened to isolate and identify the presence of the MY decolorizing mixed bacterial culture from agricultural soil and non-dye-contaminated environments. Until today, no studies have shown that mixed bacterial cultures isolated from non-dye-contaminated environments have been able to decolorize MY. Thus, this demonstrates the novelty of this research. For the first screening, 1 g of soil was weighed and grown in minimal salt media (MSM) composed of glucose (1%), $(NH_4)_2SO_4$ (0.40 g/L), KH_2PO_4 (0.20 g/L), K_2HPO_4 (0.40 g/L), NaCl (0.10 g/L), $Na_2M_oO_4$ (0.01 g/L), $MgSO_4.7H_2O$ (0.10 g/L), $Fe_2(SO_4)_3H_2O$ (0.01 g/L), $MnSO_4.H_2O$ (0.01 g/L), and yeast extract (1.00 g/L) supplemented with 50 mg/L of MY dye in a 250 mL conical flask. The cultivation of bacteria was conducted at room temperature on a rotary shaker (120 rpm). After the first screening, the secondary screening was performed by exposing the chosen mixed culture from the first screening to different dye concentrations ranging from 100 mg/L to 400 mg/L. The absorbance of the dye was measured by aseptically drawing 1.0 mL of the growth media and centrifuged it at 10,000$\times$ g for 10 min. The absorbance of the supernatant of the sample was measured at 434 using a UV—visible spectrophotometer. The mixed bacterial culture that was able to decolorize a high dye concentration was chosen for further study. The chosen mixed bacterial culture was maintained by subculturing in MSM consisting of 100 mg/L of the MY dye and 1.0% glucose in a conical flask. The decolorizing efficiency of the dye was expressed as the percentage of decolorization.

$$\%\ \text{Decolourisation} = \frac{(\text{Initial absorbance} - \text{Final absorbance})}{\text{Initial absorbance}} \times 100\%.$$

2.2. Identification of the Chosen Mixed Bacterial Culture Using Metagenomic Analysis

The metagenomic analysis is a method that is used to investigate complex microbial communities from environmental samples without culturing or isolating a single organism [14]. It is estimated that about 99% of microorganisms that are present in natural environments are not readily culturable, and this has made it impossible to investigate the functional roles of different microbes in a certain niche. Therefore, metagenomics studies have made it possible to analyze complex genomes in a microbial community [15].

In this study, the chosen mixed bacterial culture was sent to Apical Scientific Sdn Bhd for metagenomics analysis. 16s rRNA is used for the diversity analysis of bacteria communities, while the internal transcribed spacer (ITS) is used for the diversity analysis of fungi. To identify the genus of a certain microorganism, the 16s rRNA gene is used since it is the "gold standard" that is routinely used in classifying prokaryotes. For the metagenomic analysis, the steps include experimental design, sampling, sample fractionation, DNA extraction, DNA sequencing, assembly, binning, annotation, and statistical analysis [16,17].

In brief, the forward and reverse reads were merged using FLASH 2 and quality screened for sequence length and nucleotide ambiguity. All sequences that were shorter than 150 bp or longer than 600 bp (sequenced on the MiSeq platform) were removed from downstream processing. Reads were then aligned with 16S rRNA or the UNITE ITS database and inspected for chimeric errors. After these quality assessment steps, reads were clustered at 97% similarity into operational taxonomy units (OTUs).

2.3. Optimisation of Significant Parameters Using Response Surface Methodology (RSM)

The experimental process of optimizing MY dye decolorization was conducted using RSM with the BBD as an experimental matrix. In this study, RSM was used to investigate the relationships between four different significant parameters—dye concentration (A), glucose concentration (B), ammonium sulfate concentration (C), and pH (D)—and dye decolorization as a response and to optimize the relevant conditions of the variables and to predict the optimal conditions. The results obtained from the experiment were statistically analyzed using the Design Expert version 6.0.10, and a total of 29 runs were carried out. The results were analyzed using the Design Expert add-ons program, including analysis of variance (ANOVA), to determine the interaction between the variables and the response. The quality of the fit of this model was expressed by the coefficient of determination (R^2) in the same program [18]. All of the parameters were studied based on the range in Table 1.

Table 1. The upper limit and lower limit of Box–Behnken design (BBD).

Parameters	Unit	Upper Limit	Lower Limit
Dye concentration	mg/L	200.0	50.0
Glucose concentration	%	2.0	0.5
Ammonium sulphate concentration	g/L	1.0	0.1
pH		6.0	7.5

To start the experiment, 1.0% of the fresh mixed bacterial culture was grown in MSM supplemented with various concentrations of dye (A), glucose (B), and ammonium sulfate (C) and at varying pH values (D), based on Table 2. The design experiments were carried out in conical flasks and the conical flasks were incubated on a rotary shaker (120 rpm) for 24 h. After 24 h, 1.0 mL of the sample was withdrawn and centrifuged at 10,000× *g* for 10 min. The supernatant of the sample was measured at 434 nm using a UV–visible spectrophotometer, and the percentage of dye decolorization was calculated using the formula given above.

2.4. Effect of Different Initial Dye Concentrations on Dye Decolorisation and Bacterial Growth

The mixed bacterial culture was grown in MSM supplemented with 50 mg/L of MY and incubated on a rotary shaker for 24 h. Next, 1.0% of the mixed bacterial culture was inoculated in MSM supplemented with different concentrations of MY (30, 60, 90, 120, 130, 150, 180, 210, 240, 270, and 300 mg/L). This was performed in duplicates. They were then incubated at 120 rpm on a rotary shaker at room temperature. Every 2 h for the first 24 h, 1.0 mL of the culture was aseptically withdrawn and centrifuged at 10,000× *g* for 10 min. The supernatant was measured at 434 nm. After 24 h, the reading was taken every 6 h until 66 h, and then the reading was taken every 12 h until 114 h.

2.5. Effect of Heavy Metal Ions on Dye Decolorisation

In this study, the effect of heavy metals on the dye colorization of free mixed bacterial was evaluated. MSM was separately supplied with 1.0 mg/L of copper (Cu), arsenic (As), zinc (Zn), chromium (Cr), nickel (Ni), silver (Ag), lead (Pb), and mercury (Hg). The free mixed bacterial cells were prepared based on optimum conditions that were previously obtained from RSM. The mixed bacterial culture was grown in MSM supplemented with

50 mg/L of MY dye and 1.0 mg/L of the heavy metals listed. This was performed in duplicates. The mixed bacterial culture was then incubated at 120 rpm on a rotary shaker at room temperature. Bacterial cultures without metals served as controls. After 24 h of incubation, 1.0 mL of the culture was withdrawn aseptically and centrifuged at 10,000× g for 10 min. The supernatant was measured at 434 nm, and the percentage of dye decolorization was calculated using the formula above.

Table 2. Experimental runs predicted by response surface methodology (RSM).

Run	Factor 1 A: Dye Concentration (mg/L)	Factor 2: B: Glucose Concentration (%)	Factor 3: C: Ammonium Sulphate Concentration (g/L)	Factor 4: D: pH
1	50	2.00	0.55	6.75
2	50	1.25	0.55	7.50
3	125	1.25	1.00	6.00
4	125	1.25	1.00	7.50
5	200	1.25	0.55	7.50
6	200	1.25	1.00	6.75
7	200	1.25	0.55	6.00
8	125	2.00	1.00	6.75
9	50	1.25	0.55	6.00
10	50	0.50	0.55	6.75
11	125	200	0.55	6.00
12	125	1.25	0.10	6.00
13	200	1.25	0.10	6.75
14	125	0.50	1.00	6.75
15	125	0.50	0.55	6.00
16	125	0.50	0.10	6.75
17	125	1.25	0.55	6.75
18	50	1.25	1.00	6.75
19	200	0.50	0.55	6.75
20	125	1.25	0.55	6.75
21	50	1.25	0.10	6.75
22	125	2.00	0.10	6.75
23	125	2.00	0.55	7.50
24	200	2.00	0.55	6.75
25	125	1.25	0.55	6.75
26	125	1.25	0.55	6.75
27	125	1.25	0.55	6.75
28	125	1.25	0.10	7.50
29	125	0.50	0.55	7.50

3. Results

3.1. Isolation and Screening of Mixed Cultures of Bacteria from the Soil Sample

In this study, 40 agricultural soil samples were collected for the mixed culture MY degrader. After the secondary screening process, isolates, namely FN3 from oil palm estate soil collected from the Universiti Putra Malaysia (UPM), GPS 2.9876, 101.7234, were selected for further study. This was because isolate FN3 was able to decolorize up to 200 mg/L of dye compared to the other isolates, which could only decolorize up to 100 mg/L of MY dye. Isolate FN3 was able to decolorize 92.93% of 200 mg/L of MY, whereas the other isolates were only able to decolorize 100 mg/L of MY [19].

3.2. Metagenomic Analysis of Mixed Bacterial Culture

We performed a metagenomics analysis of mixed bacterial culture FN3 by using 16s rRNA for bacterial identification and ITS for fungi identification. Bacterial 16s rRNA genes have nine hypervariable regions (V1–V9) that indicate sequence diversity among the different bacteria. This is used for bacterial diversity analysis [14,20]. For fungi diversity analysis, ITS was used. ITS is located between the 18S and 5.8S rRNA genes and has a high degree of sequence variation. The chosen mixed bacterial culture was sent to Apical Scientific Sdn. Bhd, Malaysia for metagenomic analysis. For bacterial identification using 16S rRNA, out of the nine hypervariable regions, regions V3–V4 were chosen as the primer. The sequences were grouped and clustered into OTUs. In brief, the forward and reverse reads were merged using FLASH 2 and quality screened for sequence length and nucleotide ambiguity. All sequences that were shorter than 150 bp or longer than 600 bp (sequenced on the MiSeq platform) were removed from downstream processing. Reads were then aligned with the 16S rRNA database and inspected for chimeric errors. After these quality assessment steps, reads were clustered at 97% similarity into OTUs, rare OTUs with only one (singleton) or two reads (doubleton), which are often spurious, were deleted from downstream processing.

We identified 97% as Kingdom bacteria which is a major part of the microbial communities. Out of this 97%, 45% was comprised of phylum Firmicutes, 44% was comprised of phylum Proteobacteria, and 9% was comprised of phylum Bacteroidetes. This shows that the major communities come from phylum Firmicutes. From this phylum, it has been 100% identified that the class of bacteria is Bacilli, the order of bacteria was Bacillales, the family was Bacillaceae, and the genus of bacteria was 100% Bacillus, as in Figure 1.

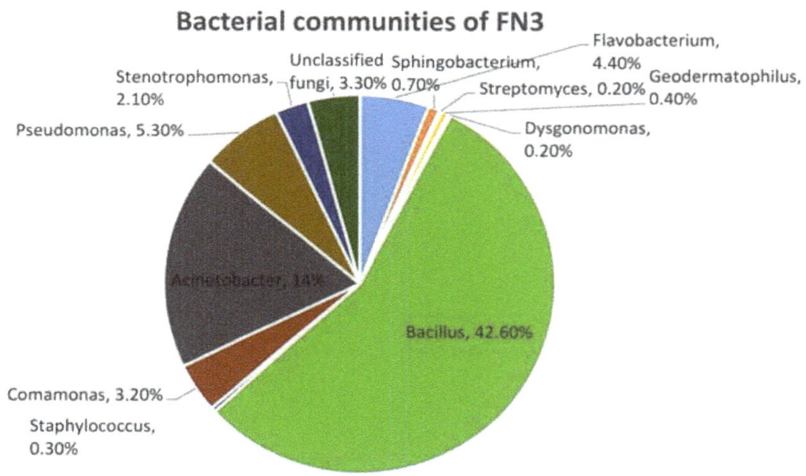

Figure 1. The bacterial communities that were identified in mixed bacterial culture FN3 by metagenomic analysis.

In Figure 2, it can be seen that a total of 115,952 reads of sequences composed of 97% unclassified fungi and 3% fungi from phylum Ascomycota were identified. In the phylum Ascomycota, the fungi genus identified was Candida. For an OTU to be categorized as "unclassified," based on the UNITE database, there is no "species" that is annotated up to a rank. This has given the perspective that it might be a novel species [21].

3.3. Optimisation of Significant Variables Using the Box-Behnken Design (BBD)

This study focused on the combined effects of four significant variables for the decolorization of the MY dye by mixed bacteria. To optimize the process variables for maximal dye decolorization, 29 experimental runs were conducted. Table 3 shows the experimental and response results together with the response predicted by RSM. Maximal and minimal dye decolorization was observed in runs 1 and 3, respectively, as shown in Table 3.

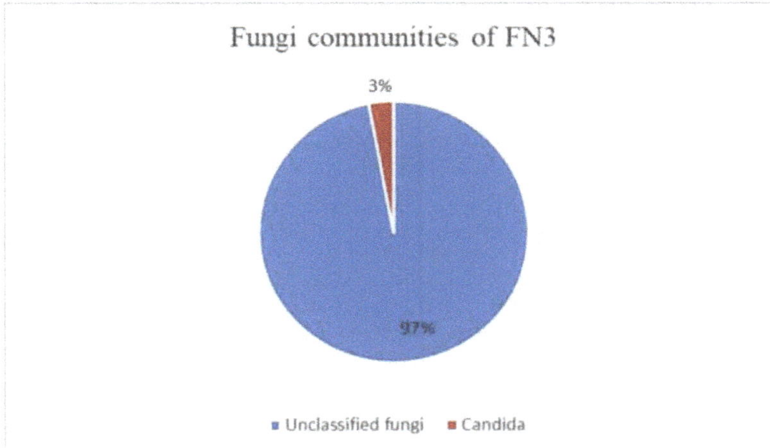

Figure 2. The fungi communities that were identified in mixed bacterial culture FN3 by metagenomic analysis.

Table 3. The BBD for the four independent variables on dye decolorization in actual and prediction values.

Run	Factor 1 A: Dye Concentration (mg/L)	Factor 2: B: Glucose Concentration (%)	Factor 3: C: Ammonium sulphate Concentration (g/L)	Factor 4: D: pH	Decolorization (%)	Prediction by RSM (%)
1	50	2.00	0.55	6.75	88.35	95.8
2	50	1.25	0.55	7.50	87.14	75.53
3	125	1.25	1.00	6.00	5.85	−4.49
4	125	1.25	1.00	7.50	45.43	40.95
5	200	1.25	0.55	7.50	38.00	27.87
6	200	1.25	1.00	6.75	17.85	20.95
7	200	1.25	0.55	6.00	2.86	13.62
8	125	2.00	1.00	6.75	29.92	40.61
9	50	1.25	0.55	6.00	23.00	32.27
10	50	0.50	0.55	6.75	14.20	22.57
11	125	2.00	0.55	6.00	39.00	27.30
12	125	1.25	0.10	6.00	0.00	10.34
13	200	1.25	0.10	6.75	9.60	9.97
14	125	0.50	1.00	6.75	0.00	6.41
15	125	0.50	0.55	6.00	8.02	−0.32
16	125	0.50	0.10	6.75	0.93	−10.61
17	125	1.25	0.55	6.75	61.08	58.77
18	50	1.25	1.00	6.75	50.35	44.98
19	200	0.50	0.55	6.75	14.86	13.27
20	125	1.25	0.55	6.75	48.02	58.77
21	50	1.25	0.10	6.75	60.36	52.25
22	125	2.00	0.10	6.75	61.19	53.98
23	125	2.00	0.55	7.50	74.48	77.81
24	200	2.00	0.55	6.75	41.29	38.78
25	125	1.25	0.55	6.75	60.50	58.77
26	125	1.25	0.55	6.75	63.77	58.77
27	125	1.25	0.55	6.75	60.49	58.77
28	125	1.25	0.10	7.50	6.20	22.41
29	125	0.50	0.55	7.50	0.00	6.69

ANOVA is a measurable investigation that is part of the analysis in RSM. It has been applied to identify the contrast between at least two groups that change in an experiment, and it is typically used to show that there is a significant outcome from the experiment. Along these lines, ANOVA was utilized to evaluate the significance of the model compared with the experimental values [22]. Table 3 shows the ANOVA of the regression parameters of the predicted response surface quadratic model for dye decolorization. The regression model is as follows:

$$\text{Dye decolorization} = 58.77 - 16.58 * A + 24.69 * B + 0.93 * C + 14.38 * D - 11.93 * AB + 4.57 * AC - 7.25 * AD - 7.59 * BC + 10.88 * BD + 8.34 * CD - 3.36 * A^2 - 12.81 * B^2 - 23.38 * C^2 - 18.09 * D^2$$

The predicted response fitted well with that of the experimentally obtained response. The adequate approximation of the selected model was measured by applying the diagnostic plots available in the Design Expert version 6.0.10 software, namely the externally studentizedresiduals plotted against the normal probability, the predicted versus studentizedresiduals, the runs versus studentizedresiduals, and the actual responses versus the predicted response values. Figure 3a shows that the externally studentized residuals plotted against the normal probability yielded a straight line showing the normal distribution of the experimental data. As shown in Figure 3b, the predicted versus externally studentized residual runs versus the externally studentized residuals and the actual responses versus the predicted responses, respectively, lie below the interval of ±4.00, indicating that the approximation of the model was good, with no data errors. Figure 3d illustrates the actual responses plotted against the predicted responses, which fit each other with correlation coefficients (R^2 and R^2 adj) of 0.913 and 0.825, respectively, for dye decolorization. Therefore, the developed model was suitable for predicting the efficiency of dye decolorization under the investigated conditions [23]. Table 4 shows that the F value of the model was 10.43 with a low probability value ($F < 0.001$), indicating that the model was significant for dye decolorization. On the other hand, a value of P that is less than 0.0001 is statistically significant for the quadratic equation of the model [24]. Values of $p > F$ less than 0.0500 indicated that the model terms were significant, while values greater than 0.1000 indicated that the model terms were not significant. The lack of fit for the F test (0.0750) was statistically insignificant, implying that the model fitted the data. The non-significant value of lack of fit (>0.05) revealed that the quadratic model was statistically significant for the response and, therefore, it could be used for further analysis. The goodness of fit of the model was evaluated using the determination coefficient (R^2). In this case, the value of R^2 was 0.913 and the value of the adjusted R^2 was 0.825, which was in agreement with the predicted R^2 (0.525), indicating that the model was adequate for predicting the dye decolorization with any combination of the variables. An R^2 value close to 1.00 showed that the model was sufficiently strong in its prediction [25].

3.4. Determination and Validation of Optimal Conditions

The maximal decolorization was accomplished by the desirability function technique. This technique incorporates the wants and needs for every one of the factors to construct a system for deciding the connection between the anticipated color decolorization for every factor and the desirability of the reactions. The optimal conditions predicted by RSM were as follows: 73 mg/L of dye, 1.934% glucose, 0.433 g/L of ammonium sulfate, and pH of 7.907, which resulted in an overall 97.551% dye decolorization with a desirability value of 1 (Table 5). To verify this optimal condition, a validation experiment was performed according to the predicted condition obtained. The experimental result was compared with the given predicted value by measuring the deviation between both values. Verification experiments performed under the predicted conditions indicated the validity and adequacy of the predicted models. The results obtained through the validation of the experiment indicate the suitability of the developed quadratic models, and it may be noted that these optimal values are valid within the specified range of process parameters.

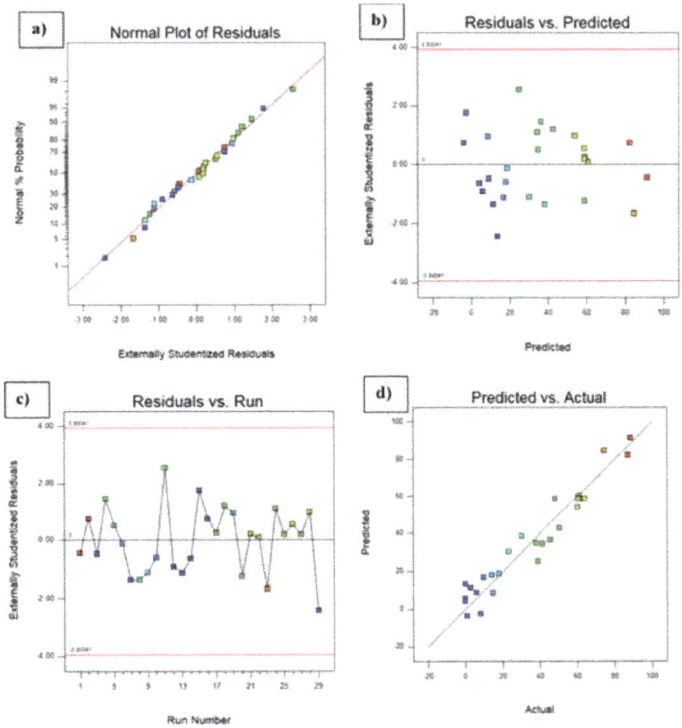

Figure 3. Diagnostic plots showing (**a**) the externally studentized residuals plotted against the normal probability, (**b**) the predicted versus the externally studentized residuals, (**c**) the run number versus the externally studentized residuals, and (**d**) the actual responses versus the predicted response.

Table 4. Analysis of variance (ANOVA) for the fitted quadratic polynomial for optimization of Metanil Yellow (MY) decolorization.

Source	Sum of Squares	df	Mean Squares	F Value	*p*-Value Prob > F	
Model	20132.04	14	1438	10.43	<0.0001	significant
A-Dye concentration	3298.09	1	3298.09	23.93	0.0002	
B-Glucose concentration	7312.19	1	7312.19	53.05	<0.0001	
C-Ammonium sulphate concentration	10.3	1	10.3	0.075	0.7885	
D-pH	2480.26	1	2480.26	17.99	0.0008	
AB	569.3	1	569.3	4.13	0.0616	
AC	83.36	1	83.36	0.6	0.4497	
AD	210.25	1	210.25	1.53	0.2372	
BC	230.13	1	230.13	1.67	0.2173	
BD	473.06	1	473.06	3.43	0.0851	
CD	278.56	1	278.56	2.02	0.1771	
A^2	73.09	1	73.09	0.53	0.4785	
B^2	1064.3	1	1064.3	7.72	0.0148	
C^2	3545.47	1	3545.47	25.72	0.0002	
D^2	2123.12	1	2123.12	15.4	0.0015	
Residual	1929.84	14	137.85			
Lack of fit	1777.99	10	177.8	4.68	0.075	Not significant
Pure error	151.85	4	37.96			
Cor total	22061.88	28				

Table 5. The optimum conditions were obtained using the desirability function technique.

Dye Concentration (mg/L)	Glucose Concentration (%)	Ammonium Sulphate Concentration (g/L)	pH	Decolorization (%)	Desirability
73	1.934	0.433	7.097	97.551	1

3.5. *Effects of Different Initial Dye Concentration on Dye Decolorisation and Bacterial Growth*

The mixed bacterial culture of FN3 was grown in optimized MSM with different MY dye concentrations ranging from 30 mg/L to 300 mg/L, as shown in Figure 4. From the graph, it is observed that a low dye concentration was decolorized at a higher rate compared to the high dye concentration by the mixed culture FN3.

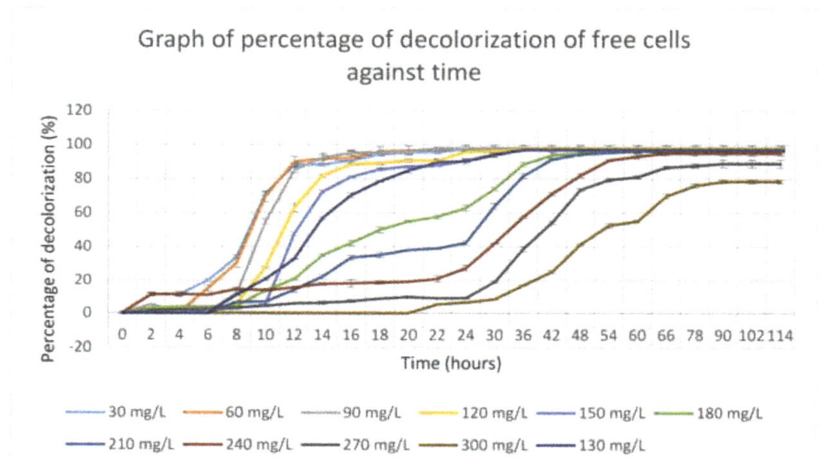

Figure 4. Dye decolorization percentages of free cells against time.

Meanwhile, mixed bacterial culture FN3 decolorized 300 mg/L of MY dye with an almost 80% decolorization rate for 114 h. Mixed bacterial culture FN3 fully decolorized MSM supplemented with 30–240 mg/L even though it took up to 66 h to decolorize Meanwhile, for MSM supplemented with a dye concentration of 270 mg/L and 300 mg/L, mixed bacterial culture FN3 able to decolorize up to 80% dye decolorization only.

The dye decolorizdecolorization percentage by the mixed bacterial culture FN3 became static after a certain incubation time, which might be due to the inhibition of the dye by the mixed bacterial culture FN3.

3.6. *Response Surface Plots of the Affecting Parameters*

The surface response of the quadratic models was applied to visualize the effects of each experimental parameter, with two parameters maintained at the optimal value and the other two varying within the experimental ranges, as depicted in Figure 5. The three-dimensional response surface plots are the graphical representations of the regression equation. The main goal of the response surface is to efficiently track the optimum values of the variables such that the response is maximized. By analyzing the plots, the best response range can be calculated. Figure 5a shows the 3D surface response of the interaction effect of the glucose concentration and dye concentration. The dye decolorization increased when the glucose concentration was high and the dye concentration was low. This indicates that with an increasing supply of glucose, the bacteria used the glucose as a substrate for growth and, thus, increased the dye decolorization. Figure 5b shows the 3D surface response of the interaction effect of the ammonium sulfate concentration and dye concentration. The dye decolorization increased when the ammonium sulfate concentration was between 0.2 and 0.5 g/L. Figure 5c shows the interaction effect of pH and dye concentration. The dye decolorization increased when the pH was between 6.0 and 7.5. The highest dye decolorization was achieved when the pH was around 7.5. This indicated that the bacteria preferred more alkaline conditions compared to acidic conditions for growth. For Figure 5d, the dye decolorization increased when the ammonium sulfate concentration was between 0.1 and 0.6 g/L, while the higher the glucose concentration, the higher the dye decolorization. It can be concluded that the bacteria need a nitrogen source also besides carbon source for decolorization of the dye. For Figure 5e, the higher the pH and glucose concentration, the higher the dye decolorization percentage. For Figure 5f, the dye decolorization increased when the pH was between 6.0 and 6.9 [26].

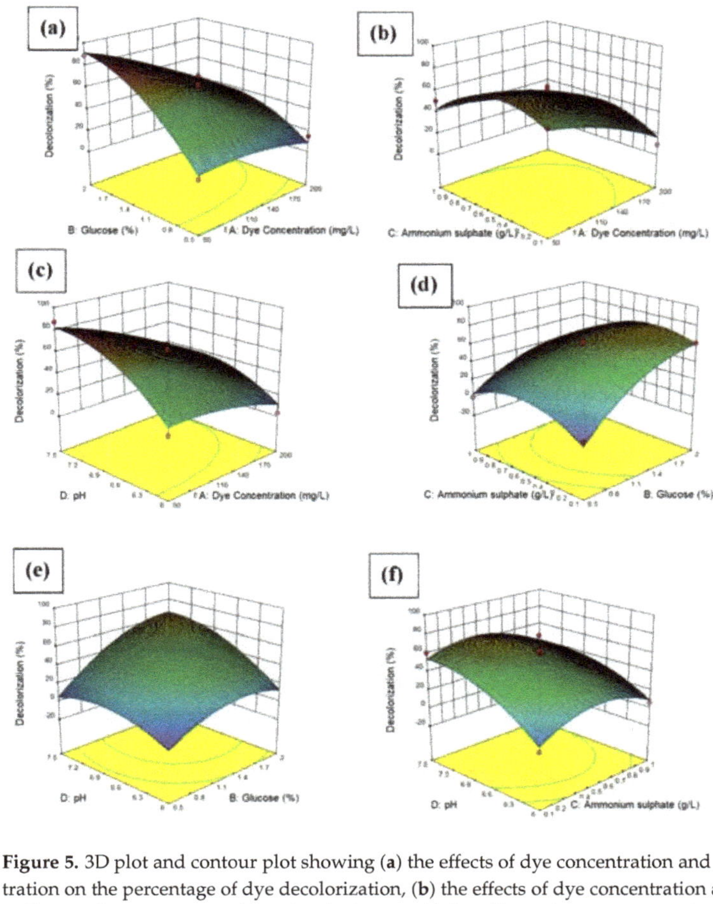

Figure 5. 3D plot and contour plot showing (**a**) the effects of dye concentration and glucose concentration on the percentage of dye decolorization, (**b**) the effects of dye concentration and ammonium sulfate on the percentage of dye decolorization, (**c**) the effects of dye concentration and pH on the percentage of dye decolorization, (**d**) the effects of glucose concentration and ammonium sulfate on the percentage of dye decolorization, (**e**) the effects of glucose concentration and pH on the percentage of dye decolorization, the (**f**) the effects of ammonium sulfate and pH on the percentage of dye decolorization.3.7 Effects of heavy metal ions on dye decolorization.

Heavy metals exist in various contaminated sites. Bioremediation of certain pollutants is often affected by the presence of heavy metals. Thus, studies to determine the effects of metal ions on bacterial bioremediation were performed. In this study, 1 mg/L of heavy metals, such as nickel (Ni), copper (Cu), lead (Pb), chromium (Cr), silver (Ag), zinc (Zn), mercury (Hg), and arsenic (As), were used.

Figure 6 demonstrates the percentage of dye decolorization of free cells with the presence of the heavy metals listed, and MSM with no addition of heavy metals served as the control. It was observed that the MY dye was decolorized in 24 h with the presence of heavy metals. This proved that the presence of heavy metals does not affect the dye decolorization. The percentage of dye decolorization (96.62%) was the highest with the addition of lead (Pb), whereas the lowest dye decolorization (56.19%) was achieved with the addition of copper (Cu).

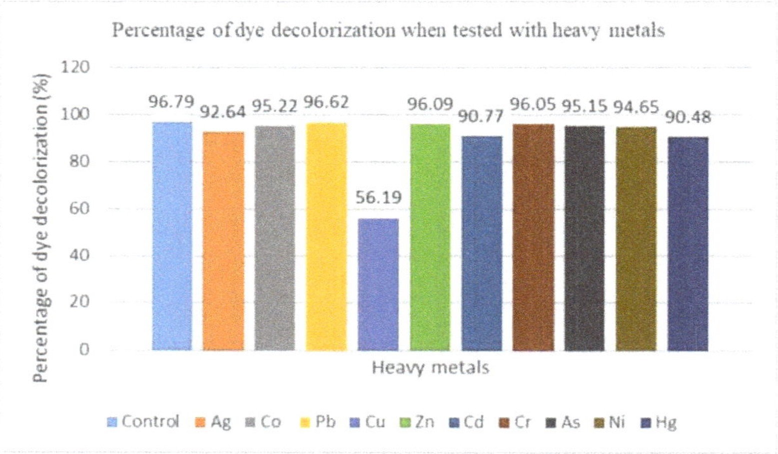

Figure 6. Bar chart showing the percentage of dye decolorization when tested with heavy metals.

4. Discussion

In this study, MY, a type of azo dye, was successfully decolorized within 24 h by a locally isolated mixed bacterial culture from Malaysian agricultural soil, namely FN3. According to a previous study, most of the microbial species that have been found to decolorize azo dyes originate from dye-contaminated wastewater or soil [27]. However, isolate FN3 has been isolated from agricultural soil. This indicates that dye-decolorizing bacteria can also be found in soil that has not been contaminated with dyes. Furthermore, in this work, a mixed bacterial culture was isolated instead of a pure culture. The advantages of mixed culture in dye decolorization have been reported in several reports [12,27,28]. The use of a mixed bacterial culture can result in a higher rate of dye decolorizdecolorization compared to the pure culture. From the metagenomics analysis of the mixed bacterial culture FN3, it was discovered that about 42.6% of the microbial community of FN3 is *Bacillus* sp. *Bacillus* sp. has been reported to decolorize azo dyes successfully [12,29]. It has been found that *Bacillus* sp. is well known and documented for its dye-decolorizing potential in previous studies [30]. About 42.6% of the microbial community was comprised of *Bacillus* sp., while another main species identified was *Acinetobacter* sp. (14%). A previous study reported the MY dye decolorization by the *Bacillus* sp. strain AK1 and *Lysinibacillus* sp. strain AK2 [2].

From the ITS diversity analysis, it was found that 97% of the mixed bacterial culture FN3 is "unclassified fungi." Based on the UNITE database, which is the reference database in this metagenomics analysis of fungi, there is no "species" that is annotated up to a rank. This is indicated as "unclassified" fungi.

MY dye decolorization involves the enzymatic breaking of an azo bond (–N=N–) by the azoreductase enzyme. This process in azo dye decolorization eventually proceeds in two stages. In the first stage, the azoreductase breaks the azo bond, and this will produce colorless aromatic amines. The second stage involves the degradation of the colorless aromatic amines. In this study, the FN3 mixed bacterial culture needs aeration for growth. The mixed bacterial culture could not survive and could not decolorize MY with no aeration. It can be concluded that the mixed bacterial culture uses an aerobic mechanism. For the aerobic mechanism in the degradation of aromatic amines, it involves the replacement of the functional groups of the aromatic rings with other groups, such as hydroxyl groups. Next, two oxygen atoms are incorporated into the aromatic rings [9].

This study also indicates the potential use of the statistical optimization tool using RSM. It helps in finding the optimum values of each parameter and correlates each parameter to optimize the dye decolorization within a short period of time and for less cost. Before the use of RSM, scientists used OFAT to optimize certain conditions. This method is tedious and not cost-effective since a number of experiments are run and the parameters are not correlated. In this study, the dye decolorization was optimized and the optimum dye concentration was 73 mg/L with 97% decolorization. The chosen mixed bacterial culture needs glucose as the carbon source to grow and decolorize the dye. However, some bacteria are able to use the dye as the sole carbon source [31]. Even though our attempts to isolate the bacteria in order to use the dye as the sole carbon source have failed, the outsourcing of a carbon source such as glucose is being introduced to start the decolorization process. Another counted parameter is pH. Bacteria can live at various pH levels. In this study, the optimum pH suitable for the bacterial culture to decolorize the dye at an optimum level was 7.097, which is a neutral condition. From previous studies, *Bacillus* sp gives maximum dye decolorization at neutral pH [12].

The effect of different dye concentrations on the percentage of dye decolorization is significant in the study in order to identify the tolerance of mixed bacterial culture towards MY. The mixed bacterial culture FN3 was grown in MSM supplemented with 30 to 300 mg/L of MY. From this study, 240 mg/L of MY was completely decolorized. The concentration of 300 mg/L of MY was not decolorized completely by the mixed bacterial culture FN3. High concentrations of dyes cause the inhibition of metabolic processes of microorganisms. This is probably due to the toxic effects of the dyes on microorganisms through the inhibition of metabolic activity [32]. When the concentration of the dye is too high, the cells are saturated with the dye molecules, thus inactivating the transport system of the dyes [33]. The active sites of the enzymes that are responsible for decolorization may have been blocked by the dye molecules, thus reducing the decolorization efficiency [34].

Heavy metals are detrimental to microbes since they affect enzymatic functions, act as redox catalysts in the production of reactive oxygen species (ROS), disrupt ion regulation, and affect DNA and protein production [35]. Based on a previous study, the effects of heavy metal ions were tested on the dye decolorization. The dye decolorization was not affected by the presence of heavy metal ions [36,37].Dye decolorization is achieved with the addition of most heavy metals, such as Ag, Co, Pb, Zn, Cd, Cr, As, Ni, and Hg. This indicates that the mixed culture FN3 was able to tolerate the toxic effect to achieve decolorization. In this study, 56.19% of dye decolorization was achieved when 1 mg/L of Cu was added. The percentage of dye decolorization was low compared with the other tested heavy metals Thus, the low dye decolorization may be related to the inhibition of enzymes and metabolic pathways caused by Cu [33].

5. Conclusions

In conclusion, MY dye was decolorized by a mixed culture of bacteria and fungi from an agricultural soil sample. The mixed bacterial culture was successfully identified by 16s rRNA and ITS for bacterial and fungi metagenomic analysis, respectively. The result showed that the highest percentage in the microbial communities was from the genus *Bacillus* and "unclassified" fungi. The MY dye decolorization by FN3 was also successfully optimized by RSM. The optimal conditions predicted by RSM were 73 mg/L of dye 1.934% glucose, 0.433 g/L of ammonium sulfate, and a pH of 7.097, with decolorization up to 97.551%. The mixed cultures of FN3 also had high tolerance towards various heavy metals, as the dye decolorization of MY is not affected when heavy metals are added into the growth medium.

Author Contributions: Conceptualization, M.I.E.H., F.N.A.M.; data curation, M.I.E.H., F.N.A.M. methodology, M.I.E.H.; supervision, M.I.E.H., S.B.A.W., S.S.A.G., U.H.Z., K.M., M.Y.A.S.; writing—Original draft, F.N.A.M.; writing—Review and editing, M.I.E.H., F.N.A.M. All authors have read and agreed to the published version of the manuscript.

Funding: This project was financed by funds from Putra Grant (GP-IPM/2017/9532800) and Yayasan Pak Rashid Grant UPM (6300893-10201).

Conflicts of Interest: The authors declare no conflict of interest.

References

1. Singh, R.P.; Singh, P.K.; Singh, R.L. Bacterial Decolorization of Textile Azo Dye Acid Orange by Staphylococcus hominis RMLRT03 *Toxicol. Int.* **2014**, *21*, 160–166. [CrossRef]
2. Anjaneya, O.; Souche, S.Y.; Santoshkumar, M.; Karegoudar, T.B. Decolorization of sulfonated azo dye Metanil Yellow by newly isolated bacterial strains: Bacillus sp. strain AK1 and Lysinibacillus sp. strain AK2. *J. Hazard. Mater.* **2011**, *190*, 351–358. [CrossRef] [PubMed]
3. Aksu, Z.; Kılıç, N.K.; Ertuğrul, S.; Dönmez, G. Inhibitory effects of chromium(VI) and Remazol Black B on chromium(VI) and dyestuff removals by Trametes versicolor. *Enzym. Microb. Technol.* **2007**, *40*, 1167–1174. [CrossRef]
4. Hao, O.; Kim, H.; Chiang, P.-C. Decolorization of Wastewater. *Crit. Rev. Environ. Sci. Technol.* **1999**, *30*, 449–505. [CrossRef]
5. Moreira, R.; Peruch, M.G.; Kuhnen, N.C. Adsorption of textile dyes on alumina. Equilibrium studies and contact time effects *Braz. J. Chem. Eng.* **1998**, *15*. [CrossRef]
6. Sarioglu, M.; Bali, U.; Bisgin, T. The removal of C.I. Basic Red 46 in a mixed methanogenic anaerobic culture. *Dye. Pigment.* **2007** *74*, 223–229. [CrossRef]
7. Wang, H.; Su, J.; Zheng, X.; Tian, Y.; Xiong, X.; Zheng, T. Bacterial decolorization and degradation of the reactive dye Reactive Red 180 by Citrobacter sp. CK3. *Int. Biodeterior. Biodegrad.* **2009**, *63*, 395–399. [CrossRef]
8. Pandey, A.; Singh, P.; Iyengar, L. Bacterial decolorization and degradation of azo dyes. *Int. Biodeterior. Biodegrad.* **2007**, *59*, 73–84 [CrossRef]
9. Alabdraba, W.; Bayati, M. Biodegradation of Azo Dyes a Review. *Int. J. Environ. Eng. Nat. Resour* **2014**, *1*, 179–189.
10. Sarkar, S.; Banerjee, A.; Halder, U.; Biswas, R.; Bandopadhyay, R. Degradation of Synthetic Azo Dyes of Textile Industry A Sustainable Approach Using Microbial Enzymes. *Water Conserv. Sci. Eng.* **2017**, *2*, 121–131. [CrossRef]
11. Moosvi, S.; Kher, X.; Madamwar, D. Isolation, characterization and decolorization of textile dyes by a mixed bacterial consortium JW-2. *Dye. Pigment.* **2007**, *74*, 723–729. [CrossRef]
12. Jain, K.; Shah, V.; Chapla, D.; Madamwar, D. Decolorization and degradation of azo dye – Reactive Violet 5R by an acclimatized indigenous bacterial mixed cultures-SB4 isolated from anthropogenic dye contaminated soil. *J. Hazard. Mater.* **2012**, *213–214*, 378–386 [CrossRef]
13. Sharma, D.C.; Satyanarayana, T. A marked enhancement in the production of a highly alkaline and thermostable pectinase by Bacillus pumilus dcsr1 in submerged fermentation by using statistical methods. *Bioresour. Technol.* **2006**, *97*, 727–733. [CrossRef]
14. Ghosh, A.; Mehta, A.; Khan, A.M. Metagenomic Analysis and its Applications. In *Encyclopedia of Bioinformatics and Computational Biology*; Ranganathan, S., Gribskov, M., Nakai, K., Schönbach, C., Eds.; Academic Press: Oxford, UK, 2019; pp. 184–193. [CrossRef]
15. Streit, W.R.; Schmitz, R.A. Metagenomics – the key to the uncultured microbes. *Curr. Opin. Microbiol.* **2004**, *7*, 492–498. [CrossRef]
16. Thomas, T.; Jack, G.; Meyer, F. Metagenomics-A guide from sampling to data analysis. *Microb. Inform. Exp.* **2012**, *2*, 3. [CrossRef]
17. Bashir, Y.; Singh, S.; Konwar, B. Review Article Metagenomics: An Application Based Perspective. *Hindawi Publ. Corp. Chin J. Biol.* **2014**, *2014*, 1–7. [CrossRef]
18. Murugesan, K.; Dhamija, A.; Nam, I.-H.; Kim, Y.-M.; Chang, Y.-S. Decolourization of reactive black 5 by laccase: Optimization by response surface methodology. *Dye. Pigment.* **2007**, *75*, 176–184. [CrossRef]
19. Pokharia, A.; Singh, S. Isolation and Screening of Dye Decolorizing Bacterial Isolates from Contaminated Sites. *Text. Light Ind Sci. Technol.* **2013**, *2*, 54–61.
20. Chakravorty, S.; Helb, D.; Burday, M.; Connell, N.; Alland, D. A detailed analysis of 16S ribosomal RNA gene segments for the diagnosis of pathogenic bacteria. *J. Microbiol. Methods* **2007**, *69*, 330–339. [CrossRef]
21. De Mandal, S.; Sanga, Z.; Nachimuthu, S.K. *Metagenomic Analysis of Bacterial Community Composition among the Cave Sediments of Indo-Burman Biodiversity Hotspot Region*; PeerJ PrePrints: Corte Madera, CA, USA; London, UK, 2014; pp. 2167–9843.

22. Sanusi, S.N.A.; Halmi, M.I.E.; Abdullah, S.R.S.; Hassan, H.A.; Hamzah, F.M.; Idris, M. Comparative process optimization of pilot-scale total petroleum hydrocarbon (TPH) degradation by Paspalum scrobiculatum L. Hack using response surface methodology (RSM) and artificial neural networks (ANNs). *Ecol. Eng.* **2016**, *97*, 524–534. [CrossRef]
23. Halmi, M.I.E.b.; Abdullah, S.R.S.; Wasoh, H.; Johari, W.L.W.; Ali, M.S.b.M.; Shaharuddin, N.A.; Shukor, M.Y. Optimization and maximization of hexavalent molybdenum reduction to Mo-blue by Serratia sp. strain MIE2 using response surface methodology. *Rend. Lincei* **2016**, *27*, 697–709. [CrossRef]
24. Toolabi, A.; Malakootian, M.; Taghi Ghaneian, M.; Esrafili, A.; Ehrampoush, M.; Askarishahi, M.; Tabatabaei, M.; Khatami, M. Optimizing the photocatalytic process of removing diazinon pesticide from aqueous solutions and effluent toxicity assessment via a response surface methodology approach. *Rend. Lincei* **2019**, *30*, 155–165. [CrossRef]
25. Gunawan, E.; Basri, M.; Rahman, M.; Bakar Salleh, A.; Rahman, R. Study on response surface methodology (RSM) of lipase-catalyzed synthesis of palm-based wax ester. *Enzym. Microb. Technol.* **2005**, *37*, 739–744. [CrossRef]
26. Sharma, P.; Singh, L.; Dilbaghi, N. Optimization of process variables for decolorization of Disperse Yellow 211 by Bacillus subtilis using Box–Behnken design. *J. Hazard. Mater.* **2009**, *164*, 1024–1029. [CrossRef]
27. Khehra, M.S.; Saini, H.S.; Sharma, D.K.; Chadha, B.S.; Chimni, S.S. Decolorization of various azo dyes by bacterial consortium. *Dye. Pigment.* **2005**, *67*, 55–61. [CrossRef]
28. Khehra, M.S.; Saini, H.S.; Sharma, D.K.; Chadha, B.S.; Chimni, S.S. Comparative studies on potential of consortium and constituent pure bacterial isolates to decolorize azo dyes. *Water Res.* **2005**, *39*, 5135–5141. [CrossRef]
29. Tony, B.D.; Goyal, D.; Khanna, S. Decolorization of textile azo dyes by aerobic bacterial consortium. *Int. Biodeterior. Biodegrad.* **2009**, *63*, 462–469. [CrossRef]
30. Patil, P.S.; Shedbalkar, U.U.; Kalyani, D.C.; Jadhav, J.P. Biodegradation of Reactive Blue 59 by isolated bacterial consortium PMB11. *J. Ind. Microbiol. Biotechnol.* **2008**, *35*, 1181–1190. [CrossRef]
31. Sudha, M.; Saranya, A.; Gopal, S.; Natesan, S. Microbial degradation of Azo Dyes: A review. *Int. J. Curr. Microbiol. Appl. Sci.* **2014**, *3*, 670–690.
32. Contreras, M.; Grande-Tovar, C.; Vallejo, W.; Lòpez, C. Bio-Removal of Methylene Blue from Aqueous Solution by Galactomyces geotrichum KL20A. *Water* **2019**, *11*, 282. [CrossRef]
33. Shah, M.P.; Patel, K.A.; Nair, S.S.; Darji, A. Microbial decolorization of methyl orange dye by Pseudomonas spp. ETL-M. *Int. J. Environ. Bioremediation Biodegrad.* **2013**, *1*, 54–59.
34. Li, H.-H.; Wang, Y.-T.; Wang, Y.; Wang, H.-X.; Sun, K.-K.; Lu, Z.-M. Bacterial degradation of anthraquinone dyes. *J. Zhejiang Univ.-Sci. B* **2019**, *20*, 528–540. [CrossRef]
35. Gauthier, P.T.; Norwood, W.P.; Prepas, E.E.; Pyle, G.G. Metal–PAH mixtures in the aquatic environment: A review of co-toxic mechanisms leading to more-than-additive outcomes. *Aquat. Toxicol.* **2014**, *154*, 253–269. [CrossRef]
36. Igiri, B.; Okoduwa, S.I.R.; Idoko, G.; Akabuogu, E.; Adeyi, O.; Ejiogu, I. Toxicity and Bioremediation of Heavy Metals Contaminated Ecosystem from Tannery Wastewater: A Review. *J. Toxicol.* **2018**, *2018*, 2568038. [CrossRef]
37. Murugesan, K.; Kim, Y.-M.; Jeon, J.-R.; Chang, Y.-S. Effect of metal ions on reactive dye decolorization by laccase from Ganoderma lucidum. *J. Hazard. Mater.* **2009**, *168*, 523–529. [CrossRef]

sustainability

MDPI

Article

Immobilization of Metanil Yellow Decolorizing Mixed Culture FN3 Using Gelling Gum as Matrix for Bioremediation Application

Fatin Natasha Amira Muliadi [1], Mohd Izuan Effendi Halmi [1,*], Samsuri Bin Abdul Wahid [1], Siti Salwa Abd Gani [2], Khairil Mahmud [3] and Mohd Yunus Abd Shukor [4]

1 Department of Land Management, Faculty of Agriculture, Universiti Putra Malaysia, Serdang 43400, Malaysia; fatin.muliadi@gmail.com (F.N.A.M.); samsuriaw@upm.edu.my (S.B.A.W.)
2 Department of Agricultural Technology, Faculty of Agriculture, Universiti Putra Malaysia, Serdang 43400, Malaysia; ssalwaag@upm.edu.my
3 Department of Crop Science, Faculty of Agriculture, Universiti Putra Malaysia, Serdang 43400, Malaysia; khairilmahmud@upm.edu.my
4 Department of Biochemistry, Faculty of Biotechnology and Biomolecular Sciences, Universiti Putra Malaysia, Serdang 43400, Malaysia; mohdyunus@upm.edu.my
* Correspondence: m_izuaneffendi@upm.edu.my; Tel.: +603-9769-4958

Abstract: In this study, the Metanil Yellow (MY) decolorizing mixed culture, namely FN3, has been isolated from agriculture soil. The mixed culture was immobilized using gellan gum. In order to optimize the immobilization process for maximal dye decolorization, Response Surface Methodology (RSM) was performed. The optimal conditions for immobilization predicted by desirability function are 130 mg/L of MY dye concentration, 1.478% of gellan gum concentration, 50 beads and 0.6 cm of beads size with the percentage of decolorization of 90.378%. The correlation coefficients of the model (R^2 and R^2 adj) are 0.9767 and 0.9533, respectively. This indicates that the established model is suitable to predict the effectiveness of dye decolorization under the investigated condition. The immobilized beads of mixed culture FN3 were able to be reused up to 15 batches of decolorization. The immobilized cells also have high tolerance towards heavy metals. This was proven by higher dye decolorization rate by the immobilized cells even with the addition of heavy metals in the media. The decolorization potential of the mixed culture indicates that it could be useful for future bioremediation of soil contaminated sites and treatment solutions of water bodies polluted with MY dye.

Keywords: immobilization; mixed culture; Metanil Yellow; response surface methodology

Citation: Muliadi, F.N.A.; Halmi, M.I.E.; Wahid, S.B.A.; Gani, S.S.A.; Mahmud, K.; Shukor, M.Y.A.. *Sustainability* **2021**, 13, 36. https://dx.doi.org/10.3390/su13010036

Received: 30 September 2020
Accepted: 18 November 2020
Published: 22 December 2020

1. Introduction

Azo dyes are extensively used in various industries such as textile industry, food, paper-making and cosmetic industries [1,2]. With the growing demand of the textile industry, approximately 40,000 different dyes and pigments are being use and 2000 azo dyes are currently in use. The production of azo dyes annually worldwide is around 7×10^5 tonnes [3]. Azo dyes are highly water-soluble. They include one or more azo (-N=N-) groups and sulfonic (SO^{3-}) groups [4]. Due to the dyeing process, the textile industry releases an enormous amount of wastewater into the environment. Typically, about 10–15% of the dye is lost in effluents, since they do not bind to the fibres [5]. This has led to water pollution.

Metanil Yellow (MY) (3-(4-Anilinophenylazo) benzene sulfonic acid sodium salt is a type of acidic azo dyes [6]. MY is commonly used as soap colouring, spirit lacquer, shoe polish, bloom sheep dip, for preparation of food stains, leather dyeing, for manufacturing of pigment lakes and paper staining [7]. MY is categorized as nonpermitted food colourants. It is widely used as alimentary dye adulterant in India [8]. However, it was found out later that MY is carcinogenic. From the toxicity investigation data, it is known that when MY is fed to the animals, it produces testicular lesions leading a decrease in the rate of spermatogenesis [7]. According to Saratale (2009), exposure to the dyes leads to potential health hazards such as asthma, rhinitis and dermatitis. Apart from the health hazards to human health, azo dyes also affect the ecosystems. When the wastewater effluents are being discharged into the open water sources, it affects the photosynthetic activity of aquatic organisms [9] and the dissolved dyes may also affect the aquatic organisms as it may be toxic to them due to their breakdown products [10].

Researchers are finding various ways of removing dyes from wastewater. As they are relatively resistant to biodegradation, the elimination of coloured effluents in wastewater treatment systems is mainly based on physical or chemical procedures such as adsorption, concentration, chemical transformation and incineration [11]. However, these methods are high-energy costs, high sludge production and formation of byproducts [12]. Therefore, biological methods for removing dyes need to be developed. The advantages of these methods is that they are environmental-friendly, economical and cost-competitive [13]. Biodegradation by different microorganisms appears to be an attractive alternative by using microorganisms as dye decolorizing agent [14].

Generally, degradation and decolorization of azo dye by bacteria proceeds in two stages. The first stage involves reductive cleavage of the dyes' azo linkages, resulting in the formation of generally colourless but potentially hazardous aromatic amines. The second stage involves the degradation of the aromatic amines [15]. Reductive cleavage of the (-N=N-) bond is the initial step of the bacterial degradation of azo dyes. It can occur by different types of mechanisms such as through enzymes which are azoreductase and laccase enzymes, low molecular weight redox mediators, chemical reduction by biogenic reductants like sulfide or a combination of all of these [14]. The use of enzymes is beneficial according to substrate specificity and may be effectively used in textile water pretreatment. During azoreductase enzymatic dye degradation, azo bond (-N=N-) is being cleaved by the enzyme and four electrons are transferred as reducing equivalent. In each stage, two electrons transfer to the azo dye which is the electron acceptor and decolorization happened when the colourless solution is formed. The resulting intermediate is toxic aromatic amine which is later degraded by the aerobic process or sometimes microaerophilically [4].

It has been reported that bacterial dye decolorization and degradation can occur by a pure culture of bacteria and also by mixed culture of bacteria. According to the previous study, it has been reported that mixed bacterial culture can give a better degradation rate than the individual strain. Individual species have limited metabolic capability to mineralize dye completely and in many cases it has been observed that mainly due to lack of catabolic pathway, aromatic amines are not further degraded. Catabolic and synthrophic interactions of indigenous species lead to the complete degradation of azo dyes [16]. Azo dyes are not readily metabolized under aerobic conditions, and as a result of metabolic pathways, it degraded into intermediate compounds but not mineralized. It can be completely degraded under coupled aerobic, anaerobic degradation. Therefore, coupled anaerobic treatment followed by aerobic treatment can be an efficient and effective degradation method of azo dyes.

Immobilization of bacteria is studied extensively as it offers many advantages in the bioremediation field. There are many ways to immobilize the bacteria such as entrapment, adsorption, encapsulation and crosslinking [17]. Entrapment and encapsulation are considered as some of the preferred methods of immobilization. By immobilization, it prevents cell washout and maintains a high density of bacterial cells in bioreactor. Aside from that, the catalytic stability of bacteria is also improved compared to free cells [18]. Immobilization of whole cells also provides a conducive microenvironment for bacterial cells [19] and the entrapped cells can better tolerate various environmental stresses [20]. There are various immobilization matrices available. One of them is gellan gum. Gellan gum is said to be a premium natural polymer for encapsulation of active microorganisms. Gellan gum gel has better rheological characteristics compared to agar and carrageenan gels at equivalent concentrations [21]. Besides that, the gels are stable in a wide range of pH which is from pH 2–10 [21,22]. The interaction between gellan gums and ions is nonspecific, making the gels able to interact with a wide variety of cations. This is unlikely with other ion-sensitive gelling polysaccharides such as alginate [21].

Conventional method of optimization which is "one factor at a time" (OFAT) approach, is laborious, time consuming and incomplete. It involves varying a single independent variable while the other variables are maintained at a constant level. Thus, response surface methodology (RSM) using Central Composite Design (CCD) and Box–Behnken design (as factorial experimental design) which involves full factorial search by examining the simultaneous, systematic and efficient variation of important components is applied to model the decolorization process, identify possible interactions, higher-order effects and determine the optimum operational conditions [23]. RSM is a collection of statistical and mathematical techniques that are useful for developing, improving and optimizing processes. In this study, decolorization of MY by newly isolated mixed culture from agricultural soil namely FN3 is optimized by a statistical study of Response Surface Methodology (RSM) for optimum decolorization. Box Behnken design matrix is chosen for optimization. The isolated mixed culture is then immobilized using gellan gum. Mixed bacterial culture FN3 consists of bacteria mostly from *Bacillus* sp and yeasts [24].The immobilized cells are tested with addition of 1 mg/L of various heavy metals to investigate the significant effects to the dye decolorization. Reusability of the immobilized cells are also being tested in this study. To the best of our knowledge, no study has reported on the immobilization of mixed culture using gellan gum to decolorize MY dye. From the previous study, many works reported on the use of pure culture of bacteria instead of the mixed culture of bacteria, as reported in this study.

2. Materials and Methods

2.1. Cultivation of Mixed Culture FN3

The MY decolorizing mixed culture FN3 from an in-house culture collection was provided by the Department of Land Management at University Putra Malaysia [24]. The mixed culture was cultured in minimal salt medium (MSM) that contained (g/L): glucose, 10; $(NH4)_2SO_4$, 0.4; KH_2PO_4, 0.2; K_2HPO_4, 0.4; NaCl, 0.1; $Na2MoO_4{\cdot}2H_2O$, 0.01; $MgSO_4{\cdot}H_2O$, 0.1; $MnSO_4.H_2O$, 0.01; $Fe(SO_4)_3{\cdot}H_2O$, 0.01; yeast extract, 1.0 and supplemented with 50 mg/L of MY dye (Sigma-Aldrich, St Louis, MO, USA, 70% purity).

2.2. Immobilization of the Mixed Culture Using Gellan Gum

For the immobilization part, the mixed culture FN3 was grown on a large scale in minimal salt medium (MSM) supplemented with 50 mg/L of MY dye. The medium also contained (g/L): glucose, 10; $(NH4)_2SO_4$, 0.4; KH_2PO_4, 0.2; K_2HPO_4, 0.4; NaCl, 0.1; $Na_2M_0O_4$, 0.01; $MgSO_4{\cdot}H_2O$, 0.1; $MnSO_4{\cdot}H_2O$, 0.01; $Fe(SO_4)_3{\cdot}H_2O$, 0.01; yeast extract, 1.0. The 5 L conical flask containing 3 L sterile MSM (pH 7.097) supplemented with 50 mg/L of MY dye was set up. A sterile glass-fibre syringe filter with 0.45 µm pores was used to filter microorganism from the incoming air provided by an oxygen pump. The pump was used to enhance the aeration in the flask and speed up the growth of the mixed culture. 300 mL of the selected mixed culture was cultured in 3 L MSM (pH 7.097) in 5 L scale-up set. After two days of incubation, the culture was centrifuged at 15,000× *g* for 10 min using a high-speed centrifuge (Beckman Coulter, Brea, CA, USA). The pellet was collected and immobilized using gellan gum for the cell immobilization study.

A method for the production of gellan gum beads, as proposed by Moslemy et al. (2003) with some small changes, was used. A 0.75% (*w/v*) of dispersed gellan gum was first dissolved in 55 mL of distilled water by continuously heating to 75–80 °C. The mixture was continuously stirred to prevent the clumping of gellan gum. Next, its pH was adjusted to 7 with 0.1 M NaOH. For microbial encapsulation, 5.25 g wet weight of microbial cells was dispersed in 150 mL gellan gum solution before emulsification. The mixture was then stirred for 1 min and emulsified dropwise using pipette through 2 mm diameter tips into

2% of $CaCl_2$. The gellan gum droplets immediately gelled in the $CaCl_2$ solution. The beads were then kept in the $CaCl_2$ solution for one hour before being transferred to sterile distilled water. The beads were kept overnight in distilled water at 4 °C before being harvested by filtration. These beads were used for the MY dye decolorization after being washed with sterile distilled water and beads containing no cells served as the control [25].

2.3. Optimization of Immobilized Beads Using RSM

Response surface methodology acts as a modelling technique in determining the optimum conditions in a multivariable system and to evaluate the relationship between controllable experimental factors and observed results [26]. In the optimisation of the medium for optimum MY dye decolorization, Box–Behnken design was chosen as the experimental matrix. The significant parameters involved were dye concentration (A), gellan gum concentration (B), number of beads (C) and beads size (D) with the percentage of dye decolorization as the response. The design of the experiment and the statistical analysis of the data was conducted using Design Expert software, 6.0.10; Stat-Ease Inc, Minneapolis, USA [27] with a total number of 29 runs. The results were analysed using Design Expert add-ons program including ANOVA to find out the interaction between the variables and the response.

The quality of the fit of this model was expressed by the coefficient of determination (R^2) in the same program. To start the experiment, the immobilized cells were prepared according to the runs predicted by RSM. The immobilized mixed culture beads were grown in MSM supplemented with MY dye in conical flasks and incubated on a rotary shaker (120 rpm) for 24 h. After 24 h, the percentage of dye decolorization was calculated using the formula as stated above.

2.4. Metabolites Analysis Using High-Performance Liquid Chromatography (HPLC)

The metabolites of the degraded Metanil Yellow were extracted three times using ethyl acetate. Firstly, the minimal salt media supplemented with 50 mg/L of MY was degraded with mixed culture FN3 for 24 h at 120 rpm rotary incubator. After 24 h, the culture broth was centrifuged at 12,000 rpm for 15 min. The metabolites were extracted from the clear supernatant with an equal volume of ethyl acetate. The mixture was vigorously mixed in order to dissolve the metabolites. The organic layer was then separated. The extracts were dried over anhydrous Na_2SO_4 and evaporated to dryness using a rotary evaporator. The crystals obtained were dissolved in HPLC grade methanol for further analysis of HPLC and FTIR [28,29]. HPLC analysis of the control dye, 50 mg/L of Metanil Yellow and the ethyl acetate extracts were performed using Waters 2690 instrument, Milfrd, USA, equipped with C_{18} column having symmetry of 250 × 4.6 mm. Methanol was used as the mobile phase with a flow rate of 1.0 mL/min for 15 min and UV detector at 434 nm [30].

2.5. Metabolites Analysis Using Fourier-Transform Infrared Spectroscopy (FTIR)

The metabolites of the degraded dye were extracted using ethyl acetate. The FTIR analysis of the metabolites was carried out on Thermo Nicolet 6700 instrument and compared with a control dye, 50 mg/L of Metanil Yellow dye in the mid-IR region of 400–4000 cm^{-1} with eight scan speed. The samples were mixed with spectroscopically pure KBr in the ratio of 5:95, pellets were fixed in the sample holder and the analysis was carried out [31].

2.6. Reusability of the Immobilized Beads

The reusability test of the immobilized beads is a critical process in bioremediation. This test was used to determine the ability of immobilized cells for Metanil Yellow dye decolorization after completing one decolorization cycle. The incubation of immobilized cells for 24 h was calculated as one cycle. The microbial beads were prepared according to the optimised parameters obtained from RSM; gellan gum concentration: 1.478%, beads number: 50 beads and beads size: 0.6 cm. The microbial beads and minimal salt media with 50 mg/L of Metanil Yellow dye were prepared in 50 mL liquid phase. The microbial beads were incubated overnight at 120 rpm rotary shaker at room temperature. After 24 h of incubation, 1 mL of the culture was withdrawn aseptically and centrifuged at 10,000× *g* for 10 min. The supernatant was measured at 434 nm wavelength and the percentage of decolorization was calculated. After the first cycle of decolorization, the beads were filtered from the growth medium and then washed with sterile distilled water. Next, the filtered microbial beads were incubated in the fresh MSM supplemented with 50 mg/L of MY dye. The process was repeated until no more dye decolorization was observed.

2.7. Scanning Electron Microscopy (SEM) of Immobilized Mixed Culture FN3

For scanning electron microscopy of the dye degrading mixed culture, the gel beads were firstly fixated with 2.5% (*w/v*) of glutaraldehyde solution overnight. After that, the dehydration process was performed. The dehydration started with increasing acetone concentration (35%, 50%, 75%, 95% and 100%). The gel beads were dehydrated subsequently to remove the remaining water. The dehydrated gel beads were then dried under critical conditions in a CO_2 atmosphere. The dehydrated gel beads were then cut in half using sterile scalpel before being coated with gold and examined under the scanning electron microscope [32].

2.8. Effects of Heavy Metal Ions on Dye Decolorization of Immobilised Cells

In this study, the effect of heavy metals on the dye decolorization for immobilised cells was also carried out. MSM was separately supplied with one ppm of heavy metals consisting of copper (Cu), arsenic (As), zinc (Zn), chromium (Cr), nickel (Ni), silver (Ag), lead (Pb) and mercury (Hg). The immobilised cells were prepared based on optimum conditions that previously obtained from RSM. The MY dye concentration used was 50 mg/L. The immobilised beads were incubated at 120 rpm on a rotary shaker at room temperature. The microbial beads without heavy metals served as controls. After 24 h of incubation, 1 mL of the culture was withdrawn aseptically and centrifuged at 10,000× *g* for 10 min. The supernatant was measured at 434 nm wavelength and the percentage of dye decolorization was calculated after that. The studies were performed in duplicates.

3. Results and Discussions

3.1. Optimization of Immobilized Beads Using RSM

This study focused on the combined effects of four significant variables for the decolorization of MY dye by mixed culture FN3. In order to optimize the process variables for maximal dye decolorization, 29 experimental runs were conducted. Table 1 shows the experimental and response results together with the response predicted by RSM. The trial run 12 showed the highest dye decolorization which is 94.37%. The maximal and minimal dye decolorization was observed at run 12 and run 7, respectively, as shown in Table 1. There are few studies that studied favourable conditions for Metanil Yellow dye decolorization. Microorganisms that able to decolorize Metanil Yellow dye with favourable conditions have been listed in Table 2.

ANOVA is a measurable investigation that is part of the analysis in Response Surface Methodology (RSM). It has been applied to identify the contrast between at least two groups that change in an experiment and is typically used to show that there is a significant outcome from the experiment. Along these lines, ANOVA was utilized to evaluate the significance of the model compared with the experimental values [37]. Table 3 showed the analysis of variance (ANOVA) of the regression parameters of the predicted response surface quadratic model for dye decolorization. The regression model was given as follows:

$$\text{Decolorization} = 17.04 - 43.14 * A - 6.08 * B + 1.07 * C + 0.17 * D - 0.87 * AB - 0.76 * AC - 2.55 * AD + 2.43 * BC + 4.97 * BD + 3.15 * CD + 23.7 * A^2 + 6.01 * B^2 + 5.06 * C^2 + 7.77 * D^2$$

Table 1. The Box–Behnken design for the four independent variables on dye decolorization in actual and predicted values.

Run	A: Dye Concentration(mg/L)	B: Gellan Gum Concentration (%)	C: Number of Beads	D: Beads Size (cm)	Decolorization (%)	Predicted RSM Decolorization (%)
1	225	0.750	50	0.45	33.12	32.82
2	225	0.750	10	0.45	45.80	35.55
3	350	1.125	30	0.30	10.75	7.74
4	100	1.125	30	0.30	88.23	88.93
5	225	1.500	30	0.60	34.31	29.87
6	225	1.125	30	0.45	17.27	17.04
7	350	1.500	30	0.45	0.00	−3.35
8	225	1.125	30	0.45	12.40	17.04
9	100	1.125	10	0.45	89.36	87.11
10	225	1.500	10	0.45	19.53	18.52
11	225	1.125	50	0.3	35.38	27.61
12	100	1.125	30	0.6	92.67	94.37
13	225	1.125	10	0.6	15.26	25.82
14	225	0.750	30	0.6	38.00	32.10
15	225	1.500	30	0.3	15.18	19.59
16	225	1.125	50	0.6	34.16	34.25
17	225	0.750	30	0.3	38.73	41.69
18	350	1.125	30	0.6	5.00	2.99
19	100	0.750	30	0.45	88.96	95.10
20	350	1.125	50	0.45	2.20	2.96
21	350	1.125	10	0.45	2.10	2.35
22	225	1.125	30	0.45	21.40	17.04
23	225	1.125	30	0.45	16.53	17.04
24	225	1.500	50	0.45	16.57	25.52
25	225	1.125	30	0.45	17.59	17.04
26	100	1.500	30	0.45	89.23	84.67
27	225	1.125	10	0.3	29.07	31.77
28	350	0.75	30	0.45	3.20	10.55
29	100	1.125	50	0.45	92.50	90.77

Table 2. Lists of microorganisms that are able to decolorize Metanil Yellow (MY) dye with the favourable conditions.

Microorganisms	Optimal pH and Temperature for Reduction	Optimal C Source	Dye Concentration Tested (mg/L)	References
Vibrio harveyi TEMS1	20 °C	Glucose	100	[33]
Bacillus sp AK1 *Lysinibacillus* sp AK2	7.2 37 °C	Metanil Yellow	200	[7]
Unknown local isolates (NII and RHG)	-	Glucose	50	[34]
Oenococcus oeni ML34	7.0 30 °C	Glucose	1000	[35]
Bacillus sp Neni-10	6.3 34 °C	Glucose	150	[36]

The predicted response fitted well with those of the experimentally obtained response The adequate approximation of the selected model was measured by applying the diagnostic plots available in the Design-Expert version 6.0.10 software, namely the externally studentized residuals plotted against the normal probability, predicted versus studentized residuals, runs versus studentized residuals and actual responses versus the predicted response values.

Figure 1a showed that the externally studentized residuals plotted against the normal probability yielded a straight line showing normal distribution of the experimental data. As shown in Figure 1b, the predicted versus externally studentized residual runs versus externally studentized residuals and actual responses versus predicted responses, respectively, lie below the interval ±4.00 indicating that the approximation of the model was good with no data error. Figure 1d illustrated the actual responses plotted against the predicted responses value which fit each other with correlation coefficients (R^2 and R^2 adj) of 0.9767 and 0.9533, respectively for dye decolorization. Therefore, the developed model was suitable for predicting the efficiency of dye decolorization under the investigated conditions [38]. Table 2 depicted that the F value of the model is 41.84 with a low probability value ($p < 0.001$), indicating that the model was significant for dye decolorization. On the other hand, the value of p less than 0.0001 is statistically significant for the quadratic equation of the model [39].

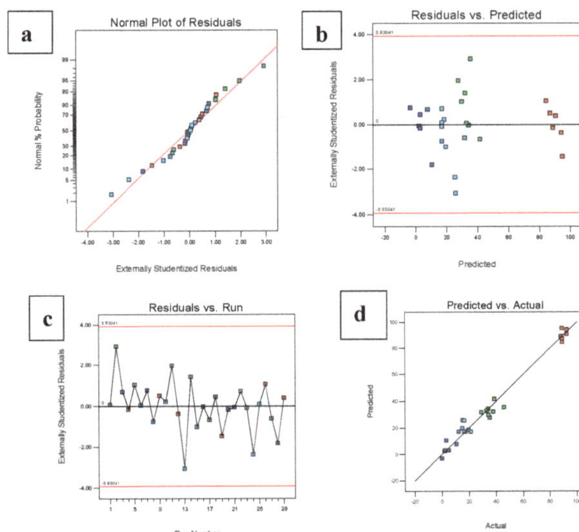

Figure 1. Diagnostic plots showing (**a**) the externally studentized residuals plotted against the normal probability, (**b**) the predicted versus the externally studentized residuals, (**c**) the run number versus externally studentized residuals, (**d**) the actual responses versus the predicted responses.

Values of $p > F$ less than 0.0500 indicated that the model terms were significant while values higher than 0.1000 indicated that the model terms were not significant. The lack of fit for the F test (5.8) was statistically insignificant, implying that the model fitted the data. The nonsignificant value of lack of fit (>0.05) revealed that the quadratic model was statistically significant for the response and, therefore, it can be used for further analysis. The goodness of fit of the model was checked using the determination coefficient (R^2). In this case, the value of R^2 was 0.9767 and the value of adjusted R^2 was 0.9533 which was in reasonable agreement with the predicted R^2 (0.8719), indicating that the model was adequate for predicting the dye decolorization with any combination values of the variables. The R^2 value close to 1.00 showed that the model was sufficiently strong in its prediction [40].

Table 3. Analysis of variance (ANOVA) for the fitted quadratic polynomial order for optimization of dye decolorization of Metanil Yellow dye by immobilized cells of mixed culture FN3.

ANOVA for Response Surface Quadratic Model						
Analysis of Variance Table [Partial Sum of Squares—Type III]						
Source	**Sum of Squares**	**df**	**Mean Square**	**F Value**	**p-Value Prob > F**	
Model	26,709.35	14	1907.81	41.84	<0.0001	significant
A: Dye concentration	22,334.7	1	22,334.7	489.87	<0.0001	
B: Gellan Gum Concentration	444.07	1	444.07	9.74	0.0075	
C: Number of beads	13.67	1	13.67	0.3	0.5927	
D: Beads size	0.35	1	0.35	7.75^{-3}	0.9311	
AB	3	1	3	0.066	0.8014	
AC	2.31	1	2.31	0.051	0.8252	
AD	25.94	1	25.94	0.57	0.4632	
BC	23.64	1	23.64	0.52	0.4833	
BD	98.62	1	98.62	2.16	0.1635	
CD	39.62	1	39.62	0.87	0.367	
A^2	3643.94	1	3643.94	79.92	<0.0001	
B^2	233.92	1	233.92	5.13	0.0399	
C^2	165.93	1	165.93	3.64	0.0772	
D^2	391.46	1	391.46	8.59	0.011	
Residual	638.3	14	45.59			
Lack of fit	597.15	10	59.71	5.8	0.0524	Not significant
Pure Error	41.15	4	10.29			
Cor Total	27347.66	28				

3.2. Determination and Validation of Optimal Conditions

The maximal decolorization was accomplished by the desirability function technique. This technique incorporates the wants and needs for every one of the factors to construct a system for deciding the connection between the anticipated colour decolorization for every factor and the desirability of the reactions. The optimal conditions predicted by RSM were as follows: 130 mg/L of dye concentration, 1.478% of gellan gum concentration, 50 beads and 0.600 cm of beads size, which resulted in an overall 90.378% of dye decolorization with desirability value of 1 (Table 4). Figure 2 showed the immobilized mixed culture beads using gellan gum after optimization process. To verify this optimal condition, a validation experiment was performed according to the predicted condition obtained. The experimental result was compared with the given predicted value by measuring the deviation between both values. Verification experiments were performed at the predicted conditions, indicating the validity and adequacy of the predicted models. The results obtained through the validation of the experiment indicate the suitability of the developed quadratic models, and it may be noted that these optimal values are valid within the specified range of process parameters.

Table 4. The optimum conditions obtained by using the desirability function technique.

Dye Concentration (mg/L)	Gellan Gum Concentration (%)	Number of Beads	Beads Size (cm)	Decolorization (%)	Desirability
130	1.478	50	0.600	90.378	1

Figure 2. The immobilized mixed culture beads using gellan gum.

3.3. Response Surface Plots of the Affecting Parameters

The surface response of quadratic models was applied to visualize the effects of each experimental parameter with two parameters maintained at the optimal value and the other two varying within the experimental ranges as depicted in Figure 3. The three-dimensional response surface plots are the graphical representations of the regression equation. The main goal of the response surface is to track efficiently for the optimum values of the variables such that the response is maximized. By analysing the plots, the best response range can be calculated. Figure 3a showed the 3D surface response of the interaction effect of the dye concentration and gellan gum concentration. The dye decolorization increased when the gellan gum and dye concentration were low. This gave a brief view that with the increasing concentration of gellan gum, the dye decolorization was affected. Optimum gelling gum concentration is needed as it is in charge of the mechanical power of the beads and the efficiency of the beads on decolorizing the dye.

Figure 3b showed the 3D surface response of the interaction effect of the number of beads and dye concentration. The dye decolorization increased when low dye concentration and high number of beads are used. By increasing the number of beads, this means more cell loading is used. This will in turn increase the capability of the beads to decolorize the dye. Figure 3c showed the interaction effect of bead size and dye concentration. The dye decolorization increased when low dye concentration and low bead size are used. The optimum bead size used is 0.6 cm. Small bead size indicates that surface area is small and this increases the biodegradation process [41]. Figure 3d showed the interaction between the gellan gum concentration and the number of beads. The highest dye decolorization was recorded when the low number of beads and gellan gum concentration were used. Figure 3e showed the interaction between the beads size and the gellan gum concentration. Low number of beads size and gellan gum concentration gave the highest dye decolorization. In Figure 3f, the interaction was between beads size and the number of beads. The optimum bead size, 0.6 cm with high number of beads, gave the highest dye decolorization.

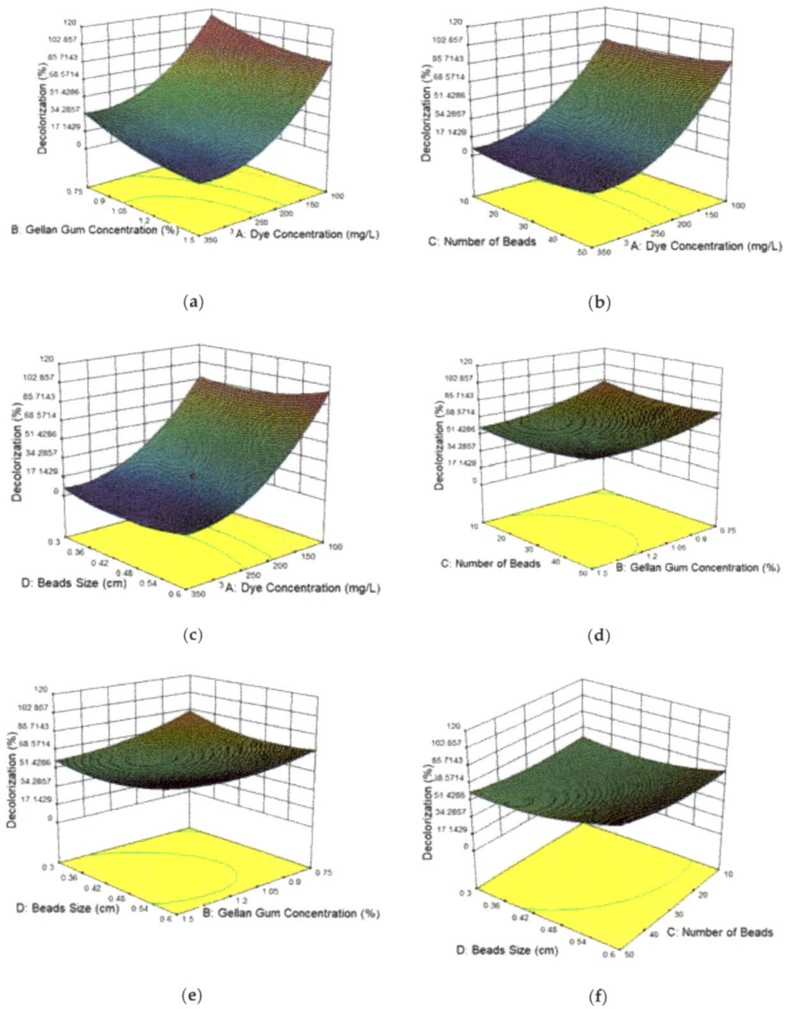

Figure 3. 3D plot and contour plot showing (**a**) the effects of dye concentration and gellan gum concentration, (**b**) the effects of dye concentration and number of beads, (**c**) the effects of dye concentration and beads size, (**d**) the effects of gellan gum concentration and number of beads, (**e**) the effects of gellan gum concentration and beads size, (**f**) the effects of number of beads and beads size.

3.4. Analysis of High Performance Liquid Chromatography (HPLC) and Fourier-Transform Infrared Spectroscopy (FTIR)

The HPLC analysis of Metanil Yellow dye (control) at 50 mg/L showed that two peaks appeared at retention time of 2.40 min and 2.60 min indicating that Metanil Yellow dye is not 100% pure (Figure 4). After the dye decolorization process, the disappearance of the peaks was seen in the case of controlled dye and formation of completely different peaks at retention time of 2.20 min, 2.30 min, 2.50 min and 2.70 min were observed in Figure 5. The appearance of new peaks in the decolorized dye products and disappearance of peaks in control dye support the Metanil Yellow dye decolorization by the mixed culture [28]. It is not confirmed whether the metabolite or the breakdown product formed is a low molecular weight nontoxic component or another toxic product which needs further analysis. From previous study of Metanil Yellow dye decolorization, peak of R_T value 2.8 indicated metanilic acid [7]. However, in this study, negative peak can be seen at R_T value of 2.8. This gave indication that in this study, the Metanil Yellow dye decolorization by mixed culture FN3 did achieve the metanilic acid formation. The negative peak could be due to the disturbance of the local equilibrium in the mobile phase and the stationary phase. The disturbance is created when a sample is injected.

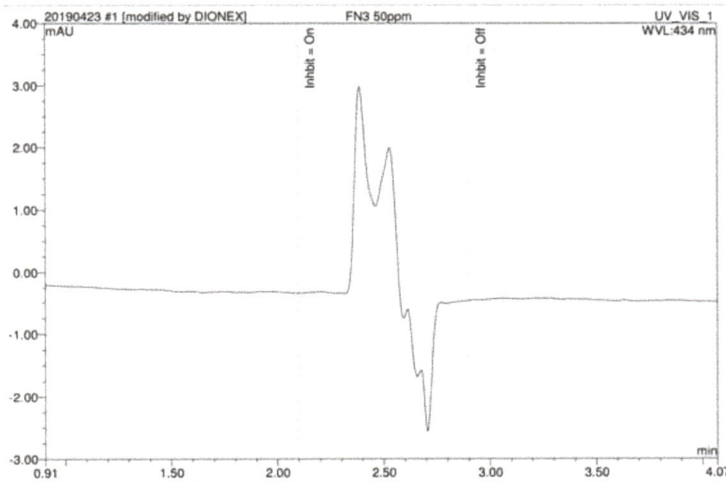

Figure 4. HPLC chromatogram of the control dye, Metanil Yellow.

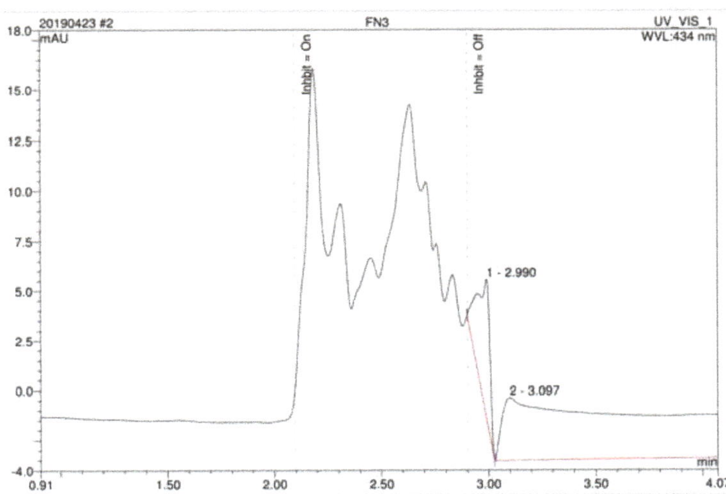

Figure 5. HPLC chromatogram of the decolorized products obtained after treatment with mixed culture FN3.

The FTIR spectrum of control dye Metanil Yellow showed peak of wavelength 1635.07 cm^{-1} that signified presence of (-N=N-) stretching as shown in Figure 6. There were variations in the peaks in the FTIR spectrum of metabolites extracted from decolorized sample of dye when compared to the control dye spectrum as in Figure 7. The absence of peaks with wavelength of 1635.07 cm^{-1} indicated the reductive cleavage of azo bond [30]. The wavelength of 1671.40 cm^{-1} indicated the aromatic stretch of C=C. It specified the stretching of benzene rings' compounds of Metanil Yellow. Other peaks at wavelengths of 2325.84 cm^{-1}, 1048.95 cm^{-1}, 979.67 cm^{-1}, 557.27 cm^{-1} and 431.08 cm^{-1} have formed and these changes were clear evidences of the decolorization of Metanil Yellow dye by the mixed culture. The peak of 1048.95 cm^{-1} indicated the C-N stretching whereas peak of 979.67 cm^{-1} indicated the bending of C=C [28,30,31]. The peak between 3500 and 3000 cm^{-1} gave indication of the stretching of (-NH) [42].

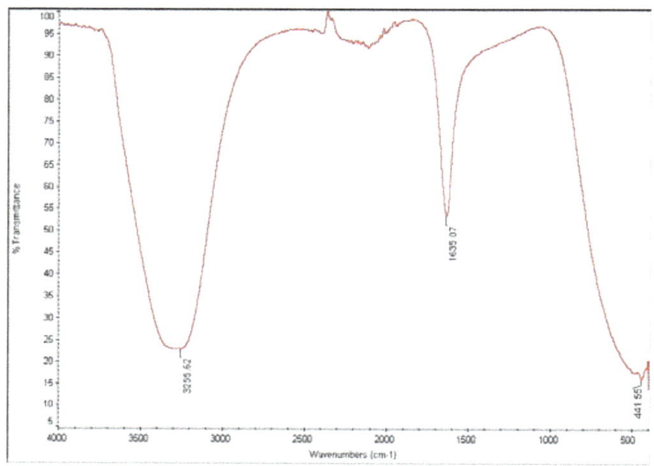

Figure 6. FTIR spectrum of the control Metanil Yellow dye.

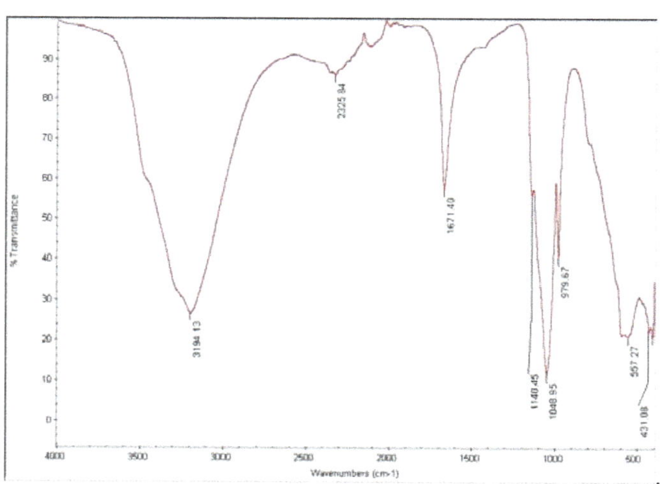

Figure 7. FTIR spectrum of decolorized products of Metanil Yellow dye decolorization by mixed culture FN3.

3.5. Reusability of the Immobilized Beads

One of the advantages of immobilization is the reusability of the immobilized beads. These experiments were carried out to test the reusability of the gellan gum immobilized beads. Figures 8 and 9 showed the decolorization profile of the immobilized beads with initial dye concentration of 50 mg/L. It is shown that the immobilized beads were able to be reused up to 15 cycles with complete dye decolorization in 24 h.

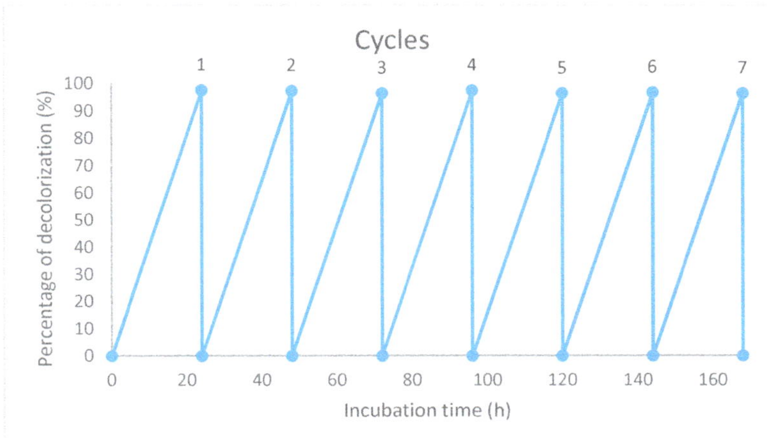

Figure 8. The repeated usage (cycles 1 to 7) of the immobilized cells with initial dye concentration of 50 mg/L. Cycles 1 to 7 took 24 h per cycle for complete Metanil Yellow decolorization.

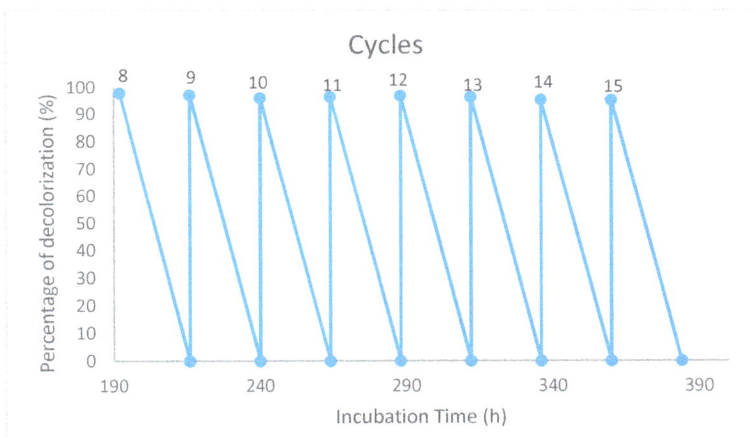

Figure 9. The repeated usage (cycles 8 to 15) of the immobilized cells with initial dye concentration of 50 mg/L. Cycles 8 to 15 took 24 h per cycle for complete Metanil Yellow decolorization.

Studies on the reusability of immobilized beads have been carried out by Chang et al. (2000). In the study, immobilized Pseudomonas luteola with calcium alginate, k-carrageenan and polyacrylamide can be reused up to four times with 75, 85 and 80% of dye decolorization [43]. Another study has shown that the gellan gum immobilized beads were able to be reused up to 20 times without substantial loss of catalytic activity [44]. No data was available on the reusability of gellan gum immobilized cells in Metanil Yellow dye decolorization. Based on this study, the immobilized beads have showed positive results on its continuous use in dye decolorization. These immobilized beads of mixed culture FN3 could be used to decolorize Metanil Yellow dye up to 15 batches.

Nowadays, immobilization method has become one of the important measures to tackle bioremediation effectively. One of the immobilization techniques, encapsulation, has emerged as a promising solution to overcome practical limitations of using free cell formulations. A defined stable, consistent and protective microenvironment is provided by the polymeric matrix of the support material, where cells can survive and metabolic activity can be maintained for extended periods without the immediate release of a large number of cells. Better toleration towards numerous environmental stresses by the entrapped cells may be released after adaptation to surrounding environmental conditions [20].

The mechanical strength of immobilized cells depends on the types of matrices of immobilization used. In this study, gellan gum was chosen to immobilize the mixed culture of FN3. Gellan gum is an ionic heteropolysaccharide. It is also believed able to be used repetitively up to 20 batches without substantial loss of catalytic activity [44]. This proves that the durability of the beads is above adequate.

3.6. Scanning Electron Microscopy (SEM) of Immobilized Mixed Culture FN3

The SEM micrographs of mixed culture FN3 immobilized in gellan gum matrix were presented in Figure 10. The microscopic observation revealed that in Figure 10a, the mixed culture FN3 had been successfully entrapped in the gellan gum. In Figure 10b, the micrograph showed the mixed culture FN3 which is the rod-shaped bacterium and Candida sp budding. There are quite a number of studies regarding dye decolorization by yeast From a study, *Candida palmioleophila* JKS4 isolated from activated sludge from wastewater treatment plants have been reported to be able to decolorize various types of azo dyes [45] As identified from the metagenomics analysis, the most abundant percentage of bacteria in mixed culture FN3 was *Bacillus* sp. We can conclude that the rod-shaped bacterium in the microscopic observation was Bacillus sp. Bacillus sp was one of the early discovered bacteria with dye decolorizing [46]. In Figure 10c, the micrograph showed the cross-section of the gellan gum before the dye decolorization was performed. It was proven that the mixed culture FN3 was successfully immobilized in the gellan gum. The micrograph was compared with Figure 10d where Figure 10d showed the cross section of the beads after dye decolorization. The mixed culture FN3 was still entrapped within the gellan gum matrix even after Metanil Yellow dye decolorization. This indicated that the immobilized beads could be reused as the mixed culture was still available in the beads. This also proved that gellan gum has strong and rigid characteristic as immobilization matrix.

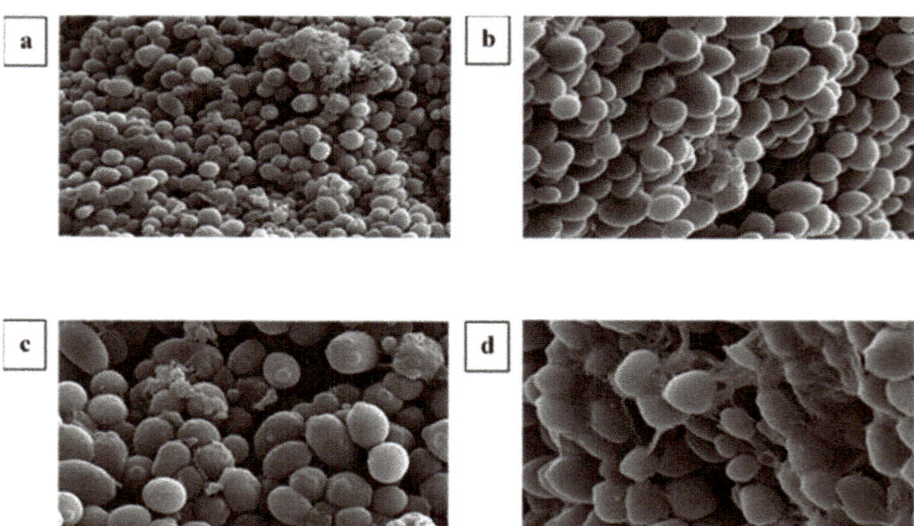

Figure 10. Scanning electron micrographs of (**a**) mixed culture FN3 successfully being entrapped in gellan gum matrix, (**b**) mixed culture FN3 showing a rod-shaped bacterium and budding of yeasts, (**c**) the mixed culture fn3 before dye decolorization and (**d**) after dye decolorization.

3.7. Effects of Metal Ions on Dye Decolorization of Immobilised FN3

Heavy metals exist particularly in various contaminated sites. Bioremediation of certain pollutants is often affected by the presence of heavy metals. Thus, to determine the effects of the metal ions on decolorization, these experiments were performed. In this study, one mg/L of heavy metals such as nickel (Ni), copper (Cu), lead (Pb), chromium (Cr), silver (Ag), zinc (Zn), mercury (Hg) and arsenic (As) were used. Figure 11 demonstrated the percentage of dye decolorization of immobilized beads with the presence of heavy metals as listed. The medium with no addition of heavy metals served as control and the percentage of decolorization was high. Medium supplemented with Cu inhibited the dye decolorization of Metanil Yellow where the percentage of decolorization was 90.09%, which was the lowest compared to the others. Medium with the addition of Ag, Pb, Co, Zn, Cd, Cr, As, Ni and Hg gave 99.91%, 98.65%, 96.67%, 96.96%, 94.38%, 96.94%, 93.85%, 96.29% and 99.32% of dye decolorization, respectively.

Heavy metals are detrimental to microbes as they affect the enzymatic functions, act as redox catalysts in the production of reactive oxygen species (ROS), destruct the ion regulation and affect the DNA and protein production [47]. From the data observed, this shows that the immobilized beads of microbes have a high tolerance towards heavy metals as the metal ions do not have significant effects on the Metanil Yellow dye decolorization. Dye decolorization is achieved with the addition of most heavy metals such as Ag, Co, Pb, Zn, Cd, Cr, As, Ni and Hg. From this study, the mixed culture FN3 was able to tolerate the toxic effect of heavy metals to achieve decolorization. In this study, 90.09% of dye decolorization was achieved by immobilized beads when one ppm of Cu is added. The percentage of dye decolorization is low compared with the others. Thus, it may be related to the Cu inhibition of enzymes and metabolic pathways [48].

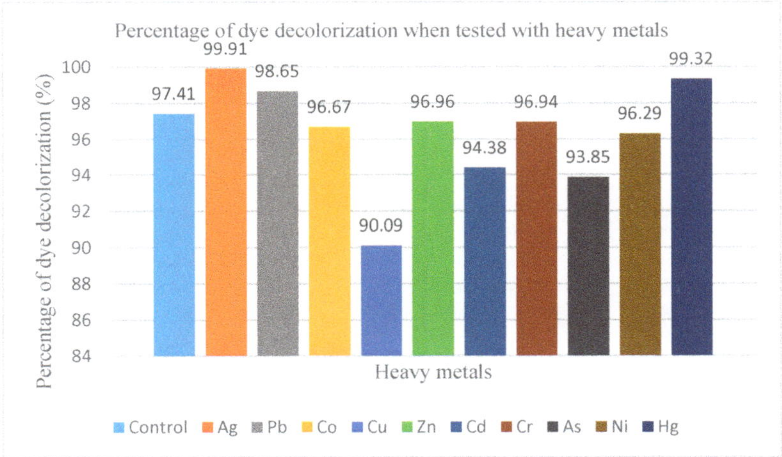

Figure 11. Effects of different heavy metals on Metanil Yellow dye decolorization by immobilized cells of mixed culture FN3.

4. Conclusions

As for the conclusion, the mixed culture FN3 has been successfully immobilized using gellan gum. The optimization of the dye decolorization by the immobilized beads has been performed using Response Surface Methodology (RSM), an optimization method that has many advantages as compared to a conventional optimization method, One Factor at a Time (OFAT). The optimum conditions obtained are 130 mg/L of dye, 1.478% of gellan gum concentration, 50 beads and 0.6 cm of beads size with decolorization up to 90.38%. The immobilized beads are also tested for the reusability purpose and it is proven that gellan gum immobilized beads are able to be reused up to 15 batches of dye decolorization without loss of decolorizing activity. The effects of heavy metals ions are also tested with the immobilized beads. From the data, most of the heavy metals do not give a significant effect on the Metanil Yellow dye decolorization. Therefore, the findings from the study established the fact that technology of immobilization could be chosen as a practical strategy to enhance the working performance of bioremediation of Metanil Yellow dye.

Author Contributions: Conceptualization, M.I.E.H.; methodology, F.N.A.M.; software, M.I.E.H. and M.Y.A.S.; validation, M.I.E.H.; formal analysis, F.N.A.M.; investigation, F.N.A.M.; resources, M.I.E.H.; data curation, M.I.E.H., and S.S.A.G.; writing-original draft preparation, F.N.A.M.; writing-review and editing, F.N.A.M. and M.I.E.H.; visualization, M.I.E.H. and F.N.A.M; supervision, M.I.E.H., S.B.A.W., M.Y.A.S., S.S.A.G., and K.M.; project administration, M.I.E.H.; funding acquisition, M.I.E.H.. All authors have read and agreed to the published version of the manuscript.

Funding: This project was financed by funds from Putra Grant (GP-IPM/2017/9532800) and Yayasan Pak Rashid Grant UPM (6300893-10201).

Institutional Review Board Statement: Not applicable.

Informed Consent Statement: Not applicable.

Data Availability Statement: The data presented in this study are available on request from the corresponding author. The data are not publicly available due to privacy.

Conflicts of Interest: The authors declare no conflict of interest.

References

1. Chang, J.-S.; Chou, C.; Lin, Y.-C.; Lin, P.-J.; Ho, J.-Y.; Hu, T.-L. Kinetic characteristics of bacterial azo-dye decolorization by Pseudomonas luteola. *Water Res.* **2001**, *35*, 2841–2850. [CrossRef]
2. Singh, R.P.; Singh, P.K.; Singh, R.L. Bacterial Decolorization of Textile Azo Dye Acid Orange by Staphylococcus hominis RMLRT03. *Toxicol. Int.* **2014**, *21*, 160–166. [CrossRef]
3. Puvaneswari, N.; Jayarama, M.; Gunasekaran, P. Toxicity assessment and microbial degradation of azo dyes. *Indian J. Exp. Boil.* **2006**, *44*, 618–626.
4. Sarkar, S.; Banerjee, A.; Halder, U.; Biswas, R.; Bandopadhyay, R. Degradation of Synthetic Azo Dyes of Textile Industry: A Sustainable Approach Using Microbial Enzymes. *Water Conserv. Sci. Eng.* **2017**, *2*, 121–131. [CrossRef]
5. Sudha, M.; Saranya, A.; Gopal, S.; Natesan, S. Microbial degradation of Azo Dyes: A review. *Int. J. Curr. Microbiol. Appl. Sci.* **2014**, *3*, 670–690.
6. Sivashankar, R.; Velmurugan, S.; Sathya, A.B.; Pallipad, S. Biosorption of Hazardous Azo Dye Metanil Yellow using Immobilized Aquatic weed. In Proceedings of the International Conference on Future Trends in Structural, Civil, Environmental and Mechanical Engineering–FTCEM, Bangkok, Thailand, 13–14 July 2013.
7. Anjaneya, O.; Souche, S.Y.; Santoshkumar, M.; Karegoudar, T.B. Decolorization of sulfonated azo dye Metanil Yellow by newly isolated bacterial strains: Bacillus sp. strain AK1 and Lysinibacillus sp. strain AK2. *J. Hazard. Mater.* **2011**, *190*, 351–358. [CrossRef]
8. Ghosh, D.; Singha, P.; Firdaus, S.; Ghosh, S. Metanil yellow: The toxic food colorant. *Asian Pac. J. Health Sci.* **2017**, *4*, 65–66. [CrossRef]
9. Aksu, Z.; Kılıç, N.K.; Ertuğrul, S.; Dönmez, G. Inhibitory effects of chromium(VI) and Remazol Black B on chromium(VI) and dyestuff removals by Trametes versicolor. *Enzym. Microb. Technol.* **2007**, *40*, 1167–1174. [CrossRef]
10. Hao, O.; Kim, H.; Chiang, P.-C. Decolorization of Wastewater. *Crit. Rev. Environ. Sci. Technol.* **1999**, *30*, 449–505. [CrossRef]
11. Moreira, R.; Peruch, M.G.; Kuhnen, N.C. Adsorption of textile dyes on alumina. Equilibrium studies and contact time effects. *Braz. J. Chem. Eng.* **1998**, *15*, 21–28. [CrossRef]
12. Sarioglu, M.; Bali, U.; Bisgin, T. The removal of C.I. Basic Red 46 in a mixed methanogenic anaerobic culture. *Dye. Pigments* **2007**, *74*, 223–229. [CrossRef]
13. Wang, H.; Su, J.; Zheng, X.; Tian, Y.; Xiong, X.; Zheng, T. Bacterial decolorization and degradation of the reactive dye Reactive Red 180 by Citrobacter sp. CK3. *Int. Biodeter. Biodegr.* **2009**, *63*, 395–399. [CrossRef]
14. Pandey, A.; Singh, P.; Iyengar, L. Bacterial decolorization and degradation of azo dyes. *Int. Biodeter. Biodegr.* **2007**, *59*, 73–84. [CrossRef]
15. Alabdraba, W.; Bayati, M. Biodegradation of Azo Dyes a Review. *Int. J. Environ. Eng. Nat. Resour.* **2014**, *1*, 179–189.
16. Jain, K.; Shah, V.; Chapla, D.; Madamwar, D. Decolorization and degradation of azo dye–Reactive Violet 5R by an acclimatized indigenous bacterial mixed cultures-SB4 isolated from anthropogenic dye contaminated soil. *J. Hazard. Mater.* **2012**, *213-214*, 378–386. [CrossRef]
17. Partovinia, A.; Rasekh, B. Review of the immobilized microbial cell systems for bioremediation of petroleum hydrocarbons polluted environments. *Crit. Rev. Environ. Sci. Technol.* **2018**, *48*, 1–38. [CrossRef]

18. Unnikrishnan, S.; Ramamoorthi, P.; Sridhar, S. Studies on decolorization of malachite green using immobilized Pseudomonas putida. *J. Chem. Pharm. Res.* **2015**, *7*, 589–596.
19. Ge, X.; Yang, L.; Xu, J. Cell Immobilization: Fundamentals, Technologies, and Applications. *Ind. Biotechnol.* **2017**, 205–235. [CrossRef]
20. Moslemy, P.; Guiot, S.R.; Neufeld, R.J. Encapsulation of bacteria for biodegradation of gasoline hydrocarbons. In *Immobilization of Enzymes and Cells*; Springer Science and Business Media LLC: Berlin/Heidelberg, Germany, 2006; pp. 415–426.
21. Moslemy, P.; Neufeld, R.J.; Millette, D.; Guiot, S.R. Transport of gellan gum microbeads through sand: An experimental evaluation for encapsulated cell bioaugmentation. *J. Environ. Manag.* **2003**, *69*, 249–259. [CrossRef]
22. Ashtaputre, A.A.; Shah, A.K. Studies on a Viscous, Gel-Forming Exopolysaccharide from Sphingomonas paucimobilis GS1. *Appl. Environ. Microbiol.* **1995**, *61*, 1159–1162. [CrossRef]
23. Sharma, D.C.; Satyanarayana, T. A marked enhancement in the production of a highly alkaline and thermostable pectinase by Bacillus pumilus dcsr1 in submerged fermentation by using statistical methods. *Bioresour. Technol.* **2006**, *97*, 727–733. [CrossRef] [PubMed]
24. Muliadi, F.N.A.; Halmi, M.I.E.; Abdul Wahid, S.; Abd Gani, S.S.; Uswatun, H.; Mahmod, K.; Abd Shukor, M.Y. Biostimulation of Microbial Communities from Malaysia Agriculture Soil for Detoxification of Metanil Yellow Dye; Response Surface Methodological Approach. *Sustainability* **2020**. (Accepted).
25. Survase, S.; Annapure, U.; Singhal, R.S. Gellan Gum as Immobilization Matrix for Production of Cyclosporin A. *J. Microbiol. Biotechnol.* **2010**, *20*, 1086–1091. [CrossRef]
26. Murugesan, K.; Dhamija, A.; Nam, I.-H.; Kim, Y.-M.; Chang, Y.-S. Decolourization of reactive black 5 by laccase: Optimization by response surface methodology. *Dyes Pigm.* **2007**, *75*, 176–184. [CrossRef]
27. Bezerra, M.A.; Santelli, R.E.; Oliveira, E.P.; Villar, L.S.; Escaleira, L.A. Response surface methodology (RSM) as a tool for optimization in analytical chemistry. *Talanta* **2008**, *76*, 965–977. [CrossRef] [PubMed]
28. Lade, H.; Govindwar, S.; Paul, D. Low-Cost Biodegradation and Detoxification of Textile Azo Dye C.I. Reactive Blue 172 by Providencia rettgeri Strain HSL1. *J. Chem.* **2015**, *2015*, 894109. [CrossRef]
29. Saratale, R.G.; Saratale, G.D.; Kalyani, D.C.; Chang, J.S.; Govindwar, S.P. Enhanced decolorization and biodegradation of textile azo dye Scarlet R by using developed microbial consortium-GR. *Bioresour. Technol.* **2009**, *100*, 2493–2500. [CrossRef]
30. Ajaz, M.; Rehman, A.; Khan, Z.; Nisar, M.A.; Hussain, S. Degradation of azo dyes by Alcaligenes aquatilis 3c and its potential use in the wastewater treatment. *AMB Express* **2019**, *9*, 64. [CrossRef]
31. Kurade, M.B.; Waghmode, T.R.; Jadhav, M.U.; Jeon, B.-H.; Govindwar, S.P. Bacterial–yeast consortium as an effective biocatalyst for biodegradation of sulphonated azo dye Reactive Red 198. *RSC Adv.* **2015**, *5*, 23046–23056. [CrossRef]
32. He, F.; Hu, W.; Li, Y. Biodegradation Mechanisms and Kinetics of Azo Dye 4BS by a Microbial Consortium. *Chemosphere* **2004**, *57*, 293–301. [CrossRef]
33. Ozdemir, G.; Pazarbasi, B.; Kocyigit, A.; Ersoy Omeroglu, E.; Yaşa, İ.; Karaboz, I. Decolorization of Acid Black 210 by Vibrio harveyi TEMS1 a Newly Isolated Bioluminescent Bacterium Enhanced Decolorization of Cibacron Red FN-2BL by Schizophyllum Commune from Izmir Bay. *World J. Microbiol. Biotechnol.* **2007**, *24*, 1375–1381. [CrossRef]
34. Johari, W.L.W.; Isa, R.I.M.; Ghazali, N.; Arif, N.M.; Shukor, M.Y.A. Decolorization of Azo Dyes by Local Microorganisms. In *Proceedings of the From Sources to Solution*; Springer Science and Business Media LLC: Berlin/Heidelberg, Germany, 2013; pp. 357–361.
35. El Ahwany, A.M.D. Decolorization of Fast red by metabolizing cells of Oenococcus oeni ML34. *World J. Microbiol. Biotechnol.* **2008**, *24*, 1521–1527. [CrossRef]
36. Mansur, R.; Gusmanizar, N.; Roslan, M.A.H.; Ahmad, S.A.; Shukor, M.Y. Isolation and Characterisation of a Molybdenum-reducing and Metanil Yellow Dye-decolourising Bacillus sp. strain Neni-10 in Soils from West Sumatera, Indonesia. *Trop. Life Sci. Res.* **2017**, *28*, 69–90. [CrossRef] [PubMed]
37. Sanusi, S.N.A.; Halmi, M.I.E.; Abdullah, S.R.S.; Hassan, H.A.; Hamzah, F.M.; Idris, M. Comparative process optimization of pilot-scale total petroleum hydrocarbon (TPH) degradation by Paspalum scrobiculatum L. Hack using response surface methodology (RSM) and artificial neural networks (ANNs). *Ecol. Eng.* **2016**, *97*, 524–534. [CrossRef]
38. Halmi, M.I.E.b.; Abdullah, S.R.S.; Wasoh, H.; Johari, W.L.W.; Ali, M.S.b.M.; Shaharuddin, N.A.; Shukor, M.Y. Optimization and maximization of hexavalent molybdenum reduction to Mo-blue by Serratia sp. strain MIE2 using response surface methodology. *Rendiconti Lincei* **2016**, *27*, 697–709. [CrossRef]
39. Toolabi, A.; Malakootian, M.; Taghi Ghaneian, M.; Esrafili, A.; Ehrampoush, M.; Askarishahi, M.; Tabatabaei, M.; Khatami, M. Optimizing the photocatalytic process of removing diazinon pesticide from aqueous solutions and effluent toxicity assessment via a response surface methodology approach. *Rendiconti Lince* **2019**, *30*, 155–165. [CrossRef]
40. Gunawan, E.; Basri, M.; Rahman, M.; Bakar Salleh, A.; Rahman, R. Study on response surface methodology (RSM) of lipase-catalyzed synthesis of palm-based wax ester. *Enzym. Microb. Technol* **2005**, *37*, 739–744. [CrossRef]
41. Beshay, U. Production of alkaline protease by Teredinobacter turnirae cells immobilized in Ca-alginate beads. *Afr. J. Biotechnol.* **2003**, *2*, 60–65. [CrossRef]
42. Pokharia, A.; Singh, S. Biodecolorization and degradation of xenobiotic azo dye -Basic Red 46 by Staphylococcus epidermidis MTCC 10623. *Int. J. Res. Biosci.* **2016**, *5*, 10–23.
43. Chang, J.-S.; Chou, C.; Chen, S.-Y. Decolorization of azo dyes with immobilized Pseudomonas luteola. *Process. Biochem.* **2001**, *36*, 757–763. [CrossRef]

44. Nallapan maniyam, M.; Latif Ibrahim, A.; Cass, T. Enhanced cyanide biodegradation by immobilized crude extract of Rhodococcus UKMP-5M. *Environ. Technol.* **2017**, *40*, 1–41. [CrossRef]
45. Jafari, N.; Kasra-Kermanshahi, R.; Soudi, M.R. Screening, identification and optimization of a yeast strain, Candida palmioleophila JKS4, capable of azo dye decolorization. *Iran J. Microbiol.* **2013**, *5*, 434. [PubMed]
46. Saratale, R.G.; Saratale, G.D.; Chang, J.S.; Govindwar, S.P. Bacterial decolorization and degradation of azo dyes: A review. *J. Taiwan Inst. Chem. Eng.* **2011**, *42*, 138–157. [CrossRef]
47. Gauthier, P.T.; Norwood, W.P.; Prepas, E.E.; Pyle, G.G. Metal–PAH mixtures in the aquatic environment: A review of co-toxic mechanisms leading to more-than-additive outcomes. *Aquat. Toxicol.* **2014**, *154*, 253–269. [CrossRef] [PubMed]
48. Shah, M.P.; Patel, K.A.; Nair, S.S.; Darji, A. Microbial decolorization of methyl orange dye by Pseudomonas spp. ETL-M. *Int. J. Environ.* **2013**, *1*, 54–59.

MDPI
St. Alban-Anlage 66
4052 Basel
Switzerland
Tel. +41 61 683 77 34
Fax +41 61 302 89 18
www.mdpi.com

Sustainability Editorial Office
E-mail: sustainability@mdpi.com
www.mdpi.com/journal/sustainability

www.ingramcontent.com/pod-product-compliance
Lightning Source LLC
Chambersburg PA
CBHW061602220926
51850CB00005B/401
* 9 7 8 3 0 3 6 5 1 8 8 2 4 *